Genetics

Genetics

John R. S. Fincham
University of Edinburgh

WRIGHT · PSG

1983   Bristol . London . Boston

© **John Wright & Sons Limited** 1983

Published by   John Wright & Sons Ltd, 823–825 Bath Road,
Bristol BS4 5NU, England.
John Wright PSG Inc., 545 Great Road, Littleton,
Massachusetts 01460, USA.

British Library Cataloguing in Publication Data
Fincham, J. R. S.
  Genetics.
  1. Genetics
  I. Title
  575.1   QH430

ISBN 0 7236 0661 7

Library of Congress Catalog Card Number: 82-42672

Printed in Great Britain by
John Wright & Sons (Printing) Ltd
at the Stonebridge Press
Bristol BS4 5NU

Preface

Writing a general book on a popular and rapidly expanding subject like genetics belongs to the same category of human folly as trying to cross the Atlantic in a rowboat. In both kinds of enterprise one runs a considerable risk of drowning. In this voyage I did my best to ignore the waves of rather good texts with broadly similar aims that started to appear as soon as my book had passed the point of no return, but I could not help worrying about the tides of new data that constantly threatened to overwhelm my system of navigation. I have tried hard to get my view of genetics accurately positioned in the present and pointing in the right direction for the future, but some of the most exciting and important aspects of contemporary genetics, such as the control of eukaryotic gene activity and its connection with chromatin structure, cannot be definitively charted at the present time. A book such as this has to be based on a blend of established fact and debatable clues. I hope that the reader will find the distinction between the two sufficiently clear.

My aim has been to provide a picture of the current state of genetics for the student or other scientifically interested reader with no previous knowledge of the subject. I have assumed some familiarity with, or at least access to, elementary biochemistry. The original concept was of a general book for the student who did not want to be distracted by a lot of references to original sources. However, I have always felt rather cheated by general books that make interesting statements without giving chapter and verse. It seemed a reasonable compromise to provide sufficient references to lead the persistent enquirer to the original sources, but to place these unobtrusively, without titles, at the back of the book. I am not sure now that this was the best plan, but there it is. Often the reference given is only one of several that might have been cited, and has been selected because of its comparatively recent date and potential value as a guide to earlier papers. The sources listed at the ends of chapters under Further Reading are of a somewhat different character; they are mostly books and reviews that will provide the reader with a great deal of additional background information.

Some readers of the book in typescript have commented that the approach is 'ahistorical', which is broadly true. Here and there I have been unable to resist a historical allusion, but I have certainly not treated the history of genetics systematically, nor given credit to all the pioneers who would have featured prominently in a historically-orientated work. This is my excuse for not providing an index of authors, even though I have tried to make the general subject index as comprehensive as possible. The rather full subject index is, in turn, my excuse for not including a glossary of technical terms. I believe that nearly all terms a reader will wish to have explained will be listed in the index and defined on the page of first reference.

I feel that I should say something about the Problems at the end of most of the chapters. By the standards of some other books there are not a great many of them, but I have tried to make them interesting. In most cases they are based on real data and require the student to think a little beyond the material presented in the preceding chapter(s). One or two reviewers have remarked that the problems are very demanding and, having just re-worked most of them in search of errors, I must admit that some (though not all) are rather difficult. However, the reader can take comfort from the fact that behind every difficult problem there is an easier one, namely looking up the answer and showing that it works. Any student who can show that an answer *does not* work should be given extra credit.

Finally, I must thank all those without whose help this book could hardly have been possible. Dr John Gillman, of John Wright & Sons, must take the credit, if credit is the right word, for having persuaded me to undertake the project. He has been a constant source of encouragement. Mrs Jane Sugarman has been a most helpful and understanding sub-editor, and a model of equanimity in trying circumstances. Among a number of reviewers who saw the draft in typescript Dr Frank Stahl provided a particularly thorough and critical appraisal. I am indebted to him for a large number of suggestions on specific points, most of which I adopted. Several of my Edinburgh colleagues were kind enough to read individual chapters. Professor Douglas Falconer, in particular, helped to make the chapter on quantitative genetics less defective than it might otherwise have been. Mr Graham Thomas was generous of his time in reading the entire book in galley proof, and he picked up a number of errors that had escaped my own scrutiny. Mr E. D. Roberts, formerly artist in the Edinburgh Department of Genetics, undertook, after his official retirement, the onerous task of producing the drawings for figures. Unfortunately he was not able to finish quite all of them, and most of the remainder were drawn by me. I am afraid that the reader will have no difficulty in distinguishing his work from mine. Most of the photographs were solicited gifts. The donors are acknowledged in the legends, and I thank them all for their generosity.

J. R. S. F.

Contents

Abbreviations *for enzymes and other molecules*

ADH	alcohol dehydrogenase
2-AP	2-aminopurine
APRT	adenine phosphoribosyltransferase
ATP	adenosine triphosphate
ATPase	adenosine triphosphatase
5-BrdU	5-bromodeoxyuridine
5-BU	5-bromouracil
cAMP	adenosine cyclic 3′ : 5′-monophosphate
cDNA	complementary (copy) DNA
CRM	cross-reacting material
CRP	catabolite repression protein (cAMP-binding protein)
ctDNA	chloroplast DNA
CTP	cytidine triphosphate
DNAase	deoxyribonuclease
dsRNA	double-stranded RNA
GDH	glutamate dehydrogenase
G6PD	glucose 6-phosphate dehydrogenase
HGPRT	hypoxanthine–guanine phosphoribosyltransferase
hnRNA	heterogeneous nuclear RNA
Ig	immunoglobulin
IGP	indoleglycerol phosphate
mRNA	messenger RNA
mtDNA	mitochondrial DNA
NTP	nucleoside triphosphate
PGA	phosphoglyceraldehyde
6PGD	6-phosphogluconate dehydrogenase
PGK	phosphoglycerate kinase
PGM	phosphoglucomutase
rDNA	DNA from which rRNA is transcribed
rRNA	ribosomal RNA
SDS	sodium dodecylsulphate
TK	thymidine kinase
tRNA	transfer RNA
TSase	tryptophan synthetase
UV	ultraviolet light

1 Cells, DNA and Chromosomes

1.1   Introduction

Living organisms are characteristically made up of cells. Most bacteria and a number of fungal and algal species consist of single free cells. All large organisms are multicellular, but even these originate from single cells—spores or fertilized eggs.

The essential property of the cell is its capacity for self-replication, that is to say its ability to propagate more cells like itself. In the case of unicellular organisms, the products of cell division, the daughter cells, are most commonly just like the mother cell which gave rise to them. In multicellular forms we see diversification of cell type as development proceeds, so that the mature organism contains many different kinds of cell each with a specialized function. Notwithstanding this differentiation, at least those cells which are destined to give rise to the next generation retain the potential for recapitulating the whole course of development characteristic of the species. Thus both in unicellular and multicellular organisms there must exist a system of information, a sort of master plan or blueprint, contained within a single cell and capable of being copied and transmitted faithfully through cell division to daughter cells. In this chapter we will be concentrating upon those structures and molecules which are particularly relevant to this self-copying, or self-replicating, property of cells.

1.2   Prokaryotic and eukaryotic modes of organization

The living world is divided into two main classes of organisms—the prokaryotes and the eukaryotes. The former class consists of the bacteria and the blue-green algae (Cyanophyta), and the latter includes all the rest.

The typical prokaryotic cell is enclosed by a protective cell wall, the composition of which varies greatly from one group to another but is always different from anything found in eukaryotes. Inside the cell wall the **protoplasm** is bounded by a double-layered selectively permeable **cell membrane** which, in striking contrast to the situation in eukaryotes, is the only membrane in the cell. The prokaryotic cell membrane is a complex structure performing many different functions. The enzyme complexes responsible for cell respiration and also (in photosynthetic bacteria) the pigments and enzymes involved in photosynthesis are associated with the cell membrane. In the much larger cells of eukaryotes all these functions are carried out by complex internal membranous structures (**organelles**).

The most conspicuous internal feature of the prokaryotic cell, easily seen in thin sections with the electron microscope, is a compact mass of intricately folded and coiled deoxyribonucleic acid (DNA). If a bacterial cell is made to burst by exposure to low osmotic pressure after removal of its cell wall with the enzyme lysozyme, the DNA is sometimes released in the form of a completely unfolded

Fig. 1.1

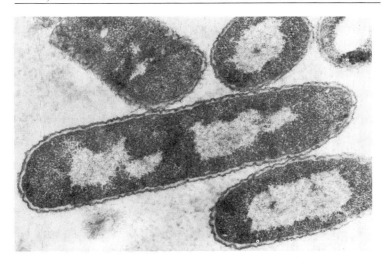

Electron micrograph of thin sections of *Escherichia coli* cells as seen with the
electron microscope. The centrally placed cell sectioned longitudinally shows
two light-coloured areas which are full of DNA. There is a double-layered
plasma membrane enclosing the cell inside the cell wall but no internal
membrane enclosing the DNA. The cell, apart from the DNA, is packed with
darkly stained ribosomes. Magnification × 32 500.
Photograph by courtesy of Dr P. Highton.

closed-loop molecule. In the best studied bacterium, *Escherichi coli* (henceforth to
be called *E. coli*), the circumference of this closed loop is 1300 μm, an enormous
length considering that the cell into which it was packed is a rod about 3 μm long
and 1 μm in diameter (*see Fig.* 1.1). In the intact cell the DNA is attached to a
specific site on the cell membrane. Prokaryotic DNA is 'naked', with the negative
charges on its acidic phosphate groups neutralized by polyamines, which are basic
compounds of small molecular weight, rather than, as in eukaryotes, being
complexed with specific proteins.

Eukaryotic Organelles

Eukaryotic differ from prokaryotic cells in their much larger size,
which would presumably make it impossible for the whole protoplasmic volume to
be 'serviced' by the bounding cell membrane. In the eukaryotic cell, respiration (i.e.
the oxidation of reduced metabolic intermediates by molecular oxygen, with the
generation and chemical storage of energy) is the prerogative of an internal
organelle bounded by its own double membrane and called the **mitochondrion**. The
inner mitochondrial membrane is invaginated to form a series of incomplete
transverse partitions, or **cristae**, and it is these which contain the complexes of
enzymes and cytochrome pigments which carry electrons from the respiratory
substrates to oxygen. In vascular plants and algae there is a second major class of
organelle, the **plastid**, which, in its mature form as a **chloroplast**, is responsible for
photosynthesis and is packed with an ordered array of membranes (**grana**)
containing chlorophyll pigments.

A vitally important fact about these eukaryotic organelles is that they contain a
part, often several per cent, of the DNA of the cell. Organellar DNA is in a form
quite different from that which is described below for the DNA of the cell nucleus.

It appears to be much more like the supercoiled 'naked' loop found in prokaryotes, though much smaller and usually present in many copies per organelle. *Table* 1.1 shows some of the available data on the dimensions and quantities of organellar DNA in various organisms.

Table 1.1   **Quantities of DNA in genomes of different organisms and viruses**

A. Viruses

Tobacco mosaic*†	6·4
φX174 (*E. coli* phage)†	5·5
Lambda (*E. coli* phage)	48
T4 (*E. coli* phage)	circa 200
Simian virus 40	5·2
Adenovirus	circa 30–33

B. Prokaryote

Escherichia coli	circa $4·5 \times 10^3$

C. Eukaryotes	Haploid nuclear genome‡	Mitochondrial genome	Chloroplast genome
i. Fungi			
Saccharomyces cerevisae (yeast)	$1·8 \times 10^4$ (18)	74	—
Neurospora crassa	$4·3 \times 10^4$ (7)	60	—
ii. Flowering plants			
Crepis capillaris	$1·3 \times 10^6$ (3)		?
Lupinus albus (lupin)	$2·7 \times 10^6$ (24)		?
Vicia lathyroides (bean)	$2·4 \times 10^6$ (6)		?
Vicia faba (broad bean)	$1·2 \times 10^7$ (6)	105	135
Secale cereale (rye)	$9·2 \times 10^6$ (7)		circa 130
Zea mays (maize)	$5·4 \times 10^6$ (10)		circa 130
Tradescantia paludosa	$1·9 \times 10^7$ (6)		?
iii. Invertebrate animals			
Nematode:			
Caenorhabditis elegans	8×10^4 (6)	?	
Insects:			
Drosophila melanogaster	$1·7 \times 10^5$ (4)	18	—
Locusta migratoria	6×10^6 (11)		—
iv. Vertebrate animals			
Amphibians:			
Rana pipiens (frog)	6×10^6 (13)	16	—
Xenopus laevis (frog)	3×10^6 (18)	18	—
Triturus cristatus (newt)	2×10^7 (12)		
Necturus maculosus	8×10^7 (19)		—
Birds:			
Gallus domesticus (fowl)	$1·3 \times 10^6$ (circa 39)§	16	—
Mammals:			
Mus musculis (mouse)	3×10^6 (20)	15	—
Rattus norwegius (rat)	3×10^6 (21)	15	—
Homo sapiens	3×10^6 (23)	15	—

All quantities are expressed in kilobase-pairs (kilobases in the case of viruses with single-stranded amino acid).
kbase-pair ≃ 617 500 daltons.
Sources of data: for *Caenorhabditis*, Sulston and Brenner;[35] for *Vicia* spp., Chooi;[36] for the fungi, Fincham et al.;[37] for the plants, King;[38] for animals, White[39] and Swanson et al.[40]
* RNA not DNA.
† Single-stranded nucleic acid.
‡ Haploid number of chromosomes in parentheses.
§ Ten fairly large, the others small, some extremely small.

Apart from the organelles, each of which is completely enclosed by its own membrane, the typical eukaryotic cell contains throughout much of its volume an open system of membranes, often in parallel layers, to which are bound the ribosomes which function in protein synthesis. This system, called the **endoplasmic reticulum**, is responsible for most of the protein synthesis of the cell, although an important part is also played by special classes of ribosomes located in the mitochondria and (where present) the plastids. These organellar ribosomes function in the synthesis of proteins the structures of which are encoded in the organellar DNA (*see* Chapter 7, pp. 176–195 and, in their somewhat smaller size and susceptibility to inhibition by certain antibacterial antibiotics such as erythromycin and chloramphenicol, they are rather similar to the ribosomes of bacteria. This is the principal basis for the idea, argued particularly by Margulis,[1] that organelles originated in evolution from symbiotic prokaryotes.

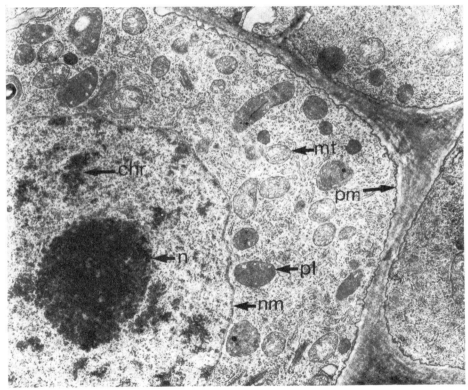

Fig. 1.2

Electron micrograph of a thin section of part of a pea root cell.
Key: pm, plasma membrane (double layered) surrounding the cytoplasm inside the thick cellulose cell wall; mt, mitochondrion with clearly visible internal membranes (cristae) which connect (though this is not very evident in the section) with the inner of the two membranes enclosing the mitochondrion; pl, plastid also enclosed by a double membrane (in a leaf cell this could develop into a chloroplast); nm, double-layered nuclear membrane enclosing the chromatin and nucleolus; n, nucleolus; chr, chromatin material, condensed in some regions and diffuse in others. Within the cytoplasm one can see ribosomes, many of which are attached to membranes (endoplasmic reticulum). Magnification ×9400.
Photograph by courtesy of Professor M. M. Yeoman.

The Eukaryotic Cell Nucleus

The structure which occupies the central position, both in the eukaryotic cell and in the science of genetics, is the cell nucleus. It is bounded by a membrane, the **nuclear envelope**, which contains pores and may have interconnections with the membranes of the endoplasmic reticulum. The nucleus contains, as a rule, by far the greatest part of the DNA of the cell, though in some green leaves the chloroplast DNA may actually be greater in amount. The nuclear DNA is packaged in a number of **chromosomes**, so called because they can be vividly stained with various dyes for observation under the microscope. The number of chromosomes is a characteristic of the individual species. Making use of criteria to be described later in this chapter, it is possible to distinguish one chromosome from another and to show that, depending on the stage of the life history (*see* Chapter 2), a nucleus is usually either **haploid**, with one chromosome of each kind, or **diploid** with a double set, each chromosome type being represented twice. As well as the chromosomes, the nucleus is often seen to contain one or more prominent spherical bodies, the **nucleoli**. These are generally formed in association with specific regions of specific chromosomes (**nucleolus organizer** regions) and they represent centres of synthesis of ribosomal precursor material.

Before proceeding further we must briefly review the structure and mode of synthesis of DNA, which, from now on, will come increasingly to occupy the centre of the stage.

1.3   Structure and replication of DNA

DNA occurs in the cell predominantly in double-stranded form, the two component strands being interwound in a double helix. Each strand is a string of deoxynucleotides condensed together, and each deoxynucleotide consists of a purine or pyrimidine base linked to a molecule of deoxyribose (forming a **deoxynucleoside**) linked in turn to orthophosphate. The deoxyribose is able to form ester bonds with phosphate through hydroxyl groups on both carbon atoms 3 and 5 (called 3′ and 5′ in the deoxynucleotide). The phosphate itself is able to bond to the deoxyribose through any of its oxygens. The backbone of the poly(deoxynucleotide) chain is a succession of deoxyribose—phosphate bridges (*Fig.* 1.3a), the sequence running phosphate—5′-deoxyribose-3′—phosphate—5′-deoxyribose . . . and so on, the chain being eventually terminated by a 3′-hydroxyl group. The bases are attached to the number 1 carbon atom of each deoxyribose as side-chains. There are four kinds of base in DNA: two purines, adenine (A) and guanine (G), and two pyrimidines, thymine (T) and cytosine (C). (In many species a proportion of the bases are methylated, e.g. to give 5-methylcytosine or N^6-methyladenine, but this does not affect the properties we are concerned with in this chapter.) Within each chain the four bases can occur in any proportions and in any order; this, of course, together with the great length of the chains, explains the suitability of DNA molecules for encoding vast amounts of genetic information. The transmission of this information through successive cell divisions depends on the self-copying property of the DNA which is inherent in its structure and mode of synthesis.

In the double helix, or **duplex**, of DNA the bases of the intertwined strands are directed inwards, and each base from one strand pairs, through weak hydrogen bonds, with a base of the other strand. The two strands are of opposite polarities; that is to say, a given direction along the duplex is $3' \rightarrow 5'$ for one strand and $5' \rightarrow 3'$ for the other. The key to the self-replication of DNA lies in the fact that the base pairing is specific. The difference in size between purines and pyrimidines means that a purine must always pair with a pyrimidine; there is no room for two purines, and two pyrimidines would occupy too little space to bridge the gap between the two backbones (*Fig. 1.3b*). Furthermore, the dispositions in the bases of the hydrogen-donating ($>NH$ and $-NH_2$) and hydrogen-accepting atoms ($>C=O$ and $\gg N$) are such that only two purine–pyrimidine pairings will fit, namely, adenine–thymine (A–T) and guanine–cytosine (G–C); these are, in fact, the only pairs found in normal duplex DNA (*Fig. 1.3c*). The former pair is held by two hydrogen bonds and the latter by three. Duplexes rich in G–C are more stable than those rich in A–T, and the temperature at which the duplex structure 'melts' into single strands can be used as an index of base composition: the higher the temperature the greater the percentage of G–C.

Fig. 1.3.

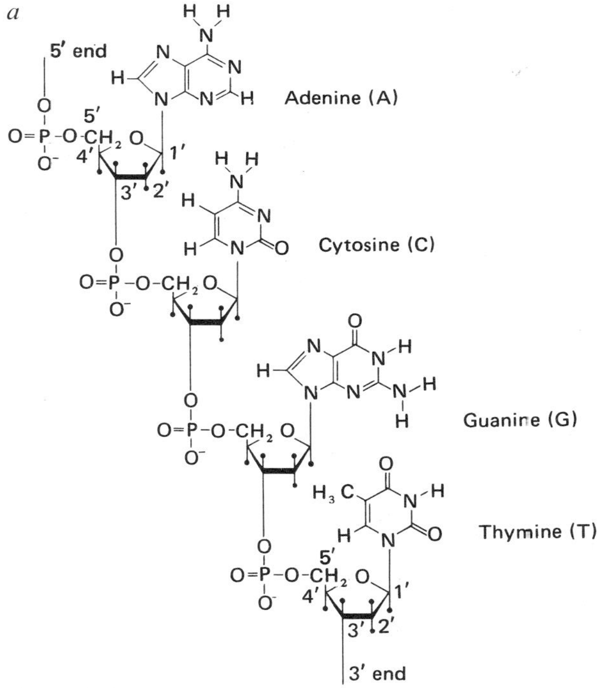

Views of DNA structure.
a. Structure of a DNA single strand. In the five-membered deoxyribose rings the carbon atoms are at the unlabelled angles and the hydrogen atoms are represented by small filled circles. The stretch of DNA sequence shown has each of the four bases present once; the bases can in fact occur in any order or (in a single strand) in any proportions. Note the distinction between the 5' and 3' ends of the sequence. The carbon atoms of the upper deoxyribose at the top of the figure are labelled $1' \rightarrow 5'$ according to the standard convention.
Ribonucleic acid (RNA) has the same structure except that there is an OH (2'-hydroxyl) group attached to the 2'-carbons of the pentose units (i.e. ribose instead of deoxyribose) and uracil instead of thymine (replacement of CH_3 by H).
Based on a representation of RNA in Stent and Calendar.[32]

Fig. 1.3   *b*

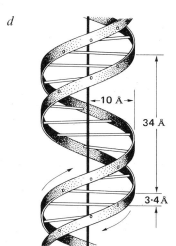

b. Simplified diagram of the structure of double-stranded DNA, with the purines (adenine and guanine) represented by longer, and the pyrimidines (cytosine and thymine) by shorter bars. The two strands are shown as parallel and straight for simplicity.

c

Adenine	Thymine	Guanine	Cytosine

c. The chemical structures of the base-pairs, with hydrogen bonds represented by dotted lines. Arrows indicate attachment to deoxyribose.

d

d. The general form and dimensions of the double helix. The helically wound ribbons represent the deoxyribose—phosphate backbone, and the rungs connecting them represent the base-pairs.
Redrawn from Stent and Calendar.[32]

Fig. 1.3   *e*

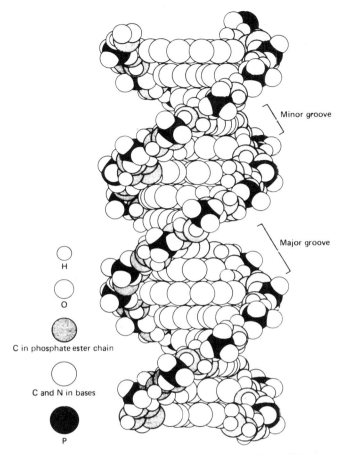

Minor groove

Major groove

H

O

C in phosphate ester chain

C and N in bases

P

e. A space-filling model of double-stranded DNA, with the different atoms shown to scale and in their correct packing. Figure by courtesy of Professor M. H. F. Wilkins.

The rule of specific base pairing means that a unique base sequence along one strand of a duplex implies a unique complementary sequence along the other. Watson and Crick, who proposed the duplex structure in 1953,[2] saw that this suggested a mechanism for faithful replication of base sequence. If the synthesis of DNA involved the separation of the two strands of each duplex and the synthesis of a new complementary strand alongside each, the result would necessarily be two duplex molecules each identical to the original one. We now know, from a great deal of biochemical evidence, that this is indeed the mechanism of DNA replication. It is called **semiconservative**, since each new duplex molecule contains one pre-existing strand and one newly synthesized one.

There was one rather perplexing problem of DNA replication which came to light as enzymes capable of catalysing the process were isolated (from *E. coli* in the first instance) and their properties investigated. These enzymes, called **DNA polymerases**, catalysed the sequential addition of nucleotide residues (supplied in the reaction mixture as nucleoside 5′-triphosphates) to the end of the growing

Box 1.1   **DNA weights and measures**

A–T deoxynucleotide pair ('base-pair') weighs 617 daltons.
G–C deoxynucleotide pair ('base-pair') weighs 618 daltons.
1000 base-pairs (1 kilobase-pair, 1 kb) weighs approx. 617 500 daltons or
  $6.175 \times 10^5/6.02 \times 10^{23}$ g $= 1.026 \times 10^{-18}$ g $= 1.026 \times 10^{-6}$ picogrammes (pg).
1 pg of DNA $= 9.75 \times 10^5$ kilobases $= 0.975 \times 10^6$ kb $\approx 10^6$ kb.
Distance between successive base-pairs $= 3.4$ Å $= 0.34$ nm.
1 kb measures 340 nm or 0.34 μm.
1 μm of double-stranded DNA $= 2.9$ kb.
NB 6.02×10^{23} is the Avogadro number, the number of molecules in a gramme-molecule.

Fig. 1.4

a

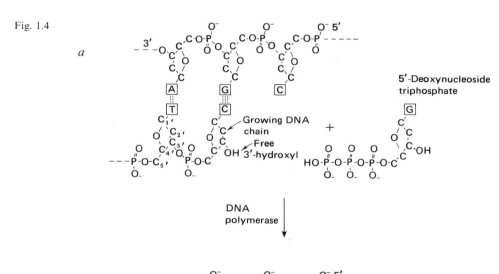

a. The synthetic reaction catalysed by DNA polymerase. A deoxynucleoside triphosphate (here shown as deoxyguanosine triphosphate) is cleaved with release of inorganic pyrophosphate and the deoxynucleoside 5'-monophosphate (i.e. deoxynucleoside 5'-monophosphate) moiety is condensed with the free 3'-hydroxyl at the growing end of the chain being synthesized. The appropriate deoxynucleoside triphosphate is 'chosen' by the enzyme to give a standard base-pair (either A–T or G–C) between the newly added base and the base opposite on the complementary (template) strand (here shown as cytosine). The bases are shown as boxes labelled with appropriate initial letters. The rest of the chemical formulae are shown in full except that the hydrogens needed to saturate the carbon valencies are omitted for simplicity. The chains are to be imagined as being extended beyond the broken lines—at both 3′ and 5′ ends for the upper (template) chain and at the 5′ end for the lower (newly synthesized) chain.

Fig. 1.4

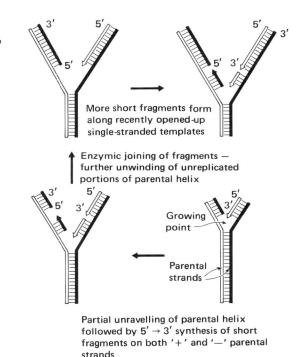

More short fragments form along recently opened-up single-stranded templates

Enzymic joining of fragments — further unwinding of unreplicated portions of parental helix

Growing point

Parental strands

Partial unravelling of parental helix followed by 5′ → 3′ synthesis of short fragments on both '+' and '−' parental strands

b. Semiconservative replication of DNA with discontinuous synthesis, in the 5′ to 3′ direction, on both strands. The short regions of new synthesis indicated by the arrows are thought to be initiated by short RNA 'primer' sequences (*see* text).
Redrawn from J. D. Watson.[33]
In principle, new synthesis can proceed continuously behind one branch of the fork while fresh priming and reinitiation is constantly necessary behind the other. Thus one might expect greater delay in synthesis of the complementary strand in one branch of a replication fork (*see Figs* 1.7 and 1.9*a* for examples).

chain. A pre-existing chain, to which the new strand will be complementary and with which it will form a duplex, has to be added to the reaction mixture as a 'template' and at least a short sequence of complementary strand has to be already present on the template to act as a 'primer'. Deoxynucleotides are added to only the 3′-hydroxyl side of the priming sequence and chain growth in the enzyme-catalysed system occurs in the 5′→3′ direction (*Fig.* 1.4*a*). This was hard to reconcile with evidence (*Figs* 1.4, 1.6, 1.8) that synthesis followed behind a **replication fork** which travelled in a direction which was $3' \rightarrow 5'$ for one strand and $5' \rightarrow 3'$ for the other. The dilemma was resolved as evidence accumulated that, *in vivo* as well as *in vitro*, synthesis was in the $5' \rightarrow 3'$ direction on *both* strands, following the fork on one strand and 'backwards' in short discontinuous pieces on the other. The first indication of this mode of synthesis was obtained by T. Okazaki, who found very newly synthesized DNA was in small single-stranded pieces ('Okazaki fragments'). An alternative explanation of this observation has been proposed (*see* Stent and Calendar[32]), but continuing experiments have supported the idea of discontinuous synthesis (*Fig.* 1.4*b*). *Figs* 1.7 and 1.9 show some electron microscope evidence from sea urchin mitochondrial and *Drosophila* nuclear DNA that synthesis is delayed somewhat on one strand adjacent to the replication fork, as would be expected if synthesis on this strand was taking place 'backwards'.

Discontinuous synthesis posed, in turn, another biochemical problem, namely the source of the primer for the 'backwards' synthesis. The known DNA polymerases cannot initiate a new chain but need a 3'-hydroxyl group already in position with which to condense the next deoxynucleotide (*Fig.* 1.4*a*). The answer to this problem appears to involve ribonucleic acid (RNA), a type of macromolecule similar to DNA except that ribose replaces deoxyribose and uracil replaces thymine, and which, as we shall see, is normally synthesized in the cell by complementary copying from DNA. RNA polymerase can initiate RNA synthesis at certain sites on a DNA template *without* a primer. Considerable biochemical evidence is now available to show that each new initiation of DNA chain synthesis is primed by the prior synthesis of a short stretch of RNA. DNA polymerase then takes over, condensing the 5'-phosphate of a deoxyribonucleotide residue with the 3'-hydroxyl terminus of the RNA chain, and then proceeding with normal DNA chain synthesis. The RNA primer sequence is subsequently removed and the gap filled by DNA.

1.4   Modes of DNA replication in prokaryotes and eukaryotes

The mechanism of DNA replication which was described above is summarized in *Fig.* 1.4*b*. The details of what happens at the replication fork are still being worked out, but there is little doubt that the main feature of the model, the travelling fork generating two semiconserved duplexes behind it, is true in both prokaryotes and eukaryotes.

The first demonstration of semiconservative replication in *E. coli* was made in 1958 by Meselson and Stahl,[3] who made the DNA heavy by growing the bacterium in a medium with ammonium salt containing heavy nitrogen (^{15}N) as nitrogen source. The DNA containing the heavy isotope could be separated from DNA of normal density by sedimentation to equilibrium in a density gradient established by subjecting a solution of caesium chloride to a high gravitational field in an ultracentrifuge. It was found that after one round of DNA replication in normal medium, the cells having previously been made uniformly heavy by prolonged growth in the heavy medium, the DNA was of exactly intermediate density (as if

Fig. 1.5

100 nm

Autoradiographic detection of the replication forks in *E. coli* DNA.
We see the transition from low-level tritium labelling at the replication origin
(arrowed) to high-level labelling at the diverging replication forks. For further
explanation *see* text.
Drawn from the photograph of Prescott and Kuempel.[4]

containing one heavy and one light strand). After one further round of replication
the DNA was separable into two equal fractions, one intermediate in density and
one wholly light (as if the heavy and light strands had been separated and a new
light strand associated with each). The intermediate-density DNA was indeed
separable, after melting, into light and heavy single strands.

Following this definitive experiment, further information was obtained, on both
bacterial and eukaryotic cells, in two ways. In one experimental plan, newly
synthesized DNA is made radioactive by forcing the cells to utilize a radioactive
precursor, usually tritiated thymidine ([³H]thymidine), and then the pattern of
radioactive labelling is determined by autoradiography. The second experimental
approach is the direct observation of replicating DNA molecules with the electron
microscope. The DNA is usually coated with the basic protein cytochrome *c* (the
technique invented by Kleinschmidt) in order to make the strands thick enough for
observation, and additional contrast can be obtained by shadowing or negative
staining with heavy metal atoms. The picture obtained can be sufficiently clear for
a distinction to be made between single and double-stranded DNA (*see Figs* 1.7,
1.9).

Fig. 1.5*a* shows a result from an experiment on *E. coli* using radioactive
labelling: virtually identical pictures can be obtained of DNA replication origins in
mammalian cells. If cells of a mutant strain of *E. coli* requiring an external supply
of a particular amino acid for growth (*see* p. 346) are deprived of their required
amino acid and thus prevented from synthesizing protein, they will complete their
current round of DNA replication but start a new round only after the amino acid
is supplied again. In the experiment illustrated, DNA synthesis was initiated by
amino acid addition and, at the same time, a low concentration of [³H]thymidine
was added to the growth medium in order to label newly synthesized DNA. After a
few minutes of synthesis a much higher concentration of [³H]thymidine was added
to the culture so that the initial low-level labelling was followed by a high level.
After several more minutes the DNA was extracted from the cells, spread on a
slide, dried and covered with radiosensitive film. After a period of exposure the film
was developed and the labelled DNA molecules could be seen as tracks of silver
grains in the film; the level of labelling could be assessed by the concentration of
grains. The pattern of labelling revealed that DNA synthesis started at one point in
the 1300-μm loop and proceeded from this point simultaneously in both directions
at approximately equal rates. The figure shows a relatively early stage of
replication, with two heavily labelled forks diverging from a more lightly labelled
origin. In other experiments, in which DNA synthesis was allowed to proceed in
the presence of label for 20–30 minutes, the two replication forks were seen to
converge on a point in the loop diametrically opposite the starting point; their
meeting completed the synthesis of two completely labelled daughter duplexes. The
whole process is summarized in *Fig.* 1.6.

Fig. 1.6

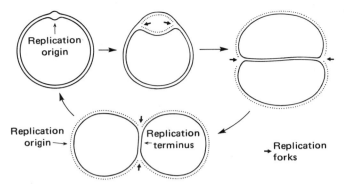

Diagram of the replication of bacterial DNA.
Pre-existing DNA chains are shown as full lines, newly synthesized chains as
dotted lines. The replication forks and their direction of travel are indicated by
arrows.

Fig. 1.7

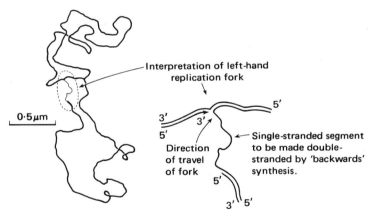

Replicating mitochondrial DNA from the oocyte of a sea urchin.
The replication of the molecule is almost completed. A single-stranded region
can be seen in one branch adjacent to the left-hand replication fork,
see interpretative diagram.
Drawn from the electron micrograph of Matsumoto et al.[5]

Fig. 1.8

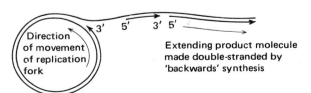

A diagram to illustrate the 'rolling circle' model of DNA replication. An
indefinite number of end-to-end copies (a concatenate) are produced
from a circular template.

Bidirectional replication from a single origin is also seen in the smaller closed-loop DNA molecules of bacterial plasmids and eukaryotic organelles (*Fig.* 1.7) as well as of a number of animal viruses such as simian virus 40 (SV40) and polyoma. Unlike the enormous DNA molecule of *E. coli*, these smaller loops can, when relieved of their supercoils (*see* p. 252), easily be seen in the act of replication with the electron microscope.

The characteristic appearance of the mode of replication which we have just been considering is the so-called *theta* (θ) form—a closed loop doubled over part of its circumference. This is not, however, the only mode of prokaryotic DNA replication. Many bacteriophages, including the *E. coli* phages P2 and lambda (λ), replicate their DNA, at least in the later stages of their intracellular growth, by a process often referred to as the **rolling-circle** mechanism. Here there is a single replication fork travelling round and round the closed duplex, generating a free-ended single strand of indefinite length which is a series of end-to-end copies (or concatenate) of the sequence of the parental molecule (*Fig.* 1.8). This single-stranded product is, either immediately or after a little delay, made double stranded by complementary synthesis, and it is subsequently cleaved into single copies for packaging into virus particles. Rolling-circle replication is also import-ant in another connection, being the probable mechanism of the **transfer replication** of transmissible plasmids, with which we shall be much concerned in Chapter 9.

Another bacteriophage variation of some importance is that seen late in the replication cycle of the *E. coli* phage ϕX174, which we describe in detail in Chapter 15. Here the replicating duplex loop uses just one of its strands as template for the synthesis of *single-stranded* progeny loops, which are packaged as such in the infective phage particles; synthesis of the complementary strand does not occur until after infection of another host cell.

Eukaryotic DNA Replication

A combination of autoradiographic and electron microscopic studies on eukaryotic cells of various kinds has shown the same essential features as found in *E. coli*—namely semiconservative replication at divergent replication forks. Autoradiographs of DNA origins in, for example, mouse or hamster cells show pictures virtually identical to *Fig.* 1.5. There is, however, one striking difference between prokaryote and eukaryote DNA replication. Whereas the bacterial DNA molecule replicates from a single origin, the DNA of eukaryote chromosomes always has multiple replication origins, more or less regularly spaced (*Fig.* 1.9). This is the case even where, as in yeast, the average amount of DNA per chromosome is less than in *E. coli*. The stretch of DNA replicated from a single origin is called a **replicon**, and replication is completed when the 'eye-shaped' loops of adjoining replicons merge. Replicon size in eukaryotic chromosomes is usually in the range of 10–100 μm—very much less than the 1500 μm which is replicated from a single origin in *E. coli*. On the other hand, the rate of fork movement, typically of the order of 1 μm/min, is only a tenth or less of the *E. coli* rate. The total rate of DNA synthesis (the product of number of replicons and the rate of synthesis in each) is fairly similar between eukaryotes and prokaryotes.

Fig. 1.9   *a*

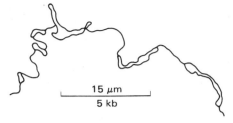

15 μm
5 kb

a. Replicating chromosomal DNA from *Drosophila melanogaster*. Note the multiple replication loops.

b

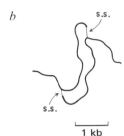

1 kb

b. One loop at a higher magnification. Note the thinner (single-stranded—arrowed s.s.) region on one side of each replication fork (cf. *Fig.* 1.7). Drawn from the electron micrographs of Kriegstein and Hogness.[6]

c

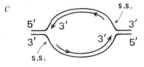

c. Diagram interpreting the structure of a replication loop. The arrowheads indicate the direction of synthesis of newly synthesized DNA fragments. The single-stranded region behind each travelling replication fork is the strand on which complementary synthesis has to occur 'backwards' by periodic fresh initiation.

1.5 DNA sequence organization

When we look at the quantities of DNA present in the cell nuclei of eukaryotes two things are strikingly apparent. The first is the enormous amounts found in many eukaryotes as compared with bacteria. For example, man, a fairly average eukaryote in this respect, has a haploid DNA complement of about 2.9×10^9 base-pairs (2.9×10^6 kilobase-pairs), which would be about a metre in length if in one piece and fully extended. The second is the very great difference in quantity between different eukaryotes, even ones which would appear to be quite closely related (*see*, for example, the two species of *Vicia, Table* 1.1). There appears to be no close relationship between DNA quantity and degree of biological complexity, which may seem to contradict the central message of this book:

Box 1.2 C_0t **values are measures of sequence complexity**

If c_0 is the initial concentration of a given sequence of single-stranded DNA (and its complement) in mol/l, and at time t this concentration has decreased to c through annealing to form duplex structure, then the rate of decrease at this time of the concentration of the single-stranded sequence will be:

$$\frac{dc}{dt} = -kc^2$$

where k is a constant almost independent of kind of DNA but somewhat dependent on fragment size.

Integrating:

$$\int_t^0 \frac{dc}{c^2} = -k\,dt \quad \text{gives} \quad {}^{c_0}\left[-\frac{1}{c}\right]_c = -k\left[t\right]_t^0$$

and

$$\frac{1}{c} - \frac{1}{c_0} = kt$$

If $t_{\frac{1}{2}}$ is the time taken for 50 per cent reassociation (i.e. $c = c_0/2$),

$$\frac{2}{c_0} - \frac{1}{c_0} = \frac{1}{c_0} = kt_{\frac{1}{2}} \tag{1}$$

If now, instead of expressing concentration in mol/l, we express it in terms of mass (C_0, g/l), then $c_0 = C_0/M_r$, where M_r is the molecular weight. In this context molecular weight is equivalent to the **complexity** of the sequence, that is the length of the sequence before it repeats.

Then, from equation (1) above,

$$C_0 t_{\frac{1}{2}} = \left(\frac{1}{k}\right) M_r$$

namely that it is the DNA that carries the genetic information which guides development. In *Table* 1.1 we see, for example, the surprising difference between the broad bean and the lupin, both leguminous plants, and between the newt and the frog, which are amphibia of apparently similar complexity; the higher member of each of these pairs far outstrips man in terms of quantitative DNA endowment. The puzzle posed by these comparisons is often referred to as the 'C-value paradox', the C-value being the constant amount of DNA found in a haploid chromosome set. One is bound to wonder whether the apparently excessive quantities of DNA present in some species as compared to others really represent a greater complexity of sequences, or whether they are to be explained by some sequences being present in repeated copies. This question can now be answered experimentally.

Most investigations of sequence repetition have been based on the method of analysis first worked out and published in 1968 by Britten and Kohne.[7] This depends on the fact that when a DNA sample is broken into fragments (usually by mechanical shearing), melted to single strands and then allowed to reassociate at a temperature below, but not too far below, the duplex melting point, each sequence will reform a double-stranded structure with its complementary sequence at a rate proportional to the product of their respective concentrations. A single-stranded fragment with a base sequence which is highly repeated will reassociate relatively quickly, while one present in only one copy per chromosome set (**genome**) will take very much longer. It can be shown that the product of the DNA concentration in g/l (C_0) and the time taken for 50 per cent reannealing ($t_{\frac{1}{2}}$) is proportional to the complexity of the sequence, that is to say the length before it repeats itself. The reasoning is summarized in *Box* 1.2. *Fig.* 1.10 shows the results obtained in the first demonstration of the $C_0 t$ analysis. Because of the great range of complexity in the different DNA samples under test, it was convenient to plot $C_0 t$ values on a logarithmic scale from 10^{-2} to 10^5. On such a scale, a DNA sample of a single degree of complexity reassociates over quite a narrow range. The relationship between the proportion of the sample remaining single stranded and the log $C_0 t$ is S shaped, with the point of inflexion corresponding to 50 per cent reassociation. The validity of the method was established by the use, as standards, of a series of prokaryotic DNA samples, each thought to consist virtually entirely of unique non-repetitious sequences. The value of $C_0 t_{\frac{1}{2}}$ determined for each was indeed closely proportional to the amount of DNA per nuclear body (in the case of *E. coli*) or per particle in the case of bacteriophages. This consistency enables the $C_0 t$ curves given by the genomes of more complex organisms to be interpreted with some confidence.

When DNA from a higher plant or animal is subjected to the same sort of experiment, a rather complex $C_0 t$ curve is always found. One generally finds reassociation proceeding over a wide range of $C_0 t$ values, with sometimes the appearance of a number of more or less well-defined steps. This indicates the presence of sequences showing widely differing degrees of repetition (reiteration). A substantial fraction of the DNA of a eukaryote is always found to reassociate at a very high $C_0 t$ value, consistent with its consisting of sequences present in single copies per genome. At the other extreme, a significant amount of DNA is often found to reform double-stranded structure virtually instantaneously. This ultra-

Fig. 1.10

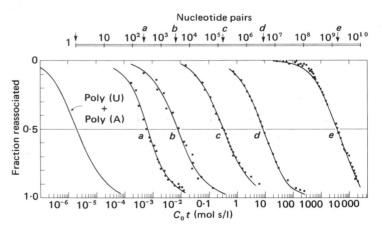

Kinetics of reassociation of double-stranded nucleic acid from various sources. The upper scale shows genome sizes in terms of nucleotide pairs. The nucleic acid samples were: (*a*) the highly repetitious mouse DNA 'satellite' fraction; (*b*) bacteriophage MS2 (a double-stranded RNA phage); (*c*) bacteriophage T4; (*d*) total *E. coli* DNA and (*e*) the non-repetitive fraction of DNA from calf. The poly(U)–poly(A) curve was obtained with synthetic polyuridylate + polyadenylate— i.e. UUUU.../AAAA..., having a repeat unit of only one since *all* base-pairs were U–A. From Britten and Kohne.[7]
The arrows on the upper scale indicate the complexities of the sequences, i.e. the length of each sequence before it repeats itself; some of these were already known from other kinds of measurement.

rapid reaction has a special explanation, namely that it is due to the occasional presence of complementary sequences on the *same* single-stranded fragment, as will occur if there are **inverted repeats** in the original duplex. The occurrence and possible significance of inverted repeats are discussed later in this book (pp. 445, 448). There is also usually a fraction, very variable in amount from one species to another, which reassociates very fast but at a $C_0 t$ value measurably greater than zero (perhaps 10^{-4} to 10^{-3}). This fraction could be explained as consisting of relatively simple, very highly reiterated sequences and that this is indeed the case is shown in several instances by evidence of a different kind.

A simple sequence, because it is simple, may by chance have a base composition sufficiently unlike the bulk of the DNA to be separable from it on the basis of buoyant density (GC is denser than AT). A minor DNA fraction forming a 'satellite' band away from the main band in a density gradient is indeed found in many species; sometimes there is more than one satellite. It is clear that 'low-$C_0 t$' DNA and 'satellite' DNA are largely the same thing. In many cases the simple sequence has been determined chemically. A guinea-pig satellite, for instance, was shown by E. Southern to consist of a sequence of six base-pairs repeated, with minor variations, many millions of times in the genome.

Distinct both from the low-$C_0 t$ satellite DNA and from the DNA with a sufficiently low reassociation rate to be single copy, there is usually found a substantial proportion of DNA with a range of intermediate $C_0 t_{\frac{1}{2}}$ values (often in the range 10^{-1} to 10^2). This is 'intermediate-repetitive DNA', and it consists partly of sequences of known functions (*see* pp. 224–225) and partly of sequences of unknown function, but about which it is most interesting to speculate (*see* pp. 444–455).

The discovery of repetitive sequences has not provided a complete answer to the *C*-value paradox. While DNA-rich species do tend to have higher proportions of repetitive sequences, there are also large differences in quantity of single-copy sequence which still do not seem to correlate very well with differences in biological complexity. We shall return to this problem in Chapter 15.

1.6 The packing of eukaryotic DNA in chromosomes

Both in prokaryotes and, to an even greater extent, in eukaryotes, there is a problem in explaining how the quantity of DNA known to be present is packed into the cell. In bacteria, where a millimetre or more of DNA duplex is packed into a cell measured in micrometres (µm), part of the answer lies in the fact that the native DNA is nearly always in supercoiled form. The double helix needs to have one 360° turn every ten base-pairs in order to be able to lie relaxed with its central axis in a straight line. Either underwinding or overwinding results in the duplex being thrown into supercoils, rather as a piece of rope or a double strand of rubber will form internal coils when one end is twisted and the other held firm. Relaxation of a DNA supercoil can occur by internal rotation if one strand of the duplex is cut by the action of an **endonuclease** enzyme. In this case one strand is able to spin around the other until the number of helical turns compatible with a relaxed extended conformation is attained. The striking effect of nicking one strand of supercoiled closed-loop DNA is shown in *Fig.* 10.1 (p. 252).

The 1300-μm DNA molecule of *E. coli*, loosely termed a 'chromosome', though it does not conform to the microscopists' original definition of that term, is extensively supercoiled. It is further compacted in a series of folds and loops, the organization of which is not well understood. In the true chromosomes of eukaryotes, on the other hand, the packing of DNA takes a special and characteristic form about which a good deal has been learnt in recent years. The most direct evidence in favour of the idea that each single chromosome contains a single duplex DNA molecule has been obtained in bakers' yeast (*Saccharomyces cerevisiae*) and *Drosophila melanogaster*. In the former species the chromosomes are so small that DNA molecules sufficient in size to account for all the DNA of a chromosome should be capable of being resolved and followed from end to end under the electron microscope. Petes et al.[8] have indeed found and measured yeast DNA molecules of the requisite sizes. The very much longer duplexes of *Drosophila* chromosomes have been sized by Kavanoff and Zim[9] using viscosity measurements, and again the molecules seemed to be large enough to correspond to whole chromatids. These two organisms are admittedly low on the scale of DNA quantity but it seems probable, nevertheless, that the single-duplex theory is valid for eukaryotes in general. Evidence of a totally different kind, on the mode of disjunction of DNA at chromosome division (*see* p. 29), points in the same direction.

The previous paragraph leads to the conclusion that an unreplicated chromosome or chromatid contains a single DNA duplex running from end to end. How, then, are we to explain how such an immensely long molecule is packed into the relatively short and thick chromosome? It is now clear that a considerable part of the necessary packing is achieved by complexing of the DNA with a set of special proteins called **histones**. These molecules, which are basic because of their high content of the basic amino acids arginine and lysine, are of five classes: H1, H2a, H2b, H3 and H4. The last four of these are astonishingly similar in structure in all eukaryotes in which they have been studied. It is as if, having evolved a set of molecules to make a perfect fit to DNA, which is always of the same structure and lateral dimensions, eukaryotic organisms have been restrained by natural selection from any further change in this respect.

A great deal is now known about the mode of association of DNA with histones, and present knowledge, based on X-ray analysis, is summarized in *Fig.* 1.11. The fundamental unit of organization is the **nucleosome**, in which a length of DNA duplex of about 140 base-pairs is wound twice around a protein core composed of two molecules each of H2a, H2b, H3 and H4. Each nucleosome is joined to the next by a DNA 'spacer' of about 100 base-pairs, the exact length being somewhat variable.

Nucleosome structure first became evident when chromosomal material (chromatin) was treated with a nuclease enzyme from *Micrococcus* spp. that cuts double-stranded DNA, and the size distribution of the fragments was examined in the ultracentrifuge. A regularly graded series of fragments was found, and turned out to be runs of one, two, three etc. nucleosomes. The nuclease cuts the DNA between the nucleosomes in a random way, but the DNA within each nucleosome is protected. Further evidence for the nucleosome model came from electron microscopy. When chromatin is extracted with saline of suitable ionic strength the H1 histone is removed and the residual material, when prepared for electron

Fig. 1.11

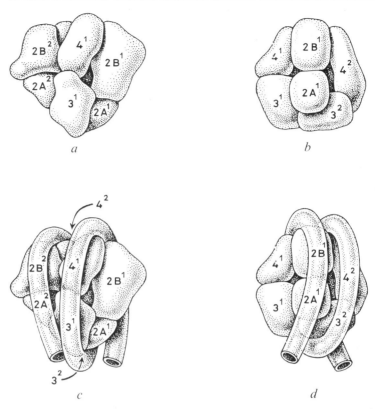

Model of nucleosome structure based on the X-ray crystallographic analysis of Klug et al.[14] The nucleosome contains eight protein subunits—two each of four different kinds of histone molecule, H2A, H2B, H3 and H4. It can be dissociated into a horse-shoe-shaped tetramer with composition H4–H3–H3–H4 and two dimers, each H2A–H2B, one on each face of the 'horseshoe'. Two views of the nucleosome, one rotated 90° with reference to the other are shown in (a) and (b); (c) and (d) show how the double-stranded DNA, about 140 nucleotide pairs in length, is wound around the nucleosome.

microscopy, shows a 'beads-on-a-string' appearance (*Fig. 1.12b*). The H1 histone seems to have the function of stabilizing the supercoiling of the DNA between the nucleosomes, pulling the nucleosomes together into a fibre of a fairly uniform 100-Å fibre which then forms higher-order helix of about 250 Å; indications of this can be seen in *Fig. 1.12a*.

Apparently, the histones remain associated with the DNA during replication. There is evidence that when the DNA replicates, all the pre-existing histone molecules are retained on one of the daughter duplexes while new histone molecules are synthesized on the other.[15]

Chromatin also contains **non-histone proteins** of various kinds, but these are about ten times less abundant than the histones. Prominent among them are the **high mobility group** proteins,[41] which are recognized by their small size (about 30 000 daltons or less), high content of charged amino acids and consequent rapid

Fig. 1.12

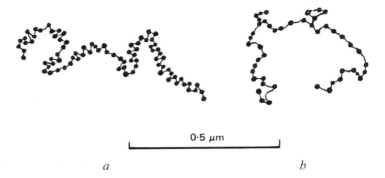

0·5 μm

a   b

a. Chromatin of rat liver spread in 0·2 M EDTA
to partly relax the fibre structure. Nucleosomes, seen as small spheres, are
helically packed to give a fibre of about 200 Å (i.e. 20 nm or 0.02 μm) diameter.
b. Similar preparation as in (a) except that histone H1, which is involved in
internucleosome interaction, has been removed by high salt concentration;
nucleosomes now appear strung out in a fibre with maximum diameter 100 Å
(0·01 μm).
Drawn from the electron micrographs of Thoma and Koller.[10]

c

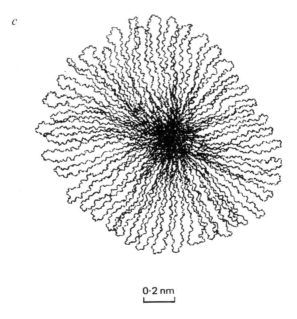

0·2 nm

c. Slightly idealized drawing of an electron microscope photograph of a transverse section of a human metaphase chromatid after treatment with EDTA.
100-Å chromatin fibrils (see (b)) are disposed in radial loops with each loop
pinned at its bases to a protein core.
Based on the electron micrograph of Marsden and Laemmli.[11]
In another paper,[12] comparable pictures were presented of chromosomes
which had been treated to remove histones and hence nucleosome structure.
The loops remained intact as naked DNA, considerably extended in length,
with no free ends—consistent with one continuous molecule per chromatid.

migration in electrophoresis. Non-histone proteins seem to be concerned in the transcriptional synthesis of ribonucleic acid (*see* p. 39).

Even two orders of coiling of the DNA duplex, the supercoiling in the nucleosome and the further coiling of the 100-Å nucleosome fibre to make a 250-Å 'thick fibre', go only a small way towards accounting for the full thickness of the eukaryote chromosome even in its most extended condition. Crick and his colleagues[16] have estimated that the largest human chromosome, for example, contains a DNA duplex which would measure about 8×10^4 µm if fully extended, and that it required a linear contraction of some 1300-fold to fit into the longest observed chromosome thread. Superimposed on the $7 \times$ contraction in the nucleosome string, and a further $6 \times$ in the formation of the thick fibre, a yet higher-order helix (or 'solenoid') was seen in some electron microscope preparations. This gives a further contraction of $40 \times$ and a chromatid thickness of 4000 Å or 0·4 µm, close to the borderline of visibility with the light microscope.

The 'solenoid' model is, however, not the only one available for explaining the higher-order contraction of chromosomes. Laemmli and his colleagues have studied, with the electron microscope, the appearance of human metaphase chromosomes (*see Fig.* 6.7) after they have been subjected to various treatments to relax their tightly compact structure. After treatment with ethylenediamine tetraacetic acid (EDTA), an agent which removes divalent cations such as magnesium, the chromosomes showed a pattern of looped nucleoprotein fibres (strings of nucleosomes) apparently pinned at intervals to a much shorter protein core running along the axis of each chromatid (*Fig.* 1.12*c*). Removal of the histones prior to fixation and electron microscopy did not break the looped structure; naked DNA molecules, visualized by the Kleinschmidt technique (cf. p. 13), were seen to be still attached in extended loops to what was evidently a non-histone protein axis. Whether the gathering of the nucleosome strings into loops is an alternative to packing them in 'solenoids', or whether both modes of contraction can coexist in highly condensed chromatin, is not at present clear.

1.7 Mechanism of disjunction of DNA

Every organism has a mechanism to ensure that, at cell division, each daughter cell receives a complete set of DNA sequences. The process of equal distribution of DNA at cell division is called **disjunction**.

Prokaryotes

In prokaryotes the problem of disjunction is simplified by the complete set of essential DNA sequences being present in a single enormous molecule. Like many other things in its bacterial cell, the DNA is attached to the cell membrane, probably at or close to the origin of replication. It seems likely that, as a round of DNA synthesis is initiated, the membrane attachment site splits into two, one part joined to each of the two daughter duplexes formed between the diverging replication forks. As the cell elongates in preparation for the next division, the membrane at the equator of the cell grows lengthwise, carrying apart

Fig. 1.13

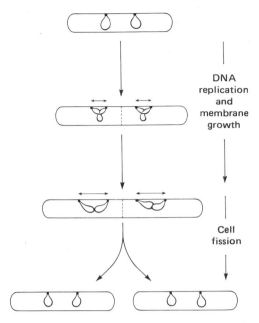

DNA
replication
and
membrane
growth

Cell
fission

Postulated mode of division of a rod-shaped bacterium such as *Escherichia coli*. The DNA is a closed loop, much longer and more elaborately supercoiled than shown in the diagram, attached to the cell membrane at a point close to the DNA replication origin (cf. *Fig.* 1.6). After DNA replication (or perhaps while it is proceeding), the two daughter molecules (genomes) are supposed to be segregated by growth of the cell membrane between them.
Based on Ryter.[13]

the two DNA attachment sites and separating the daughter duplex loops as they are formed. In rapidly growing cultures of *E. coli* and other rod-shaped bacteria a further round of DNA synthesis usually starts before cell fission is completed by septum formation; indeed, replication may start again at the origin even before the replication forks from the previous initiation have reached the terminus. Each cell contains either two or four nuclear bodies, depending on whether cell fission is about to occur or has just occurred. In *Fig.* 1.13 a mechanism is summarized which was first proposed by Jacob and Cuzin and supported by electron microscope observations of membrane–DNA association and membrane growth in *Bacillus*.[13] Plasmids, the separate relatively small DNA loops present in many bacterial cells, are probably distributed at cell division in much the same fashion. The restriction of many of them to one or a few copies per cell may reflect a requirement for attachment to specific membrane sites which are in limited supply.

Mitosis in Eukaryotes

Disjunction of chromosomes in eukaryotic cells is an altogether more complicated process. The total process of cell division is called **mitosis**. *Fig.* 1.14 shows the cycle of events, as it affects the nucleus, in *Tradescantia paludosa*, a representative of a plant group with particularly large and clearly visible chromosomes. Measurements of quantity of DNA (by photometry of stained preparations) show that the amount of DNA doubles some time, usually hours, before the microscopically observable changes in the nucleus prior to cell division. This period of DNA synthesis is called S-phase, the interval between S and mitosis is G2 and the interval between mitosis and the next round of DNA synthesis is G1. Mitosis itself is heralded by a progressive condensation of the chromatin. As this stage, called **prophase**, proceeds, the chromosomes, hitherto too finely spun-out to be visible as separate entities, become increasingly apparent. In species with reasonably large chromosomes the lengthwise division of each chromosome into half-chromosomes (**chromatids**) can be made out at middle to later prophase.

At the conclusion of prophase, when the chromosomes are maximally condensed and usually obviously split into chromatids, a series of dramatic changes marks the transition to **metaphase**. Except in the fungi and certain protozoa (*see below*), the nuclear envelope disappears, so that there is no longer any apparent barrier between the chromosomes and the cell matrix (the cytoplasm). The chromosomes become enmeshed in a system of protein fibres, called the **spindle**, occupying a space the shape of a cylinder sometimes attenuated towards its ends (the **poles**). The spindle fibres are in parallel alignment and in fungi (*see Fig. 1.15*),

Fig. 1.14

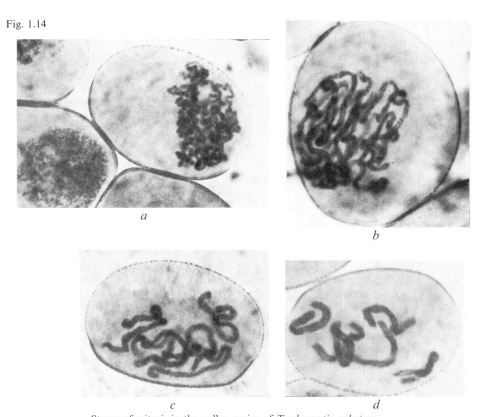

a

b

c

d

Stages of mitosis in the pollen grains of *Tradescantia paludosa*.
a. Early prophase (pre-prophase lower left). *b*. Mid-prophase—note chromosomes visibly divided into chromatids, with paired chromomeres. *c*. Late prophase—six divided chromosomes visible. *d*. Pro-metaphase. Staining with acetic-orcein.

algae and animals (though not in higher plants and some protozoa), they converge at each pole to join a **spindle pole body**, sometimes called a centrosome (cf. *Fig.* 1.15). The spindle pole body is an interesting example of an apparently self-maintaining structure which probably does not contain nucleic acid. During early prophase there is a single body attached to the nuclear membrane and, as prophase proceeds, it appears to divide into two, the products taking up their new polar positions as the spindle is formed. Some spindle fibres may stretch from pole to pole, but others join to the chromosomes. At metaphase, each chromosome is held in balance on the equator of the spindle between fibres extending to opposite poles. Most eukaryotes have on each chromosome a specific **spindle fibre attachment site**, sometimes called a **kinetochore** or, as in this book, a **centromere**. Centromeres can often be recognized as constrictions, points of minimal thickness, and also as points at which the chromosomes appear undivided when the regions to each side (the chromosome arms) are visibly split into chromatids. A few taxonomically scattered groups, including some insects, arachnids and flowering plants, are said to have 'diffuse centromeres', which means that spindle fibres can attach virtually all the way along the chromosomes rather than at one special site.

In the great majority of eukaryotes, there are localized centromeres and the chromosomes at metaphase tend to become arranged with their centromeres

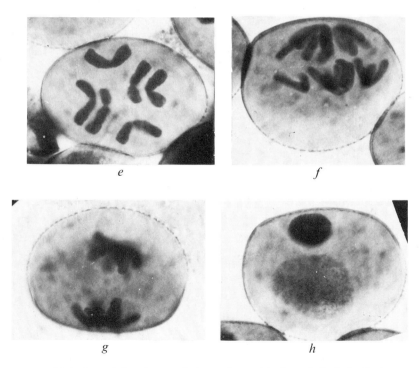

e. Metaphase. *f*. Anaphase. *g*. Late anaphase—telophase. *h*. End of mitosis. Note differentiation of large diffuse 'vegetative' nucleus which will not divide again and the dense germinal nucleus which will divide again in the pollen tube to produce two nuclei fertilizing the egg and endosperm respectively (cf. *Fig.* 2.5). Magnification × 1100.

Fig. 1.15

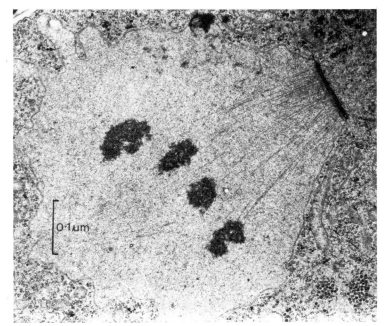

Electron micrograph of a section of an ascus of the fungus *Ascobolus stercorarius,* showing four metaphase bivalent chromosomes, a part of the spindle, and one of the two plaque-like spindle pole bodies (the other is out of the plane of the section). This is the first metaphase of meiosis (cf. *Fig.* 3.1), but the spindle in mitosis is similar. Spindle fibres can be seen connecting the pole body to chromosome centromeres; others run from one pole to the other. Note that in fungi, though not in higher plants and animals, the nuclear envelope (a double-layered membrane) remains intact throughout nuclear division; the spindle pole body is attached to it. Note the mitochondria and the small densely stained ribosomes in the cytoplasm of the cell.
The photograph, kindly supplied by Dr Denise Zickler,[25] is reproduced here by permission of the publishers.

towards the centre of the spindle equatorial plane and the arms radiating towards the periphery. A mitotic metaphase sectioned or squashed in the equatorial plane gives the best picture of the **karyotype**, that is the whole array of chromosomes with their characteristic sizes and centromere positions. A selection of karyotypes is shown in *Fig.* 1.16; the great range of chromosome sizes (reflecting DNA quantity) found among eukaryotes is very evident.

At the onset of the next stage of mitosis, **anaphase**, the division of the chromosomes is completed by the lengthwise splitting of the centromeres. The chromatids now become chromosomes in their own right, each with its individual centromere. The pairs of daughter chromosomes appear to be pulled apart towards opposite poles by the spindle fibres, the centromeres leading the way. Depending on whether the chromosomes have centromeres sited centrally (**metacentric**), off-centre (**acrocentric**) or terminally (**telocentric**), they appear at anaphase as V, J or I shaped (*see Fig.* 1.14*e*). At **telophase**, the movement is completed, the two separated groups of chromosomes are clustered at the poles, the spindle disappears, and a new nuclear envelope is formed to enclose each chromosome group. At this stage the metaphase contraction of the chromosomes begins to relax, and the chromosomes gradually revert to the finely spun-out condition

Fig. 1.16

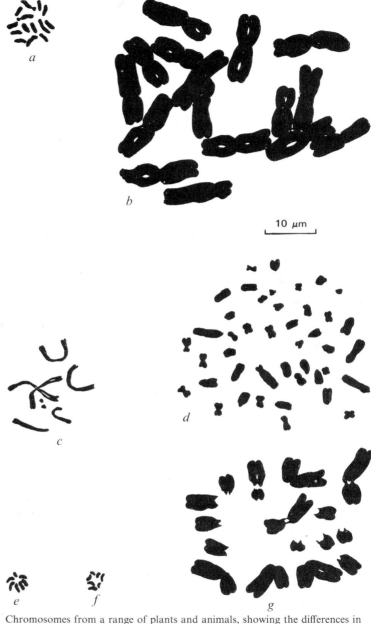

10 μm

Chromosomes from a range of plants and animals, showing the differences in size. All are seen at metaphase and are drawn at the same magnification (*see* scale bar).
a. Medicago murex, a leguminous plant; from Lesins et al.[26]
b. Allium cepa, onion; from Bennett and Rees.[27]
c. Drosophila melanogaster, fruit fly; from Gall et al.[28]
d. Rattus rattus, rat; from Young and Dhaliwal.[29]
e. Dictyostelium discoideum, slime mould; from Robson and Williams.[42]
f. Neurospora crassa, Ascomycete fungus; from Singleton.[30]
g. Myrmelliotettix maculatus, grasshopper; from Hewitt and John.[31]

characteristic of non-dividing nuclei. At the same time, the nucleolus, which usually dwindles and disappears during prophase, is quickly reformed—a sign that the nucleus, which appears to be inert during mitosis, is once again metabolically active. The completion of division of the cell into two new cells, each with a completely enclosing cell membrane, and equipped with a nucleus containing the same set of chromosomes as the mother cell, usually follows closely after telophase.

In fungi there is a major variation on the above scenario. Instead of disappearing at the end of prophase, the nuclear envelope remains intact throughout mitosis. The spindle is formed inside the envelope which assumes a dumb-bell shape as mitosis proceeds, eventually pinching apart into two spherical ready-made envelopes for the two daughter groups of chromosomes.

The spindle mechanism has the effect of ensuring that each new cell receives the same essential set of chromosomes as was present in the mother cell. It is not at present clear whether there is any parallel mechanism for distribution of the equally essential DNA of mitochondria and (in plants) of plastids. It is commonly assumed that, since there are usually many of these organelles per cell, a random apportionment will nearly always result in each daughter cell being seeded with at least one.

1.8 Evidence that replication of chromosomal DNA is semiconservative

If each chromatid really contains only one DNA duplex (the one-thread or **unineme** hypothesis), and if DNA replication is semiconservative, then the replication of the chromosome must also be semiconservative, at least so far as its DNA is concerned. In 1957, J. H. Taylor published evidence that this was indeed the case. In his experiments, the DNA synthesized in one round of replication was made radioactive by supplying the cells (of bean roots in the first trial) with [^{3}H]thymidine. He followed the distribution of the radioactive label through subsequent rounds of replication in the absence of the radioactive precursor. The method involved the application of radiosensitive film to microscope slides carrying spread metaphases, and noting which chromatids were radioactive as shown by silver grains in the film above them. The results were reasonably clear: the first cells to come to metaphase after labelling had radioactivity equally distributed between the two chromatids of each chromosome, but after one more round of chromosome replication each chromosome had one labelled and one unlabelled chromatid. This was exactly as predicted by the unineme hypothesis with semiconservative replication. The picture was a little complicated by the occurrence of apparent switches of radioactivity from one chromatid to the other at one or more points in some chromosomes, as if sister chromatids were occasionally exchanging corresponding segments by some kind of breakage and reunion (**sister-strand exchange**). A limitation of the technique was the scatter of the silver-grains, which sometimes made it difficult to distinguish an

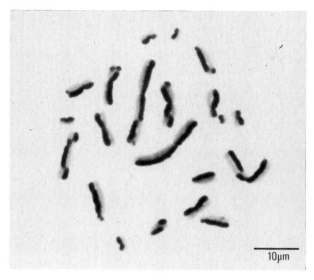

a

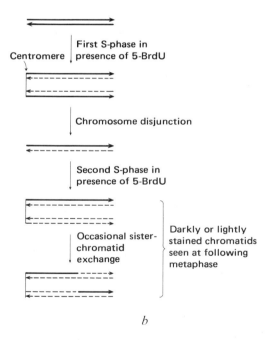

Centromere

First S-phase in presence of 5-BrdU

Chromosome disjunction

Second S-phase in presence of 5-BrdU

Occasional sister-chromatid exchange

Darkly or lightly stained chromatids seen at following metaphase

b

a. Demonstration of semiconservative chromosome replication in cultured Chinese hamster cells by differential Giemsa staining of chromatids which have incorporated 5-bromodeoxy-uridine (5-BrdU).[18] The metaphase chromosomes each have 5-BrdU incorporated into one strand of the DNA duplex in one chromatid and into both strands of the other chromatid. The effect of the incorporated BrdU is to reduce the staining intensity. Note the occasional exchanges between chromatids (sister-chromatid crossing over). Photograph by courtesy of Professor H. J. Evans.
b. Diagram to show the interpretation of (*a*). 5-BrdU-substituted DNA strands are shown as interrupted and non-substituted strands as continuous lines.

unexpected labelling of both chromatids at a particular point (**isolabelling**) from closely placed sister-strand exchanges. During the past decade the development of a new and elegant method of labelling newly synthesized DNA has removed this sort of ambiguity.

The newer technique depends on the fact, which is unexplained but convenient, that DNA in which the thymine (i.e. 5-methyluracil) is partly replaced by its analogue 5-bromouracil stains less strongly with the Giemsa reagent (cf. p. 35). Thus, if one essentially repeats the Taylor experiment, but supplying 5-bromo-deoxyuridine (5-BrdU) instead of [³H]thymidine as a labelling precursor, it is possible to locate the label with a precision limited only by the resolving power (less than 1 μm) of the light microscope, rather than the several micrometres scatter of radiation following the disintegration of a tritium atom. *Fig.* 1.17 shows an example of the brilliantly clear demonstration of semiconservative chromosome replication which is possible by the 5-BrdU/Giemsa technique. It also shows very clear sister-strand exchanges. Some of these appear to be induced by the presence of the 5-BrdU (or tritium) which is necessary for their visualization, but other evidence suggests that they also occur naturally at some frequency.

1.9   Chromosome landmarks

Although all chromosomes have a common basic architecture, they do possess individual features which are of great use to the geneticist in attempts to correlate functions and particular chromosome regions. The first characteristic of a chromosome is its length, and this by itself may sometimes distinguish it from other members of the haploid set. The second is the position of the centromere, which can be seen as a constriction at metaphase. Other constrictions can also be seen in some chromosomes at metaphase, most notably the nucleolus organizers, of which there is at least one per haploid set. The lack of condensation of nucleolus organizers is probably connected with their continued attachment to the nucleolus at the time when contraction of the rest of the chromosomes commences.

Heterochromatin

Another distinguishing feature of some chromosomes is the presence of regions called **heterochromatin**, which remain heavily condensed at times when the other regions, the **euchromatin**, are highly extended, that is in the non-dividing nucleus and early prophase. In many species heterochromatin tends to be concentrated in the regions flanking the centromeres, but it also occurs sporadically elsewhere. Sex chromosomes are often largely or entirely hetero-chromatic. The Y-chromosome of *Drosophila* spp. or of mammals is usually permanently condensed; the complex and interesting situation with regard to the mammalian X-chromosomes is reviewed in Chapter 17 (p. 509).

With the development of methods (*see* p. 19 *above*) for fractionating DNA sequences according to their different degrees of repetitiveness, it has become possible to show that heterochromatin is rich in very highly repetitive ('satellite') sequences. The technique is to treat a chromosome preparation spread on a microscope slide in such a way as to separate the complementary strands of the DNA duplexes and then to immerse the slide, under conditions favouring the reformation of duplex molecules, in a solution containing radioactively labelled

Fig. 1.18

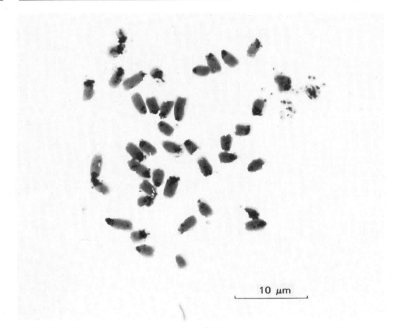

In situ hybridization of tritium-labelled (³H-labelled) mouse satellite DNA to mouse metaphase chromosomes, revealed by silver-grains formed in a superimposed radiosensitive film. The radioactive satellite finds complementary sequences to bind to mainly in the centromeric regions at the chromosome ends.[19]
Photograph by courtesy of Dr K. W. Jones.

single-stranded satellite DNA. The chromosome regions containing satellite corresponding, and complementary, to the labelled 'probe' sequences will assimilate some of the label into stable hybrid duplexes and will remain radioactive after rinsing off excess probe. Exposure of the resulting preparation, after drying, to radiosensitive film reveals which chromosome regions contain the satellite sequence. The result of one of the first experiments on the distribution of mouse satellite DNA is shown in *Fig.* 1.18; it will be seen that it occurs around the centromeres of all the chromosomes. That heterochromatin DNA consists largely of highly repetitive relatively simple sequences is probably true for all species. For example, the heterochromatin at the base of the *Drosophila* X-chromosome has been shown by Peacock[17] to consist of blocks of different satellite sequences arranged in a characteristic pattern. Heterochromatin is usually, and perhaps always, inert genetically; that is to say it is responsible for no indispensable functions and is not transcribed into RNA (*see* p. 450).

Chromomeres and Polytene Chromosome Bands

In addition to the major blocks of heterochromatin, smaller condensed regions, in a characteristic pattern for each chromosome, can be made out at prophase in many species. These are called **chromomeres**, and they serve as

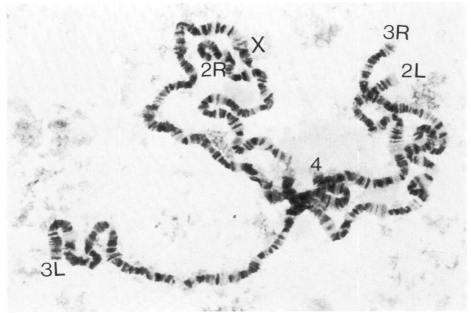

Fig. 1.19   A squashed and stained salivary gland nucleus of *Drosophila melanogaster* showing the five long and one very short polytene chromosome arms radiating from the centromeres fused in the chromocentre. Note that chromosomes II and III have medial centromeres and so each is represented by a left (L) and a right (R) arm. Each arm consists, in fact, of two closely paired homologues. Photograph by courtesy of Dr Lalji Singh. Magnification × 600.

important landmarks in studies of chromosome rearrangements. They are especially evident at the pachytene stage of meiosis (*see* p. 60).

Chromomeres can be seen in the finest detail in the hypertrophied (**polytene**) chromosomes of the nuclei of the cells of salivary glands and certain other protein-secreting organs of *Drosophila* and other insects of the order Diptera. Polytene (i.e. many-stranded) chromosomes consist of about a thousand parallel chromatin strands formed by some ten successive replications without separation. The individual strands remain extended as in an ordinary non-dividing nucleus. As a result of the lateral multiplication, the chromomeres appear as bands, ranging from prominent to very faint. About 5000 bands can be distinguished in all, and experts can distinguish each small section of each chromosome by its characteristic pattern of darker and fainter bands. We shall see in Chapter 15 (pp. 431–435) the value of polytene band analysis in working out the functions of specific chromosome loci.

Polytene nuclei show a number of peculiarities not seen in ordinary nuclei, in addition to chromosome hypertrophy. The chromosomes are paired, homologue to homologue, as they are in the pachytene of meiosis (p. 60) but not usually in somatic cells. The centromere regions are aggregated in an apparently sticky mass called the chromocentre. Thus a well-spread preparation of a salivary gland nucleus shows a haploid number of chromosome arms (each in reality a pair of closely associated homologues) radiating from the chromocentre (*Fig.* 1.19). Another interesting feature is that the heterochromatic centrometric regions, rich in satellite DNA sequences, do not share in the general magnification and are thus greatly diminished relative to the euchromatic chromosome arms. This lent some plausibility to the proposal that, if centromeric heterochromatin has a function at all, it is connected with chromosome movement, but confirmation of this idea has been hard to get. A terminally differentiated cell, such as that of a salivary gland, will never divide again but needs to be very active with respect to certain DNA-dependent metabolic functions; selective magnification of euchromatin and not heterochromatin can be seen as an adaptation to its special role.

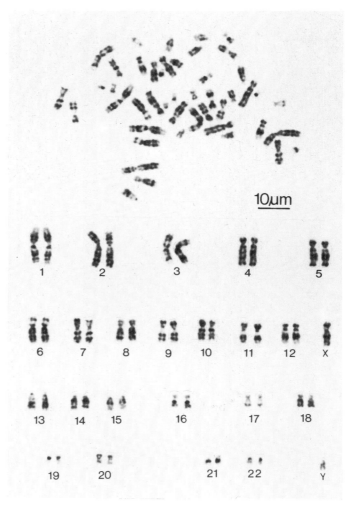

a. A normal human male karyotype (diploid chromosome set) stained by the G-banding technique. The upper part of the figure shows a metaphase set of chromosomes in their original positions after flattening on the microscope slide. The lower part shows the chromosomes cut out of the photograph and arranged in matching pairs. The banding patterns permit much more accurate matching than would otherwise be possible.

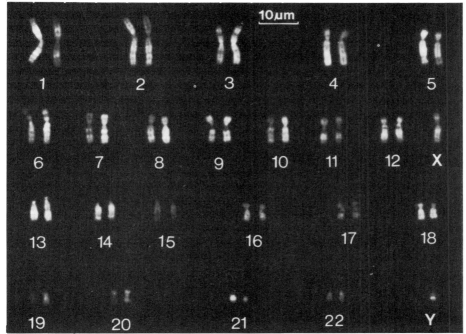

Fig. 1.20*b*   *b*. Another normal male karyotype stained by the Q-banding technique. The bands show fluorescence induced by ultraviolet light.
Photographs by courtesy of Professor H. J. Evans.

Chromosome Banding Techniques

Within the past decade a new set of staining techniques has been used to reveal many hitherto concealed chromosomal landmarks. These techniques are known collectively as chromosome banding,[20] and are of four main types:

1. *Q-banding*, which was the first to be developed (by T. Caspersson and his associates in Sweden), involves staining the chromosomes with the fluorescent dye quinacrine. Certain chromosomal blocks and bands fluoresce much more brightly than others in ultraviolet light and can be seen with a fluorescence microscope. The pattern is reproducible and can be used to distinguish chromosomes of otherwise similar size and form. There is evidence, that the Q-bands, as the more highly fluorescent regions are called, are rich in A–T base-pairs.

2. *G-banding* is obtained by staining with the complex **Giemsa** dye mixture. After appropriate pretreatments, for example with acetic acid and salt, Giemsa staining reveals a detailed pattern of more and less darkly stained bands. The banding seems to be related to the chromomere pattern, which can be seen in prophase chromosomes (*Fig.* 1.14*b*) following staining by traditional procedures. It probably represents the different densities of packing of DNA. Examples of the application of the Q and G-banding techniques to human chromosomes are seen in *Fig.* 1.20.

3. *R-banding* is that seen when chromosome preparations are stained after treatment with phosphate buffer at pH 6·5 and 87 °C to partially convert ('melt') part of the DNA to single-stranded form, thus reducing its stainability. It is thought that R-bands represent those chromosome regions richest in G–C base-

pairs and thus most resistant to melting. As would be expected on this hypothesis, R-banding gives about the same pattern as Q-banding but in a negative form; Q-bands correspond to R-interbands and Q-interbands to R-bands. R-bands can be stained either with Giemsa reagent or with acridine orange, which is a dye giving either green or orange fluorescence depending upon whether the DNA is double stranded or single stranded.

4. *C-banding* is obtained by Giemsa staining following alkaline extraction of a large part of the DNA from the chromosomes. What is left tends to be the highly compacted DNA of the heterochromatin, and the C-bands (often considerable blocks of material) usually indicate centromeric and other heterochromatic regions.

The various banding techniques find many applications in modern cytogenetics, some of which are referred to later in this book (pp. 237, 241).

1.10   How DNA exercises control

A central achievement of the science of genetics, which will be documented in the following chapters, has been the analysis of the hereditary transmission of biological traits in terms of what were at first abstract entities, genes and linkage groups (*see* Chapter 3), but which turn out to correlate perfectly with DNA and chromosomes. The clinching evidence of the role of DNA is the phenomenon of **transformation** in which a heritable trait can be transferred to one cell by treating it with purified DNA isolated from another. Transformation with free DNA molecules was first demonstrated in various species of bacteria (*see* pp. (275–276) and it has now been shown to work, under appropriate conditions, with yeast and animal cells (pp. 242–245).

Working in parallel with the geneticists, biochemists have, during the past 25 years, shown by their own methods how the chemical make-up of a cell is ultimately determined by its DNA.

In the briefest summary, DNA is **transcribed**, by a complementary copying process analogous to DNA replication, into ribonucleic acid (RNA) which then, after some further **processing**, functions in protein synthesis. A part of the RNA consists of **messenger** (mRNA) molecules which, through a process of **translation**, code for the polypeptide chains of proteins. The proteins, acting as specific chemical catalysts (**enzymes**) and as structural components, are in direct control of cell metabolism and growth.

1.11   Summary and perspectives

This chapter has been concerned with biochemistry and cytology, that is the microscopical study of cells, rather than with genetics as such. Both kinds of study point to the genetical importance of DNA. Biochemistry tells us that DNA is synthesized by a self-copying process, which is one of the essential properties of the genetic material, and that, through transcription into RNA and the control exercised by the RNA over protein synthesis, it is capable of determining the macromolecular composition, and hence the structure and metabolism of the rest of the cell. Cytology tells us that all cells, of both prokaryotes and eukaryotes, possess mechanisms which ensure the equal distribution of the DNA at cell division. In eukaryotes the problem of equal distribution of an enormous length of DNA is solved through the packing of the DNA into chromosomes and the separation of the latter at anaphase of mitosis by means of the spindle mechanism. Even without genetics, classically defined as the study of the inheritance of variation, we might well be convinced that DNA must be the substance in which heredity is encoded.

What biochemistry and cytology cannot easily tell us, however, is how the DNA is locally differentiated into functional units, how these units are arranged, and how they act and interact to control the development of the organism. To answer these questions one needs to use the classical approach of genetics which, roughly speaking, is to find out where the functional units are and what they do by studying their variation. We can investigate how genetic variation is transmitted, which leads us to genetic mapping, and how it affects the biochemistry and morphology of the organism, which leads to knowledge of the functions of the genetic units or genes.

Genetic variation will be the central concern of the remainder of this book.

An Outline of Transcription and Translation from DNA

Readers who have had some introductory molecular biology can now proceed to Chapter 2. For others, this appendix attempts to provide, in skeletal form, the biochemical background necessary for the understanding of later parts of the book.

I. Transcription

We have already seen how DNA is replicated by the transfer of successive deoxynucleotide residues from deoxynucleoside triphosphates to the 3'-hydroxyl end of the growing chain, and how a pre-existing DNA chain is used as a template to determine which residue shall be added at each step. A very similar mechanism is used for the synthesis of ribonucleic acid (RNA). DNA is again the template, but a difference from DNA replication is that, usually, only one strand of a DNA duplex is used for complementary copying of an RNA strand (**transcription**) at any particular point in the duplex.

All RNA (except in certain viruses where it is self-replicating—*see* pp. 280, 426) is transcribed from DNA. This is confirmed by the finding that all cellular RNA molecules will bind in stable DNA/RNA heteroduplexes when allowed to associate with single-stranded DNA from the same species. Such stable molecular hybridization requires a close complementary matching between the associating strands and is thus consistent with the RNA having been transcribed from the DNA.

The immediate precursors of RNA synthesis are ribonucleoside triphosphates (*Fig. 1.4a*), just as those for DNA synthesis are deoxyribonucleoside triphosphates. Essentially the same base-pairing rules apply as in DNA synthesis, but with the difference that uracil in RNA replaces thymine (which is 5-methyluracil) in DNA. Uracil pairs with adenine in the same way as thymine does. Again as in DNA synthesis, RNA chain growth is in the 5'→3' direction, with the template oppositely orientated 3'→5'. It is believed that, when it is being used for RNA transcription, the DNA duplex opens up locally to allow the RNA-synthesizing enzyme (one of a family of **RNA polymerases**) access to the template strand. The opening point travels along the duplex to allow transcription to proceed along the template; the transcribed RNA strand is peeled off as it is synthesized and the DNA duplex closes up behind it.

In prokaryotes there may be only one type of RNA polymerase. In eukaryotes, on the other hand, three species have been distinguished, responsible for transcribing different DNA sequences. Polymerase I synthesizes the main RNA molecules of the large and small ribosome subunits, polymerase II makes all the different kinds of **messenger RNA** (i.e. coding for proteins) and polymerase III is responsible for the synthesis of the small RNA molecules involved in protein synthesis, including transfer RNAs and 5-S RNA. These kinds of RNA are reviewed below (p. 40). The RNA polymerases are complex molecules each containing several different polypeptide chains, some common to all three enzymes and some present in only one.

Transcription of DNA in eukaryotes seems to occur without loss of nucleosome structure, even though the DNA binding to histones might have been expected to get in the way of RNA polymerases. However, some change in nucleosome structure is associated with transcription, the effect being to make the DNA susceptible to DNA-degrading enzymes including deoxyribonuclease I. Presumably there is a partial unpacking of nucleosome components which gives access to RNA polymerase, and at the same time exposes the DNA to nuclease attack. Sensitivity to DNAase I is correlated with the presence of high mobility group proteins,[41] which may have some essential *general* function in rendering chromatin transcribable.

Fig. 1.21

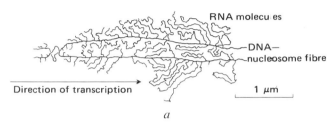

Direction of transcription   1 μm

a

Visualization of transcription of RNA with the electron microscope.
a. Paired DNA–nucleosome fibres immediately after replication in the early *Drosophila* embryo. Both DNA duplexes are being transcribed in an identical way with RNA molecules growing from left to right. The longest transcripts here are of the order of 5 kilobases in length. Nucleic acid molecules coated with cytochrome *c* for better visibility.
Drawing based on the photograph of McKnight and Miller.[21]

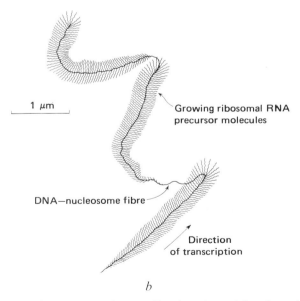

1 μm

Growing ribosomal RNA
precursor molecules

DNA–nucleosome fibre

Direction
of transcription

b

b. A single DNA–nucleosome fibre from *Drosophila* embryo showing transcription of tandemly repeated genes for ribosomal RNA.
Drawing based on the electron micrograph of Laird and Chooi.[22]
Note the high frequency of transcription initiation (close spacing of transcripts) as compared with the unknown gene shown in (*a*).

(For a recent review *see* Ref. 41.) Other non-histone proteins are less well characterized.

Electron microscopy allows transcription of DNA into RNA to be directly visualized (*Fig.* 1.21). In *Drosophila* abundantly transcribed polytene chromosome bands can be identified by their hybridization to their radioactively labelled transcripts (*see* p. 433). The most transcriptionally active bands sometimes appear as 'puffs' of diffusely unfolded chromatin strands. In the 'lampbrush' chromosomes of the oocytes of newts and other amphibia (*see* p. 68) the lateral loops can likewise be shown to be coated with recently synthesized RNA. Both polytene bands and lampbrush loops appear to be units of coordinated transcription.

Not all DNA sequences are transcribed. 'Satellite' (highly repetitive) DNA sequences appear never to be transcribed at all. Furthermore, the set of sequences that is transcribed in one type of cell is usually not the same as (though it may overlap) the set transcribed in another type of cell in the same organism. The problem of cell differentiation, with which we deal in Chapter 17, is largely the problem of control of transcription.

II. Processing and Functions of RNA Molecules

The sizes of the RNA molecules produced by transcription are determined by starting and stopping signals in the sequence of the template DNA. There is, however, much more to RNA synthesis than the primary transcription. A considerable amount of further processing is involved as we shall see. Further consideration of the controls of RNA transcription and processing is deferred to Chapter 14.

The function of RNA in general is in protein synthesis. The most abundant kinds of RNA in cells of all kinds are **ribosomal RNA** (rRNA) molecules: these are structural components of the ribosomes, the engines of protein synthesis (*see below*). Many other kinds of RNA molecules, mostly present in small amounts, serve as **messenger RNA** (mRNA). They provide the coded programmes which tell the ribosomes, which have no preference of their own, which particular polypeptide chains to synthesize. Finally, there is **transfer RNA** (tRNA), which is essentially a set of keys for decoding the information for the messenger molecules.

Ribosomal RNA molecules, and the DNA sequences from which they are transcribed, provide some of the best-known examples of gene structure in relation to gene product. Here it will suffice to say that the two major RNA molecules contained in the ribosomes, the one from the large ribosomal subunit about twice as large as the one from the small subunit, are derived from a single primary RNA transcript by a process of cutting and trimming. In prokaryotes the two rRNA species are called 23 S and 16 S after their respective sedimentation coefficients in the ultracentrifuge, and are about 3·0 and 1·5 kilobases (kb) long. In the somewhat larger ribosomes of eukaryotes the two major rRNA molecules are 28 S and 18 S (about 5 and 2 kb, respectively). In association with the large ribosomal subunit there is a much smaller rRNA species which sediments at 5·8 S, and which is also cut out of the same primary transcript. In both pro- and eukaryotes there is another small RNA component (5 S, about 120 nucleotides) which is transcribed separately. Ribosomes also contain about 50 specific protein molecules, one of each kind per ribosome. The DNA sequences which are transcribed into rRNA sequences are present in multiple copies in all organisms (*see* p. 225).

Fig. 1.22

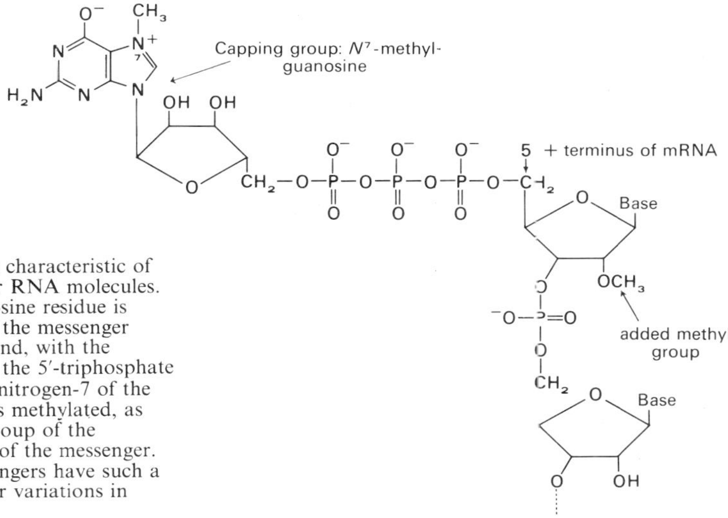

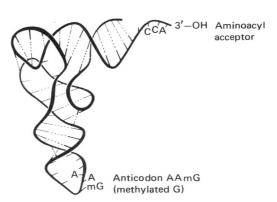

Capping group: N^7-methyl-guanosine

5 + terminus of mRNA

added methyl group

The 'capping' group characteristic of eukaryote messenger RNA molecules. Note that the guanosine residue is joined to the end of the messenger the 'wrong' way round, with the 5'-carbon linking to the 5'-triphosphate of the mRNA. The nitrogen-7 of the capping guanosine is methylated, as is the 2'-hydroxyl group of the terminal nucleotide of the messenger. All eukaryote messengers have such a cap, with only minor variations in structure.

Fig. 1.23

3'—OH Aminoacyl acceptor

Structure of a tRNA molecule (tRNA^{Phe}) from *Escherichia coli*. After Rich and Raj Bhandary.[23]

Anticodon AAmG (methylated G)

Messenger (mRNA) molecules are, at least in eukaryotes, the products of elaborate processing of primary RNA transcripts. Most kinds of eukaryotic mRNA are synthesized in the nucleus and have to be exported across the membrane of the nuclear envelope before they can associate with ribosomes to promote protein synthesis. Before this export takes place the mRNA molecules are modified by the addition of a 'capping' group, a methylated guanosine joined 'back to front' through its 5' group to the 5' terminus of the RNA (*Fig.* 1.22). A second modification is the addition of a 'tail', in the form of a sequence of 25–50 adenylate residues, to the 3' terminus. Messengers coding for histone proteins seem to be exceptional in not having poly(A) tails. Most remarkably, many eukaryotic mRNAs are spliced together from discontinuous segments of the primary transcripts (*see* Chapter 14).

Transfer (tRNA) molecules are also the products of secondary processing of the primary transcripts, and two or more different ones are known, in at least some cases, to be cut out from a common larger precursor. Many of their bases are modified after transcription, by methylation or in other ways. An essential feature of tRNA molecules is their secondary folded structure, which is stabilized by hydrogen-bonded base pairing in a series of hairpin loops (*Fig.* 1.23).

III. Translation into Protein

The code through which mRNA is translated into the specific amino acid sequences of polypeptides is written in terms of adjacent (non-overlapping) sequences of three bases, each such sequence being called a **codon**. The code, which has now been established with great certainty by a variety of experiments both biochemical and genetical, and which applies to prokaryotes and eukaryotes alike (except for a few variations in eukaryote mitochondria—*see* p. 404) is shown in *Table* 1.2. Of the 64 possible codons, AUG uniquely specifies the amino acid methionine and also serves as the signal for the initiation of polypeptide synthesis. Three other codons, UAA, UAG and UGA, code for no amino acid ('nonsense' codons) and are polypeptide chain-termination signals. The other 60 codons all code for one or other of the other 19 amino acids common in proteins. An amino acid may be coded for by 1, 2, 3, 4 or 6 alternative codons, a feature referred to as **degeneracy** in the code.

The decoding and translation of the mRNA depends on the tRNA molecules in partnership with the ribosomes. The tRNA molecules, which in most organisms are of about 40 or more different kinds, are the decoding keys. Each is a sequence of about 80 nucleotides secondarily folded as a result of self-complementary base pairing. In an exposed position at the end of one of the loops is a sequence of three bases, the **anticodon**, which is complementary to, and therefore capable of specifically pairing with, one or more of the codons for a particular amino acid. It should be noted that an anticodon may pair with more than one codon because, as indicated in *Table* 1.2, there is some latitude ('wobble') in the pairing in the third position (3' end of the codon, 5' end of the anticodon, since they pair in inverted orientation). At the other end of the tRNA molecule, through the free 3'-hydroxyl of its terminal nucleotide, an ester linkage can be formed with a specific amino acid, the reaction being catalysed by a specific **aminoacyl-tRNA synthetase**. Each amino acid is brought into position for addition to the carboxyl end of the growing polypeptide chain as an aminoacyl residue attaches to the 3' terminus of the appropriate tRNA molecule, the anticodon of the tRNA pairing with the mRNA codon next in line for translation.

The role of the ribosome is to stabilize the association between mRNA and tRNA and to catalyse peptide bond formation. Its large subunit carries two binding sites for tRNA, and each is aligned with a binding site on the smaller subunit for a messenger codon. The initiating methionyl-tRNA (in prokaryotes, at least, this is in the substituted form of *N*-formylmethionyl-tRNA) binds to a site (the P or peptide site) opposite the AUG codon, and a second aminoacyl-tRNA, of a kind determined by the codon following the AUG, binds to the other (A or amino acid) site. Let us suppose that the second amino acid is valine, corresponding to the codon GUC. A transfer (or **translocation**) reaction then occurs whereby the methionyl is transferred from its linkage to tRNA to the amino group of the valine, forming methionyl-valyl-tRNA. At the same time the initiating tRNA, having shed its methionine, is ejected from the P site and the methionyl-valyl-tRNA moves across to take its place, leaving the A site vacant; the messenger is

Fig. 1.24

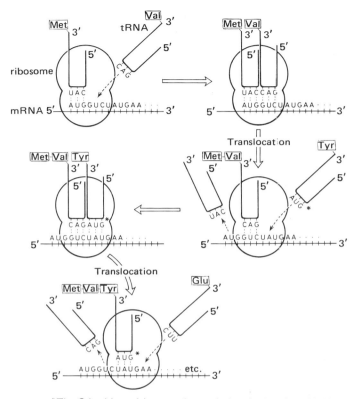

*The G in this position may be methylated—it pairs with U as a result of 'wobble' in the third position.

Mechanism of polypeptide synthesis by the ribosome–mRNA–tRNA complex.

also shifted along by three bases so that the third codon is now at the A ribosome site. Supposing the third codon to be UAU for tyrosine, a molecule of tyrosyl-tRNA will now be bound at the A site, and will accept the methionyl-valyl to form methionyl-valyl-tyrosyl-tRNA which shifts to the P site. So the process continues, with a new codon being brought into register on the A site of the ribosome as each new peptide link is formed. The growth of the polypeptide chain is terminated when a 'nonsense' codon (UAA, UAG or UGA) is brought into register. At this point there is normally (but *see* p. 367) no aminoacyl-tRNA to donate a further amino acid residue, and the completed polypeptide chain is released from the ribosome. The N-terminal methionyl group is usually, especially in eukaryotes, cleaved from the polypeptide chain before the completion of protein synthesis. The whole process is sketched in *Fig.* 1.24.

For the formation of functional proteins, polypeptide chains have to be folded and often packed together in very specific ways. Several kinds of folding and packing can be distinguished. Hydrogen bonding within a polypeptide chain can stabilize helix formation (α-structure) or, alternatively, the parallel or antiparallel alignment of sections of chain in 'pleated sheets' (β-structure). Superimposed on these forms of **secondary structure** there is often a higher-order folding called

Table 1.2 Coding for amino acids by messenger RNA sequences (codons) and the decoding (anticodon) sequences of transfer RNAs

Codon	Anticodon	Amino acid	Codon	Anticodon	Amino acid
5'UUU3' UUC	3' A^{m}AG 5'[2] AAG[1]	Phenylalanine	3'UAU5' UAC	AΨG[1] AUQ[2]	Tyrosine
UUA UUG	AAN[3] AAC[2]	Leucine	UAA UAG	— —	Termination
CUU CUA CUG	GAG[1] GAU[2] GAC[1]	Leucine	CAC CAU	GUQ[1]	Histidine
AUC AUU AUA	UAG[1]	Isoleucine	CAA CAG	GU^{mn}S[1]	Glutamine
AUG	UAC[1,2]	Methionine	AAC AAU	UUG[?]	Asparagine
GUC GUU GUA GUG	CAG[1] CAU*[2] CAI[2] CAV[1,2]	Valine	AAA AAG	UU^{mc}S[2] --UUC[1]	Lysine
UCC UCU UCA UCG	AGI[2] AGV[1]	Serine	GAC GAU	CUQ[1] CUG[2]	Aspartic acid
			GAA GAG	CU^{mn}S[1], CU^{mc}S[2]	Glutamic acid
			UGC UGU	ACG[2]	Cysteine
			UGA	—	Termination
			UGG	ACC[1] A$_{m}$CC[2]	Tryptophan

Codons	Amino acid	Anticodons
CCU, CCC, CCA, CCG	Proline	GGG[?], GGV[1]
ACC, ACU	Threonine	UGG[1]
ACA, ACG		UGU[?]
GCC, GCU, GCA, GCG	Alanine	CGI[2], CGV[1]

Codons	Amino acid	Anticodons
CGC, CGU, CGA, CGG	Arginine	GCI[1,2], GCC[?]
AGC, AGU	Serine	UCG[1]
AGA, AGG	Arginine	mcUCU[2]
GGC, GGU, GGA, GGG	Glycine	CCG[1,2], CCU[1], CCC[1]

Abbreviations for unusual nucleosides:

N = unidentified nucleoside.

^{m}G = methylguanosine.

I = inosine (i.e. hypoxanthine nucleoside).

V = 5-carboxymethoxyuridine.

U* = unidentified uridine derivative.

ψ = pseudouridine.

Q = unidentified guanosine derivative.

^{mm}S = 5-(methylaminomethyl)-2-thiouridine.

^{mc}S = 5-(methoxycarbonylmethyl)-2-thiouridine.

[1] Anticodons determined in *E. coli*.

[2] Anticodons determined in *Saccharomyces cerevisiae*.

[3] Anticodons determined in bacteriophage T4.

? Anticodons not known from these three sources, i.e. conjectural.

Data abridged from Sodd.[34]

tertiary structure, and the resulting compacted single chains (monomers) are often packed together to form oligomers (dimers, tetramers, hexamers etc., depending on the protein). In a number of well-known examples (e.g. haemoglobin, *see* pp. 356–357) more than one kind of folded chain are present in a **hetero**-oligomer, but **homo**-oligomers, with all the chains similar, are more common. All these complex steps of folding and packing occur spontaneously in normal intracellular conditions of temperature, pH and ionic strength, and no additional information need be provided beyond what is present in the amino acid sequence (primary structure) of the polypeptide chains, themselves encoded in the DNA. Since the proteins, in their various roles, determine virtually everything that goes on in the cell, it is easy to see, in principle, how the DNA exercises control over all cellular processes.

Chapter 1

Selected Further Reading

Lilley D. M. J. and Pardin J. F. (1979) Structure and function of chromatin. *Annu. Rev. Genet.* **12**, 471–512.

Stent G: S. and Calendar R. (1978) *Molecular Genetics—An Introductory Narrative*, 2nd ed., 773 pp. New York, Freeman.

Swanson C. P., Merz T. and Young W. J. (1981) *Cytogenetics—the Chromosome in Division, Inheritance and Evolution*, 572 pp. Englewood Cliffs, New Jersey, Prentice Hall.

Watson J. D. (1975) *The Molecular Biology of the Gene*, 3rd ed., 662 pp. New York, W. A. Benjamin.

Chapter 1

Problems

1.1 The base composition (expressed as fraction $G+C$) of double-stranded DNA is reflected both in its buoyant density in caesium chloride and in the temperature (T_m) at which it is half 'melted' to single strand. The following empirical relationships have been established:

Buoyant density $= 1.660 + (0.098 \times$ fraction $G+C)$.

Fraction $G+C = 2.44\ (T_m - 69.3)$, T_m being determined in standard saline. DNA samples from various sources were found to have the following buoyant densities:

Rat	1.702
Xenopus laevis	1.723
Drosophila melanogaster	1.698
Neurospora crassa	1.713
Saccharomyces cerevisiae	1.699

Estimate the fraction $G+C$ of the DNA of each species, and the expected melting temperature of each.

1.2

The sketch shows a hypothetical autoradiograph of a complete relaxed replicating DNA molecule released from a thymidine-requiring *E. coli* cell which had been grown in the presence of [³H]thymidine from the start of one round of DNA replication to about half way through the next round. Explain how the observed pattern of labelling could have arisen.

1.3

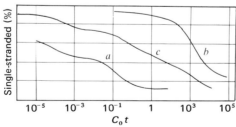

The C_0t curves shown were obtained in experiments by Davidson and Britten (*Cell* 1975, **6**, 29) on the DNA of the mollusc *Aplysia californica*. Curve *a* was obtained with total DNA averaging 2 kb in length, curve *b* was a rerun of the most slowly reannealing fraction separated from the rest by passage through hydroxyapatite, and curve *c* was the result obtained with the total DNA sheared to fragments of about 400 bases. What can you conclude about the organization of single-copy and repetitive sequences, and the relative abundances of different degrees of repitition? What could be the explanation of the observation in curve *c* that about 7 per cent of the total DNA had already reannealed before the first measurement could be taken?

1.4   Callan (*Proc. R. Soc. Lond. B Biol. Sci.* 1972, **181**, 19) estimated from fibre autoradiographs (cf. *Figs* 1.5, 1.9), of chromosomes labelled for different periods with tritiated thymidine, that in *Xenopus laevis* chromosomes replication origins were spaced on average 57 μm apart and that the replication forks travelled at a rate of about 9 μm/h. In another amphibian, *Triturus cristatus*, he obtained corresponding estimates of about 200 μm and 20 μm/h. DNA synthesis (the S-phase of mitosis) has been reported to last 13 hours in *Xenopus* and about four times as long in *Triturus*. Assuming that the initiations of replication from the different origins are spread evenly over the S-phase, and that the fibre represents a 40-fold contraction as compared with the fully extended double-stranded DNA, and taking the *C*-values listed in *Table* 1.1, calculate for each species (*a*) the total number of replication origins per haploid chromosome set, (*b*) the mean rate of DNA synthesis during S-phase and (*c*) the average number of replication units which are active at any one time.

1.5   From *Figs* 1.14*e* and 1.16*f* obtain estimates of the total lengths of the haploid sets of metaphase chromosomes of *Tradescantia paludosa* and *Neurospora crassa*. Using the *C*-values listed in *Table* 1.1, calculate the factor by which the DNA is linearly contracted in each species.

1.6   Miller et al. (*Chromosoma* 1976, **55**, 1) grew human lymph cells in culture for two replication cycles in the presence of 5-bromodeoxyuridine (BrdU). The amount of BrdU incorporated was considerably less during the second S-phase than during the first because of depletion of the drug in the growth medium. The cells were then allowed to go through a third cycle in the absence of BrdU, and their chromosomes were examined at metaphase by Giemsa staining. Ignoring sister-strand exchanges, half of the chromosomes were seen to have one darkly stained and one moderately stained chromatid, and the other half to have one moderately stained and one very lightly stained chromatid. Explain how this pattern comes about. One of the points of the experiment was to show that sister-chromatid exchanges occurring in the first, second or third mitosis could be distinguished by their different effects. Using diagrams, work out what these different effects should be.

2 The Sexual Cycle in Eukaryotes

2.1   Introduction

In the last chapter we concentrated on the capacity for self-replication which living systems obviously possess, and considered its likely molecular basis. This is one approach to genetics, but it is not the traditional one. The whole chromosome theory of inheritance in eukaryotes was built on the basis of the study of the inheritance through sexual reproduction of *differences* between individuals. By studying the *variations* in the characteristics of a species, and their modes of inheritance, it was possible to deduce in great detail the formal pattern of the genetic mechanism even without knowledge of its molecular basis. The great power of contemporary genetics derives from our present ability to link the two approaches. As we shall see in later chapters, inherited differences can now be attributed in a totally convincing way to variation in DNA.

All genetic analysis depends on cycles of mixing and segregation of sets of genetic determinants. From a study of the kinds and relative frequencies of new combinations which are segregated from the mixture, deductions can be made about the kinds, numbers and internal structures of the reassortable units which together make up the complete complement of genetic material—the **genome**. In this section, including Chapters 2–8, we consider the various kinds of segregational analysis used in experimental genetics. The first, which during the first half-century following the rediscovery of Mendel's principles of heredity in 1900 was the only method available, exploits the sexual reproductive cycle of eukaryotic organisms which is the subject of this chapter.

2.2   Types of sexual cycle

The Principle of Haploid–Diploid Alternation

The essence of sexual reproduction in eukaryotes is the fusion of nuclei (**karyogamy**) at one stage of the life cycle, resulting in a double or **diploid** set of chromosomes, balanced at another stage by the restoration of the single-set (**haploid**) condition by the process of **meiosis**. The cells which fuse to permit karyogamy are usually called **gametes**, and are most often specialized for their sexual function.

Meiosis can be summed up as *two* divisions of the nucleus accompanied by only *one* division of the chromosomes. Two nuclear divisions are necessary to complete the halving of chromosome number because the one division of the chromosome occurs in two separate steps: the chromosome arms split at the beginning of the first division, while the centromeres do not divide until the second. There are always four products of meiosis, referred to as a **tetrad**, though in seed plants and animals only one of the four products on the female side is functional. In animals the meiotic products develop directly into gametes, and are the only haploid cells in the life cycle. In plants (including fungi) the meiotic products are **spores**; these

Fig. 2.1

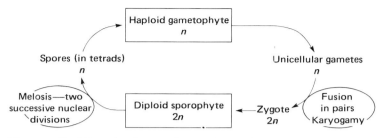

The generalized life cycle of a sexually reproducing plant.

develop into a haploid **gametophyte** generation of greater or lesser duration, which in turn eventually produces gametes.

The general eukaryote sexual life cycle is shown in diagrammatic form in *Fig.* 2.1, and its various stages in a variety of important organisms are summarized in *Table* 2.1.

Equal Haploid–Diploid Alternation

Perhaps the simplest and, at the same time, most versatile sexual system is shown by the yeast *Saccharomyces cerevisiae*, different strains of which are used by bakers and brewers. The yeast grows as free uninucleate cells which multiply by budding. Most naturally occurring strains are stably diploid under good nutritional conditions, but can be induced to undergo meiosis by starvation. The four nuclei resulting from meiosis are enclosed in four spores (**ascospores**) which are formed inside the mother cell wall. This formation of meiotic spores *within* the mother cell, or ascus, is the defining feature of the Ascomycete group of fungi.

In many yeast strains (though not all, *see* p. 485) pure haploid cultures grown from single ascospores will remain haploid indefinitely. Under the microscope the haploid cells appear very similar to the diploid, but smaller. If all the ascospores

Fig. 2.2

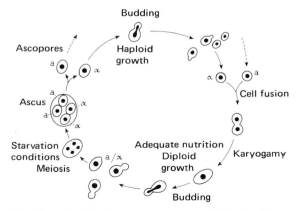

The life cycle of *Saccharomyces cerevisiae*—bakers' and brewers' yeast.

Table 2.1   A summary of some eukaryote life histories

Organism	2n Diploid phase	n Products of meiosis	n Haploid phase	u Gametes	n | n Dikaryon	2n Zygote
Neurospora crassa (filamentous fungus)	Ascus (within fruit body— perithecium)	Ascospores Four spore pairs	Filamentous mycelium bearing asexual spores (conidia)	Ascogonium (in protoperithecium) Conidium (opposite mating type)	Ascogenous hyphae within perithecium	Ascus initial
Schizophyllum commune (bracket fungus)	Basidium (borne on gills of fruit body)	Basidiospores, borne externally on basidium in tetrads	Monokaryotic mycelium	Mycelial cells of compatible mating types	Mycelium—the major growth phase—forming fruit bodies (brackets)	Young basidium
Saccharomyces cerevisiae (yeast)	Budding cells	(Induced by starvation) Tetrads of ascospores	Budding cells	Ordinary haploid cells of opposite mating type	(Karyogamy immediately ‾‾‾‾‾‾ ↑ follows gamete fusion)*	Budding diploid cell ↑
Chlamydomonas reinhardii (unicellular motile green alga)	Dormant zygote	Eight swarmers (tetrad of pairs) formed on zygote germination	Motile single cells	Motile cells— opposite mating types, morphologically identical	(Karyogamy immediately ‾‾‾‾‾‾ follows gamete fusion)	Dormant zygote (zygospore) ↑
Ulva (multicellular green marine alga— 2-layered sheet of cells)	Sheet-like thallus	Zoospores formed in tetrads	Sheet-like thallus (similar to diploid)	Motile cells— opposite mating types, morphologically identical	(Karyogamy immediately ‾‾‾‾‾‾ follows gamete fusion)	Germinates to give haploid →thallus

Any moss	Capsule borne on haploid plant	Spores, initially in tetrads, liberated from capsule	Leafy plant bearing archegonia and antheridia at fertile shoot apices	Egg (in archegonium); sperm (liberated from antheridium)	(Karyogamy immediately follows gamete fusion)	Fertilized egg, developing into →diploid capsule
Typical fern	Plant with fronds bearing sporangia	Spores liberated from sporangia	Small green prothallus bearing archegonia and antheridia	Egg (in archegonium); sperm (liberated from antheridium)	(Karyogamy immediately follows gamete fusion)	Fertilized egg, developing into →fern plant
Pinus (Gymnosperm tree)	Pine tree from seed	Megaspores and microspores (pollen) in separate female and male cones	Nutritive endosperm with embedded archegonia; Pollen tube (4 nuclei)	Egg (in archegonium); Pollen tube nucleus	(Karyogamy immediately follows gamete fusion)	Fertilized egg developing into →dormant embryo in seed
Zea mays (Angiosperm herbaceous plant)	Maize (corn) plant	Megaspores in ovules; microspores (pollen) in anthers of flowers	Embryo sac (8 nuclei including egg nucleus); Pollen tube (3 nuclei including 2 gamete nuclei)	Egg (in embryo sac); Pollen tube nucleus	(Karyogamy immediately follows gamete fusion)	Fertilized egg develops into →dormant embryo in maize kernel†
Drosophila melanogaster (fruit fly)	Larval stages leading through pupal stage to adult fly	♀ eggs............♂ spermatozoa		♀ eggs ♂ spermatozoa	(Karyogamy immediately follows gamete fusion)	Fertilized egg developing into →larva

* Except in mutant strain—see p. 176.

† The nutritive endosperm of the kernel is *triploid* tissue, stemming from fusion of two embryo sac nuclei and the second pollen tube gamete nucleus.

Note: The multicellular green algae, mosses, ferns and gymnosperms are not at present very much used in experimental genetics but are included here to illustrate types of life cycle intermediate between the almost entirely haploid (e.g. *Neurospora, Chlamydomonas*) and almost entirely diploid (flowering plants and animals) types.

from an ascus are allowed to germinate and bud together, the whole culture very soon reverts to the diploid condition by cell fusion and karyogamy. The stability in the haploid state of cultures grown from single ascospores is due to the existence of two self-incompatible **mating types** called *a* or *α*. All diploids are the product of the fusion of *a* and *α* haploid cells. All ascospores are either *a* or *α* and each meiotic tetrad has two of each kind—our first example of simple genetic segregation. The life cycle is shown in *Fig.* 2.2.

Other simple eukaryotes, notably some of the simpler algae (*see* the example of *Ulva* in *Table* 2.1), show a regular and obligatory alternation of morphologically similar haploid and diploid generations. Here the growth and life of the haploid (gametophyte) generation is limited; eventually it has to produce gametes which can only fuse in pairs to produce the diploid (sporophyte) generation. This, in turn, can only reproduce by producing haploid spores by meiosis, and these grow into gametophytes. The gametes all look the same but, as in yeast, are of two mutually compatible but self-incompatible mating types.

Predominantly Haploid Organisms

Most fungi and many algae have, in the normal course of events, no mitosis in the diploid phase; the diploid nucleus formed by karyogamy undergoes meiosis directly to restore the haploid condition.

A simple example is the unicellular green alga (*Chlamydomonas reinhardi*) which has been much used in experimental studies of plastid heredity (*see* pp. 193–195). The free cells, which are motile by virtue of paired flagella, divide mitotically by equal fission. Members of the same cell lineage cannot initiate sexual reproduction since they will be of one or other of the two self-incompatible mating types, called + and −. Mixed cultures of + and − cells, however, can produce zygotes by fusions of +/− pairs of gametes, which are very similar to vegetative cells. The

Fig. 2.3

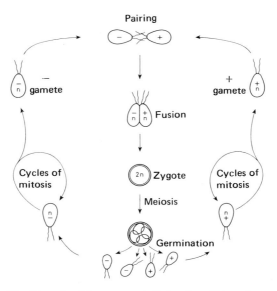

The life cycle of the green unicellular alga, *Chlamydomonas reinhardi*.

zygotes are initially motile, using all four flagella derived from the parent gametes, but soon become transformed, without further nuclear division, into resistant dormant **zygospores**. Under suitable conditions, of which the most critical is light, the zygospores germinate by meiosis to form a tetrad of four haploid nuclei (two + and two −) and then, by a further mitotic division, eight motile haploid cells which re-establish the haploid phase. The cycle is illustrated in *Fig.* 2.3.

In the two groups of fungi which, apart from yeast, have been the most investigated by geneticists, i.e. the Pyrenomycete section of the Ascomycetes and the Hymenomycete ('mushroom') section of the Basidiomycetes, there is an extra complexity in what is still essentially a haploid life cycle. There is no true multiplying diploid phase, but the fusion of haploid cells of compatible mating type does not at once lead to karyogamy. Rather there is a brief (Pyrenomycetes) or indefinitely prolonged (Hymenomycetes) **dikaryotic** phase in which the mutually compatible haploid nuclei associate and divide synchronously in pairs. When the

Fig. 2.4

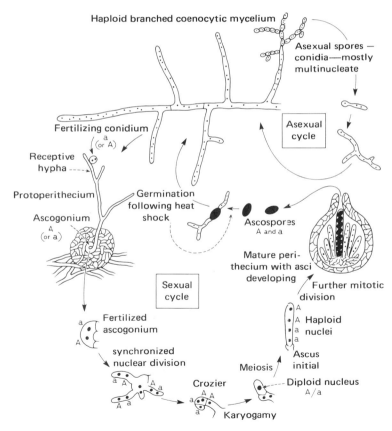

The life cycle of the Ascomycete fungus *Neurospora crassa*. For the sake of simplicity, the microconidia, which in practice are not of great importance, are now shown. The protoperithecium and perithecium are pictured as if sectioned vertically.

paired nuclei eventually fuse in the fungal fruit body, the diploid nucleus which results immediately undergoes meiosis to form haploid spores either directly (as in the Hymenomycete tetrad of basidiospores) or after a further mitotic division (as in the eight-spored Pyrenomycete ascus).

These fungal life histories are summarized in *Table* 2.1 and that of the Pyrenomycete *Neurospora crassa* is illustrated in *Fig.* 2.4. Because of the importance of the latter species, both in its own right and as a representative of the eight-spored Ascomycetes in general, some further description is in order.

The vegetative phase of *Neurospora* consists of a **mycelium** of branching filaments or **hyphae**. The mycelium is not truly cellular, with one nucleus per cell, but rather **coenocytic**, with many nuclei present together in each of the compartments into which the hyphae are subdivided by cross-walls (**septa**). The most conspicuous feature of the fungus is the great abundance of mitotically produced pink-orange spores, or **conidia**, which are produced in chains at the ends of aerial hyphae and contain varying numbers of nuclei. The conidia are carried in the air and provide the principal means of spread of the fungus. As in yeast and *Chlamydomonas*, there are two mating types, called in this case *A* and *a*. Unlike these unicellular organisms, however, *Neurospora* has distinct male and female sexual structures, both borne on the same hermaphroditic mycelium. The female organ is the **protoperithecium**, which consists of a coiled cell, the **ascogonium**, extending into a receptive hypha and enclosed by a small knot of hyphae. The male function of fertilization of the ascogonium by a strain of the other mating type can be carried out (most commonly) by a conidium or by a suitably placed hypha.

A nucleus from the fertilizing element passes via the receptive hypha into the ascogonium where it associates with a nucleus of the opposite mating type to initiate a brief dikaryotic phase. This consists of a small mass of **ascogenous hyphae** which fills the flask-shaped **perithecium** which develops from the fertilized protoperithecium. Fusion of pairs of nuclei, descendants of the original pair in the fertilized ascogonium, eventually occurs in the penultimate cells of the ascogenous hyphae, which become ascus initials. Karyogamy is followed immediately by meiosis, the ascus initial meanwhile growing rapidly to form an ascus, which in *Neurospora* and other Pyrenomycetes is an elongated cylindrical sac. The meiotic division spindles are oriented lengthwise and the second division spindles do not overlap. The result is a row of four haploid nuclei arranged along the length of the ascus. A further mitosis converts the primary tetrad of nuclei into four pairs, which are enclosed in eight ascospores. This is an example of an **ordered** tetrad, with the arrangement of the ascospores reflecting the two divisions of meiosis (*see* Chapter 3, pp. 77–80). As the ascospores mature they develop thick black cell walls which make them highly resistant and dormant. The ripe ascus eventually ruptures at its tip and the eight ellipsoidal spores are discharged through the neck of the perithecium. Germination of the ascospores is induced by heat shock (30 or 45 minutes at 60 °C is used in the laboratory), and this no doubt accounts for the common occurrence of the fungus on burnt vegetation in the tropical and subtropical countries in which it naturally occurs.

Less need be said about the Hymenomycetes. Here the dikaryotic phase is extensive and long lived, and the large characteristic fruit-bodies ('mushrooms' or brackets in the case of those species growing on trees) are built from dikaryotic mycelium. Nuclear fusion occurs in numerous special cells (**basidia**) borne on the gills on the underside of the fruit body, and meiosis, which follows immediately, leads to the formation of a tetrad of haploid basidiospores budded off from each basidium. The haploid mycelia produced by germination of the basidiospores take the first opportunity to fuse in compatible combinations to form dikaryons, in which there are two nuclei in each cell. The systems of self-incompatibility and cross-compatibility found in the Hymenomycetes are very complex; there are many mating types, all self-incompatible but mutually compatible in all combinations (*see Box* 18.1, p. 519).

The Transition from Mainly Haploid to Mainly Diploid Life Cycles

The mosses and ferns will not be prominent in the remainder of this book because they have not been extensively used in experimental genetics. They are, nevertheless, worth considering briefly since they can be regarded as transitional between the predominantly haploid fungi and simpler algae and the predominantly diploid seed plants and animals.

In both mosses and ferns the haploid stage (gametophyte), arising from meiotically produced spores, is a free-living green plant with its own simple roots. In mosses it is equipped with stems and leaves, while in the ferns it is a small and inconspicuous pad of cells, the **prothallus**, growing flat on the ground. In both cases the gametophyte produces female and male gametes—eggs in archegonia and sperm cells in antheridia. The zygotes formed by gamete fusion develop, in the case of moss, into a capsule within which a large number of cells undergo meiosis to form haploid spores. Though it is a substantial structure, green and capable of photosynthesis, the capsule is never free living but is borne at the apex of a gametophyte shoot.

In a typical fern, the zygote develops into a conspicuous fern plant which quickly becomes independent of the prothallus which gave it birth. Meiosis takes place in sporangia borne on the undersides of the fern fronds; the spores produced germinate to produce prothalli once more.

Extreme Reduction of the Haploid Phase: Seed Plants

In seed plants the haploid gametophyte generation is extremely reduced and concealed within the diploid sporophyte. Unlike the situation in mosses and most ferns, there are separate female and male gametophytes arising respectively from **megaspores** and **microspores**.

In the flowering plants (Angiosperms), the megaspore is the sole survivor of a tetrad of cells formed within the ovule, in the flower ovary. The megaspore develops into an **embryo sac** which develops in a number of different ways in different flowering plants. In the commonest pattern, seen for example in *Zea mays*, three rounds of mitotic nuclear divisions give a mature embryo sac containing eight nuclei, one of them being the nucleus of the single egg.

The microspores are the pollen grains, formed in tetrads by meiosis within the anther. The primary pollen grain nucleus undergoes mitosis to give two haploid nuclei in the mature pollen grain. After deposition on the receptive surface (stigma) which surmounts the ovary, the pollen grain germinates to form a pollen tube, within which one of the two pollen grain nuclei divides again to give two gamete nuclei.

Fertilization occurs when the pollen tube penetrates the ovule, and one of the two gamete nuclei fuses with the egg nucleus to give the zygote from which the next

58 CHAPTER 2

diploid generation will develop. The other gamete nucleus fuses with *two* other nuclei of the embryo sac to give a *triploid* nucleus (two sets of chromosomes from the maternal and one from the paternal parent). While the zygote divides to form the embryo of the developing seed, the triploid nucleus divides and forms the cells of the **endosperm**, which acts as a store of nutrients to be used by the embryo when the seed germinates. In maize (*Zea mays*), a plant much favoured by geneticists, the endosperm accounts for the bulk of the seed and is important in displaying much inherited variation. Whilst the endosperm has no genetic future itself, it is genetically identical to the embryo (except for its double maternal component) and often serves to indicate the type of plant which will grow from the seed. The main features of the sexual life cycle of *Zea mays* are illustrated in *Fig. 2.5*.

The other great group of seed plants, the Gymnosperms, which consists mainly of coniferous trees, differs from the Angiosperms in having a somewhat less reduced female gametophyte. The megaspore, produced by meiosis within the female cone, develops into quite an extensive haploid tissue which has a nutritive function comparable to that of the triploid endosperm of flowering plants. Embedded in this haploid tissue are a number of female organs, archegonia, each containing an egg. The pollen grains, which are produced as microspores in the male cones, are not very different in form or function from Angiosperm pollen grains; gamete nuclei from pollen tubes fertilize the archegonia. The life cycle of a

Fig. 2.5

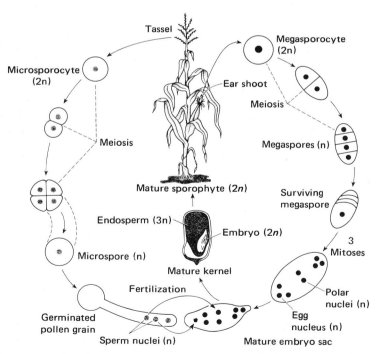

The place of meiosis and of karyogamy in the life cycle of a flowering plant such as *Zea mays*. *n*, 2*n* and 3*n* indicate haploid, diploid and triploid. Based on Srb et al.[8]

typical Gymnosperm (*Pinus*) is included in *Table* 2.1. The relatively substantial female gametophyte tissue permits certain kinds of haploid genetic analysis to be carried out in Gymnosperms more readily than in flowering plants (*see* p. 95).

Animals—entirely Diploid except for the Gametes—Separation of the Sexes

A fundamental respect in which animals differ from flowering plants is the lack of any haploid nuclei other than the immediate products of meiosis. In the testis, meiosis gives tetrads of spermatocytes which all develop into motile spermatozoa. In the ovary, only one product of each meiosis survives to become an egg; one of the products of the first meiotic division is extruded from the oocyte and one of the second-division products is likewise discarded. In neither type of gamete is there any mitotic division of the haploid nucleus. The only possible future for haploid cells in animals is to fuse in pairs to restore the diploid phase.

In the great majority of animal species each individual is either male or female. Sexual differentiation is controlled by a variety of different chromosome mechanisms, briefly reviewed at the end of this chapter. However, in one important experimental animal, the small nematode worm *Caenorhabditis elegans*, most individuals are hermaphrodite, producing both eggs and sperm and capable of self-fertilization. A few males are also produced (of XO chromosome constitution—*see* p. 71) and these can be crossed to hermaphrodites to give both males and hermaphrodites in the following generation. With its short reproductive cycle (3·5 days) and ability both to cross and to self-fertilize, *C. elegans* has become the organism of choice for certain kinds of investigation.[7] Its simple and precisely reproducible structure and pattern of development make it particularly suitable for studies of cellular differentiation (*see* Chapter 16, p. 462).

2.3 Meiosis

The process of meiosis is a remarkably uniform process across a great range of eukaryotic organisms. There are, indeed, important minority variations and we shall deal with some of these at the end of this section, but it is possible to describe a standard process which occurs in at least the great majority of fungi, seed plants and animals and probably also in algae and other groups of lower plants.

Synapsis and the Synaptonemal Complex

The first feature of the first division of meiosis, which sets it on a different path from that of mitosis, is the commencement, early in prophase, of a progressive side-by-side pairing (**synapsis**) of homologous chromosomes. This stage is called **zygotene**, and it follows an earlier stage, **leptotene**, in which a diploid number of extended chromosomes are apparently unpaired. Pairing is complete at **pachytene**, at which stage the nucleus can be seen to contain a haploid number of

Fig. 2.6

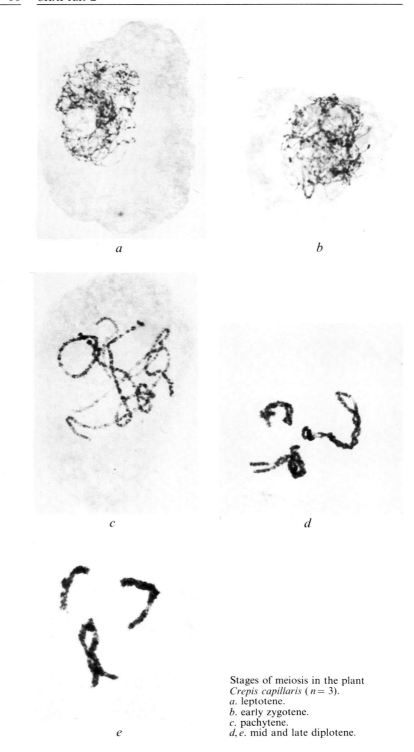

Stages of meiosis in the plant
Crepis capillaris ($n = 3$).
a. leptotene.
b. early zygotene.
c. pachytene.
d, e. mid and late diplotene.

Fig. 2.6 continued

f. diakinesis.
g. metaphase I.
h. anaphase I.
i. telophase I.
j. interkinesis.

f

g

h

10 μm

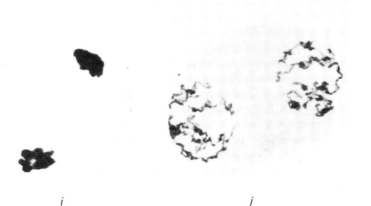

i

j

Fig. 2.6 continued

k

l

m

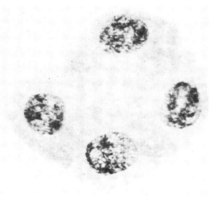

n

k. metaphase II (two in same cell).
l, m. anaphase II. (two in same cell
n. telophase II with four pollen gra
nuclei.
Photographs provided by
Dr Gareth Jones.

moderately extended but rather thick threads. The thickness is due, in part, to the fact, which can be confirmed by careful focusing at the highest power of the light microscope, that each thread is in reality two homologous chromosomes closely associated point to point along their whole length. This association depends on the homology of the chromosome segments paired at each point, as shown by the fact that structural rearrangements in one haploid set compared with the other result in interchanges of partners at pachytene (*see Fig.* 2.8). Moreover, the association is exclusively pairwise. When, as in polyploids (cf. p. 162), there are more than two chromosomes of each kind present, association at any point is confined to two chromosomes at a time.

The mechanism of the mutual recognition and attraction of homologous pairs of chromosomes is still something of a mystery. At leptotene the homologues are too far away from each other for any known type of molecular interaction to operate. At least a part of the answer to the problem probably lies in the fact that, as shown

Fig. 2.7

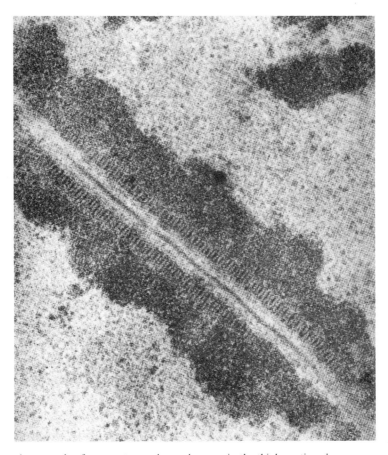

An example of a synaptonemal complex seen in the thinly sectioned young ascus of *Neottiella rutilans*. The tranverse banding of the lateral elements is characteristic of fungi, but is not usually seen in higher eukaryotes.
Magnification ×59 000.
Photograph by courtesy of the late M. Westergaard.[1] (Previously published in *Fungal Genetics*, reference on p. 94. Reproduced by permission.)

by reconstruction of the complete three-dimensional structure of the zygotene nuclei of diverse species by serial sectioning and electron microscopy, the chromosome ends **(telomeres)** are anchored to the inner side of the nuclear envelope. If this anchoring is to a specific part of the nuclear envelope for each kind of chromosome, the homologous telomeres may find each other readily and serve as starting points for a zippering-up process which eventually pulls the homologues together all along their lengths at pachytene.

Fig. 2.8

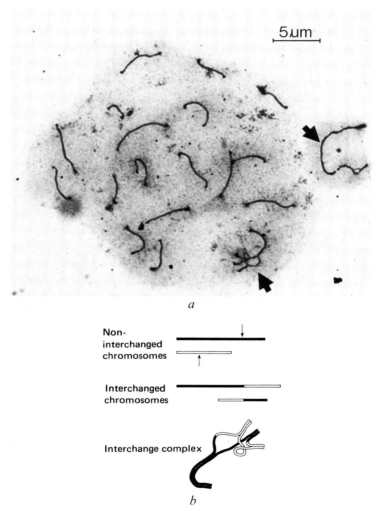

a. A preparation of a spread flattened mouse spermatocyte nucleus stained with silver at the pachytene stage.[6] Each pachytene autosomal bivalent is represented by one synaptonemal complex; a segmental interchange results in one association-of-four with a branched complex (bottom arrow). The X- and Y-chromosomes are associated end to end; each has its own silver-stained lateral element (top arrow).
b. Interpretation of the interchanged synaptonemal complexes indicating heterozygosity with respect to a chromosomal segmental interchange (cf. p. 150). Photograph by courtesy of Dr Anne Chandley.

The pachytene association of a pair of homologous chromcsomes is stabilized by a ribbon-like structure, present in essentially the same form in all eukaryotes from lower fungi to higher animals, called the **synaptonemal complex**.[1] Running along the centre of the ribbon, parallel to and equidistant between the paired chromosomes, is a **central element**. It is not known to contain any DNA (though it may do so at a very low concentration) and it is apparently composed predominantly of protein. Flanking the central element, and at the inner margins of the synapsed chromosomes, are two parallel **lateral elements** about 0·1 μm apart. These structures are readily stained with silver ions, probably due to a high concentration of proteins containing —SH groups. Although the chromosomes undergoing pairing have already replicated, each is associated with only a single lateral element which must, therefore, be shared by two chromatids.

An example of a synaptonemal complex as seen with the electron microscope is shown in *Fig.* 2.7. With the exception of the banded appearance of the lateral or central elements, which are seen in some groups of eukaryotes and not in others, the synaptonemal complex is a remarkably uniform structure in all meiotic cells. In organisms with small chromosomes, synaptonemal complexes may be more easily distinguished and counted than the chromosomes themselves. *Fig.* 2.8 shows an example of the very clear pictures which can be obtained of spread preparations of whole nuclei stained with silver ions.

One particularly intriguing feature of the synaptonemal complex, so far mainly studied in fungi and insects, is the presence of occasional **nodules** of electron-dense material adjacent to the central element. These have been called **recombination nodules** by Carpenter,[2] the idea being that they represent incipient chiasmata— points of crossing-over between chromatids (*see below*). The evidence for this interpretation is as yet inconclusive, but the correspondence between the number of nodules seen and the number of genetic crossovers required to explain genetic data is certainly very suggestive (for example in *Neurospora*[3]).

Diplotene and the Nature of Chiasmata

At the end of pachytene, which is often a relatively prolonged stage, a dramatic change occurs in the appearance of the meiotic nucleus (*Fig.* 2.6*d* and *e*). The paired homologues fall apart along the greater part of their length, remaining in juxtaposition only at a limited number of points (one or more per chromosome pair) called **chiasmata** (chiasma in the singular). Such joined pairs of homologous chromosomes are called **bivalents**. At the same time, each chromosome which has in fact been double, so far as its DNA content is concerned, since before leptotene, now becomes visibly divided into chromatids. The configuration of the chromatids at the chiasmata, while not easy to make out in all organisms, is strikingly clear in such diverse forms as grasshoppers (including locusts; *Fig.* 2.9) and newts. The appearance is as if a pair of chromatids, one from each of the paired chromosomes, has broken at exactly corresponding points and rejoined crosswise at each chiasma. Careful observation also shows that, where there are two or more chiasmata joining the same chromosome pair, the chromatids which cross over at one chiasma are not necessarily the same two as do so at the next.

Historically, there were two alternative interpretations of chiasmata. The breakage–reunion hypothesis (Janssen's **chiasmatype** theory) was not the only possibility since it was equally consistent with microscopic appearance to suppose that separated sister chromatids simply exchanged partners without any breakage.

Fig. 2.9

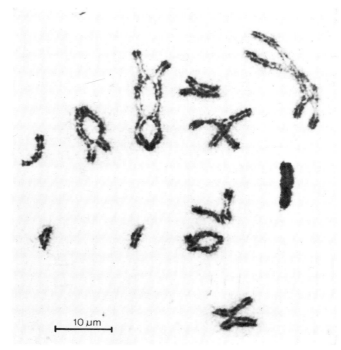

A particularly clearly analysable diplotene from locust (*Locusta migratoria* spermatocyte). Note the structure of the chiasmata with only two of the four chromatids exchanging at any one chiasma. There are 11 bivalents plus the unpaired X (at 12 o'clock).
Photograph by courtesy of Dr Gareth Jones.

The strong reason for favouring the breakage–reunion hypothesis was the explanation which it gave for genetic crossing-over (*see* p. 103). There are now, in addition, a number of lines of cytological evidence, the most graphic and direct of which depends on differentially labelling chromatids with 5-BrdU. As we saw in the previous chapter, chromatin containing a substantial amount of 5-bromouracil in place of thymine stains much less heavily with Giemsa reagent. In the case shown in *Fig.* 2.10, 5-BrdU was incorporated during the S-phase of the last premeiotic mitoses in grasshopper spermatocytes. As a result, the chromosomes entering meiosis each contained the analogue in one DNA strand. The S-phase immediately preceding meiosis, which took place after the injected 5-BrdU had been largely depleted, resulted in the segregation of labelled (more faintly staining) and unlabelled (more heavily staining) DNA into different chromatids. It can be clearly seen in the photograph that labelled and unlabelled chromatids are recombined reciprocally at the chiasma. In fact this is seen only at about half of the chiasmata, since it is just as likely for two similarly stained chromatids to undergo exchanges as for two dissimilarly stained ones to do so.

Another kind of evidence, several examples of which have been presented over the years, is available when a pair of homologues differ visibly in length as, for example, when one of them has suffered a terminal deletion in one arm. In such cases, a chiasma formed **proximal** to the deletion (i.e. between the deletion and the

Fig. 2.10

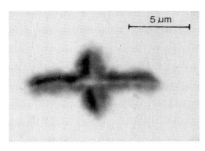

A locust diakinesis bivalent with one chiasma in which one chromatid of each chromosome is more lightly stained because it contains a proportion of 5-bromodeoxyuridine in place of thymidine.[4] Note that at the chiasma differently stained chromatids have recombined. Photograph by courtesy of Dr Gareth Jones.

Fig. 2.11

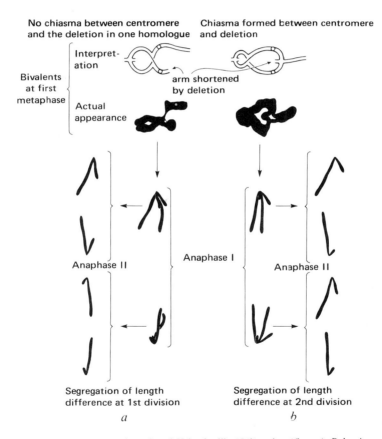

Metaphase I and anaphase I and II in the lily (*Lilium longiflorum*). Behaviour of a pair of homologous chromosomes in which one homologue is shortened by a terminal deletion. Note that chiasma formation proximal to the deletion leads to the joining of a larger and a shorter chromatid to the same centromere.
Based on Brown and Zohary.[5]

centromere) is expected, if the chiasmatype theory is correct, to lead to a longer and a shorter chromatid becoming attached to the same centromere. This can be clearly seen at anaphase I of meiosis (described below); moreover, the frequency of this appearance tallies well with the frequency of chiasma formation in the relevant chromosome arm at diplotene (*Fig.* 2.11). We may note (anticipating our description of the later stages of meiosis) that this is the first example in this book of chiasma formation leading to **second-division segregation** (cf. pp. 78–80) of a chromosome marker. Chromatids of different lengths **distal** to a chiasma (distal in the sense of more distant from the centromere) remain joined to the same centromere at the end of the first meiotic division but will be separated by centromere division at the second.

It should be emphasized that chiasmata are variable in both number and position. As a general rule every homologous chromosome pair is joined at diplotene by at least one chiasma and, indeed, the formation of bivalents is usually necessary for the regular separation (**disjunction**) of chromosomes at first meiotic anaphase. Short chromosome pairs may seldom have more than one chiasma, but longer ones will usually have two, three, four or even more, the mean number increasing with chromosome length. Again, as a general rule, chiasmata can occur anywhere along the chromosomes, although the *probability* of chiasma formation may vary from one chromosome region to another. In the fruit fly *Drosophila melanogaster*, as well as in numerous other organisms, there is little chiasma formation in the centromeric heterochromatin, and there are many other examples of chiasmata occurring preferentially either in proximal or in distal regions of chromosome arms.

In the oocytes of newts and other amphibians, diplotene is a time of intense synthetic activity, as well as being the stage at which chiasmata first become visible. The diplotene ('lampbrush') chromosomes in these cells are very extended and are characterized by paired lateral loops pinned at the base to the chromosome axis, which consists of the closely aligned chromatids. The two loops of each pair display a characteristic transcriptional activity, reflected in the quantity and distribution of the RNA transcripts which they carry. Each loop probably represents a gene, or a group of transcriptionally coordinated genes.[3]

Diakinesis and First Metaphase

At the end of diplotene the bivalents undergo a striking condensation. The component chromosomes become longitudinally contracted, to a greater extent even than in mitotic metaphase, and the division of each into chromatids is no longer evident under the microscope except, sometimes, at the chiasmata. Depending on the number and disposition of the chiasmata, the bivalents assume the forms of crosses or single or multiple loops. This stage, the last of the first meiotic prophase, is called **diakinesis** (*Fig.* 2.6*f*) and it is shortly followed by the first meiotic metaphase (*Fig.* 2.6*g*).

Just as in a mitotic division, the first metaphase of meiosis is marked by the disappearance (except in fungi) of the nuclear envelope and the appearance of a division spindle. The situation is, however, fundamentally different from mitosis in that, instead of there being a diploid or haploid number of separate chromosomes each with one centromere and two chromatids, there is a haploid number of **bivalents** each with two centromeres and four chromatids. The two centromeres of each bivalent take up symmetrical positions one on each side of the spindle equatorial plane. The bivalents have the appearance of being held in balance on the equator by equal and opposite pulls towards the two poles exerted on the centromeres by the spindle fibres.

First Anaphase and Telophase

At metaphase I the bivalents, though apparently under tension, are held together at their chiasmata. At the onset of anaphase I the paired chromatids distal to each chiasma are pulled apart and each bivalent becomes two obviously split chromosomes (called **dyads**) which separate towards the opposite poles (*Fig.* 2.6*h*). As *Fig.* 2.11 shows, dyads are very peculiar chromosomes in that, in regions distal to chiasmata, their chromatids are non-sisters (having been derived in part from different homologous chromosomes) and may be genetically different. Any difference between the two homologues forming the bivalent will be segregated at the *first* anaphase if there is no chiasma between the locus of the difference and the centromere, but if there is such a chiasma segregation will be delayed until the *second* division. The occurrence of chiasmata, in fact, is the explanation of why two divisions of meiosis are necessary to produce genetically 'pure' products. For reasons which we consider in Chapter 5, segregation may, comparatively rarely, be further delayed until the first postmeiotic mitosis, but this complication need not concern us here.

After the dyads have reached the spindle poles each group is enclosed by a new nuclear membrane (telophase I, *Fig.* 2.6*i*). In many organisms the second division of meiosis occurs almost at once; in others (as seen in *Fig.* 2.6*j*) there is a distinct **interkinesis** with a despiralization and elongation of the chromosomes prior to a new condensation at prophase II.

Second Division

Metaphase II (*Fig.* 2.6*k*) is much like two synchronized mitotic metaphases in appearance, differing only in the frequently wide separation of the chromatids attached to each centromere—a legacy of the forcible resolution of bivalents into dyads at anaphase I. The centromeres come to lie on the spindle equator at metaphase II, the centromeres split at the onset of anaphase II (*Fig.* 2.6*l* and *m*) and each haploid set of dyads becomes separated into two haploid sets of single chromosomes as the daughter centromeres disjoin to opposite spindle poles. Non-identical chromatids which, because of chiasma formation (*Fig.* 2.11), were joined in the same dyad at the first division, are now segregated into different meiotic products. Note that, because of chiasmata and the consequent mixture of first and second-division segregation of chromosomal differences, all four products of meiosis may, and generally will, inherit different combinations of the material of each of the chromosome pairs present in the original diploid. The potential of meiosis for generating new variants of the basic haploid chromosome set, given some degree of difference between homologous chromosomes in the original diploid, is clearly enormous. To independent assortment of non-homologous chromosome pairs is added crossing-over of homologues. Not only is the genetic pack shuffled and redealt, but the individual cards are split and pieced together in new ways.

The Fate of the Products of Meiosis

Depending on the organism and (in higher plants and animals) on the sex of the meiotic cell, either all four products of meiosis may be functional or only one of them. In fungi and lower plants all four spores of each meiotic tetrad are capable of germination to give a haploid generation; in most Ascomycete fungi there is an additional mitotic division within the ascus to convert the tetrad into an octad. Again, in the male organs (anthers) of flowering plants and in the testes of male animals, all four products are potentially viable—as pollen grains (microspores) in the one case and as spermatids, developing into flagellated spermatozoa, in the other. On the female side in flowering plants and animals, however, only one product of each meiosis is used. In the ovules of flowering plants, four meiotic products are formed initially as a row of four cells, but only one of these (usually the one at the end of the row nearest to the point of entry for the fertilizing pollen tube) grows to form the embryo sac (megaspore); the other three develop no further. If centromeres pass to one pole or the other at each division of meiosis purely at random (which is generally though not invariably the case), the meiotic product to develop into the embryo sac is, in effect, picked at random from the original tetrad. Non-random megaspore development occurs in some special cases, for example in species of *Oenothera* (evening primrose; *see* p. 158).

In female animals, meiosis in the oocyte involves the discarding from the cell of one product of the first division and then of one product of the single surviving second division. These two successive losses occur through the extrusion from the cell of the first and second **polar bodies**. The choice of nucleus for inclusion in a polar body depends on its position in the oocyte, but if, as is usually true, centromere disjunction is at random, the choice will be made without regard to chromosomal constitution. Later in the book we shall encounter important exceptions: so-called **meiotic drive** (where a particular chromosome finds its way into the surviving egg nucleus disproportionately frequently, *see* p. 545) and the behaviour of chromosomes with inverted segments in *Drosophila* (*see* p. 155).

Apart from the exceptions just mentioned, which involve special structural characteristics of chromosomes, the general expectation is that all products of meiosis will have an equal chance of contributing to the next generation. This expectation, though it may be upset by natural selection (Chapter 18), works well enough as a baseline for genetic analysis.

2.4   Sex chromosomes

Sex, that is to say the differentiation of male and female gametes, does not necessarily imply **sexual dimorphism**, that is the separation of the two sexual functions between morphologically distinct male and female individuals (**dioecism** as it is called in plants). Hermaphroditism, with both types of sex organs in each individual, prevails in most of the plant kingdom and also occurs in some groups of animals. Nevertheless, separation of the sexes is the rule in animals (as well as some plants), and the maintenance of such a system depends on a special chromosomal mechanism.

In all mammals, a few flowering plants and many insects (including that

Fig. 2.12

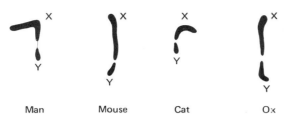

Man   Mouse   Cat   Ox

XY sex bivalents at metaphase I of meiosis in various mammals. In each case
the homology and pairing between the X and Y is confined to relatively short
regions near the chromosome ends, where a chiasma is formed. The bivalents
are seen in side view, with centromeres being pulled apart vertically upwards
and downwards.
Redrawn from White.[9]

favourite genetic object the fruit fly *Drosophila melanogaster*) females have a pair
of chromosomes, called X-chromosomes, only one representative of which is
present in the male. In place of the second X-chromosome in the male sex there is a
Y-chromosome, which is homologous to the X only to a very limited extent. At
meiosis in the male sex the X and the Y pair in their small region of homology and
usually become joined by a single chiasma in this region (*see Fig.* 2.12). The
position of the chiasma is nearly always distal to or on the opposite side of the
centromere from the large segments in which the X and Y differ and so, with rare
exceptions, the X and Y differential segments segregate from one another at the
first meiotic division. The second division leads to the formation of two X-bearing
and two Y-bearing spermatids (or pollen grains in the case of dioecious plants). In
female meiosis, the X-chromosome pair form chiasmata and disjoin like a non-sex
chromosome (**autosome**) pair, and all eggs receive a single X-chromosome. The sex
of the next generation depends on whether the egg is fertilized by an X-carrying or
by a Y-carrying male nucleus. The 2:2 X:Y ratio in every male meiotic tetrad
tends to ensure a 1:1 sex ratio, though this ratio may be disturbed somewhat by
different efficiencies of the two kinds of sperm in effecting fertilization (in man
there is about a 1 per cent excess of males at conception).

There are many variations on the well-known XX female/XY male system just
described. The flies (Diptera) as a group show a peculiarity which we must note
because of the great importance of *Drosophila melanogaster* in experimental
genetics. In female Diptera, meiosis and chiasma formation are perfectly normal,
but in the male there is no chiasma formation either by the X- and Y-chromosomes
or any of the autosomal pairs. In most eukaryotes, chiasma formation seems to be
necessary to stabilize bivalent associations at metaphase I and ensure regular
anaphase disjunction (unpaired univalents tending to distribute at random), but,
in the male fly, chromosome pairs at this stage associate with and disjoin from each
other regularly in spite of the lack of any apparent contact between them. XO
males, producing X-bearing and non-X-bearing spermatozoa in equal numbers,
are also found in some nematode worms, including the experimentally important
Caenorhabditis elegans. In this species XX animals are hermaphrodite and produce
X-bearing eggs and X-bearing sperm. Thus self-fertilization of hermaphrodites
gives only hermaphrodites in the following generation, and crosses of hermaphrodites with males give hermaphrodite and male progeny in approximately
equal numbers.

In grasshoppers and other insects of the order Orthoptera, which as a group are renowned for the clarity of their meiotic divisions under the microscope (*Fig.* 2.9), there is usually no Y-chromosome. The females are XX and the males just X (called X0 or X-zero). At anaphase I in the male the unpaired X passes to one pole and its centromere divides (giving two X-bearing products out of four) at the second division.

In the insects of the order Hymenoptera (bees, wasps, ants) the males are *haploid*, and develop **parthenogenetically**, i.e. from unfertilized eggs. Meiosis in the haploid male is replaced by single nuclear division which amounts to an ordinary mitosis. This system of male haploidy has the consequence that, far from being tied down to a 1 : 1 sex ratio, the Hymenoptera are able to vary the relative numbers of males and females by adjusting the frequency of fertilization of eggs—a facility of which these insects take full advantage in their elaborate systems of social organization.

Other insects, notably the aphids (order Hemiptera), take versatility a stage further by alternating between parthenogenesis and normal sexual reproduction. During their phase of rapid multiplication these insects produce diploid eggs (by an aberrant form of meiosis in which the second division is aborted) and these develop without fertilization. Once in each annual cycle, however, there is a reversion to normal meiosis and fertilization.

Among orders of vertebrates other than mammals one finds a number of other variations. In birds, reptiles and at least some amphibia, what is essentially an XX/XY system is found, but with the sexes reversed; the male is the sex with the similar chromosome pair (called WW), while the female has the dissimilar pair (WZ). Thus in these groups it is the female which is the **heterogametic** sex, producing two kinds of eggs, W and Z, while the male is homogametic, producing only W-carrying sperm. The fish, or many of them (and the same applies to some amphibia), appear to have sex chromosomes which are not visibly differentiated. The switch which channels development into either a male or a female direction seems to depend on minimal differences between a pair of chromosomes which look identical and seem fully homologous by the criteria of pairing and chiasma formation. In *Lebistes*, the guppy, Ø. Winge[6] showed that some strains could have the female as the heterogametic sex and other strains the male; he was even able to convert one system into the other by selection, suggesting that the developmental switch could be thrown one way or the other by any one of a number of small chromosomal differences.

For a much fuller and more authoritative account of the great variety of sex chromosome mechanisms in the animal world the reader is referred to M. J. D. White's authoritative book *Animal Cytology and Evolution*.

The sex chromosomes provide us with our first, and most directly visible, example of the first principle of eukaryotic genetics. The mutual exclusiveness of the X- and Y-chromosomes in spermatogenesis was unknown to Mendel, but is nevertheless a prime example of Mendel's First Law of segregation of unit differences, with which the following chapter is mainly concerned.

2.5 Summary

We have reviewed a number of representative types of sexual life history, all of which involve an alternation between haploid and diploid phases. We have seen how homologous sets of chromosomes are brought together by sexual cell fusion and segregate away from each other at meiosis, which consists of two divisions of the nucleus accompanied by only one division of the chromosomes; the chromosome arms split at the first division but the centromeres remain undivided until the second. Chromosome segregation has its most obvious consequence in the regular production of equal numbers of male and female-determining germ cells by the heterogametic sex in those organisms which have separate sexes. It can, however, be seen microscopically wherever a homologous pair of chromosomes differ visibly. The difference can segregate at either the second or the first division of meiosis depending upon whether or not a chiasma is formed between the point of difference and the chromosome centromere.

Chapter 2 Selected Further Reading

Darlington C. D. (1965) *Cytology*, 768 pp. London, Churchill.
Swanson C. P., Merz T. and Young W. J. (1981) *Cytogenetics—The Chromosome in Division, Inheritance and Evolution*, 2nd ed. 577 pp. Englewood Cliffs, New Jersey, Prentice-Hall.
White M. J. D. (1973) *Animal Cytology and Evolution*, 3rd ed., 961 pp. Cambridge, Cambridge University Press.

Chapter 2 Problems

See Problems at end of Chapter 3.

3 Chromosomal Segregation in Eukaryotes

This chapter gives an account of the experiments and reasoning that leads us to attribute most genetic variation to the chromosomes.

3.1 Distinguishing nuclear from extranuclear heredity

A simple and rather convincing, if not absolutely conclusive, argument in favour of the predominant role of the cell nucleus in the determination of inherited differences depends on the generally equal importance in heredity of male and female gametes. Eggs and spermatozoa in animals, and eggs and fertilizing pollen tubes in flowering plants, contribute very unequally to the fertilized egg in every respect except the nucleus. The egg contributes virtually all the cytoplasm and organelles as well as its own nucleus, whereas the male fertilizing cell often contributes very little except its nucleus.

The relative contributions of male and female gametes to the determination of a particular trait in the next generation can be assessed by means of **reciprocal crosses**. We can illustrate the principle with two simple examples. In the house mouse, *Mus musculis*, there are true-breeding *black* strains which differ from the standard type in having uniformly black hairs in their fur rather than the banded brown-grey hairs ('agouti' pattern) seen in the wild type. A cross between mice from true-breeding agouti and black strains will give only agouti mice in the resulting litter regardless of whether the female parent was black and the male agouti or whether the situation was the other way round. Agouti is **dominant** in the first generation (F_1) and black **recessive**, regardless of whether the germ cell transmitting the dominant trait is an egg, with a nucleus and much else besides, or a sperm cell, with virtually only a nucleus. For a second example, showing a somewhat different type of result, we can take the difference between true-breeding strains of snapdragon (*Antirrhinum majus*) with, respectively, magenta and white flowers. A cross between magenta and white yields seed which can be grown to give F_1 plants which all show **incomplete dominance** of magenta over white—all the flowers are slightly paler than those of the parental magenta strain. Exactly the same shade of slightly dilute magenta is obtained irrespective of which strain is used as the seed parent (female ♀) and which as pollinating parent (male ♂). Again, equal contributions of male and female gametes to the constitution of the F_1 generation are indicated, and again the nucleus would seem to be the only material contributed in equal measure by each.

In each of these two examples, the results of further breeding from the F_1 generation are the same whichever way round the original cross is made.

The rule of equal results from reciprocal crosses between true-breeding strains holds in the great majority of instances. There are two kinds of exception to the rule. The first, which is the concern of Chapter 7, is due to the determination of a genetic trait by a genetic element (as we now know, usually a DNA molecule of one kind or another) carried outside the nucleus. For example, in many different flowering plants some kinds of chlorophyll deficiency (shown as yellow or variegated foliage) are inherited exclusively or almost exclusively from the seed parent and hardly or not at all from the pollen parent. Thus a cross of yellow ♀ × green ♂ may give entirely yellow F_1 plants, and a further cross (backcross) of

F_1 yellow ♀ × green ♂ will again give all-yellow progeny seedlings. The reciprocal cross of green ♀ × yellow ♂ will in such a case usually give entirely green seedlings in the F_1 and again entirely green following the backcross of F_1 ♀ to yellow ♂. Such a result indicates strongly that whatever it is that determines the chlorophyll deficiency is transmitted virtually exclusively with the egg, presumably outside the nucleus since the nature of the pollen nucleus makes no difference. In fact the determinant in this sort of case is almost certainly the DNA of the chloroplasts, which in many species are only rarely transmitted via the pollen tube.

The second kind of exception to equality of reciprocal crosses involves sex linkage, and turns out to be a special case of nuclear heredity (*see below* p. 84).

The remainder of this chapter is concerned with the rules governing nuclear heredity. We return to extranuclear heredity in Chapter 5.

3.2 | The segregation of gene differences at meiosis

Ordered Tetrads in Ascomycete Fungi

The first principle of formal genetic analysis, sometimes called **'Mendel's First Law'**, is that, when a diploid is formed by a cross between the organisms differing with respect to a single genetic determinant, meiosis in that diploid results in the segregation of the two kinds of determinant, so that half of the haploid meiotic products carry one and half the other. More precisely, each meiotic tetrad (with infrequent exceptions considered below) shows a 2:2 segregation, a fact unknown to Mendel but one which can be readily demonstrated in an organism, such as an Ascomycete fungus, in which the recovery of meiotic products in their original tetrads is possible.

A good example is the inheritance of ascospore colour differences in *Sordaria brevicollis*. In this, as in other related species of genetical importance such as *Neurospora crassa*, the ascus is an ordered one, with the linear arrangement of the eight ascospores reflecting their descent through the two meiotic and one postmeiotic nuclear divisions. As *Fig.* 3.1 explains, the spore pairs 1+2, 3+4,

Fig. 3.1

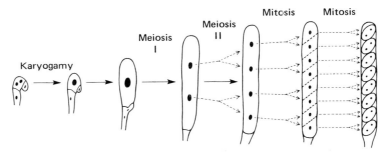

Origin and arrangement of nuclei in the ascus of *Sordaria* spp. or *Neurospora crassa.*

$5+6$, $7+8$, numbering from the tip to the base, each represent one product of meiosis. Spore pairs $1+2$ and $3+4$ stem from one product of the first division of meiosis and spore pairs $5+6$ and $7+8$ from the other, provided that the second division meiotic spindles do not overlap (which, in fact, they often do in *S. brevicollis*, though not in *Neurospora*).

Sordaria brevicollis resembles *N. crassa*, and differs from some other *Sordaria* species (such as *S. fimicola, see* p. 132), in being heterothallic, with two mating types, both of which must be present for the fruit bodies to be produced. Wild-type strains of *Sordaria brevicollis* and *Neurospora crassa* have, respectively, dark brown and black ascospores, but many mutant strains with pale ascospores are known. When a wild-type and a pale-spored strain of different mating type are grown together they constitute a sexual **cross** in which hybrid perithecia are formed, and these contain asci each with two dark and two pale spore pairs (*Fig.* 3.2). When isolated and germinated, each spore of a given colour produces a culture which, crossed with a culture of opposite mating type, will produce ascospores of that same colour again. In other words the appearance (**phenotype**) of each haploid ascospore reflects its breeding potential, genetic constitution or **genotype**.

The $2:2$ ratio of differently coloured spore pairs reflects a $2:2$ segregation at meiosis of genetic determinants of the two colours. These determinants act as strict alternatives, i.e. a haploid meiotic product can carry one or the other but not both, and such mutually exclusive factors are called **alleles**. They can also be referred to as allelic *to each other*. Allelles of the same series are alternative forms of a genetic determinant (**gene**) with a specific function, in this case in spore pigment synthesis. The theory of genes and alleles, and its experimental basis, is explored in detail in Chapters 5 and 13.

There may be many different alleles of a particular gene (an allelic series), all more or less different in their phenotypic effects, but only two of them can be present and segregating at meiosis in any particular diploid. Each series of alleles (i.e. each gene) is given a gene symbol, usually a letter or a two, three or four-letter abbreviation. Different members of each series may be distinguished if necessary by superscript numbers or letters; the normal allele typical of wild populations, if

Table 3.1	Patterns of segregation in 8-spored asci of a fungus such as *Neurospora crassa* formed in a cross between wild-type $(+)$ and mutant (m) strains					
Spores (numbered from tip to base of ascus)	*First-division segregation*		*Second-division segregation*			
1	$+$	m	$+$	m	$+$	m
2	$+$	m	$+$	m	$+$	m
3	$+$	m	m	$+$	m	$+$
4	$+$	m	m	$+$	m	$+$
5	m	$+$	$+$	m	m	$+$
6	m	$+$	$+$	m	m	$+$
7	m	$+$	m	$+$	$+$	m
8	m	$+$	m	$+$	$+$	m
Ascus pattern (Fig. 3.2)	1	2	3	4	5	6

Fig. 3.2

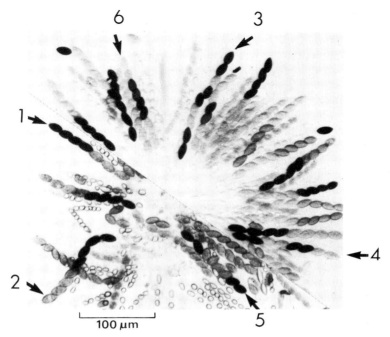

Asci from a cross between *Neurospora crassa* mutant strain carrying an allele determining pale ascospores and a wild-type strain of opposite mating type. Note 4:4 segregation of dark and pale spores in each ascus, with examples of first-division and second-division segregation patterns numbered as in *Table* 3.1. Note that the labelled type 3 ascus is lying across a type 4 ascus.

there is one, is usually given a + superscript. When there is no doubt about which gene is being referred to, the + symbol for normal or wild type is often used alone.

It will be immediately obvious that the 2:2 segregation of alleles at meiosis parallels the 2:2 segregation of the four chromatids derived from each pair of homologous chromosomes. The chromosome theory of inheritance, proof of which will accumulate as this chapter progresses, states that pairs of alleles are carried on (or are part of) homologous pairs of chromosomes. This theory also explains another feature of allelic segregation, seen in *Fig.* 3.2, which shows asci from a cross between wild-type *N. crassa* and a mutant (*per*) with both pale perithecia and pale ascospores. About 35 per cent of asci show two different-coloured spore pairs in each half of the ascus, there being four such patterns in statistically equal frequencies as set out in *Table* 3.1. In these asci the two alleles (*per* and *per*$^+$) determining the spore colour difference did not segregate from one another until the second division of meiosis; the two nuclei at the end of the first division must have carried both *per* and *per*$^+$. In the other 65 per cent, the spores in one-half of the ascus are all dark and those in the other half all light, indicating that the two alleles segregated at the first division. It is possible to draw this conclusion because it is known that in *Neurospora crassa* the spindles of the second division of meiosis are in opposite ends of the ascus with no overlapping.

The frequency of about 35 per cent second division segregation is characteristic of the *per* gene; allelic differences in other genes show different frequencies varying, depending on the gene, from zero to about 67 per cent.

If we recall the behaviour of centromeres and the nature of crossovers in the first division of meiosis, we can see how a characteristic frequency of second division can be explained as a consequence of the position on the chromosome of the gene (called the gene **locus**) in relation to the centromere. We will expect first-division segregation if no chiasma is formed between the gene and the centromere, since in this case the two chromatids carried by the (as yet undivided) centromere to each pole of the first-division spindle will carry copies of the same allele. If, on the other hand, a chiasma does occur between the gene and the centromere, the chromatids attached to each centromere will, at the locus in question, be non-sisters and will carry different alleles which will only segregate from one another at the second division when the centromere divides (*Fig.* 3.3).

The probability of a chiasma being formed between homologous chromosomes in a given segment depends on the length of that segment; thus alleles of a gene at the centromere are expected to show no second-division segregation. When the gene locus is a small distance from the centromere we expect a low frequency and, when it is far from the centromere, a high frequency of second-division segregation. In fact, percentage second-division segregation can be used as a genetic measure of distance from the centromere and makes an important contribution to genetic mapping in organisms in which tetrads can be analysed.

Fig. 3.3

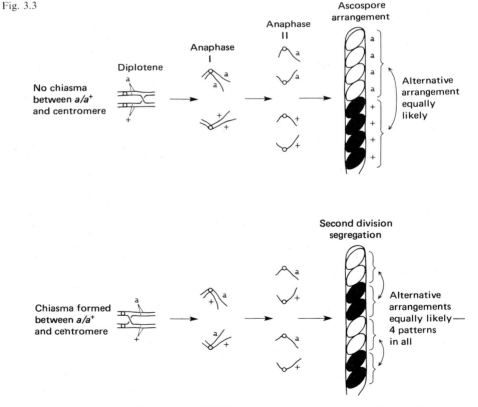

Explanation of second-division segregation in terms of crossing-over between the segregating locus ($a/+$) and the centromere.

Segregation at Meiosis seen in Randomized Meiotic Products

In higher plants and animals, in which there is no free-living haploid phase in the life cycle, it is generally not possible to observe genetic segregation at meiosis directly. One usually has to deduce the gamete genotypes from the phenotypes of the succeeding diploid generation. There are just a few situations in which a segregating difference can be observed in the meiotic products themselves. For example, in stained preparations of grasshopper testis, 50 per cent of grasshopper spermatids can be seen to contain the X-chromosome, visible as a deeply staining blob, and the other 50 per cent not. In maize, the geneticist's favourite flowering plant, there are several instances of allelic differences which can be made visible in the pollen grain. The best known of these is the difference between 'waxy' starch (amylopectin, without any long-chain amylose, determined by the mutant allele *wx*) which stains red-brown with iodine–potassium iodide reagent, and 'non-waxy' (much amylose) which is determined by the corresponding normal allele and stains blue-black. *Fig.* 3.4 shows another example—the presence or absence in the pollen grain of the enzyme alcohol dehydrogenase, for which there is a specific staining procedure. The pollen grains have been released from their original tetrads by the time they can be observed, so here the original precise 2:2 ratio in each tetrad becomes a statistical 1:1 ratio in the whole population of randomly mixed meiotic products.

Where statistical sampling error is involved, as it constantly is in genetic analysis, it is necessary to perform statistical tests to decide whether the observed numbers conform reasonably to the theoretical ratio. This book will not go deeply into statistical methods, but application of the χ^2 (chi squared) method is explained

Fig. 3.4

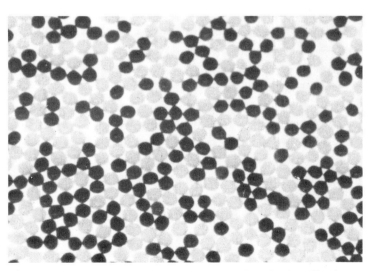

Segregation of a single gene difference in pollen grains of maize. The plant in which they were formed was a heterozygote with respect to two alleles determining, respectively, presence and absence of the enzyme alcohol dehydrogenase. The grains have been stained by a procedure which gives a deep purple colour to cells containing alcohol dehydrogenase and leaves the others colourless.
Photograph by courtesy of Dr M. Freeling.[3]

on p. 83. For an expected $1:1$ ratio the appropriate value of χ^2 is simply the square of the difference between the two observed numbers divided by their sum. If this value exceeds 3·8, the probability (P) of getting such a bad fit to the $1:1$ expectation merely by bad luck in the sample is less than 1 in 20 or 0·05. The deviation from theoretical expectation is then said to be significant at the 5 per cent level.

The Consequences of Meiotic Segregation in Diploid Organisms

Segregation of alleles at meiosis, giving $2:2$ segregation in tetrads and $1:1$ ratios among randomized meiotic products, is universal in eukaryotes, but where, as is usual in higher plants and animals, the haploid products of meiosis cannot be scored directly, one has to rely on genetic ratios in the subsequent diploid generation. Let us return to our simple example from the garden snapdragon, *Antirrhinum majus*.

We saw on p. 76 that if a true-breeding variety with white flowers (*incolorata* or *inc*) was crossed to one with dark magenta flowers (*inc*$^+$), the resulting seedlings (the F$_1$ generation) all had flowers of a rather more dilute magenta colour. Now if we **backcross** an F$_1$ plant, either as seed parent or as pollen parent, to the original white strain, the backcross progeny fall into two classes, white and dilute-magenta, in statistically equal numbers. If the backcross is made in the other direction, that is F$_1$ × full magenta, the progeny are 50 per cent full magenta and 50 per cent dilute magenta.

These results can be simply explained on the basis that at meiosis in the F$_1$ plants, both in embryo sac and in pollen grain formation, there is a $1:1$ segregation of *inc* and *inc*$^+$ products which, in the backcross to white (which contributes only *inc*), gives 50 per cent *inc*/*inc* and 50 per cent *inc*/*inc*$^+$ seeds; the backcross to full magenta (which contributes only *inc*$^+$) gives 50 per cent *inc*/*inc*$^+$ and 50 per cent *inc*$^+$/*inc*$^+$.

At this point it is necessary to introduce some essential terminology and concepts of diploid genetics. A diploid which has received the same allele from each parent (*inc*$^+$/*inc*$^+$ or *inc*/*inc* in our example) is called a **homozygote**, and one which has received differing alleles (like *inc*/*inc*$^+$) is called a **heterozygote**. Note that these terms are used in reference to specific genes, so an individual may be heterozygous with respect to one gene and homozygous with respect to another.

Another term, and basic concept, illustrated by this *Antirrhinum* example is **dominance**, as opposed to **recessivity**. One would say that dominance of one allele over the other was absent if the phenotype of the heterozygote was exactly intermediate (on some quantitative measure) between the phenotypes of the homozygotes. Dominance would be complete if the heterozygote had the same phenotype as one of the homozygotes—that carrying the dominant allele. The *inc*$^+$/*inc* heterozygote shows some degree of dominance of the *inc*$^+$ allele, in that its level of pigmentation is closer to that of *inc*$^+$/*inc*$^+$ than to the zero level in *inc*/*inc*. But dominance is not complete, since the heterozygote is somewhat paler than the *inc*$^+$ homozygote.

Incomplete dominance is not uncommon and is useful experimentally in that it allows the heterozygote to be distinguished from both homozygotes. However, complete dominance, in which the heterozygote is phenotypically virtually in-

Box 3.1   **The χ^2 (chi-squared) test for goodness of fit to theoretical ratios**

General case: Suppose we have a sample of N individuals classified into n classes. If the theoretical ratios predict, for a sample of size N, the numbers

$$e_1, e_2, e_3 \ldots e_n$$

and the numbers actually observed are

$$o_1, o_2, o_3 \ldots o_n$$

$$(\Sigma e = \Sigma o = N)$$

then

$$\chi^2_{n-1} = \frac{(e_1 - o_1)^2}{e_1} + \frac{(e_2 - o_2)^2}{e_2} + \frac{(e_3 - o_3)^2}{e_3} \ldots + \frac{(e_n - o_n)^2}{e_n}.$$

$n-1$ is the number of **degrees of freedom**, i.e. the number of variables if the total size of the sample is fixed.

The value of χ^2 obtained corresponds to a particular probability of obtaining as poor a fit to the theoretical ratio by chance, i.e. by drawing an unlucky sample from a population which as a whole really *did* conform to the theoretical ratio. This probability can be obtained by looking up tabulated values of χ^2 for various degrees of freedom. The greater the number of degrees of freedom, the larger χ^2 can be before P becomes unacceptably small.

Conventionally, if $P < 0.05$ the deviation from the expected ratio is regarded as 'significant'. If $P < 0.01$ the deviation is 'highly significant'. But remember that you may simply have the one-in-twenty or one-in-a-hundred 'bad luck' sample.

Special cases: If there are two classes only (i.e. one degree of freedom), in expected ratio $1:1$,

$$N = o_1 + o_2,$$

$$e_1 = e_2 = \frac{o_1 + o_2}{2}$$

and

$$\chi^2_1 = 2\left\{ \frac{\left[o_1 - \left(\frac{o_1 + o_2}{2} \right) \right]^2}{\frac{o_1 + o_2}{2}} \right\} = \frac{(o_1 - o_2)^2}{o_1 + o_2}$$

If the expected ratio is $3:1$

$$e_1 = \tfrac{3}{4}(o_1 + o_2), \quad e_2 = \tfrac{1}{4}(o_1 + o_2),$$

$$\chi^2_1 = \frac{[o_1 - \tfrac{3}{4}(o_1 + o_2)]^2}{\tfrac{3}{4}(o_1 + o_2)} + \frac{[o_2 - \tfrac{1}{4}(o_1 + o_2)]^2}{\tfrac{1}{4}(o_1 + o_2)}$$

which reduces to

$$\chi^2_1 = \frac{(3o_2 - o_1)^2}{3(o_1 + o_2)}$$

For $n = 1$ $P < 0.05$ if $\chi^2 > 3.8$
$P < 0.01$ if $\chi^2 > 6.6$

For $n = 2$ $P < 0.05$ if $\chi^2 > 6.0$
$P < 0.01$ if $\chi^2 > 9.2$

For $n = 3$ $P < 0.05$ if $\chi^2 > 7.8$
$P < 0.01$ if $\chi^2 > 11.3$

distinguishable (at least superficially) from one homozygote, is more frequently encountered.

If, instead of backcrossing the F_1 dilute-magenta plants to one or other parental variety, we allow the F_1 flowers to self-pollinate, we will find in the following generation (called the F_2), all the colours, full magenta, pale magenta and white, in a ratio conforming within acceptable statistical limits to $1:2:1$. This is again easily explained on the basis of $1:1$ segregation at meiosis. If there are two kinds of embryo sac, inc and inc^+ in equal frequencies, and if the same applies to the pollen grains, there will be four kinds of equally probable pollination. These will be (writing the egg genotype first, as is the convention), $inc^+ \times inc$, $inc^+ \times inc^+$, $inc \times inc$ and $inc \times inc^+$ the overall result being 1 inc^+/inc^+ (homozygous magenta) : 2 inc^-/inc (heterozygous pale magenta) : 1 inc/inc (homozygous white). This interpretation can be confirmed by backcrossing to white; the two homozygous classes give no segregation (wholly pale-magenta and wholly white progeny, respectively) while the heterozygous class behaves like the F_1 and gives a $1:1$ ratio of pale-magenta to white seedlings.

The results of further breeding tests in our other simple case, that of agouti and black mice (p. 76), are closely analogous and receive a similar explanation. In this case agouti (b^+) is completely dominant to black (b). The backcross of the F_1 (either sex) to black gives approximately 50 per cent black and 50 per cent agouti in both sexes of the backcross progeny. The backcross to pure-breeding agouti gives, genotypically, 50 per cent b^+/b^+ and 50 per cent b^+/b (as can be shown by further breeding tests) but these two classes are not distinguishable because of the complete dominance of b^+. Self-fertilization is, of course, impossible in a mammal, but brother–sister mating in the F_1 is genetically equivalent ($b^+/b \times b^+/b$) and gives a $3:1$ agouti : black ratio in the F_2, the 50 per cent of heterozygotes being indistinguishable by eye from the 25 per cent of homozygotes for the dominant allele. This is the famous $3:1$ ratio discovered by Gregor Mendel in his experiments on garden peas. We see, as he did, that it is a secondary result of the $1:1$ segregation in germ-cell formation.

3.3   Sex-linked inheritance

One important type of exception to the rule of equal results of reciprocal crosses (p. 76) is not due to genetic factors outside the nucleus but rather to a special kind of chromosomal heredity. It could have been confidently predicted from the microscopically observable distribution of sex chromosomes. Suppose, in an animal with an XX/XY (female/male) sex chromosome mechanism, that the X-chromosome (but not the Y) carries a genetic locus with a phenotypic effect other than sex determination. A male carrying a recessive X-linked allele will then always exhibit the corresponding phenotype since there will be no second X-chromosome present to supply the dominant allele. A male can be neither homozygous nor heterozygous with respect to an X-linked gene, but only **hemizygous**, without the possibility of concealment by dominance. An X-linked allele will be transmitted by a hemizygous father to all his daughters but to none of his sons (who receive his Y-chromosome instead); the same allele in a heterozygous mother will be transmitted to 50 per cent of both sons and daughters.

This is just what is seen in very many cases in mammals, including man, and in *Drosophila*. We may take as a typical case the inheritance of the *white eye* (*w*) mutant trait in *Drosophila melanogaster*, which was the first example of sex linkage in *Drosophila* or any other organism. If, starting with pure-breeding wild-type (*w*$^+$, red-eyed) and white-eyed (*w*) strains, we cross white female × red male, all the male progeny are white-eyed and all the female red-eyed. In the reciprocal cross, red female × white male, all F$_1$ progeny of both sexes are red-eyed. In each case the red-eyed female progeny can be shown to be heterozygous *w*/*w*$^+$; crossed to white-eyed males they produce a 1 : 1 ratio of red to white in both sexes, and crossed to red-eyed males they produce the 1 : 1 ratio in the male progeny only, the female progeny all being red-eyed. All this is just what one would expect if *w* were a recessive mutant allele on the X-chromosome (*see Table 3.2a*).

Proof of the postulated connection between *w* and the X-chromosome was provided in 1914 by Bridges, who investigated an unusual white-eyed stock of *Drosophila* in which some of the females gave a small proportion of exceptional progeny when crossed to wild-type males. The majority (about 96 per cent) of the progeny consisted, as expected, of approximately equal numbers of white-eyed males and red-eyed females, but there were also about 2 per cent each of red-eyed males and white-eyed females. Bridges postulated that the exceptional female progeny had inherited both X-chromosomes from their mother and a Y from their father. This is a plausible suggestion, since XXY flies in *Drosophila* are normal females; unlike the situation in mammals, *see p. 569*, the sex depends on the ratio of X-chromosomes to autosomes rather than on the presence or absence of the Y. He also postulated that the exceptional males had received an X from their father and a Y from their mother. Both of these anomalies could arise if the mothers of the exceptional families were themselves XXY. In such exceptional females one

---

Box 3.2   Segregation of a unit difference

Cross *AA* × *aa*
F$_1$ heterozygote *A*/*a*
Products of meiosis 50 per cent *A* : 50 per cent *a*
1. *Aa* self-fertilized, or *Aa* × *Aa*

		Female gametes		
		A	*a*	
Male gametes	*A*	*AA*	*Aa*	Overall ratio
	a	*aA*	*aa*	1*AA* : 2*Aa* : 1*aa*
				3 : 1 phenotypic ratio if
				A dominant

2. Backcross of F$_1$ to homozygous recessive parent, *Aa* × *aa*:

		F$_1$ gametes (female or male)		
		A	*a*	
a/*a* gametes	*a*	*Aa*	*aa*	1*Aa* : 1*aa*
(male or female)				1 : 1 phenotypic ratio if
				A dominant

NB In the convention used here, customary in maize but not in *Drosophila* genetics, the **dominant** allele of a pair is represented by a capital and the **recessive** allele by a lower-case letter, without any prejudice as to which, if either, is the 'wild type'.

Table 3.2   **Sex-linked inheritance of w (white eye) in *Drosophila***

a. Normal case

White ♀   ×   Normal ♂
X X   X Y
w w w^+ −

		Sperm	
		$\frac{1}{2}$ X $\quad w^+$	$\frac{1}{2}$ Y $\quad$ −
Eggs all X $\quad w$		$\frac{1}{2}$ X X $\quad w^+$ w Heterozygous ♀ red-eyed	$\frac{1}{2}$ X Y $\quad w$ − Hemizygous ♂ white-eyed

b. Result with exceptional white females crossed to normal males

XXY × X Y
w w − × w^+ −

Eggs			Sperm	
			X w^+	Y −
	circa 48%	X w	XX $\quad$ Normal red- w w^+ $\quad$ eyed hetero- zygous ♀♀	XY $\quad$ Normal hemizygous w − $\quad$ white-eyed ♂♂
	circa 48%	X Y w −	XX Y $\quad$ Heterozygous w w^+− $\quad$ red-eyed ♀♀ with extra Y	XYY $\quad$ White-eyed ♂♂ w − − $\quad$ with extra Y
	circa 2%	X X w w	XXX $\quad$ Triple X red- w w w^+ eyed ♀♀ poorly viable	XXY $\quad$ *Exceptional* white- w w − $\quad$ eyed ♀♀ with extra Y
	circa 2%	Y −	X Y $\quad$ *Exceptional* w^+− $\quad$ red-eyed ♂♂	YY $\quad$ Not viable − −

might expect some anomalies in meiosis. Sometimes the two X-chromosomes might form a bivalent and disjoin to opposite poles at the first division, the Y going to one or other pole more or less at random. Sometimes also, one X might pair with and disjoin from the Y, the second X being randomly distributed. The first eventuality (seemingly the more frequent alternative) must result in X and XY eggs, and the second may give XX and Y eggs. These latter, on fertilization by normal X and Y sperms, would give the exceptional progeny observed, as explained in *Table* 3.2*b*. Bridges confirmed his hypothesis by examining the chromosomes of females of the exceptional broods. As predicted, the exceptional white-eyed females had two X-chromosomes plus a Y, and half of the red-eyed females were also XXY. The XXY karyotype must be the result in the first case of a fertilization of the type XX egg × Y sperm, and in the second case by fertilization of XY egg × X sperm. The exceptional inheritance of the w/w^+ difference can thus be explained by the exceptional sex chromosome constitution, and the whole analysis provides compelling evidence for the location of the w locus on the X-chromosome.

In man, a number of well-known recessively inherited traits show the 'criss-cross' (i.e. mother-to-son, father-to-daughter) transmission typical of X-linkage. The commonest type of red–green colour blindness is one example and the rare but notorious bleeding disease, haemophilia, is another (cf. p. 565, Chapter 19). In both of these examples the defective phenotype is far more common in males than in females. Females need to inherit a recessive X-linked allele from both parents in order to show its effect; for males it is sufficient to inherit it from the mother. If the allele frequency in the whole population is p, the proportion of affected males will be p, and proportion of affected females p^2. For example, if, as is approximately the case, the frequency of colour-blind men is about 10 per cent ($p = 0.1$), we expect, and in fact we find, only about 1 per cent ($p^2 = 0.01$) of similarly colour-blind women.

3.4 Independent assortment of two allelic pairs

Random Meiotic Products

It often happens that a diploid is heterozygous for two or more phenotypically scorable allelic differences at the same time. In such cases the most common finding is that each allelic difference segregates at meiosis independently of the others. For a simple example we can turn again to the garden snapdragon, *Antirrhinum majus*. This plant produces two kinds of conspicuous flower pigment: magenta anthocyanin, already mentioned, and a bright yellow aurone, which in the wild type (dominant allele *sulf*$^+$) is confined to a small patch on the lip of the corolla but in plants of the homozygous recessive constitution *sulf/sulf* is spread over the whole flower. The *sulf*$^+$/*sulf* difference segregates at meiosis in just the same way as does *inc*$^+$/*inc* (magenta/white), and *sulf*$^+$ is completely dominant to *sulf*. The two pigments when present together over the whole flower give a bronze colour. Thus *inc*$^+$ with *sulf/sulf* gives bronze, *inc*$^+$ with *sulf*$^+$ gives magenta, *sulf*$^+$ with *inc/inc* gives white and *sulf/sulf inc/inc*, the double recessive homozygote, is yellow. (We ignore here the small difference between *inc*$^+$/*inc*$^+$ and *inc*$^+$/*inc* in intensity of magenta pigment.)

If we cross a true-breeding magenta strain to yellow, the F_1 seedlings all produce magenta flowers. Backcrossing an F_1 plant, either as seed parent or as pollen parent, to yellow gives, among the backcross progeny, nearly equal numbers of magenta, bronze, yellow and white. The obvious and simple interpretation is that meiosis in the doubly heterozygous (*inc*$^+$/*inc sulf*$^+$/*sulf*) F_1 results in random assortment of the two allelic differences; each meiotic product has a 50 per cent chance of receiving *inc*$^+$ as opposed to *inc* and a 50 per cent chance of receiving *sulf*$^+$ as opposed to *sulf*, and the two chances are independent of each other.

This will result in the four types of eggs and pollen grains, *inc*$^+$ *sulf*$^+$, *inc*$^+$ *sulf*, *inc sulf*$^+$ and *inc sulf*, being formed in statistically equal numbers, with reassorted types (*inc*$^+$ *sulf*, *inc sulf*$^+$) equalling the parental combinations (*inc*$^+$ *sulf*$^+$, *inc sulf*). When united with *inc sulf* gametes from the yellow doubly homozygous recessive, these four types of gamete will give the four phenotypes observed in the backcross progeny.

The principle of independent assortment can easily be generalized to more than two pairs of allelic differences. Thus if we have a triple heterozygote $\frac{a}{+} \frac{b}{+} \frac{c}{+}$, independent assortment will generate eight types of meiotic products in equal frequencies: $a\,b\,c, + + +, a\,b+, + + c, a + +, + b\,c, a + c, + b +$. With four independent heterozygous genes there will be sixteen equally frequent meiotic products. Each additional heterozygous gene will multiply the number of types of meiotic product by two; with n such genes the number will be 2^n, all equally frequent if the allelic differences all segregate independently.

If, instead of backcrossing to the double-recessive parent we allow the F_1 plants to self-pollinate, the F_2 seedlings again show the phenotypic proportions expected if the inc/inc^+ and $sulf/sulf^+$ allelic pairs were segregating independently. As we saw above, we expect a $3:1$ ratio in an F_2 generation when one allele of a pair is dominant. Ignoring the incomplete dominance of inc^+ for the moment, we expect, from the segregation of inc^+/inc, to find a $3:1$ ratio of plants with to plants without the magenta anthocyanin pigment; we also expect, from the segregation of

---

Table 3.3   **Independent assortment in the determination of flower colour in *Antirrhinum majus***

Magenta   Yellow
$inc^+/inc^+\ sulf^+/sulf^+ \times inc/inc\ sulf/sulf$

F_1 generation: All magenta
$inc^+/inc\ sulf^+/sulf$

		F_1 eggs			
		$inc^+\ sulf^+$	$inc^+\ sulf$	$inc\ sulf^+$	$inc\ sulf$
F_1 pollen grains	$inc^+\ sulf^+$	Magenta $\dfrac{inc^+\ sulf^+}{inc^+\ sulf^+}$	Magenta $\dfrac{inc^+\ sulf}{inc^+\ sulf^+}$	Magenta $\dfrac{inc^+\ sulf^+}{inc\ \ sulf^+}$	Magenta $\dfrac{inc^+\ sulf^+}{inc\ \ sulf}$
	$inc^+\ sulf$	Magenta $\dfrac{inc^+\ sulf^+}{inc^+\ sulf}$	Bronze* $\dfrac{inc^+\ sulf}{inc^+\ sulf}$	Magenta $\dfrac{inc^+\ sulf^+}{inc\ \ sulf}$	Bronze* $\dfrac{inc^+\ sulf}{inc\ \ sulf}$
	$inc\ sulf^+$	Magenta $\dfrac{inc^+\ sulf^+}{inc\ \ sulf^+}$	Magenta $\dfrac{inc^+\ sulf^+}{inc\ \ sulf}$	White $\dfrac{inc\ sulf^+}{inc\ sulf^+}$	White $\dfrac{inc\ sulf^+}{inc\ sulf}$
	$inc\ sulf$	Magenta $\dfrac{inc^+\ sulf^+}{inc\ \ sulf}$	Bronze* $\dfrac{inc^+\ sulf}{inc\ \ sulf}$	White $\dfrac{inc\ sulf^+}{inc\ sulf}$	Yellow $\dfrac{inc\ sulf}{inc\ sulf}$

F_2 generation: 9 Magenta:3 Bronze:3 White:1 Yellow (*Bronze results from Magenta mixed with Yellow)

Backcross of F_1 to yellow: $\dfrac{inc^+\ sulf^+}{inc\ \ sulf} \times \dfrac{inc\ sulf}{inc\ sulf}$

$$\frac{inc^+\ sulf^+}{inc\ \ sulf} : \frac{inc^+\ sulf}{inc\ \ sulf} : \frac{inc\ sulf^+}{inc\ sulf} : \frac{inc\ sulf}{inc\ sulf}$$

1 Magenta:1 Bronze:1 White:1 Yellow

sulf⁺/sulf, a 3:1 ratio of white to yellow. If the segregation of each pair is independent of the other these ratios will be combined randomly. That is to say, among the three-quarters with magenta pigment there will be a 3:1 ratio of white to yellow, and among the quarter without magenta pigment there will again be a 3:1 ratio of white to yellow. Thus we expect a 9:3:3:1 ratio of magenta-on-white (magenta), magenta-on-yellow (bronze), white and yellow, and this result is in fact found within reasonable statistical error (*see Table* 3.3).

Independent Assortment in Tetrads

Returning to the conveniently visible segregation of determinants of ascospore colour in *Sordaria brevicollis*, let us consider the situation in which two different pale-ascospore strains are crossed together. Suppose that one of these strains has *grey* ascospores because it carries an allele *g*, corresponding to *g⁺* in the wild type, and that the other has *yellow* ascospores because of an allele *y*, corresponding to *y⁺* in the wild type. The constitution of the hybrid diploid ascus will be $\dfrac{g^+ \ \ y^+}{g \ \ \ y}$. It was explained on p. 78 (*Table* 3.1), how. in ordered asci, segregation of a single pair of alleles gives distinct patterns representing second or first-division segregation depending on whether a chiasma has or has not formed

Fig. 3.5

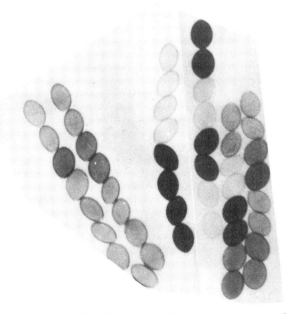

Composite picture of asci from a cross between two ascospore colour mutants in *Sordaria brevicollis*. The two single mutants both have grey ascospores, one a little paler than the other. Parental ditype (PD) asci have four grey spore pairs, with the 2:2 segregation of shades of grey just discernible. Non-parental ditype (NPD) asci have two wild-type (black) and two double-mutant (white) spore pairs. Tetratype (T) asci have two grey spore pairs (one of each parental type) and one double-mutant and one wild-type recombinant spore pair. The ascospores are approximately 20 μm long. The asci shown here are, from left to right, PD, PD, NPD, NPD, T, PD.
Photographs by courtesy of Dr D. J. Bond.

Table 3.4	Independent assortment in tetrads

Cross: $a^+ b^+ \times a b$ a/a^+ segregates   at 2nd division with frequency p
at 1st division with frequency $(1-p)$

b/b^+ segregates   at 2nd division with frequency q
at 1st division with frequency $(1-q)$

First or second-division segregation of the two pairs of alleles

	Both first		a/a^+ 1st a/a^+ 2nd b/b^+ 2nd b/b^+ 1st		Both second		
Frequency	$\dfrac{(1-p)(1-q)}{2}$	$\dfrac{(1-p)(1-q)}{2}$	$q(1-p)$ $p(1-q)$	$\dfrac{pq}{4}$	$\dfrac{pq}{2}$	$\dfrac{pq}{4}$	
Type of tetrad resulting	$\begin{matrix} a^+ b^+ \\ a^+ b^+ \end{matrix}$	$\begin{matrix} a^+ b \\ a^+ b \end{matrix}$	$\begin{matrix} a^+ b^+ \\ a^+ b \end{matrix}$	$\begin{matrix} a^+ b^+ \\ a\ b^+ \end{matrix}$	$\begin{matrix} a^+ b^+ \ a^+ b^+ \\ a\ b \ \ a\ b \end{matrix}$	$\begin{matrix} a^+ b \\ a\ b^+ \end{matrix}$	Separated at 2nd division
	$\begin{matrix} a\ b \\ a\ b \end{matrix}$	$\begin{matrix} a\ b^+ \\ a\ b^+ \end{matrix}$	$\begin{matrix} a\ b^+ \\ a\ b \end{matrix}$	$\begin{matrix} a^+ b \\ a\ b \end{matrix}$	$\begin{matrix} a^+ b^+ \ a^+ b \\ a\ b \ \ a\ b^+ \end{matrix}$	$\begin{matrix} a^+ b \\ a\ b^+ \end{matrix}$	Separated at 2nd division
	PD	NPD	T	T	PD T	NPD	

Separated at 1st division

Separated at 1st division

Totals:

$$PD = NPD = \frac{(1-p)(1-q)}{2} + \frac{pq}{4} = \frac{2(1-p-q+pq)+pq}{4} = \tfrac{1}{2}(1-p-q+\tfrac{3}{2}pq)$$

$$T = q(1-p)+p(1-q)+\frac{pq}{2} = p+q-\tfrac{3}{2}pq$$

Note: If a third allelic pair c^+/c, independent of both a^+/a and b^+/b, and showing second-division segregation frequency r, is taken into account, it is possible to evaluate p, q and r even with unordered tetrads by solving the three simultaneous equations:

Freq. $T_{ab} = p+q-\tfrac{3}{2}pq$;   Freq. $T_{bc} = q+r-\tfrac{3}{2}qr$;   Freq. $T_{ac} = p+r-\tfrac{3}{2}pr$

The solution for p is:

$$p = \tfrac{2}{3}\left[1 \pm \sqrt{\frac{4-6T_{ab}-6T_{ac}+9T_{ab}T_{ac}}{4-6T_{bc}}} \right]$$

with analogous solutions for q and r.

Method of Whitehouse.[1]

between the locus in question and the centromere. If g/g^+ and y/y^+ segregate at the second division with frequencies p and q, respectively, and if their segregation is independent, both will segregate at the first division with frequency $(1-p)(1-q)$, both at the second division with frequency pq, and one at the first and one at the second with frequency $p(1-q)+q(1-p)$.

If both segregate at the first division, the outcome must be a tetrad of spore pairs in which there are only two genotypes (i.e. a **ditype** tetrad). There are two kinds of ditype. In **parental ditypes** (PD), the two parental genotypes are represented—two grey ($g y^+$) and two yellow ($g^+ y$). In **non-parental ditypes** (NPD) the two types each have new combinations of alleles—$g^+ y^+$ (wild-type dark brown) and $g y$ (double mutant, which in this case turns out to be almost colourless). If segregation of the two allele pairs is really independent these two outcomes should be equally likely; PD and NPD asci will be statistically equal in frequency (*Fig.* 3.5).

It will be clear that if one allelic difference segregates at the first division and the other at the second the outcome *must* be tetratype (T) tetrad in which each of the four possible meiotic product genotypes is represented once. It is also easy to work out, by writing down all the possibilities that second-division segregation of both allelic differences will give PD, NPD and T tetrads in the ratio 1:1:2. The possibilities are set out in *Table* 3.4.

As first pointed out by H. L. K. Whitehouse[1] and explained in *Table* 3.4, it is possible to work out second-division segregation frequencies even in unordered tetrads by taking three independently segregating genes and determining the frequency of tetratype asci obtained with each of the three pairwise combinations of these genes. The three resulting simultaneous equations can be solved to provide estimates of the three unknowns—the second-division segregation frequencies of the three genes. This method is especially useful in yeast, where tetrad analysis is easy but the tetrads, unlike the asci of *Neurospora* and *Sordaria*, are unordered.

The interpretation of unordered tetrads is made much easier if one can find a gene so close to its centromere that its alleles segregate virtually always at the first division. A number of such marker loci are indeed available in *Saccharomyces*, and the second-division segregation frequency of any other gene can be obtained by reference to one of these centromere markers. Obviously if p is zero, the equation

$$\text{Freq. } T_{ab} = p + q - \tfrac{3}{2}pq$$

reduces simply to

$$\text{Freq. } T_{ab} = q.$$

The tetratype frequency is equal to the second-division segregation frequency of the second gene.

3.5 Exceptions to simple Mendelian segregation

There are many reasons why, in particular cases, the supposedly fundamental principle of 2:2 segregation of unit genetic differences in meiotic tetrads may not hold. As we shall see in Chapter 6, changes in chromosome number, polyploidy and aneuploidy, and structural chromosomal rearrangement such as insertions and segmental interchanges can result in distorted segregation ratios which can be fully understood in terms of abnormal distribution of chromosomal material at meiosis. More trivially, Mendelian segregation ratios are often not achieved simply because all the phenotypes are not equally viable.

Tetrad analysis also reveals, however, that there are exceptions to Mendelian segregation, due not to any chromosome aberration or failure of viability, but rather to the normal process of meiosis.

Gene Conversion and Postmeiotic Segregation

The phenomenon of gene conversion has been most obvious in tetrad analysis in the yeast *Saccharomyces*, where it is too common to be described as exceptional. Virtually every allelic difference in yeast shows something other than a 2:2 segregation in a substantial minority of the tetrads in which it is segregating. The frequency of 'aberrant segregations' varies from less than 1 per cent up to 10 per cent or somewhat more, depending on the gene.

In *Sordaria* the frequency of such aberrations is much less than in *Saccharomyces*, being usually between 0·1 and 1·0 per cent. Their lesser frequency is compensated for by their greater ease of detection, at least where ascospore colour differences are concerned. Simply by scanning asci under the microscope one can readily pick out the small minority of asci which depart from the usual 4:4 segregation patterns. Because there are eight spores in the ascus of *Sordaria*, a greater range of aberrant ratios can be more easily made out than in yeast, with its four-spored ascus.

Asci from wild-type × pale-spore mutant crosses in *Sordaria* reveal two kinds of aberrations. Firstly, there are asci showing 6:2 and 2:6 ratios of dark and light spores. These can be interpreted as cases of **conversion** of a whole meiotic product (stemming from one of the four chromatids of a diplotene bivalent) from one allelic type to the other. The equivalents in *Saccharomyces* are 3:1 and 1:3 ratios. In *Sordaria* (and also in an 8-spored Ascomycete with unordered asci, *Ascobolus*) the frequencies of mutant-to-wild conversions (6:2 asci, adopting the convention of writing the wild allele first) and of wild-to-mutant conversions (2:6 asci) are not necessarily the same. The occurrence of this kind of inequality, and its direction, probably depends on the molecular nature of the mutation in each case. In *Saccharomyces*, on the other hand, the general rule is that 3:1 and 1:3 tetrads occur with nearly equal frequencies. Thus in yeast, the 1:1 Mendelian ratio is a good fit to observation taking all asci together since the 3:1 and 1:3 tetrads balance one another.

Fig. 3.6

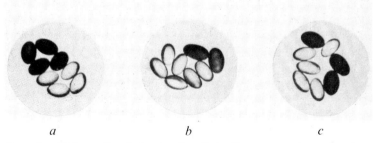

| *a* | *b* | *c* |

Examples of 'aberrant' asci, showing the effects of gene conversion and postmeiotic segregation from an *Ascobolus immersus* wild-type × pale ascospore cross. In this species, the eight ascospores from each ascus are discharged as a coherent group and can be collected on a suitable surface for observation.
a. A normal 4:4 segregation.
b. A 2:6 segregation resulting in conversion of one whole meiotic product from wild to mutant.
c. A 3:5 segregation resulting from the conversion of *one-half* of a meiotic product from wild to mutant, with segregation of wild from mutant in the spore pair formed at the first postmeiotic mitosis. The spores are about 53 μm in length.
From Fincham.[2]

A second violation of elementary genetic principle is illustrated by other aberrant types of asci, readily found in *Sordaria* and *Ascobolus*, in which segregation of an allelic difference is delayed until the mitotic division following meiosis (**postmeiotic segregation**). One such type (readily detectable only in the ordered asci of *Sordaria*) is **aberrant** 4:4 segregation in which *two* pairs of adjacent spores, each pair arising by mitosis from a single meiotic product, are mismatched with respect to an allelic difference distinguishing the parents of the cross. The effect is as if there has been a crossover between **half-chromatids**, i.e. between single DNA strands rather than between the double-stranded DNA molecules which are presumed to be involved in the normal whole-chromatid exchanges which explain segregation at the second division of meiosis.

Generally more common than aberrant 4:4 ascus ratios in *Sordaria* and *Ascobolus* are asci which show 5:3 or 3:5 segregations. These asci show segregation in only one of the four postmeiotic mitoses in the ascus. It is as if not a whole chromatid, but a half-chromatid, presumably corresponding at the molecular level to a single DNA chain, has undergone conversion. In similar terms, aberrant 4:4 asci could be interpreted as the result of *reciprocal* half-chromatid conversion, and whole chromatid conversion (seen as 6:2 and 2:6 ascus ratios) as simultaneous conversion of both strands of the double helix of the chromatid.

Postmeiotic segregation is not so readily detected in species, such as *Saccharomyces* and other yeasts, which do not have a postmeiotic mitosis preceding spore formation. It occurs nevertheless and results in the occurrence of genetically mixed colonies growing from single ascospores. The equivalent of a 5:3 segregation in yeast is the germination of the four spores of a tetrad to give two colonies of one parental type, one of the other type, and one mixed.

The various non-Mendelian meiotic ratios, though often referred to as 'aberrant', occur in every species so far as is known, though at widely different frequencies. They reflect part of the normal process of meiosis, the nature of which will be considered in Chapter 5.

3.6   Summary

Most inherited variation in eukaryotes can be accounted for by differences with respect to units (genes) each present only once in a haploid cell. Alternative forms (alleles) of the same gene can be brought together in a diploid cell by sexual fusion (to give a heterozygote) but they are segregated from one another when the chromosomes are reduced to the haploid number at the following meiosis. When meiotic products are analysed in tetrads they generally show 2:2 segregation with respect to each allelic difference. When two or more differences are segregating from the same diploid they frequently do so independently of one another. Independent assortment is readily explained on the basis that the several allelic differences are associated with different chromosome pairs. Segregation at the second, rather than the first division of meiosis is explicable in terms of crossing-over between the locus of the allelic difference and the chromosome centromere. Exceptions to 2:2 segregation are rare in higher eukaryotes but not uncommon in some fungi including *Saccharomyces*. They are attributed to gene conversion, the significance of which is reviewed in Chapter 5.

Chapter 3   # Selected Further Reading

Fincham J. R. S., Day P. R. and Radford A. (1979) *Fungal Genetics,* 4th ed. Oxford, Blackwell Scientific Publications.

Problems for Chapters 2 and 3

1   Gregor Mendel found that hybridization between true breeding round-seeded and wrinkled-seeded pea varieties gave seeds that were all round, but that when these F_1 seeds were sown and the resulting plants self-pollinated, round and wrinkled seeds were formed, often in the same pod, in an overall ratio of 3 : 1. What does this result tell one about the nature of the pea seed?

2   The fungus *Phycomyces blakesleeanus* exists in two mating types, *plus* and *minus*; confrontation of *plus* and *minus* mycelia results in the formation of zygospores which, after a period of dormancy, each germinate to give a stalked sporangium containing numerous spores. The wild-type fungus is yellow and a white mutant strain is known. In one experiment (Eslava et al., 1975, *Genetics* **80**, 445) zygospores were formed at the junction of *plus* yellow and *minus* white mycelium. Spores from one sporangium produced from a zygospore yielded 147 yellow and 135 white colonies. Of 34 yellow colonies taken at random and tested for mating type 14 were *plus* and 20 were *minus*. Of 26 white colonies, 13 were *plus* and 13 *minus*. What can you conclude about the haploid/diploid status of the different stages of the life history of the fungus? Where does meiosis occur?

3   The plant parasitic fungus *Phytophthora cactorum* (like *Phycomyces*, classed in the Phycomycete group of fungi) propagates itself asexually through motile zoospores formed in sporangia borne on the mycelium, and also sexually by oospores, which are formed through fusions of female and male gametangia, both of which may be formed on the same mycelium. Elliott and McIntyre (*Trans. Br. Mycol. Soc.* 1973, **60**, 311) studied a culture grown from a single oospore following treatment with the mutagenic chemical N-methyl-N'-nitro-N-nitrosoguanidine (cf. p. 334). Oospores formed on this culture by self-fertilization were of two types; about one-quarter germinated to give methionine-requiring colonies, while the other three-quarters resembled the parent strain and the wild-type fungus in being independent of methionine in the culture medium. Analysis of a further generation of oospores showed that all the methionine-requiring cultures produced only methionine-requiring oospores, while the first-generation methionine-independent cultures fell into two classes; about two-thirds gave an approximately 3 : 1 ratio of methionine-independent and methionine-requiring oospores, while the remaining third bred true for methionine independence. Zoospores in every case formed colonies resembling those of the culture in which they were formed, whether methionine requiring or methionine independent. What can you conclude about the haploid/diploid status of the different stages of the life history of the fungus? Where does meiosis occur?

4   Guries and Ledig (*Heredity* 1978, **40**, 27) showed that, in a species of pitch pine (*Pinus rigida*), the major malate dehydrogenase enzyme existed in the population in two distinct electrophoretic forms, *fast* (F) and *slow* (S) (cf. Chapter 18). Immature seeds from one cone were dissected so as to separate the embryos from the surrounding nutritive tissue (endosperm), and embryos and endosperms were analysed separately for malate dehydrogenase (MDH) type. The endosperms contained either S or F with approximately equal frequencies, but never both. The embryos were of three types: F, S, and F and S together. An embryo always contained the same MDH type as was present in the associated endosperm and sometimes, but not always, contained the other type as well. What does this tell you about the status of the endosperm in the life history of the pine tree? Note that the embryos were, to a large extent, the result of pollination by other trees. What ratio of embryo types would have been expected had the tree examined been entirely self-pollinated?

5   Two pure-breeding strains of maize (corn, *Zea mays*) have white and purple seeds, respectively. The colour seen in the purple-seeded strain is in the endosperm, the seed coat in this case being colourless. The cross of purple (as seed parent) × white (as pollen parent) gave cobs with purple seeds not quite so dark as those of the purple-seeded parent. The reciprocal cross (white as seed parent, purple as pollen parent) gave cobs with seeds all of a still paler purple. Bearing in mind the triploid nature of maize endosperm tissue, explain these results. What proportion of differently coloured seeds would you expect to find following self-pollination of plants grown from the coloured seeds produced from the two reciprocal crosses?

6   The mating of a pair of black rabbits resulted in a litter of eight; six (A–F) were black and two (G and H) white. A, B and C were females and D, E and F males. Three second-generation families were obtained from matings of A × D, B × E and C × F. On the basis of the simplest hypothesis regarding the origin of the white rabbits G and H, calculate the probability of finding one or more white rabbits among the second-generation families, assuming a litter size of eight in each case.

7   A cross was made between a wild-type female *Drosophila melanogaster* and a male homozygous with respect to the two recessive mutations *dumpy* (wings) and *hairy*. All of the progeny were wild in phenotype. F_1 males were crossed to *dumpy hairy* female homozygotes, and the following numbers were found among the progeny:

Phenotype	Males	Females
Wild	64	68
Dumpy	50	55
Hairy	59	52
Dumpy hairy	46	50

What ratio of the different phenotypes would you expect among the F_2 progeny from intercrosses of F_1 males and females?
What ratios might you find, and with what probabilities, among the offspring of sib matings of hairy with dumpy flies from the F_2?

8   Two pure-breeding strains of maize, with red and white seeds, respectively, were crossed. All of the cobs formed were entirely purple seeded, whichever way round the cross was made. Self-pollination of a plant grown from one of the F_1 purple seeds gave a total of 1146 seeds: 631 purple, 295 white and 220 red. Plants were grown from the F_2 purple and white seeds and crosses made between them. One such purple × white cross gave 342 purple, 333 red and 684 white seeds. Another, involving the same plant grown from the purple seed but a different plant grown from a white seed, gave 468 purple, 142 red and 550

white seeds. Explain these results. Assign genotypes to the plants involved in the various crosses.

NB The gene difference segregating to give coloured versus white in this example (C/c) is different from that segregating in the example used in question 5 (R/r). Unlike R, C shows complete dominance over its recessive allele.

9 Two pure-breeding mutant strains of *Drosophila melanogaster*, one with *vermilion* (v) rather than the wild-type brown-red eyes, and the other with *vestigial* (vg) wings, were crossed, the *vermilion* strain providing the females. In the F_1 generation all of the males had *vermilion* and all the females wild-type eye colour, and all of both sexes had normal wings. An F_2 generation obtained by mating F_1 males and females consisted of the following:

103 wild type both for wings and eyes (52 females, 51 males)
98 *vermilion* with wild-type wing (53 females, 45 males)
30 *vestigial* with wild-type eye (18 females, 12 males)
34 *vestigial vermilion* (14 females, 20 males)

Explain these results. What results would be expected if, in the first cross, the *vestigial* strain had provided the females?

10 Females from the *vermilion* strain mentioned in question 9 were crossed to males from a strain true-breeding for another bright-red eye colour mutation *scarlet* (st), indistinguishable by eye from *vermilion*. In the F_1 generation all the females were phenotypically wild type and all the males had bright-red eyes. An F_2 generation obtained by mating F_1 males and females comprised:

Females: 90 wild type, 131 bright-red
Males: 83 wild type, 140 bright-red

Explain these results.

11 Females of the *scarlet* strain mentioned in question 10 were crossed to males of a true-breeding *brown*-eyed strain. All F_1 flies had wild-type red-brown eye colour. An F_2 progeny consisted of the following:

Wild type	156 (75 females, 81 males)
Scarlet	48 (27 females, 21 males)
Brown	52 (29 females, 23 males)
White	18 (10 females, 8 males)

Propose a hypothesis to explain these results and the appearance of the new phenotype in the F_2 generation. How would you confirm the postulated genetic basis of the white phenotype?

12 A cross was made between two *Neurospora crassa* strains, one of mating type A and resistant (as the result of a genetic mutation) to sulphanilamide (sfo) and the other of mating type a and (as the result of another mutation) unable to form asexual spores (conidia)—the latter phenotype is called *fluffy* and the mutation determining it *fl*. Ascospores were dissected in order and germinated from a number of asci of which 48 gave complete germination. The cultures obtained were scored by eye for *fluffy* and tested for mating type and sulphanilamide resistance. In the following classification each ascus is shown as four spore pairs (the two members of each pair being identical) and as divided into two half-asci (spores 1–4 and spores 5–8); no account is taken of different sequences of spore pairs *within* each half-ascus and asci which are of the same type except for top-half/bottom-half reversal are lumped together.

A	sfo fl	A	sfo +	A	sfo fl	a	sfo fl	a	sfo +	a	sfo fl
A	sfo fl	A	sfo +	A	sfo +	a	sfo fl	a	sfo +	a	sfo +
a	+ +	a	+ fl	a	+ fl	A	+ +	A	+ fl	A	+ fl
a	+ +	a	+ fl	a	+ +	A	+ +	A	+ fl	A	+ +
	3		2		15		5		2		13

A	sfo fl	A	sfo +	A	sfo fl	A	sfo +	A	sfo fl	A	sfo +
a	sfo fl	a	sfo +	a	sfo +	a	sfo fl	a	sfo +	a	sfo fl
A	+ +	A	+ fl	A	+ fl	A	+ +	A	+ +	A	+ fl
a	+ +	a	+ fl	a	+ +	a	+ fl	a	+ fl	a	+ +
	3		1		1		2		1		2

The symbol + refers to the wild-type alternative in each case. What can you deduce about the positions of the loci of the three segregating differences in relation to their respective centromeres?

Linkage, Recombination and Genetic Mapping in Eukaryotes

4.1 Introduction

If, as the phenomenon of sex linkage first clearly indicated, genetic determinants (genes) are located on chromosomes, we expect to find only as many independently reassortable groups of genes as there are chromosome pairs in the diploid set. If there are more gene loci than chromosomes, some of them, one would suppose, must be linked together. This is indeed found to be the case; the total ensemble of segregating genetic differences in any particular species can be classified into **linkage groups.** Those in different groups reassort independently of one another at meiosis, while those in the same group show a greater or lesser degree of linkage, the nature of which is explained in this chapter. In well-studied species, where all the linkage groups have been identified, the number of linkage groups equals the number of chromosomes in the haploid genome.

It will be instructive to start by seeing how linkage shows itself in meiotic tetrads, and then to extend the analysis to random meiotic products.

4.2 Linkage studied in meiotic tetrads

Test for Linkage

As a simple example of linkage we may take two visible mutant traits in the Ascomycete fungus *Neurospora crassa*. The normally orange asexual spores (conidia) are changed to white by the *albino* (*al*) mutation, while the normal rapidly spreading growth habit is made more compact, and the conidia-bearing hyphae (conidiophores) much shorter by the *crisp* (*cr*) mutation. The corresponding wild-type alleles are represented by al^+ and cr^+. *Table* 4.1 shows the results of an ordered tetrad analysis of the cross of albino ($al\,cr^+$) × crisp ($al^+\,cr$).

Three points are at once apparent. First, the frequency of PD tetrads greatly exceeds that of NPD tetrads. In other words, there are far fewer recombinant (*al cr*, $al^+\,cr^+$) than parental-type ($al^+\,cr$, $al\,cr^+$) meiotic products. Classical Mendelian reassortment would give equal numbers of recombinants and parentals, and equal numbers of PD and NPD tetrads. The fraction of recombinants is, in this example, 25·4 per cent, a number obtainable either by simply adding up parental and recombinant products regardless of tetrad associations, or from the formula

$$R = \frac{\text{Tetratypes} + (2 \times \text{Non-parental ditypes})}{2 \times \text{Total asci}}$$

which amounts to the same thing.

An example of linked segregation in the tetrads of another Ascomycete fungus, *Sordaria brevicollis* is shown in *Fig.* 4.1. In this case the two segregating differences affect ascospore colour, and the different tetrad (ascus) types can be distinguished by visual inspection.

Table 4.1   **Linkage in a *Neurospora crassa* cross: *cr⁺ al* × *cr al⁺***

Ascus classification

	1	2	3	4	5	6	7
Spore pairs (half-asci bracketed)	⎧ cr + ⎩ cr +	cr al cr al	cr al cr +	cr + + al	cr al + +	cr al + +	cr al + al
	⎧ + al ⎩ + al	+ + + +	+ al + +	cr + + al	cr al + +	cr + + al	cr + + +
Ascus type	PD	NPD	T	PD	NPD	T	T
Number of asci	38	2	40	13	0	4	3

Interpretation

centromeres

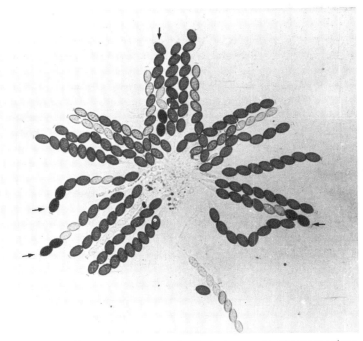

* The two crossovers in this case could just as well have a four-strand relationship, rather than two-strand as shown.

Fig. 4.1

Asci from a cross between *Sordaria brevicollis* strains carrying the spore colour mutant alleles *buff* (*b*) and *yellow* (*ylo*), respectively. The two markers are linked, and the array of asci in the photograph include numerous parental ditypes (4 buff : 4 yellow with the yellow slightly paler than the buff but the colours barely distinguishable in the more mature asci) and a few tetratypes (arrowed). The tetratype asci each have one yellow, one buff, one wild-type (black) and one double-mutant (white) spore pair.

The Nature of Crossovers

The second important point made by *Table* 4.1 is that, where recombinant products occur, they do so in reciprocally constituted pairs. Wherever the *cr al* double-mutant recombinant appears, so also, in the same ascus, does the reciprocal + + wild-type recombinant. This suggests that the recombinants are generated by reciprocal and equal exchanges, or **crossovers**, thus:

The third fundamental feature of recombination seen in the table is that, with infrequent exceptions due to double crossovers, only two members of each tetrad show the effects of crossing-over. This is a typical result; crossovers in general give tetratype tetrads, *not* non-parental ditypes. We must, therefore, modify our simple diagram of crossing-over to show that it occurs after the chromosomes are divided into chromatids, and involves only two of the four chromatids at any one point:

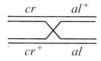

The structure of chromosome pairs (bivalents) at diplotene of meiosis, as seen in those organisms with chromosomes large enough for study with the light microscope (*Fig.* 2.10c), is exactly what is needed to explain the pattern of recombination seen in tetrads. We need only assume that the genetic determinants showing the allelic differences are carried at different loci on a particular chromosome and that crossing-over is the genetic consequence of the formation of chiasmata between chromatids after chromosome replication. This, of course, is the same assumption that we needed to make in order to explain meiotic segregation at the second division (*see* p. 80).

Centromere Positions

Accepting that *cr* and *al* are mutant alleles of genes linked on the same chromosome, we can deduce something about their relative positions with respect to the chromosome centromere. Since *al* clearly shows a much higher frequency of second-division segregation than does *cr*, we can conclude that there is more chiasma formation in the interval between *al* and the centromere than between *cr* and the centromere. This is strongly suggestive of *al* being more distant from the centromere than *cr*, but it does not in itself indicate whether the two loci are in the same or different chromosome arms. The fact, however, that in 20 asci showing second-division segregation for *cr*, as many as 17 show second-division segregation for *al* also, as compared with 42 out of 82 asci with first-division segregation of *cr*, suggests that *al* is distal to *cr* in the same arm. Such an arrangement means that a chiasma between the centromere and *cr* will bring about second-division segregation of both loci unless a second chiasma is formed between the two. (Note that, in cytogenetics, **distal to** means further from the centromere in the same chromosome arm, and **proximal to** means closer to the centromere in the same arm.)

Crossover Analysis and Sequence of Marker Loci

The simplest interpretation of the data in terms of the positions of crossovers, and the chromatids involved in them, is shown at the foot of *Table* 4.1. It will be seen that the less common ascus patterns can be explained as the results of various kinds of double crossing-over. Note that it has to be supposed that all four chromatids can take part in successive crossovers, even though only two of them are involved in any one. This fits well with the microscopic evidence reviewed in Chapter 2 that either of the two chromatids of each chromosome of a bivalent can participate in chiasma formation.

Asci of class 2, the relatively infrequent non-parental ditypes (NPDs), have to be explained as due to two crossovers which between them involve all four chromatids (**four-strand** doubles). Class 6 must result from the second crossover involving one of the same strands as the first crossover and one different one (**three-strand** doubles). **Two-strand** doubles, with the same two chromatids involved twice, cannot be unequivocally identified from data with only two markers on the same

---

Table 4.2 The data of *Table* 4.1 with the addition of a third marker—mating type A/a

```
Cross: a + al  ×  A cr +
   A        cr          +            Constitution of
   A   0    cr          +            bivalents
      I  II      III
   a        +          al
   a   0    +          al
```

Ascus types and simplest interpretation of each
(The genotypes shown are those of spore pairs; half-asci are separated by - - - - -
No account is taken of the order of the spore pairs within each half-ascus.)

A cr +	A cr +	A cr +	A cr +	A cr +	A cr +	A cr al
A cr +	a cr +	A cr al	A + al	A + al	a + al	A cr al
a + al	A + al	a + +	a cr +	A cr +	a cr +	a + +
a + al	a + al	a + al	a + al	a + al	A + al	a + +
28 asci	10	29	11	1	1	2

				2-str.	3-str.	4-str. double within III
No cross-over	Single in I	Single in III	Single in II	double I and II		

A cr al	a cr al	A cr al	a cr al	A cr al	a cr al	A cr al	a cr al	A cr al
a cr +	A cr +	a cr +	A cr +	A + al	a + al	A + +	a + +	a + +
A + al	a + al	a + al	A + al	a cr +	A cr +	a cr +	A cr +	a cr +
a + +	A + +	A + +	a + +	a + +	A + +	a + al	A + al	A + al
3 asci	2	4	2	1	2	2	1	1
4-str.	2-str.	3-str.*	3-str.*	4-str.	2-str.	3-str.*	3-str.*	3-str., 3-str., triple I, II, III

| | | double I and III | | | | double II and III | | |

* Note two distinguishable kinds of 3-strand double crossover.

Box 4.1   **Calculations based on the data of *Table* 4.2**

Deduction of map from the data

Interval	Tetrad types PD	NPD	T	Percentage recombination $\left[\dfrac{T+(2\times NPD)}{2\times total}\right]$ (estimate of map distance)
A–cr	60	0	40	20·0
A–al	33	7	60	37·0
cr–al	51	2	47	25·5

	First or second-division segregation 1st	2nd	Percentage second	Estimated map distance from centromere
A	76	24	24·0	12·0
cr	80	20	20·0	10·0
al	43	57	57·0	28·5

Best fit to a linear map:

Analysis of double-crossover frequencies
Frequency of crossing-over/tetrad (twice map distance): I 0·24, II 0·20, III 0·51

Frequency of doubles (including the triple):

	Observed	Expected	Coefficient of coincidence
I and II	0·03	0·24 × 0·20 = 0·048	0·03/0·048 = 0·62
II and III	0·07	0·20 × 0·51 = 0·10	0·07/0·10 = 0·70
I and III	0·12	0·24 × 0·51 = 0·12	0·12/0·12 = 1·0

Strand relationships in double crossovers*

	2-strand	3-strand	4-strand
I and II	1	2	0
II and III	2	4	1
I and III	2	7	3
Total	5	13	4 (compatible with a 1 : 2 : 1 ratio)

*The triple crossover ascus is counted as two doubles, I–II and II–III
NB The two 4-strand doubles within III are not counted since 2- and 3-strand doubles *within* an interval are not distinguishable; the inclusion of the 4-strand doubles would therefore bias the data.

side of the centromere (class 7 in the table could have been due to either two-strand or four-strand double crossovers, or both). We shall see more evidence on strand relationships of successive crossovers below.

The _cr + × + al_ example is an example of a two-point cross. It goes some way, with the analysis of ordered tetrads, towards being a three-point cross, with the centromere as the third point. The limitation of the centromere as a reference point in linkage analysis is that one cannot distinguish between homologous centromeres—hence the ambiguity as regards two and four-strand doubles exemplified by the class 7 asci of _Table_ 4.1. Most linkage analysis is based upon crosses with distinguishing alleles at three or more linked genes. In fact, the _cr + × + al_ cross in _Neurospora_ can be made into a three-point analysis by scoring the progeny cultures for mating type. The _A/a_ mating type difference turns out to be determined in a gene in the same linkage group as _cr_ and _al_, about the same distance from the centromere as _cr_ but in the opposite chromosome arm. _Table_ 4.2 and _Box_ 4.1 summarize the fuller analysis.

An allelic difference occurring at a certain position (**locus**) on a chromosome makes it possible to follow the transmission of that locus through successive generations, and is often referred to as a **marker**.

Interference

Data from crosses involving groups of linked markers can be used to answer two questions concerning the relationships between simultaneous crossovers in the same linkage group. First, we can ask whether there is interference between crossovers; in other words whether the occurrence of one crossover affects the probability of another crossover being formed nearby or at least in the same linkage group. The numbers in the table are not really large enough to answer the question decisively, but we see that the number of observed double crossovers is less than would be predicted on the assumption that crossing-over in one interval should occur independently of crossing-over in the adjacent interval. The degree of interference is expressed by the **coefficient of coincidence**, which is the observed number of doubles divided by the number expected on the hypothesis of no interference. A value of significantly less than unity indicates interference; a value significantly in excess of unity would imply negative interference, i.e. a tendency to clustering of crossovers, but it is not clear that this is ever found. (_See_, however, the association of crossovers with gene conversion, Chapter 5, p. 142. Positive interference is usually very evident within chromosome arms, though it tends to fade out with distance. Coefficients of coincidence between relatively close intervals within an arm, say within a distance giving 20 per cent recombination, may be as low as 0·2–0·3. It seems to be generally true, however, that there is little or no interference and across the centromere.

The second question about double crossing-over is whether the choice of chromatids for involvement in a second crossover is at all prejudiced by their involvement or not in the first. A non-random participation of chromatids in successive crossovers is called **chromatid interference**, but in fact there is doubt about whether the phenomenon exists. Random participation implies a ratio of 2-strand, 3-strand and 4-strand doubles of 1 : 2 : 1. The data of _Table_ 4.2 are too meagre to provide a good test of this prediction, but at least they are consistent with it. Much more extensive data are available from _Neurospora_ and yeast, and they show a reasonably close fit to the 1 : 2 : 1 ratio. Note that, in genetical analysis, only tetrad data (or, less efficiently, half-tetrad data, _see_ p. 127) provide information on this point. However, microscopical evidence from observations of diplotene bivalents in organisms with large chromosomes confirms the random involvement of chromatids in chiasmata.

4.3 Linkage studied in randomized meiotic products

In higher plants and in animals, meiotic products cannot be recovered in their original tetrads and so one has to analyse linkage by sampling random meiotic products. In higher eukaryotes, lacking a free-living haploid phase, the meiotic products are gametes or immediate precusors of gametes, and their genotypes must be determined retrospectively by examination of the phenotypes of the succeeding diploid generation. The efficient procedure is to cross a strain heterozygous at several linked loci to a tester strain homozygous for all the recessive alleles. The phenotypes of the progeny of such a **test-cross** are a direct reflection of meiotic segregation in the heterozygous parent. The constant genotype of the gametes from the homozygous recessive parent provides a uniform 'background' against which the different genotypes of the gametes from the multiple heterozygote can be expressed in the phenotypes of the test-cross progeny.

The rules governing linked segregation in tetrads apply, in a reduced form, to randomized meiotic products. In two-point tests, recombinant frequencies of significantly less than fifty per cent are taken as evidence of linkage. Reciprocal recombinants are expected to occur with statistically equal frequencies; departures from equality are attributed to viability differences. Map order from a three-point test-cross is determined acording to the principle of minimizing the number of crossovers needed to explain the data. Thus the least frequent pair of classes, of the eight classes generated by three segregating linked loci, is taken to be the product of double crossing-over. The other six classes will consist of two reciprocally related pairs of single crossover products and one pair (often, though not necessarily, the most frequent) representing the non-crossover (parental type) combinations.

Interference can be assessed by calculation of the coefficient of coincidence just as in tetrad analysis except that we are looking at crossovers affecting the same genetic strand (chromatid) rather than at all crossovers along a bivalent, whether they affect the same chromatid(s) or not. Provided that chromatids are involved in crossovers at random (no chromatid interference), the coefficient of coincidence will be the same whether it is calculated on a single-strand or whole-tetrad basis.

Boxes 4.2 and 4.3 show two worked examples of the use of three-point test-crosses to establish linkage relationships. The first example, from *Drosophila*, is straightforward; the second, from *Antirrhinum*, is identical in principle, but has the complication of epistasis operating between two of the markers.

An important fact to remember in considering *Drosophila* data, is that in flies there is no crossing-over in the male; in spermatogenesis all linked genes remain strictly in their parental combinations. There are, in fact, no visible chiasmata formed during meiosis in the male, though they are present in typical form in the female. This peculiarity is not relevant to the example of *Box* 4.2 (a male could not in any case show crossing-over for X-linked markers since it cannot have two X-chromosomes), but where autosomal markers are concerned, *Drosophila* test-crosses always have to be made with the female as the heterozygous parent and the male as the multiple recessive tester. *Drosophila* is exceptional; the general rule in higher plants and animals is similar frequencies of crossing-over in the two sexes.

Box 4.2   **Analysis of a three-point test-cross in *Drosophila melanogaster***

w = white eye; m = miniature wing; f = 'forked' (shortened and twisted) thoracic bristles. All are sex linked.

$$\text{Cross:} \quad \frac{w\ m\ f}{+ + +} \times \frac{w\ m\ f}{Y}$$
$$\qquad\qquad \text{female} \qquad \text{male}$$

The progeny phenotypes all reflect the genotypes of the maternal eggs; the male parent contributes only recessive alleles to the female progeny and only the Y-chromosome to the male progeny.

Progeny phenotypes (males and females together)	No. found	
$w\ m\ f$	110	} 235
$+ + +$	125	
$w\ m\ +$	40	} 83
$+ + f$	43	
$w + +$	32	} 68
$+ m\ f$	36	
$w + f$	2	} 6
$+ m\ +$	4	
	392	

Total $w^+ = 208$, $w = 184$: χ^2 (for a 1 : 1 ratio) = 1·5 ⎫ $P > 0.05$ in all cases;
$m^+ = 202$, $m = 190$: χ^2 (for a 1 : 1 ratio) = 0·37 ⎬   no significant viability
$f^+ = 201$, $f = 191$: χ^2 (for a 1 : 1 ratio) = 0·26 ⎭   effects of mutant alleles
The $w + f$ and $+ m +$ classes are obviously the least frequent, and so must be attributed to double crossing-over. Thus m must be the middle marker and the sequence of loci must be as written.

$$\text{Percentage recombination} \quad w\text{–}m = \frac{(68 + 6) \times 100}{392} = 18 \cdot 9$$

$$m\text{–}f = \frac{(83 + 6) \times 100}{392} = 22 \cdot 7$$

Expected number of double crossovers with no interference
= $0 \cdot 189 \times 0 \cdot 227 \times 392 = 16 \cdot 8$
Coefficient of coincidence = $6/16 \cdot 8 = 0 \cdot 36$

Map: w 18·9 m 22·7 f

| Box 4.3 | Analysis of a three-point test-cross in *Antirrhinum majus* |

El/el (Eluta): pigment diluted over much of flower (the dominant wild type) versus pigment uniformly intense over whole flower (the recessive allele of garden varieties).

Pal/pal (pallida): presence versus absence of anthocyanin purple pigment.

Div/div (divaricata): normal 'snapdragon' flower versus corolla lobes divergent.

Cross: $\dfrac{el\ Pal\ div}{El\ pal\ Div} \times \dfrac{el\ pal\ div}{el\ pal\ div}$

Progeny phenotypes and numbers

el Pal div   2914

$\left.\begin{array}{l} \textit{El pal Div} \\ \textit{el pal Div} \end{array}\right\}$ 3792*

El Pal div   844

el Pal Div   254

$\left.\begin{array}{l} \textit{El pal div} \\ \textit{el pal div} \end{array}\right\}$ 293*

El Pal Div   7

—————
        8104

The numbers are consistent with reciprocally consituted products being equally frequent. Thus 2914 + 844 = 3758 which is not significantly different from 3792; likewise 254 + 7 = 261, which is not significantly different from 293.

* *El/el* cannot be scored in homozygous *pal* plants, i.e. *pal* is epistatic to *El/el*.

Recombination frequencies

El/el and *Pal/pal* (information from *Pal* plants only):

Recombinant *El Pal* = 851
Total *Pal* = 4019
Percentage recombination = 21·2

El/el and *Div/div* (*Pal* plants only):

Recombinants *el Pal Div* + *El Pal div* = 1098
Total *Pal* = 4019
Percentage recombination = 27·3

Pal/pal and *Div/div*: Recombinants: *Pal Div* 261 + *pal div* 293 = 554
Total = 8104
Percentage recombination = 6·8

The map sequence is *El–Pal–Div*, since highest recombination frequency is between *El* and *Div*; the small number of *El Pal Div*, is consistent with this being a double-crossover class.

Expected number of double crossover products with no interference is

$8104 \times 0.068 \times 0.212 = 117$

of which half should be *El Pal Div*.

Coefficient of coincidence = 7/58·5 = 0·12.

Owing to the low coefficient of coincidence (i.e. high degree of crossover interference–low frequency of double crossovers), the recombination frequencies are in this case nearly additive along the map:

```
        21·2        6·8
El          Pal          Div
←—————— 27·1 ——————→
```

Data of Harrison and Fincham.[2]

4.4 Map units and the theory of linkage maps

It is conventional to represent the linkage relationships of genes by maps in which each linkage group is represented by a series of points (or loci) along a line, with distances between loci proportional to the frequency of recombination between them. One map unit (sometimes called a centimorgan after T. H. Morgan, one of the main originators of the chromosome theory) corresponds to 1 per cent recombination among total meiotic products.

The Effect of Double Crossovers

The trouble with percentage recombination as a measure of map distance is that it is additive only over short distances. When distances between loci become sufficiently great for an appreciable frequency of double crossovers to occur, the recombination frequency underestimates the true distance. The reasons for this become clear when we consider the consequences of double crossing-over for the formation of recombinants. A three-strand double within an interval gives the same result as a single—namely two recombinant products out of four in each meiotic tetrad. Four-strand doubles give four recombinant products out of four, but they are balanced by the generally equal number of two-strand doubles which give none (the second crossover merely cancelling the effect of the first). Thus, overall, double crossovers give the same result as singles—50 per cent recombination. Adding a further crossover to a row of crossovers already in place will not further increase the yield of recombinants since, on average, it will cancel as many recombinants as it creates. In short, as the average number of crossovers increases to the point where there is virtually always *at least one* per bivalent in the interval being measured, the frequency of recombination will approach 50 per cent and never exceed this value.

Map Distance Proportional to Mean Crossover Frequency

A much more satisfactory (because additive) measure of map distance is the *mean total number* of crossovers. This will not be expected to bear a constant relationship to physical distance along chromosomes, since some chromosome regions are more prone to form crossovers than others, but it will, at least, increase as real distance increases.

If we build up our linkage map by piecing together a succession of short intervals, within each of which the chance of more than one crossover being formed is negligible, we can have a composite map distance which is proportional to total crossover frequency. It will be equal to 100 times the mean number of crossovers per meiotic product and 50 times the mean number per tetrad (remember that each crossover affects only two chromatids out of four). Thus a section of a linkage map representing a fairly long chromosome with a mean chiasma frequency of 3·0 would be 150 centimorgans in length, despite the fact that recombination measured directly between two terminal markers would (as explained below) not exceed 50 per cent. One could build up the 150 units by adding together recombination frequencies over a number of small contiguous intervals.

A Simple Mapping Function

Making assumptions about crossover interference, one can construct theoretical curves relating true map distance (that is, mean number of crossovers × 50) to frequency of recombination. The simplest assumptions are complete interference on the one hand and no interference on the other. With complete interference, that is to say with the number of crossovers per bivalent limited to one, map distance is the same as percentage recombination. With no interference, the different numbers of crossovers will be distributed about the mean number according to the Poisson distribution. This distribution gives the frequencies of 0, 1, 2, 3 etc. events if each event occurs independently of any other, and the mean number of events is m. The Poisson terms are as follows:

$$
\begin{array}{cccccc}
0 & 1 & 2 & 3 & \ldots & n \\
e^{-m} & m\,e^{-m} & \dfrac{m^2}{2}e^{-m} & \dfrac{m^3}{6}e^{-m} & & \dfrac{m^n}{n!}e^{-m}
\end{array}
$$

We saw in the previous section (p. 109) that, provided chromatids are involved in

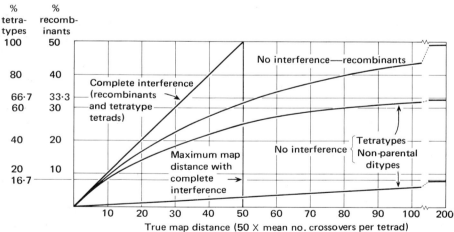

Fig. 4.2

The relationships between true map distance (mean no. of chiasmata × 50) and the frequencies of recombinant meiotic products and different tetrad types: under conditions of (*a*) no crossover interference (i.e. numbers of crossovers distributed according to the Poisson series) and (*b*) complete interference (i.e. never more than one crossover). In real situations the curves fall somewhere between these two extremes, depending on the degree of interference. Note that 66·7 per cent ($\frac{2}{3}$) and 16·7 per cent ($\frac{1}{6}$) are the frequencies which tetratype and non-parental ditype asci, respectively, approach as the map distance becomes large—these limits correspond to the 50 per cent limit for recombinant products, i.e. ($\frac{1}{2} \times \frac{2}{3}$)+(1 × $\frac{1}{6}$) = $\frac{1}{2}$.

crossovers at random, any class with one or more crossovers will, on average, give 50 per cent recombination. Thus the only significant determinant of recombination frequency is the zero term e^{-m}; all the other terms totalling $(1-e^{-m})$ will give the same recombinant frequency of $0 \cdot 5$. Hence, with no interference:

$$\text{Percentage recombination} = 50(1-e^{-m})$$

The curve given by this equation is shown in *Fig.* 4.2 together with the corresponding curves relating map distance to frequencies of the different types of tetrad

Calculation of Centromere Distances

We can now go back to the interpretation of the tetrad data of *Table* 4.2 and see why, in *Box* 4.1, the centromere distance of a locus close to its centromere is calculated as one-half the percentage second-division segregation of the locus. Neglecting double and multiple crossovers, each second-division segregation corresponds to a chiasma in the diplotene bivalent between the marker and the centromere. It is helpful to remember that second-division segregation, when an interval is bounded by a genetic marker at one end and the centromere at the other, is equivalent to a tetratype tetrad when the interval is bounded by a marker at each end—both arise from a single crossover in the interval. Since only two of the four chromatids are involved in each crossover, the percentage of tetratypes or of second-division segregations has to be divided by two in order to correspond to the percentage of crossover products.

Second-division segregation frequency, like recombination frequency, under-estimates map distance when intervals are long enough for double crossovers to become important. In fact the underestimation is even greater in the case of locus–centromere distances, since double crossovers actually give fewer second-division segregations on average than do singles. Only three-strand doubles result in distal sister alleles becoming connected to different centromeres, the prerequisite for second-division segregation. Four-strand doubles switch sister alleles from one centromere to the other and, since the two centromeres cannot be distinguished, the effect is the same as that following a two-strand double or no crossover at all—namely first-division segregation. Thus, whereas single crossovers as a class give 100 per cent second-division segregation, double crossovers give only 50 per cent (provided two, three and four-strand doubles are in the proportions $1:2:1$). It is easy to show that triple crossovers will, overall, give 75 per cent, quadruple crossovers $62 \cdot 5$ per cent (five-eighths) and quintuple crossovers $68 \cdot 8$ per cent (eleven-sixteenths). The addition of each further crossover to a row of crossovers already in place will have the effect of converting all the tetrads that would otherwise have been first-division segregations to second-division segregations, and one-half of those which would otherwise have been second-division segregations to first-division segregations. As the number of crossovers becomes large the limiting frequency of $66 \cdot 7$ per cent (two-thirds) second-division segregation is approached, which, converted naively into map units, will give a distance of only $33 \cdot 3$ centimorgans from the centromere. Thus, even more than recombination fre-quencies, second-division segregation frequencies are only to be trusted to give reasonably accurate values of map distance over relatively short intervals, say up to 20 or 30 per cent second-division segregation, within which multiple crossovers are rare.

4.5   Correlations between linkage maps and chromosomes

In all organisms which have been intensively studied genetically, allelic differences (genetic markers) can be assigned to a limited number of linkage groups which agrees with the haploid chromosome number. Among the best examples are *Drosophila melanogaster* with four linkage groups, *Zea mays* with ten, *Neurospora crassa* with seven and *Saccharomyces cerevisiae* with seventeen. Mouse (20) and man (23) are rapidly acquiring a similarly complete inventory of linkage groups equal in number to the chromosome set. In *Drosophila*, the correlation extends beyond mere number to relative size. The two longest linkage groups correspond to the second and third chromosomes, each of which has two relatively long arms flanking the centromere. A somewhat shorter linkage group of sex-linked markers corresponds to the one-armed X-chromosome, while the fourth group, consisting of a few very closely linked loci, corresponds to the tiny fourth chromosome.

Where the diplotene stage of meiosis is sufficiently clear for counts of chiasmata to be made, it is possible to verify the expected relationship between chiasma number and total map length. As we saw above, an average of one chiasma in a chromosome region should correspond to a map distance of 50 centimorgans in the linkage map, and the total map length should approximate to 50 times the mean total chiasma count. In maize, to cite the organism most favourable for such comparisons, the map length summed over all ten linkage groups is 1120 centimorgans, and this agrees well with the mean total chiasma count of 27, especially when one allows for the possibility that some of the chromosome ends may not yet be completely mapped for want of terminal markers (*Table* 4.3).

Maps of most of the genetically well-worked eukaryotic species can be found in the *Handbook of Genetics* edited by R. C. King. Such maps provide the framework within which studies of the structures and functions of particular parts of the genome are conducted. *Figs* 4.3–4.5 show the linkage maps of *Drosophila melanogaster, Zea mays* and *Saccharomyces cerevisiae*, respectively and some of their features are summarized in *Table* 4.4. It is remarkable how much these diverse organisms differ in frequency of recombination per unit of DNA. In terms of map units, the yeast genome is nearly 5 times as long as that of maize, but its DNA content is nearly 500 times less.

More detailed correlations between particular genetic loci and visually defined chromosome segments or chromomeres are described in later chapters (pp. 431, 563) but may be mentioned briefly here. Firstly, mutations leading to loss of a gene function, especially when they have recessive lethal effects as well, are often deletions of short chromosome segments, and these can often be recognized and their limits defined microscopically. The gene whose function is lost can be assumed to lie within the deleted segment, and when a number of different deletions are available, all eliminating the same function, the gene can sometimes be assigned to a very narrowly defined region—that which is lacking in all the deletions. This type of analysis is most powerful in *Drosophila*, where advantage can be taken of the polytene chromosomes of the salivary glands. As well as the high degree of fine detail that can be seen in these giant chromosomes, there is the

Table 4.3	Chiasma frequency and genetic map distance in maize chromosomes			
Chromosome no.	Relative length*	Mean chiasma frequency	Predicted map length	Map length known up to 1976
1	100	3·65	183	161
2	86	3·25	163	165
3	78	3·00	150	128
4	76	2·95	148	143
5	76	2·95	148	87
6	53	2·20	110	68
7	61	2·45	122	112
8	61	2·45	122	28†
9	53	2·20	110	138
10	44	1·95	97	99

* Measured at pachytene.
† Only three markers in this linkage group.
Table slightly adapted from Swanson et al.[1]

Table 4.4	Different frequencies of crossing-over per unit DNA in different organisms			
Organism	No. of chromosomes	Total map length	Haploid DNA quantity (kb)	No. of kb per map unit (centimorgans)
Maize	10	circa 1350	$5·4 \times 10^6$	circa 4000
Drosophila	4	280	$1·6 \times 10^5$	570
Yeast	17	> 2500	$1·8 \times 10^4$	< 10

advantage of their somatic pairing which allows, in a heterozygote, an exact side-by-side comparison of normal and deleted chromosomes.

A more recent method, which can lead to an assignment of a particular gene to a single polytene chromosome band, involves the hybridization of the DNA of the gene (or part of it) with a radioactively labelled 'probe' nucleic acid sequence known to be specifically complementary to the gene. The probe is most usually RNA transcribed from the gene, or a DNA copy of it. Some of the spectacular results which have been obtained by this technique of *in situ* **hybridization** are outlined in Chapter 16.

Fig. 4.3

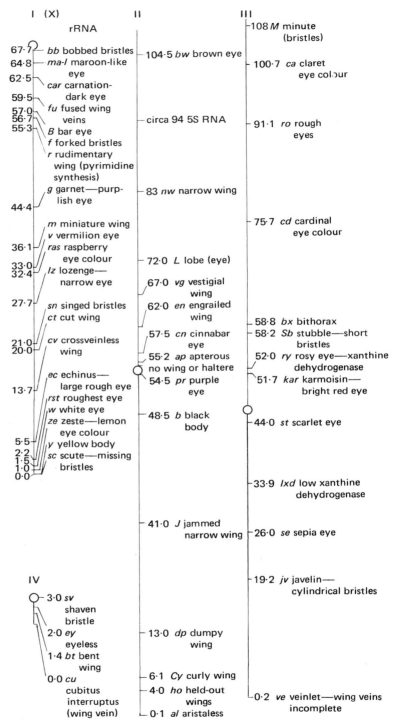

The linkage map of *Drosophila melanogaster*. Only a selection of the known markers is given here; those referred to elsewhere in this book or of special interest for other reasons are included. Centromeres are shown as circles. Data from King.[3]

Fig. 4.4

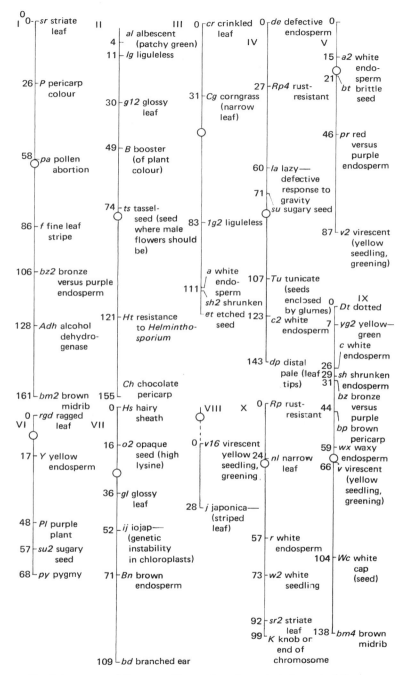

The linkage map of *Zea mays*. The markers shown include most of those
affecting the endosperm of the seed. Several of the genes affecting endosperm
colour also affect the colour of other parts of the plant. *sh2* (linkage group
III) codes for the enzyme sucrose synthetase, and is the first maize gene
to be isolated as a molecular clone (cf. p. 458).
Data from Coe and Neuffer.[4]

Fig. 4.5

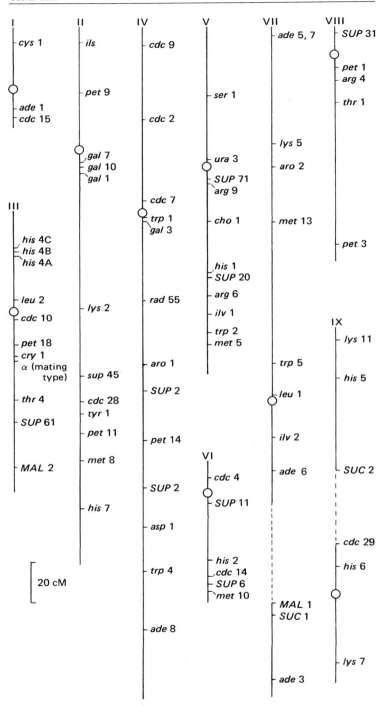

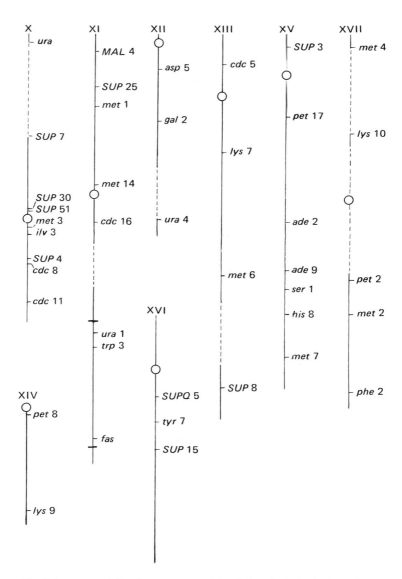

The linkage map of *Saccharomyces cerevisiae*. Only a limited selection of markers is shown: amino acid auxotrophies (abbreviations are all standard except for *ilv*, meaning isoleucine plus valine, and *aro*, meaning a mixture of aromatic amino acids); auxotrophy for adenine (*ade*) or uridine (*ura*); ability to utilize certain sugars (*MAL* maltose, *SUC* sucrose, *gal* galactose); super-suppressor loci coding for tRNA molecules (*SUP*); petite (cf. Chapter 8) *pet*; temperature-sensitivity for specific stages of the cell cycle *cdc*; fatty acid synthesis *fas*; choline requirement *cho*.
Modified after Fincham et al.[5]

4.6   The fine structure of crossovers

Tetrad data appear to tell us that crossing-over is a symmetrical process, with each crossover product accompanied in the same tetrad by a strictly reciprocal product. This is certainly the result which one sees when linked markers are relatively widely spaced, so that nearly all the observed recombination is due to crossovers distant from the markers being recombined. When, however, recombination within a very short interval is investigated, as when markers are within the same gene (*see* pp. 140–141), a very different picture emerges. As we shall see in the next chapter, crossovers are often, and perhaps always, accompanied by non-reciprocal transfer of genetic material in their own immediate vicinity. Furthermore, it appears that the primary event in recombination is an association between homologous chromatids which results in a local exchange of genetic information without *necessarily* leading to a crossover at all. The simple concept of recombination by perfectly reciprocal crossing-over, while adequate for most practical mapping, only appears valid when the crossovers are looked at from a distance. In the following chapter we explore the additional complications encountered when recombination is investigated at close quarters.

4.7   Summary and perspectives

We have seen in this chapter how genetic markers in sexually reproducing organisms can be mapped in linkage groups which clearly correspond to the chromosomes visible in the cell nucleus. The classical genetic method of making test-crosses can establish the order of the markers within each linkage group. The same linkage analysis also tells us a good deal about the nature of the crossing-over which accounts for recombination within linkage groups. It shows that crossing-over is confined to two chromatids at any one cross-over point, but that all four chromatids can participate on an equal footing in a series of multiple crossovers. It also shows that there is some interference between crossovers in that there are usually fewer multiple crossovers than one would expect if all crossovers were independent events. All of this fits with what one can see in meiotic bivalents under the microscope, assuming chiasmata to represent crossovers.

The limitation of linkage mapping is that, although it can establish the sequence of markers, it can give no certain information about their physical spacing. There is not only the problem of the non-equivalence, except at very short distances, of recombination frequences and true map units; there is also the difficulty posed by the very different frequencies of crossovers per unit length of chromosome or DNA between organisms and even between different chromosome regions in the same organism. To find out exactly how far apart genetic markers are in physical terms it is necessary to relate genetic maps to physical maps of chromosomes and DNA segments. We shall deal with physical mapping in later chapters.

Chapter 4	Selected Further Reading

King R. C. (ed.) (1975) *A Handbook of Genetics*, Vols 2–4. New York, Plenum Press.

Chapter 4

Problems

4.1   In maize a pure-breeding strain with coloured and normally formed endosperm was crossed with another strain pure breeding with respect to colourless and shrunken endosperm. All the seeds formed from the cross were coloured and non-shrunken. F_1 plants grown from these seeds were backcrossed to the colourless shrunken strain. A count of seeds from several cobs gave the following result:

Coloured non-shrunken	1056
Coloured shrunken	123
Colourless non-shrunken	136
Colourless shrunken	1102

Plants grown from coloured shrunken and colourless non-shrunken seeds were crossed together. What kinds and proportions of seeds do you expect? If coloured non-shrunken seeds from this cross are grown the plants crossed to the original colourless shrunken strain, what kinds and proportions are now expected?

4.2   Where two recessive mutations a and b are linked with recombination fraction p (percentage recombination = $100 \times p$), there are two kinds of double heterozygote, *cis* (sometimes called 'coupling') $\dfrac{a\ b}{+\ +}$ and *trans* (sometimes called 'repulsion') $\dfrac{a\ +}{+\ b}$. By means of appropriate 'Punnett squares', calculate formulae for predicting the phenotypic ratios among the progenies obtained by self-fertilizing (or sib-mating) each of these kinds of double heterozygote. Why do such progenies provide a less efficient way of calculating recombination frequencies than progenies of crosses of double heterozygotes to double recessive homozygote?

4.3   In the mouse, black (as opposed to the wild-type agouti) is determined by the recessive b, the white-belted pattern by the recessive bt, and wavy ('undulated') whiskers by the recessive un.
Black white-belted females (homozygous with respect to b and bt) were mated to wavy-whiskered (homozygous for un) males. The resulting litters were all phenotypically wild type. Brother–sister mating among the F_1 mice gave 200 F_2 mice as follows (wild-type characteristics in brackets):

74 wild type (agouti, non-belted, normal whiskers)
22 belted (agouti, normal whiskers)
36 wavy-whiskered (agouti, non-belted)
17 wavy-whiskered belted (agouti)
38 black (non-belted, normal whiskers)
13 black belted (normal whiskers)

What can you conclude about the linkage relationships of the three recessive markers?

4.4   In maize a strain homozygous for two recessive mutations, *liguleless* (*lg*) and *glossy* (*gl*) seedling, was crossed to another strain homozygous with respect to a dominant allele, *Booster* (*B*), that causes strong red pigmentation in stem and leaves. All of the F_1 plants showed the effect of *Booster* but had normal ligules and were not glossy. An F_1 plant was backcrossed to the *liguleless glossy* parent strain. Seeds formed from the cross gave the following seedling phenotypes and numbers:

lg	*gl*	+	172
+	+	*B*	162
lg	*gl*	*B*	56
+	+	+	48
lg	+	*B*	51
+	*gl*	+	43
lg	+	+	6
+	*gl*	*B*	5

Make a linkage map showing the relative positions of the three gene loci. Is there evidence for crossover interference? Calculate the coefficient of coincidence.

4.5   In *Drosophila melanogaster* the recessive sex-linked mutations white-eye (*w*), miniature-wing (*m*) and forked-bristle (*f*) map in the order *w–m–f*, with 35 per cent recombination between *w* and *m* and 20 per cent between *m* and *f*. A cross was made between wild-type males and females from a strain homozygous for all three mutations. What proportions of different phenotypes do you expect to find (*a*) in the F_1 generation and (*b*) in an F_2 generation obtained by allowing the F_1 flies to mate together. In calculating the answer to (*b*) assume (i) that there is no interference between crossovers and, more realistically, (ii) that there is interference with a coefficient of coincidence of 0·5.

4.6   In *Drosophila melanogaster* the eye colour mutations *brown* (*b*) and *cinnabar* (*cn*) are linked 45 centimorgans apart on chromosome 2. *Cinnabar* resembles *vermilion* and *scarlet* in colour and, like these mutants, interacts with *brown* in the homozygous double mutant to give a white eye. Supposing that you have flies of each of the two doubly heterozygous types, *cis* $\dfrac{cn\ bw}{+\ +}$ and *trans* $\dfrac{cn+}{+bw}$, what phenotypes (and in what proportions) would you expect from the following crosses (female written first in each case): (i) *cis* × white double homozygote, (ii) *trans* × white double homozygote, (iii) *cis* × *trans*, (iv) *cis* × *cis*, (v) *trans* × *trans*? Bear in mind that there is no crossing-over in the *Drosophila* male.

4.7   In *Neurospora crassa* a cross was made between a single mutant strain with a nutritional requirement for arginine and a double-mutant strain requiring both histidine and methionine. Fifty asci were dissected into spore pairs (no regard being paid to the order of the spore pairs in each ascus) and the spores were germinated and classified with respect to nutritional requirement(s). The following types and numbers of tetrad were found:

his met arg	*his met* +	*his met arg*	*his met arg*	*his met* +
his met arg	*his met* +	*his met* +	*his* + *arg*	*his* + +
+ + +	+ + *arg*	+ + *arg*	+ *met* +	+ *met arg*
+ + +	+ + *arg*	+ + +	+ + +	+ + *arg*
9	6	22	1	3

his met arg	*his met* +	*his met arg*	*his met* +
his + +	*his* + *arg*	*his* + +	*his* + *arg*
+ *met arg*	+ *met* +	+ *met* +	+ *met arg*
+ + +	+ + *arg*	+ + *arg*	+ + +
4	2	1	2

What do you conclude about the linkage relationships of the three segregating markers?

4.8 A cross was made between proline-requiring and adenine-requiring strains of *Neurospora crassa*, asci were dissected with the spore pairs in order, and the germinated ascospores were classified with respect to proline and adenine requirement. The following types and numbers of asci were obtained, account being taken only of whether different spore pairs came from the same half-ascus or from different half-asci:

pro +	*pro* +	*pro* +	*pro ad*	*pro ad*
pro +	*pro ad*	+ *ad*	+ *ad*	+ +
+ *ad*	+ +	*pro* +	*pro* +	*pro* +
+ *ad*	+ *ad*	+ *ad*	+ +	+ *ad*
58	19	20	1	2

Construct a linkage map showing the positions of *pro* and *ad* in relation to the centromere.

4.9 Suppose, in an organism for which tetrad analysis is possible, you have the triple heterozygote $\dfrac{a\ b\ c}{+\ +\ +}$, the map distances between a and b and between b and c each being 10 centimorgans. Assuming no chiasma (crossover) or chromatid interference, calculate the expected frequencies of the different tetrad types which should be formed at meiosis.

5 Genetic Fine Structure and the Mechanism of Eukaryote Recombination

We saw in the last chapter how linkage analysis permits the mapping of genetic markers in linkage groups, corresponding to chromosomes. In the examples considered, the markers were relatively widely spaced—very far apart indeed in molecular terms. We now turn to the special problems encountered in **fine-structure mapping**, that is when the markers involved are tightly linked—separated, in terms of DNA, only by some hundreds or thousands of base-pairs.

5.1 Mutation, the nature of alleles and the definition of the gene

In their studies of genetic variation, geneticists constantly refer to the **wild type** of their experimental organism. This is not a very satisfactory concept. In a long-domesticated organism like *Zea mays*, for example, a true wild type is not easy to identify, and even in truly wild populations of plants and animals there is considerable variation between individuals, as we shall see in Chapter 18. Natural populations do not as a rule have an absolutely standard normal allele of each gene. Nevertheless, one can usefully distinguish between alleles which are compatible with the normal range of variation shown in wild populations and alleles which take the phenotype clearly outside the normal range and thus constitute clear genetic markers. It is often convenient to distinguish mutant from wild type in this sense, notwithstanding the likelihood that different wild-type alleles will not be absolutely the same.

Mutant alleles arise by abrupt and usually stably inherited change (**mutation**) from wild-type alleles. Mutation, which is considered at greater length in Chapter 13, occurs at low frequency in all populations, and can be greatly increased in frequency by mutagenic treatments, including ultraviolet or ionizing radiation and exposure to certain chemicals. Most mutations with effects which are sufficiently strong to be seen clearly in the phenotype result in loss or drastic impairment of some normal function (*see* Chapter 12 for the biochemical basis for this statement). The effect of a defective mutant allele is very often more or less concealed when it is present in a diploid heterozygote with a wild-type allele to supply the normal function. Mutations, in other words, are commonly recessive, though there are many exceptions.

The Classical Gene

The term 'gene' was coined in 1909 by the Danish geneticist Johanssen to denote the presumed physical basis of the Mendelian factor-pair. As chromosome mapping was developed, the gene came to be viewed as a unit or particle residing at a specific point on a chromosome—the gene **locus**. It was taken for granted that crossing-over did not occur within genes, only between them.

A second property attributed to the gene was that it was a single unit of **function**. Different mutations in the same gene were expected to have effects on, generally impairments of, the same function. Thus, given two uncharacterized mutations, the answers to the two questions 'Do they map at the same locus?' and 'Do they affect the same function?' were expected to be the same.

Complementation Tests for Functional Allelism

Let us take some examples from the well-worked area of eye colour in *Drosophila melanogaster*. The normal wild-type colour is a rather brownish red, due to a mixture of red and brown pigments. Many recessive mutations have been found which, when homozygous, eliminate the brown component, giving a scarlet eye. Their phenotypic similarity is not, by itself, good evidence that they all share the same functional defect since there are several different enzyme functions acting in sequence in the synthesis of brown pigment. The simple test is to cross together individuals from true-breeding homozygous mutant stocks and to observe the phenotype of the F_1 flies (both sexes, if the mutations are autosomal, or females if they are sex linked) which carry both mutations. If this phenotype is wild type, with red-brown eyes, it is evident that each mutant type is able to contribute to the F_1 what is lacking in the other—the two are complementary or, in the usual terminology, show **complementation**. In other words, the two mutations must have eliminated different functions in pigment synthesis and, by the functional criterion, should be in different genes. If, on the other hand, the F_1 flies have scarlet eyes like the parents the two mutants are shown to be non-complementary and deficient in the same function; they are thus taken to be **allelic**, that is mutant in the same gene.

On the basis of the complementation criterion the scarlet-eyed mutants of *Drosophila* fall into three clear groups, *scarlet* (*st*), *cinnabar* (*cn*) and *vermilion* (*v*). Members of different groups complement each other completely, and they also turn out to map at clearly distinct chromosomal loci. The F_1 generation from the cross between the autosomal mutants *st* and *cn* can be represented as $\dfrac{st \quad cn^+}{st^+ \quad cn}$; the phenotype is wild type since both dominant wild-type genes st^+ and cn^+ are present. If the F_1 flies are allowed to mate, the F_2 generation exhibits something close to a $9:7$ ratio of wild-type to scarlet-eyed flies. This, as the reader should be able readily to work out, is what one would expect if the two gene differences were segregating independently, both sexes producing $st\ cn^+$, st^+cn, st^+cn^+ and $st\ cn$ gametes in a $1:1:1:1$ ratio. In fact, many detailed studies show that *st* and *cn* are located, respectively, on the second and third chromosomes. The third type of scarlet-eyed mutant (*v*) is on the X-chromosome (first chromosome) and shows sex linkage.

A contrasting result is given by the eye-colour mutants *white* and *apricot*. The former lacks completely both eye-colour pigments and the latter shows a very marked dilution of them. Both are sex linked and recessive in the female. In the cross between flies of the pure-breeding white and apricot strains the F_1 males show the phenotype of the maternal parent, since no X-chromosome is inherited from the father. The females receive an X-chromosome from both parents and are *pale apricot* in phenotype, as if the effect of the *apricot* chromosome was merely being diluted by the *white* rather than being complemented by it. We thus conclude in this case that the two mutations are in the same functional gene (i.e. are allelic to one another). This is recognized by their both being given the same basic gene symbol, *white* being represented by *w* and *apricot* by w^a; wild-type alleles are shown as w^+.

Complementation tests of the kind just considered can be made in any organism which has a stable and observable diploid phase in its life cycle. In *Saccharomyces* (yeast) species, for example, complementation between haploid mutants can be readily assessed, provided that the strains are of different mating types, by allowing the cells to fuse to form a diploid culture.

In some of the best studied filamentous fungi, such as *Neurospora*, there is no stable diploid phase but complementation tests can still be easily carried out through the use of **heterokaryons**. When two strains of the same mating type, similar with respect to a number of other genes which control compatibility, are inoculated close together, fusions of hyphae occur resulting in the mixing of two kinds of nuclei in a common cytoplasm. If the two strains are mutant in different functional genes, the heterokaryon, which is usually established within 24 hours, will generally grow like wild type since each component will supply the normal function which is defective in the other. It does not usually seem to matter that the complementary chromosomes are in different nuclei. This test has been used in *Neurospora crassa* to classify a large number of **auxotrophic** mutants, each of which has lost the ability to synthesize a particular metabolite and thus requires that metabolite as a supplement in the growth medium. For example, some hundreds of arginine-requiring mutants fall into twelve different **complementation groups**, each group corresponding to a different functional gene and a different step in arginine synthesis (*see* pp. 350–351, Chapter 12).

5.2 The divisibility of the functional gene (cistron): attached-X analysis in *Drosophila*

Mutations attributed to the same functional gene on the basis of complementation tests, always map close together. Indeed, except in very large scale analyses, they generally appear to be inseparable by recombination. For example, pale-apricot w/w^a females mated to w males give white and apricot progeny in a 1 : 1 ratio with the apricot male offspring significantly darker than the pale-apricot females (the phenomenon of **dosage compensation**—*see* p. 509, Chapter 17). There are usually no w^+ red-eyed recombinants unless the sample is very large. From this result it is concluded that w and w^a are not only defective in the same function—they are also *almost* inseparable by recombination.

They are not quite inseparable, however, as E. B. Lewis[1] first showed. When hundreds of thousands of offspring of w/w^a females are screened, a few red-eyed individuals do appear, due not to mutation (since spontaneous mutation is far less frequent even than this), but to the separability by recombination of the w and w^a mutational sites. Lewis demonstrated the recombinational origin of the exceptional w^+ progeny by the ingenious use of **attached-X** chromosomes, a method which requires a little explanation.

In an attached-X *Drosophila* stock, the female flies have their pairs of X-chromosomes joined at the centromeres. The way in which the genetic system is maintained is shown in *Fig.* 5.1. It will be seen that the attached-X pair is transmitted from mother to daughter, while the father's X-chromosome is inherited by his sons. The explanation of this bizarre reversal of the usual state of affairs is that, in the attached-X ($\widehat{XX}$) female, the two Xs cannot separate at the first division of meiosis but instead pass together to the same spindle pole; the Y-chromosome, which is also present in females of attached-X stocks, passes to the other pole. Thus the eggs formed are 50 per cent $\widehat{XX}$ and 50 per cent Y; fertilization by normal X and Y-carrying sperm leads, as the figure explains, to only two kinds of fully viable fly—$\widehat{XX}$Y females and XY males. Note that, in *Drosophila*, it is the

Fig. 5.1

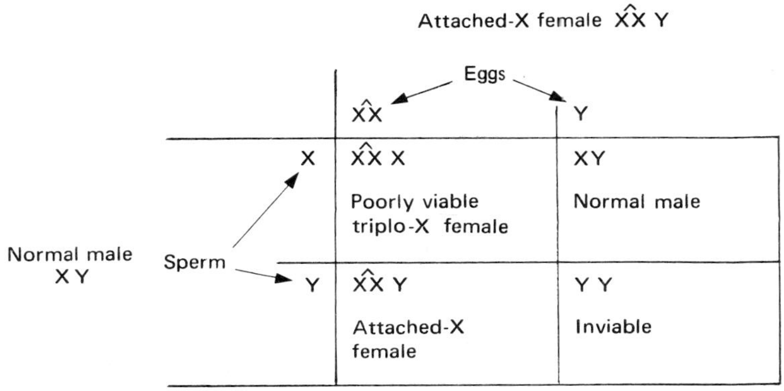

Inheritance of the attached pair of
X-chromosomes in attached-X stocks
of *Drosophila melanogaster*.

number of X-chromosomes in proportion to the number of autosomes that determines the sex. The Y has no effect except on sperm function; a fly with one X and no Y is a normal-looking but sterile male.

In attached-X females the four X-chromatids present at first meiotic prophase are not distributed to all four meiotic product nuclei but to only two of them. The eggs which receive an attached-X pair are, so far as the X-chromosome is concerned, equivalent to **half-tetrads**. They provide an opportunity of recovering the two reciprocal products of crossing-over. When a single crossover occurs between chromatids of the mutually attached X-chromosomes there is a 50 per cent chance that the two participating chromatids will still be attached together following the division of the fused centromere at the second division of meiosis (this is explained in *Fig.* 5.2). The half-tetrad is thus the next best thing to complete tetrad analysis, which is available only in lower eukaryotes, particularly fungi.

Before use can be made of half-tetrad analysis there are two obstacles to be overcome. The first is the necessity of introducing markers to differentiate the attached-X pair so that their recombination can be detected. This can be done without too much difficulty by using triplo-X females ($\widehat{X}X$ X), which are rather inviable but nevertheless obtainable. By crossing-over between the free X and one of the attached pair, markers can be transferred from the former to the latter. The second problem is to separate the reciprocally recombined chromosomes for separate analysis after the desired recombination between the attached pair has taken place. Separation, in fact, occurs spontaneously with a low frequency and is recognized when a detached X-chromosome is transmitted from $\widehat{X}X$ mother to XY son. The frequency can be greatly increased by X-rays, and detached X-chromosomes can be obtained fairly readily by this means.

In his analysis of the w/w^a situation, Lewis constructed an attached-X stock carrying w on one X and w^a on the other. The two X-chromosomes were also distinguished by various other markers, the most relevant of which were *scute* (*sc*) and *split* (*spl*), both affecting bristles and mapping close to the w locus to the left and the right, respectively. Another feature of the experimental design was the presence in the stock of long heterozygous inversions in the second and third chromosomes; these effectively reduce recombination in the autosomes (*see* Chapter 6) and, for reasons not entirely understood, increase recombination in the X by a factor varying between two and six. Lewis thus increased his chances of finding recombination between closely linked sites on the X. He screened about

Fig. 5.2

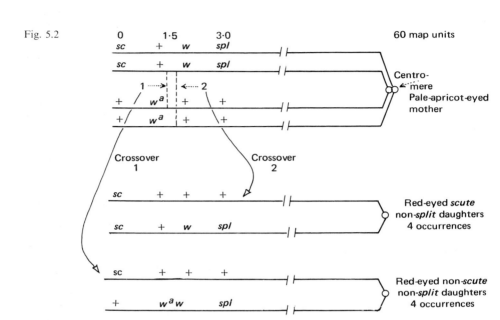

Demonstration of the origin by crossing-over of red-eyed (w^+) recombinants from pale-apricot w/w^a heterozygotes in an attached-X stock of *Drosophila melanogaster*. For the sake of simplicity only the closest flanking markers *scute* (*sc*) and *split* (*spl*), both affecting bristles, are shown. The constitutions of the recombined X-chromosomes were demonstrated by further test-crosses following detachment of the attached-X pair in each case. Numbers above the gene symbols are map units, measured from the chromosome end.
Simplified after E. B. Lewis.[1]

Notes: 1. There were also 4 red-eyed female progeny with attached-X constitutions explicable by *two* crossovers, one between w^a and w and the other between *spl* and the centromere; there were thus 12 w^+ recombinants in total.

2. 40 000 (approx.) other female progeny all had pale-apricot eyes, as expected for no recombination between w and w^a. Hence the frequency of w^+ recombinants was 12 among 80 000 X-chromosomes or 0·015 per cent. Assuming an equal number of the reciprocal $w\,w^a$, double-mutant recombinant type (not phenotypically distinguishable from w single mutant), we estimate 0·03 per cent total recombination, a frequency which would be lower without the inversion heterozygosity present in the autosomes (*see text*).

3. In this example all of the recombination between allelic sites could have been due to reciprocal crossing-over, in contrast to the examples from fungi reviewed later in the chapter (pp. 130–142). This result is not altogether typical even of *Drosophila*. In a similar attached-X analysis of recombination within the *maroon-like* (*mal*) locus, Hilliker and Chovnick[2] found no evidence for reciprocal crossing-over. Virtually all the recombination within the locus appeared to be due to conversion without crossing-over. The average *Drosophila* situation is probably much like that in the fungi, with partial association between conversion and crossing-over (cf. p. 132).

40 000 daughters of $\dfrac{sc\,w\,spl}{+\,w^a\,+}$ attached-X females and found twelve with red eyes. These latter all proved to carry a recombinant X-chromosome of constitution $sc\,w^+\,+$; the appearance of w^+ was always accompanied by a crossover joining material left of w with material right of w^a in the parental chromosomes. This suggested strongly that w and w^a were mutations at different sites, and that the w chromosome had a normal site corresponding to w^a to the left and the w^a chromosome a normal site corresponding to w to the right: $\dfrac{sc\,+\,w\,spl}{+\,w^a\,+\,+}$. Furthermore, in about half of the exceptional red-eyed flies the reciprocally recombined chromosome $+\,w^a\,w\,spl$ was also shown to be present as the other member of the attached pair. The double mutant $w^a\,w$ has the same phenotype as the single mutant $+\,w$ and had to be distinguished by an additional genetic test. By a detachment of the X-chromosomes and a cross to *scute*, females of the genotype $\dfrac{+\,w^a\,w\,spl}{sc\,+\,+\,+}$ were produced. Among their progeny, an apricot-eyed male of the constitution $sc^+\,w^a\,+\,spl^+$ was identified; the flanking markers were recombined just as would be expected if the separation of w^a from w had occurred through crossing-over between the w^a and w sites.

Lewis was the first to point out the paradox, as it appeared at that time, that females of the constitution $\dfrac{w^a\,w}{+\,+}$ (called the *cis* arrangement of the two heterozygous sites) had wild-type red eyes while the *trans* arrangement of the same components, $\dfrac{w^a\,+}{+\,w}$, gave pale-apricot eyes. This is the **cis-trans position effect**, the essence of which is that the two wild-type ($+$) sites only complement one another when adjacent on the same chromosome; their joint function fails when they are separated on different homologues. Sites bearing this relationship to one another are said to be in the same **cistron**, a term first proposed by Benzer in 1957 to denote

---

Box 5.1   **Terms relating to genetic units**

Gene = Cistron: The smallest unit of independent function. At the molecular level, the DNA containing the information for a single polypeptide or single species of RNA. Defined by complementation tests and/or molecular analysis.

Allele: A particular form of a gene. Alleles of the same gene are said to be members of the same **allelic series**, allelic *to each other*, or simply **allelic**.

Site (within a gene): the smallest part of a gene capable of independent mutation and recombination. At the molecular level a DNA base-pair. Alleles of the same gene may differ from one another in single sites or in several.

Mutation: The *process* by which a new heritable variant arises. Also used to mean the persisting and heritable *change* in the genome, e.g. a mutant site may be referred to as a mutation.

Mutant: (Noun) An organism carrying a mutation (in the second sense above). (Adjective) Derived from a pre-existing type by mutation. Thus one may refer to a **mutant allele** or a **mutant site**.

Locus: A *position* in a chromosome or linkage map more or less precisely defined by genetic mapping or cytological observation. Sometimes used instead of **gene**, especially when the allelic relationships of the mutations defining the locus are unclear. Thus a locus may turn out to contain a number of different genes, possibly but not necessarily functionally related or, conversely, could conceivably be only a part of a large and complex gene.

Marker: A mutation, more or less precisely located on a chromosome or linkage group, with a phenotypic effect which enables its transmission to be followed.

the genetic unit of function. As Benzer first established through his analysis of *rII* mutants of bacteriophage T4 (*see* Chapter 11, p. 293), the cistron is subdivisible by recombination into a large number of sites, mutation of any one of which can modify, reduce or eliminate the function of the whole cistron. In Chapter 12, we shall review some of the evidence bearing on the molecular basis of cistron function. For the time being it will be sufficient to bear in mind that the most usual function of a cistron is to code for a single type of polypeptide chain, and that individual mutations within the cistron most usually alter single nucleotide pairs in the DNA. A guide to the terminology of genetic units is provided in *Box* 5.1.

We now know that the cistron concept, though it brought considerable clarification when first put forward, is fraught with complications, some of which will be considered in Chapter 12. It is often convenient to use the older term **locus**, to mean a short chromosome segment within which mutations generate a limited spectrum of more or less related effects, without prejudging the questions of biochemical function and whether the region can properly be called a cistron or single gene.

5.3 Conversion, postmeiotic segregation and their association with crossing-over

The example of *w* and *w^a* in *Drosophila* might suggest that recombination within genes was no different from recombination between them except for shorter distances and lower frequencies. In both cases, it would appear, recombination occurs by crossing-over with the formation of reciprocally constituted products. However, investigation of recombination within further *Drosophila* genes[2] and, especially, a great accumulation of data from tetrad analysis in fungi, showed that the situation was not so simple.

In general (and the *white* analysis is rather exceptional in this respect), recombination between extremely closely linked sites is characterized by *non-reciprocal* events in the immediate vicinity of the point of recombination. Indeed, there is now good reason to suppose that all crossovers involve local non-reciprocal exchange of genetic information, and that the only reason why this is not usually apparent is that, in most chromosome mapping with fairly widely spaced markers, the latter are nearly always far away (on the molecular scale) from the recombination events.

An Example from *Neurospora*

One of the pioneering studies was the work of M. E. Case and N. H. Giles[3] on *pan-2* mutants of *Neurospora crassa* requiring the vitamin pantothenic acid. Almost any two *pan-2* mutants of independent origin gave *pan-2^+* recombinants when crossed together though the frequency was low (10^{-5}–10^{-3}). The best information on the origin of these recombinants came from analysis of asci formed in crosses between two strains carrying different *pan-2* mutations (numbers *3* and *5*) and distinguishing markers at two loci flanking *pan-2–ylo* (yellow conidia) 3·6 map units away to the left and *trp-2* (tryptophan requirement) 7·7 map units to the right. One of the crosses examined was *ylo pan-2³ +* × *+ pan-2⁵ trp*, mutants *3* and *5* having the property, unexpected in alleles (*see* p. 125), of being complementary in heterokaryons. The phenomenon of allelic complementation is considered in Chapter 12. In the present context its significance is that it provides a ready means of distinguishing between the two mutants; *pan-2³* and *pan-2⁵* segregants form pantothenate-independent hetero-karyons when paired with *pan-2⁵* and *pan-2³* testers, respectively, while the double mutant *pan-2³,⁵* will complement neither single mutant.

Table 5.1 shows the constitutions of 11 asci, out of 856 dissected and analysed, which showed something other than the normal segregation of two *pan-2³* and two *pan-2⁵* meiotic products. Though the numbers are small, the data illustrate many of the features which have been found to be typical of recom-bination between alleles both in *Neurospora* and in other fungi.

Least typical are the two asci (the last class listed in the table) which could have arisen through crossing-over of the classical reciprocal type between the two *pan-2* sites, with *5* on the left (*ylo*) and *3* on the right (*trp-2*) side. This is the type of result we have already seen in the case of the *w* locus in *Drosophila*, but in general it is

Table 5.1		Constitutions and interpretation of eleven *Neurospora crassa* asci showing recom-bination within the *pan-2* gene (data of Case and Giles[3])				
		Parental chromosomes	*ad-1*	*pan-2* 5 +	+	
			+	+ 3	*trp-2*	
Ascus class	No. of asci		Constitutions of spore pairs*			Interpretation
1	3		*ad-1*	5 +	+	Double-site
			ad-1	+ 3	+	conversion
			+	+ 3	*trp-2*	5 + → + 3
			+	+ 3	*trp-2*	No crossing-over
2	4		*ad-1*	5 +	+	Conversion 5 → +
			+	+ +	+	Crossing-over
			ad-1	+ 3	*trp-2*	adjacent to 5
			+	+ 3	*trp-2*	
3	2		*ad-1*	5 +	+	Conversion 3 → +
			ad-1	5 +	+	No crossing-over
			+	+ +	*trp-2*	
			+	+ 3	*trp-2*	
4	2		*ad-1*	5 +	+	Crossing-over
			ad-1	5 3	*trp-2*	between 5 and 3
			+	+ +	+	
			+	+ 3	*trp-2*	

There were 845 other asci each of which had two *5 +* and two *+ 3* spore pairs.
* The parental-type spore pairs are shown in lines 1 and 4 and the recombined spore pairs in lines 2 and 3, without regard to their actual positions in the ascus.

much less common than the type exemplified by classes 2 and 3, in which $pan-2^+$ recombinants have been formed *without* the simultaneous formation of the double mutant $pan-2^{3,5}$. Here the effect is as if, in one of the two chromatids participating in the recombination event, the mutant site had been unilaterally **converted** to its wild-type counterpart without any transfer of information in the opposite direction.

In the three asci of class 4, conversion seems to have involved both sites simultaneously (**coconversion**); a meiotic product which, on simple Mendelian principles, should have had the constitution $\underline{5\,+}$, emerges as $\underline{+\,3}$. Coconversion is most common where extremely closely linked sites are involved; the closest pairs are much more commonly converted together than separately. Note that coconversion does *not* result in recombination between the sites converted; it merely replaces one parental allele by the other.

Four of the nine asci displayed in *Table* 5.1, that show conversion in *pan-2*, also show crossing-over of the flanking markers, suggesting an association between conversion and crossing-over which is amply confirmed by the much more extensive tetrad data available from studies on *Sordaria* and yeast.

Conversion Analysed in *Sordaria fimicola*

The great usefulness of ascospore colour mutants in *Sordaria* and *Ascobolus* species for the detection of comparatively rare aberrant segregation was noted in Chapter 3 (p. 92). One of the most informative analyses was carried out by Kitani and Whitehouse[4] on a *Sordaria fimicola* cross between two spore colour mutants carrying different alleles of the *grey* (*g*) gene. One of these, g^{4b}, had colourless ascospores while the other, g^5, had light-grey ones, the wild-type colour being dark-brown. The cross was also segregating with respect to two markers flanking *g*; g^5 was in coupling with the morphological mutation *mat*, 0·4 map units to the left, while g^{4b} was coupled to another morphological mutation *corona* (*cor*), 3·4 units to the right.

Kitani recorded a total of 85 199 asci from this cross. The great majority had two colourless and two light-grey spore pairs, as would be expected from regular segregation of g^{4b} from g^5, but 209 asci showed other patterns, sometimes including dark-brown phenotypically wild-type spores. All of the spores from these unusual asci were dissected out, germinated, grown into cultures and tested for genotype by appropriate crosses. The analysis revealed not only g^+ recombinants but also colourless ascospores of the double-mutant ($g^{5,4b}$) type. The latter were distinguished from g^{4b} single mutants by their failure to give any g^+ recombinants in crosses to g^5.

The genotypes of the most numerous and/or informative asci are set out in *Table* 5.2. The first point to note is that, although orthodox crossing-over between sites within *g* is rare (only 7 asci), *all* the different kinds of aberrant segregation of the *g* sites are associated with a relatively high frequency of crossing-over between the flanking markers, amounting overall to nearly 50 per cent tetratype asci with respect to *mat* and *cor*. Since, in unselected asci, *mat* and *cor* show only 3·8 per cent recombination (i.e. about 7·6 per cent tetratype asci) it is clear that there must be some kind of correlation between aberrant segregation of the *g* marker sites and crossing-over in the vicinity.

Two other points should be added here. First, in asci which show both aberrant segregation at *g and* crossing-over of the flanking markers, it is generally the same

Table 5.2   **Some of the commoner patterns of aberrant segregation of markers in the *grey* (*g*) gene of *Sordaria fimicola*, and their interpretation in terms of hybrid DNA formation and mismatch correction. (Data from Kitani and Whitehouse[4])**

$$\text{Linkage map:} \quad \frac{mat \qquad g^5 \; g^{4b} \qquad\quad cor}{\qquad\quad 0\cdot4 \qquad 3\cdot4}$$

(*mat* and *cor* are mutants recognized by mycelial morphology; 5 and *4b* are, respectively, pale-grey and colourless ascospore mutants)

$$\text{Cross:} \quad \frac{mat \qquad 5 \qquad + \qquad\qquad +}{+ \qquad + \qquad 4b \qquad\quad cor}$$
$$\underbrace{\qquad\qquad\qquad\qquad}_{g}$$

Among 85 200 asci examined, 206 showed visibly aberrant segregation of the *g* markers. The selection listed here includes the commonest classes. The classes omitted, numerous but individually infrequent, mostly require more complex explanations, e.g. hybrid DNA covering both *g* sites with two or more separate correction events.

Ascus class no.	Number of occurrences	Constitution of the two relevant ascospore pairs*		Simplest model for origin	Notes
1	1	mat 5 + + mat + 4b +	+ 5 + cor + + 4b cor		†
2	13	mat 5 + + mat + + +	+ 5 4b cor + + 4b cor		†
3	7	+ 5 + + + + + +	mat 5 4b cor mat + 4b cor		†
4	14	mat 5 + + mat + + +	+ + 4b cor + + 4b cor		‡
5	0	mat 5 + + mat + + +	+ 5 4b cor + 5 4b cor		†
6	12	+ 5 + + + + + +	mat + 4b cor mat + 4b cor		‡

Table 5.2 (continued)

Ascus class no.	Number of occurrences	Constitution of the two relevent ascospore pairs*		Simplest model for origin	Notes
7	14	mat 5 + + / mat 5 + +	+ + + cor / + + 4b cor		†
8	7	mat 5 + cor / mat 5 + cor	+ + + + / + + 4b +		
9	7	mat + + + / mat + + +	+ + 4b cor / + + 4b cor		†
10	3	+ + + + / + + + +	mat + 4b cor / mat + 4b cor		
11	5	mat 5 + + / mat + 4b +	+ + + cor / + + 4b cor		†
12	4	+ 5 + + / + + 4b +	mat + + cor / mat + 4b cor		†
13	2	mat 5 + + / mat + + +	+ 5 + cor / + 5 + cor		§
14	2	mat + + + / mat + 4b +	+ + + cor / + + 4b cor		§

Table 5.2 (continued)

Ascus class no.	Number of occurrences	Constitution of the two relevent ascospore pairs*	Simplest model for origin	Notes
15	7	mat 5 4b cor + + + + mat 5 4b cor + + + +	(diagram)	
16	3	mat + 4b cor + + + + mat 5 4b cor + + 4b +	(diagram)	§

Diagram for class 15:

```
              5 ↓  +
     mat      5    +    +
              +  ⤬  4b
     +        +    4b   cor
                   ↑
```

Diagram for class 16:

```
                 ┌ 5   4b ┐↓
     mat         │ +   4b │  +
                 └    +    ⤬
     +          (+)    4b   cor
                 ┄┄┄       ↑
```

Key to diagrams: Boxes enclose regions of hybrid DNA.

Rings round site symbols indicate origin by correction of mismatch.

Dotted rings indicate possible correction, but very likely no hybrid DNA in the first place.

⋁⋁ indicates cutting of crossing strands and sealing the cut ends to reconstitute non-crossover chromatids.

⊥ indicates cutting of the 'outside' non-crossed strands and their crosswise rejoining to complete a whole-chromatid crossover.

*Each spore pair derives from one chromatid of a diplotene bivalent. Only the two which participated in the interaction at g are shown; the other two, in general, are of the two parental constitutions with only occasional coincidental crossing-over between mat and cor.

†The crossed single-strands could equally well have been on the left of the hybrid segments.

‡The circled site could have been produced by correction $^5/+ \rightarrow ^+/-$ but was probably never hybrid—cf. absence of the class which would have resulted from $^5/+ \rightarrow ^5/5$ (class 5).

§Hybrid DNA of different extents is a likely explanation in this case. But hybrid with $^5/+ \rightarrow ^+/+$ correction at the circled site also possible.

Note the absence of evidence for coconversion (cocorrection) of sites 5 and 4b, in contrast to the frequency of this kind of event in the Neurospora example in Table 5.1. This is a consequence of the relatively wide separation of 5 and 4b; they are not, however, too far apart to be included in a common tract of hybrid DNA.

---

two chromatids which are involved in both kinds of event. Second (a point which emerges from more detailed study of the data of *Table* 5.2), the associated crossover most commonly occurs, as nearly as one can tell, adjacent to the site(s) within g which exhibits the conversion or postmeiotic segregation. It will be argued at the end of this chapter that conversion and postmeiotic segregation are the symptoms of events which occur commonly, very probably always, in the immediate vicinity of crossovers.

We will now consider in more detail some of the more frequent patterns of aberrant segregation (*Table* 5.2) and what they may reveal about mechanisms. Each spore-pair in the ascus derives, so far as the g linkage group is concerned, from one chromatid present at the first meiotic anaphase. If, as was argued in

Chapter 1 (p. 20), each chromatid contains just one double-stranded DNA molecule, postmeiotic segregation within a spore pair can only mean that the corresponding chromatid at the end of the first prophase of meiosis had hybrid DNA over the segregating g site(s), with one DNA single strand derived from one parent of the cross and the complementary strand from the other parent. The occurrence of postmeiotic segregation of the g alleles in *two* spore pairs in the same ascus (i.e. an aberrant 4:4 segregation) is most simply explained as due to a reciprocal exchange of homologous segments of single-stranded DNA between two chromatids of different parental origins. The segments involved in postmeiotic segregation are variable in extent and position since they sometimes include both mutant sites (ascus type 1—*Table* 5.2) and sometimes one site but not the other (e.g. ascus types 2 and 3).

Having been forced to postulate hybrid DNA as the cause of postmeiotic segregation, we can also use it as a plausible explanation for conversion. We need only suppose that some mechanism exists for correcting mismatches in DNA base-pairs. Suppose the base-pair at a wild-type site is A–T, and that it changes by mutation to G–C, then hybrid DNA formed at that site in a wild/mutant heterozygote will contain either an A–C or a G–T mismatch. Correction would entail the removal of a mismatched base, or a stretch of single strand containing it, by nuclease activity, followed by the filling of the gap by repair synthesis using the surviving strand as a template. Depending on which of the mismatched bases was removed in any particular case, and there is no *a priori* reason to expect any preference for one over the other, the mismatched pair could be repaired either to wild type or to mutant. Such correction occurring on both participating chromatids will give one of three ratios of wild type to mutant: (i) 6:2 where correction is to wild type on both chromatids, (ii) 2:6 where correction is from wild type to mutant on both chromatids and (iii) 4:4, indistinguishable from a normal segregation, where correction is in one direction on one chromatid and in the opposite direction on the other. Correction on one chromatid only will give 5:3 or 3:5 if hybrid DNA has been formed on both.

There are two arguments for regarding the mismatch correction hypothesis as likely; one depends on analogy and the other on experiment. The analogy is with the repair of damage to DNA caused by irradiation with ultraviolet light (UV). The effect of UV on DNA is to cause cross-links between adjacent pyrimidine residues: giving **pyrimidine dimers**. These interfere with DNA replication as well as with transcription into RNA and they are lethal unless removed. One way in which both prokaryotic and eukaryotic cells can remove pyrimidine dimers is by making a single-strand incision adjacent to and on the 5′ side of each dimer and then removing a tract of DNA containing the dimer by 5′→3′ exonuclease action. The process is discussed in more detail in Chapter 13 (*see Fig.* 13.7). It is only a small step to postulate an analogous system for removing one member of a mismatched base-pair, with the incision enzyme recognizing a mismatch rather than a dimer.

Experimental evidence of mismatch correction has been obtained by M. Meselson and his colleagues (Wagner and Meselson[5]) using the DNA of bacteriophage lambda (cf. Chapter 11). It is possible to separate the two complementary DNA strands of lambda DNA and then to reanneal the 'plus' strand of the wild type with the 'minus' strand of a mutant phage or the other way round. Such fabricated hybrid DNA (heteroduplex) was introduced into cells of the *Escherichia coli* host by the method outlined on p. 275. It was shown that a high proportion of cells infected with single heteroduplex molecules released either purely wild-type or purely mutant progeny phages rather than the 50:50 mixture which would be expected if semiconservative replication of the lambda DNA occurred without any correction of the mismatch. It was thus concluded that correction, in one direction or the other, occurred with a high probability before replication. The results of experiments in which the artificial heteroduplex had two different mismatches could be interpreted by supposing that correction could be initiated at any mismatch and involved the excision of a tract of single strand up to a few kilobases in length, starting on the 5' side of the mismatch and running in the 5'→3' direction. The enzymes responsible for this kind of correction in *E. coli* have not been identified at the time of writing, and direct evidence that a comparable process occurs in meiotic nuclei of eukaryotes is still lacking. Nevertheless, the *possibility* of conversion by mismatch correction has been demonstrated.

Several ascus classes (e.g. 13 and 14 in *Table* 5.2) showed postmeiotic segregation at one site within *g* (aberrant 4:4, 5:3 or 3:5 ratio) while the other showed whole-chromatid conversion (6:2 or 2:6 ratio). Such cases can readily be interpreted as arising from hybrid DNA covering both sites and correction acting on one only. Whether or not correction initiated at one mismatch affects another mismatch within the same tract of hybrid DNA will depend on several factors: the length of the excised segment (which may be assumed to be variable), the distance separating the two mismatched sites and probably also the DNA strand attacked. If excision proceeds with a constant chemical polarity (e.g. 5'→3'), it will, if on one strand, travel towards the other site and, if on the other strand, travel away from it. In the cross under consideration, *5* and *4b* are evidently too far apart for frequent coconversion (in contrast to the *pan-2* sites shown in *Table* 5.1).

The Initiation of Hybrid DNA: One-Chromatid Versus Two-chromatid Heteroduplex

The key ideas of hybrid DNA formation and mismatch correction were first put forward by R. Holliday,[6] who envisaged a process of equal and reciprocal exchange of single DNA strands of the same polarity between chromatids, as pictured in *Fig.* 5.3A. Starting with hybrid DNA of similar length and position on both participating chromatids, all the aberrant ascus segregations which are actually found can be generated by correction (occurring from wild to mutant or from mutant to wild) on either chromatid, on neither or on both. Correction on neither gives aberrant 4:4, on one but not the other 5:3 or 3:5, and on both together 6:2, 2:6 or apparently normal 4:4 depending on the directions of correction. Yet there are good reasons for rejecting the hypothesis of equal and reciprocal exchange of DNA single strands as the *sole* kind of initiating event.

A strong and simple argument depends on the data from *Saccharomyces cerevisiae*. In this organism one finds plenty of 5:3 and 3:5 segregations but practically no aberrant 4:4 segregations—very hard to explain if there are mismatches to be corrected on *both* chromatids in those asci destined to give aberrant segregation. In *Sordaria* it is clear that there must be some asci with hybrid DNA covering *g* on both participating chromatids, since aberrant 4:4 segregations of *g* alleles do occur nearly as often as 5:3 and 3:5 segregations. But there is a good argument for concluding that even here single-strand transfer is not always reciprocal. It depends on the use of the flanking markers, in non-crossover

Fig. 5.3

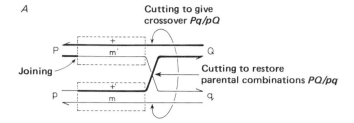

A. The interaction between diplotene chromatids postulated by Holliday[6] to explain the occurrence of conversion and postmeiotic segregation in the vicinity of single-strand DNA exchanges (potential crossovers). The dotted boxes enclose regions of hybrid DNA. If the hybrid regions contain the site of a base-pair difference between the interacting chromatids, they will have base-pair mismatches which may or may not be corrected. Correction leads to conversion, failure of correction to postmeiotic segregation. Arrows indicate polarities of strands. P, Q are flanking markers; +/+', m/m' wild-type and mutant base-pairs.

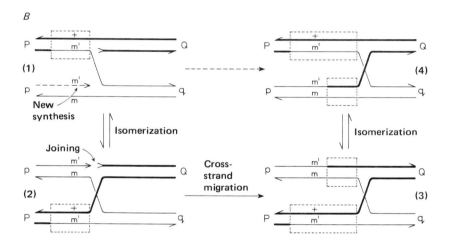

B. The modification of the Holliday model proposed by Meselson and Radding.[7] (1) One-way transfer and assimilation of a single DNA strand onto recipient chromatid; the strand displaced in the recipient is destroyed by nuclease action and the gap in the donor is filled by new synthesis (broken line). (2) Isomerization (by 180° rotation to the left of the crossing strand), bringing the free ends of the single strands into alignment for joining; this results in a structure like that postulated by Holliday (A above) but with hybrid DNA on only one chromatid. (3) Extension of hybrid DNA along both chromatids by migration of the crossed-strand position; this can occur by spinning each duplex about its long axis, unwinding on one side of the crossing-point and winding up on the other. Such spinning will occur to some extent merely as a result of thermal agitation. (4) A second isomerization converting the half-chromatid (single-strand) crossover to a whole-chromatid (double-strand) crossover. This step was also postulated by Holliday. It seems more problematical than the isomerization shown in (2) since, to occur easily, it appears to require cutting and resealing of one of the crossing strands. The chromatids are eventually separated by cutting and healing of the crossed strands giving crossing-over of flanking markers P and Q in (2) and (3) but not in (1) and (2).

Fig. 5.3. (*cont.*)

C

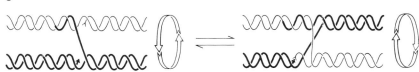

C. Proposed isomerization (1)⇌(2) with the DNA molecules drawn more realistically as double helices.

asci, for distinguishing the two participating chromatids. In an ascus with a 5 : 3 or 3 : 5 ratio, we can ask whether the participating chromatid not showing postmeiotic segregation has retained its original *g* allele or whether it has been converted to the *g* allele of the other parent. If this chromatid had hybrid DNA covering *g* there is no general reason for expecting it to be corrected back to rather than away from its original allele. Yet in many crosses, and not just the one analysed in *Table* 5.2, one finds a large and significant excess of 5 : 3 and 3 : 5 asci in which the complete original genotype of one of the participating chromatids is retained (compare, for example, classes 4 and 5 in *Table* 5.2). Rather than postulating some inherent bias in the correction mechanism favouring the resident strand, a bias which there is no independent reason to expect, it seems altogether simpler to suppose that the non-segregating chromatid never had hybrid DNA at the *g* locus and that, at least so far as this part of the chromatid was concerned, single-strand DNA transfer was one-way from a donor to a recipient chromatid. This alternative implies that the displaced tract of single strand in the recipient is not transferred reciprocally but merely lost, while the gap left in the donor DNA duplex is filled by repair synthesis using the intact complementary strand as template.

We are left with what may seem a slightly awkward conclusion—namely that hybrid DNA formation in *Sordaria* meiosis is partly reciprocal and partly one-way. It turns out, however, that this fits very neatly into a general model for recombination which is attractive for a number of other reasons. This model was devised by M. Meselson and C. M. Radding[7] (*Fig.* 5.3*B*). It postulates that interaction between the two participating chromatids starts with invasion of the DNA duplex of one chromatid by a short length of single strand from the other, forming hybrid DNA on the recipient chromatid only. This could be followed, as explained in the figure, by the extension of hybrid DNA along *both* chromatids, so that the final result is hybrid DNA on each, but to different extents. The relative average lengths of reciprocal and one-way hybrid DNA would depend on the relative rates of the different parts of the process, and it would not be at all surprising to find variation between different organisms, e.g. reciprocal hybrid DNA relatively very restricted in yeast and more extensive in *Sordaria*. Indeed, it is perhaps a fault of the model that, by suitable adjustment of the several parameters which it contains, it could be made to fit almost any data. Its virtue is in suggesting physical states of DNA and kinds of enzyme activity likely to be involved in recombination, and so stimulating experiments to test for the presence in meiotic cells of such structures and enzymes. Indeed, some relevant DNA structures have already been found in yeast meiotic cells.[10]

5.4   The origin of crossovers

An important postulate of the recombination model put forward by R. Holliday. and also of the Meselson–Radding model derived from it, is that reciprocal single-strand (half-chromatid) bridges between chromatids lead, in a proportion of cases, to crossing-over of whole chromatids. Given the association, which we have already remarked upon, between aberrant segregation and (by inference) hybrid DNA formation on the one hand and crossing-over on the other, it is indeed attractive to postulate that single-strand DNA transfer is a prerequisite for both hybrid DNA formation and for crossing-over.

A way in which the reciprocal single-strand exchange could be effected was proposed by Meselson and Radding[7] and is shown in *Fig.* 5.3C. This looks an easy and certainly very possible step. How the complementary single strands become exchanged to convert the half-chromatid crossover to a whole-chromatid crossover is much more controversial and the exact mechanism for the completion of crossovers is still very much an open question.

The relationship of the mechanism to chromatin structure and the synaptonemal complex (cf. p. 63) is particularly problematical. At the present time, the most that one can say with reasonable assurance is that crossovers tend to occur very close and, most probably, immediately adjacent to the regions of hybrid DNA which we need to postulate in order to explain postmeiotic segregation and gene conversion. A single DNA strand exchange very likely potentiates a crossover and at the same time creates hybrid DNA; the crossover may abort but the hybrid DNA remains.

Do all crossovers arise in association with hybrid DNA? The best answer to this question is based on the yeast data. In *Saccharomyces cerevisiae*, the frequency of aberrant segregation is very high in comparison with most other organisms of which we have knowledge. It varies a good deal from one gene to another, but we can take 2 per cent as a rough average for the total of all types of aberrant segregation for single markers. Neglecting the possible minority of hybrid DNA regions which, although they include a marker, leave no trace because correction restores a normal 4:4 segregation, we can take 2 per cent as a rough estimate of the probability that hybrid DNA will cover any particular site in the genome. This, in other words, will be the proportion of the genomic DNA involved in hybrid DNA formation at the first prophase of meiosis. There are about 1.5×10^4 kilobase-pairs in the haploid genome of *S. cerevisiae* and so our 2 per cent of heteroduplex amounts to 300 kb. We have no precise estimate of the average length of a heteroduplex region, but the known length, in terms of genetic map distance and of coconversion segments combined with the calibration of map distance (in a few genes—*see* p. 143) in terms of physical length, suggests that coconversion commonly extends for as much as 1 kb. This can be taken as a minimum average length for heteroduplex regions. On this basis there may be as many as 300 such regions distributed over the seventeen yeast bivalents. If, say, 40 per cent of these become the positions of crossovers, which is approximately what the genetic data indicate, we might expect about 120 crossovers, corresponding to a total genetic map length (recall the argument on p. 109) of 6000, an average of 330 per linkage group. The present yeast genetic map (*Fig.* 4.5) shows rather more than

3000 map units. Bearing in mind that the mapping is still incomplete, we can regard the agreement as reasonable and well within the likely error due to the various approximations used in the calculation.

The only other organism for which a similar calculation might be attempted is *Neurospora crassa* where, however, information on conversion frequencies is not nearly so good. Here, the aggregate map length, which is likely to be closer to the true final value than is the case in yeast, is about 650 map units, corresponding to an average frequency of 13 chiasmata. Taking these to originate from about 30 heteroduplex regions, and guessing that these each extend for about 1 kb (the same as we assumed for yeast), we are led to an estimate of 30 kb for the amount of DNA in heteroduplex during meiosis, which is a little less than one-thousandth of the genome. About one aberrant segregation in a thousand is very roughly what has been found in those few studies in which sufficient asci have been analysed (cf. *Table* 5.1).

The purpose of these very approximate calculations is to show that it is quite reasonable to postulate that *all* crossovers arise from regions of hybrid DNA which, if they happen to include marker sites, can also result in postmeiotic segregation and/or conversion. Putting the matter the other way round, aberrant segregation may be apt to occur to genetic markers whenever they happen to be close to a potential crossover (chiasma). On this view, the reason why the classical 2:2 segregation in tetrads is so much the rule is that most of the time the markers one is looking at are (in molecular terms) very distant from the nearest incipient crossover.

5.5 | Criteria for mapping very closely linked sites

Having looked at the special features of recombination between very closely linked genetic sites, we are in a better position to appreciate the different ways in which maps of such sites (**fine-structure** maps) may be made. There are three main methods, in ascending order of reliability.

Recombination Frequency

Firstly, one can attempt to place sites of mutation within a gene or two adjacent genes in a linear order based on the frequencies of recombination between them. Whether recombination is by reciprocal crossing-over between sites, or whether it is due predominantly to conversion of one site independently of another, its probability must tend to decrease as the sites become closer together. A difficulty is, however, that we know that conversion accounts for most recombination at the fine-structure level and, in so far as this depends on recognition of mismatches in hybrid DNA, its frequency very probably depends on the kind of mismatch and hence on the *nature* of the mutations which gave rise to the markers—not merely on their distances apart. Such 'marker effects' may be the reason for the rather unreliable and internally inconsistent maps which have often been obtained in yeast and *Neurospora*.

Flanking Markers

Much more reliable, both in theory and in practice, is the method depending on the use of flanking markers. Suppose our heterozygote is $\dfrac{A\ 1 + B}{a + 2\ b}$, where A/a and B/b are markers fairly close to (but well out of coconversion range of) the locus under study, and 1 and 2 sites within the locus. If now we select for $+ +$ recombinants, we expect the $a + + B$ flanking marker crossover type to greatly outnumber the $A + + b$ type. The opposite result would lead us to conclude that we had written the sites within the locus in the wrong order and that 2 should really be to the left of 1. The reasoning is obvious if the hypothesis is that the $+ +$ recombinants are being generated by classical reciprocal crossing-over; $a + + B$ can arise from a *single* crossover with the order as written, while $A + + b$ requires a *triple*, which will be very rare if the intervals between the locus and the flanking markers are short. The same expectation holds, indeed, even if some or most of the $+ +$ recombinants arise through conversion of either 1 or 2 to its $+$ counterpart, *provided* that crossovers associated with the conversion are located immediately adjacent to the converted segment. For instance, if 1 is converted to $+$, and there is a crossover adjacent to site 1 (and it makes no difference to the result whether the crossover is to the left of 1 or between 1 and 2), then the associated crossover will generate $a + + B$. Ordering closely linked sites on the basis of the relative frequencies of the two reciprocal classes of flanking marker recombinants gives convincingly consistent results, not only in fungi but also in *Drosophila* (*see*, for example, *Fig. 5.2*). The method can, however, still fail if the intralocus recombination is mainly due to conversion either not associated with crossing-over at all, or associated rather frequently with crossing-over *not* immediately adjacent to the converted segment—for example conversion of 1 and crossing-over to the *right* of 2. Each of these confusing circumstances has been encountered in particular cases.

There are examples in fungi, especially in *Neurospora*, where flanking markers have to be used in a different way. In these cases, the most consistent ordering of sites is obtained on the basis of the inequality, not of the crossover classes $a + + B$ and $A + + b$, but of the non-crossover classes $A + + B$ and $a + + b$. The class $A + + B$ can be plausibly ascribed to conversion of site 1, and the class $a + + b$ to conversion of site 2, without associated crossing-over in either case. In some cases it turns out that the sites can be arranged in a consistent sequence on the basis that the flanking marker combination originally associated with the site to the left is always in excess over that originally associated with the site to the right (or vice versa). Such a relationship could arise if there was a consistent gradient of conversion frequency of individual sites from one end of the map to the other. The reason for such an apparent gradient is not clear. It could be due to a gradient of probability of heteroduplex formation, but other explanations may be possible. At all events, the method, which may be called **conversion polarity**, sometimes works where other methods fail.[8]

Deletion Mapping

Undoubtedly the soundest method of fine-structure mapping is that of **overlapping deletions**. It is applicable in cases where one has an extensive collection of mutants within a short chromosome segment which include not only 'point mutations', i.e. changes involving single DNA base-pairs or very short sequences of base-pairs, but also deletions in each of which a more extensive sequence has been excised and lost from the DNA. It is obvious that two mutants with overlapping deletions will not be able to give any wild-type recombinants when crossed together, since neither will be able to supply what is missing in the other. On the other hand, non-overlapping deletions should be able to recombine since, between them, they have all the sequence which is necessary to reconstitute a normal chromosome. A deletion mutant should also be able to form recombinants with any point mutant whose lesion falls outside the deletion, but unable to do so with a set of point mutants (which may, however, recombine with *each other*) which all fall within the deletion.

Through the use of these simple rules, it has been possible, in a number of cases, to construct detailed and completely unambiguous maps of sites within genes in fungi as well as in bacteria and bacteriophages (*see* Chapter 17). The beauty of the method is that it does not depend on accurate counting of recombinants. All one needs from each cross is a simple yes-or-no answer to the question: do recombinants occur or not? To be sure, it is necessary to complicate this question somewhat, and enquire whether such apparent recombinants as are formed could have been due to new mutation. But spontaneous mutation is usually so much less frequent than recombination, even of the most closely linked sites, that ambiguity does not arise.

A deletion map of the gene of *Saccharomyces cerevisiae* which determines the amino acid sequence of the major cytochrome *c* (respiratory pigment protein) of the organism is shown in *Fig. 5.4*. This is one of the few cases[8] where it is possible to compare the gene map with the positions of the mutational alterations in the polypeptide specified by the gene. The comparison shows very clearly that the map

Fig. 5.4

Deletion mutants	Point mutants													
	74	91	31	239	134 130	6	120	210	36	17	19	155	200	137 39
392	—	—	—	—	+	+	+	+	+	+	+	+	+	+
400	+	—	—	—	—	—	—	—	—	—	—	+	+	+
402	+	+	—	—	—	—	—	+	+	+	+	+	+	+
415	+	+	+	—	—	—	—	—	—	—	—	—	—	—
417	+	+	+	+	—	—	—	—	—	—	—	—	—	—
422	+	+	+	+	+	—	—	—	+	+	+	+	+	+
416	+	+	+	—	—	—	—	+	+	+	+	+	+	+
425	+	+	+	+	+	+	+	—	—	—	+	+	+	+
429	+	+	+	+	+	+	+	+	—	—	—	—	—	—
433	+	+	+	+	+	+	+	+	+	—	—	—	—	—
435	+	+	+	+	+	+	+	+	+	+	+	—	—	—
436	+	+	+	+	+	+	+	+	+	+	+	+	—	—
437	+	+	+	+	+	+	+	+	+	+	+	+	+	—

A deletion map of the *cyc1* gene of yeast (*Saccharomyces cerevisiae*). The matrix shows the results of crosses between each of the deletion mutants, *392, 400* etc., with each of the single-site ('point') mutants. + indicates recombinants formed; — indicates no recombinants. The point mutants are arranged in a linear order such that each deletion removes a continuous segment of the map. This order turned out to correspond to the sequence of amino acid residues in cytochrome *c* (*see* p. 357).
Data from Sherman et al.[9]

corresponds to physical reality—it represents the linear code controlling the amino acid sequence.

The limitation of deletion mapping, so far as higher eukaryotes are concerned, is the difficulty of selecting a sufficient number of overlapping deletion mutants. The use of deletions for defining the spatial relationships of different genes, and their correspondence to visible bands on polytene chromosomes, has, however, been of great importance in *Drosophila* genetics (*see* pp. 431–432).

5.6   Summary and perspectives

Fine-structure genetic mapping is important in two ways.

First, it has led to new and totally unexpected findings about the nature of the recombination process itself—most importantly the association of crossovers with gene conversion and the exchange of genetic information by gene conversion even in the absence of crossing-over. Detailed studies on the recombination of closely linked sites using tetrad analysis have led to the formulation of a plausible general theory linking conversion and crossing-over. This theory makes predictions about the nature of the physical intermediates and enzymes involved in recombination, and these predictions may soon be tested.

Secondly, fine-structure mapping techniques have led to the establishment of the sequences of the mutational sites of many genes, some of which have well-defined biochemical functions and are open to physical study. As we see in later chapters, a combination of genetic and physical mapping of the same gene can be very powerful in the elucidation of gene function.

Chapter 5   Selected Further Reading

Fincham J. R. S., Day P. R. and Radford A. (1979) *Fungal Genetics*, 4th ed. Oxford, Blackwell Scientific Publications.

Chapter 5 | **Problems**

5.1 The *lozenge* (*lz*) locus of *Drosophila melanogaster* maps at 27·7 units on the X-chromosome, flanked by *singed* (*sn*, 21·0) and *cut* (*ct*, 20·0) on one side and *vermilion* (*v*, 33·0) on the other. Flies homozygous or hemizygous for the recessive *lz* mutant alleles lz^{46}, lz^9 or lz^{BS} have eyes smaller than the wild type with somewhat disrupted organization of facets and somewhat reduced pigmentation. A more extreme allele, lz^s, when homozygous or hemizygous, causes a more extreme reduction in eye size, with no true facets and markedly reduced pigmentation. These four alleles are non-complementary in all heterozygous combinations. Green and Green (*Proc. Natl Acad. Sci. USA* 1949, **35**, 587–591) bred females of the three following X-chromosome constitutions:

$$(1) \quad \frac{+ \; lz^{BS} \; +}{sn \; lz^{46} \; v}, \quad (2) \quad \frac{+ \; lz^{BS} \; +}{ct \; lz^9 \; v}, \quad (3) \quad \frac{+ \; lz^{46} \; +}{ct \; lz^9 \; v}.$$

They tested for X-chromosome recombination in these females by crossing them to males carrying lz^9 and appropriate recessive flanking markers. They obtained the following results:

Female type	No. X-chromo-somes tested	Kinds of meiotic products with non-parental *lz* alleles	No. found
1	20 554	$sn \; lz^+ \; +$	9
		$+ \; 'lz^{s'} \; v$	5
2	16 255	$ct \; lz^+ \; +$	13
		$+ \; 'lz^{s'} \; v$	5
3	16 098	$ct \; lz^+ \; +$	4
		$+ \; 'lz^{s'} \; v$	5

The new alleles designated 'lz^s' conferred more or less the same phenotype as the already known lz^s allele. What do you suppose they really were and how could your conjecture be tested? (Note that they were only distinguished in male test-cross progeny, presumably because they were recessive to the other *lz* alleles in heterozygous females.) This study was carried out at a time when it was taken for granted that the unit of genetic function was also an indivisible unit of recombination. What new thinking do you think it provoked?

5.2 Ballantyne and Chovnick (*Genet. Res.* 1971, **17**, 139) studied the fine structure of the *Drosophila melanogaster rosy* (*ry*) locus by means of attached third chromosome arms. Note that the mechanics of transmission of attached autosome arms is more complicated than that of attached-Xs, but that their effect is the same in allowing the inheritance from one parent of two of the four chromatids present at the prophase of meiosis (a 'half-tetrad'). Some tens of thousands of progeny were obtained from attached-3R females of constitution

$$\left\langle \frac{+ \quad ry^5 \quad +}{kar \; ry^{41} \; l26} \right.$$

mated to males which were homozygous with respect to another *rosy* mutation. (All of the *ry* mutations were distinguished, when homozygous, by their eye colour and also by their deficiency in the enzyme xanthine dehydrogenase, a deficiency which is lethal in the presence of high concentrations of purine.) The markers *kar* and *l26* are, respectively, 0·3 cM to the left and 0·2 cM to the right of *ry*; ry^5 and ry^{41} were independently obtained mutations and were thought to be at different sites. The progeny larvae were cultured in medium containing purine so as to select for those which carried a recombinant ry^+ gene. From the few survivors the following kinds and numbers of attached-3R chromosome were recovered and characterized by further breeding tests:

$$\left\langle \frac{+ \; ry^* \; l26}{kar \; ry^+ \;\; +} \right. \qquad \left\langle \frac{+ \; ry^5 \;\; +}{kar \; ry^+ \;\; +} \right. \qquad \left\langle \frac{kar \; ry^{41} \; l26}{kar \; ry^+ \;\; +} \right. \qquad \left\langle \frac{+ \; ry^5 \;\; +}{kar \; ry^+ \; l26} \right.$$
$$\qquad\qquad 6 \qquad\qquad\qquad\qquad 3 \qquad\qquad\qquad\qquad 2 \qquad\qquad\qquad\qquad 10$$

$$\left\langle \frac{+ \; ry^+ \;\; +}{kar \; ry^{41} \; l26} \right.$$
$$\qquad\qquad 2$$

What can you deduce about the order of the ry^5 and ry^{41} sites in relation to the flanking markers and about the possible mechanism(s) through which the ry^+ recombinants are produced? What may be the nature of the *rosy* allele marked *?

5.3   In *Aspergillus nidulans*, adenine-requiring mutations map at a locus situated between the markers *yellow* (*y*), 0·2 centimorgans to the left and biotin requirement (*bio*), 6 centimorgans to the right. Pritchard (*Heredity* 1955, **9**, 343) made crosses between a number of pairs of non-complementing *ade* mutants; in each case the strains crossed differed also with respect to the flanking markers. Ascospores from each cross were plated in known numbers on to medium supplemented with biotin but lacking adenine. The adenine-independent colonies which grew up were counted and classified for *y* and *bio*. From three crosses the following results were obtained:

Cross	Percentage frequency ade^+	*y +*	*+ bio*	*y bio*	*+ +*
$y \; ade^8 + \; \times \; + \; ade^{16}bio$	0·28	31	6	102	0
$y \; ade^8 + \; \times \; + \; ade^{11}bio$	0·18	43	10	136	0
$y \; ade^8 + \; \times \; + \; ade^{20}bio$	0·32	12	7	59	1

What can you deduce about the mechanism of origin of the ade^+ spores? What is the order of the *ade* mutational sites in relation to the flanking markers?

5.4   Stadler and Towe (*Genetics* 1963, **48**, 1322) dissected and germinated the ascospores from 1651 asci from the *Neurospora crassa* cross $lys\, cys^{17} + \times + cys^{64}\, ylo$, cys^{17} and cys^{64} being two non-complementing cysteine-requiring mutants and *lys* (lysine requiring) and *ylo* (yellow conidia) being marker mutations flanking the *cys* locus closely to left and right, respectively. Fourteen asci were found containing at least one cys^{+} ascospore. Their detailed constitutions, written as tetrads of spore pairs (except in one case* of postmeiotic segregation) were as follows:

lys 17 +	lys 17 +	lys 17 +	lys 17 +	lys 17 +
lys + +	+ + +	lys 64 +	lys 17 +	+ + +
+ 64 ylo	lys 64 ylo	+ + ylo	+ + ylo	lys 17 ylo
+ 64 ylo	+ 64 ylo	+ 64 ylo	+ 64 ylo	+ 64 ylo
5	*4*	*1*	*1*	*1*

lys 17 +	lys 17 +
+ 17 +	lys 17 +
+ + ylo	+ +/64*ylo
lys 64 ylo	+ 64 ylo
1	*1*

*One spore pair segregating post-meiotically for site *64*.

What light do these asci throw on the mechanism of recombination within the *cys* gene and the order of the mutant sites?

5.5   Fogel and Mortimer (*Proc. Natl Acad. Sci. USA* 1969, **62**, 96) dissected and analysed large numbers of asci from crosses between different non-complementing *arg4* mutants of *Saccharomyces cerevisiae*. Among the arginine-requiring ascospores, single mutants were distinguished from each other and from double mutants. In the following tabulation of results, *a* represents, in each cross, the *arg4* mutant site mapping nearer to the centromere end of the gene, and *b* represents the site mapping nearer the centromere-distal end.

a +	a +	a +	a +	a +	a +	a +	a +	
a +	+ +	a +	a +	a b	+ b	a +	a b	
+ b	+ b	a b	+ +	+ b	+ b	a +	+ +	
+ b	+ b	+ b	+ b	+ b	+ b	+ b	+ b	
639	3	5	18	20	2	1	9	$\begin{cases} a = \text{mutant site } 4 \\ b = \text{mutant site } 17 \end{cases}$
508	1	3	3	2	14	13	0	$\begin{cases} a = \text{mutant site } 2 \\ b = \text{mutant site } 17 \end{cases}$

Calculate the recombination frequency between *17* and each of the other two sites, and also the frequency of coconversion in each case. Consider the relationship between recombination and coconversion.

Chromosome Variations

The classical rules of chromosomal inheritance outlined in Chapters 3 and 4 were devised for eukaryotic organisms with the diploid phase characterized by two identical, or at least highly homologous, sets of chromosomes. By simple and logical extensions, the chromosome theory can also explain the markedly altered modes of inheritance shown by diploids with structurally dissimilar chromosome sets and by polyploids with more than two sets. In this chapter we review the consequences of such structural and numerical variations.

6.1 Consequences of structural change

Segmental Interchanges

One of the remarkable features of chromosomes, and one which is not yet understood, is the tendency of broken ends to rejoin to form repaired chromosome sequences fully capable of normal replication and, provided that they still possess the centromere, of stable transmission through cell division. It seems that chromosome ends formed by breakage are in some way 'sticky' in a way that normal ends (**telomeres**) are not. The joining of broken ends is quite unspecific and, when there are two breaks in a nucleus at the same time, the four broken ends can become joined in any pair-wise combination. In this way chromosome segments can become interchanged.

As for the origin of chromosome breaks, they occur apparently spontaneously with a very low frequency, but are found in abundance following the treatment of cells with ionizing radiation such as X-rays, or with any one of a variety of injurious chemicals, including alkylating agents. Most or all of these chemicals will also induce gene mutation, and the action of some of them is described in Chapter 12, pp. 333–338.

The only stable segmental interchanges are those in which the two reconstituted chromosomes each have a single centromere. The other possibility, of two fragments with centromeres joining to give a dicentric chromosome and the two

Fig. 6.1

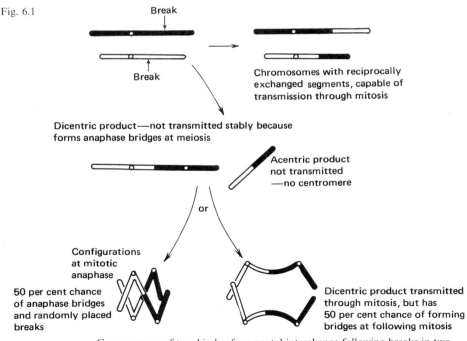

Consequences of two kinds of segmental interchange following breaks in two non-homologous chromosomes.

without centromeres forming an acentric chromosome, cannot survive mitosis. The acentric product tends to get lost at anaphase, while the dicentric product will form, with a 50 per cent probability at each mitotic anaphase, a double bridge leading to chromosome breakage at arbitrary points. The different possibilities are shown in *Fig.* 6.1.

A diploid with a pair of exchanged chromosome segments in one haploid set and the two corresponding normal chromosomes in the other is called an **interchange heterozygote**. Such heterozygosity has distinctive cytological and genetical consequences. Let us consider what is likely to happen at pachytene of meiosis when the four chromosomes, two normal and two with interchanged segments, undergo pachytene pairing. As *Fig.* 6.2 shows, full pairing of all the segments can occur only through the association of all four chromosomes in a cross-shaped configuration. (*See also Fig.* 2.8, p. 64.) At diplotene there will be the possibility of chiasma formation in all four limbs of the cross.

Suppose, first, that no chiasmata are formed between the centromeres and the interchange points. In *Fig.* 6.2*A* it is assumed that chiasmata are present in all four arms of the cross distal to the exchange points to give a ring of four chromosomes, but it makes little difference to the outcome if there are only three chiasmata, giving an open chain of four, or only two chiasmata, provided that these join two pairs of centromeres. The important variable is the orientation of centromeres at metaphase I. If this is **alternating**, with a resulting figure-of-eight arrangement of chromosomes in the case of a ring of four, or a zig-zag in the case of a chain of four, a normal chromosome complement will pass to one anaphase I pole and a balanced interchange set to the other. If, on the other hand, **adjacent** members of the ring or chain of four pass to the same pole (one possibility is shown in *Fig.* 6.2*B*), the meiotic products will be unbalanced and almost certainly inviable, since they will have one of the interchanged segments in duplicate and will totally lack the other. In the case where there are only two chiasmata and two associations of two, balanced or unbalanced products will arise with equal probabilities through independent disjunction of the two pairs. If the formation of only two chiasmata results in an association of three and one unpaired chromosome, the latter will be distributed at random at anaphase I, giving a high risk of unbalanced and inviable products. Thus, in summary, interchange heterozygotes will have reduced sexual fertility unless (i) the interchanged and normal pairs of chromosomes regularly form associations of four and (ii) these show regular alternate disjunction.

There is another hazard attendant upon meiosis in an interchange heterozygote. If a chiasma is formed between a centromere and the interchange point (in what is called an **interstitial** region), the result will be as shown in *Fig.* 6.2*C*. Two of the four meiotic products will receive balanced genomes, one normal and one with the reciprocal exchange, while the other two will be unbalanced. The unbalanced (and usually inviable) genomes will be those carrying the crossover products if the centromere orientation is as shown in the figure, but if the orientations of the two centromeres on the right are imagined as reversed with respect to the two on the left, it will be the non-crossover products which will tend to be lost. The first type of orientation may well be more probable, and, in any case, inviability caused by crossing-over in interstitial regions will, in effect, reduce recombination frequencies in such regions.

The main genetical consequence of heterozygosity with respect to a segmental interchange is the tying together of two different linkage groups. Since the viable meiotic products are those containing the two normal or the two interchanged chromosomes, markers on the two different chromosomes involved in the

Fig. 6.2

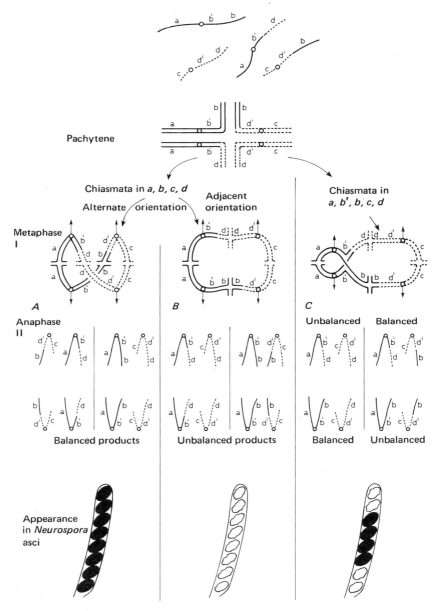

Modes of disjunction of an interchange quadrivalent at meiosis. When no chiasma is formed in either of the interstitial segments b′ or d′, alternate orientation of centromeres as in (A) will lead to balanced meiotic products being formed. Adjacent orientation as in (B) will lead to all products being unbalanced, with duplications and deficiencies with respect to the segments b and d distal to the exchange point. When a chiasma is formed in either b′ or d′, two out of the four meiotic products must be unbalanced.

Foot of figure: Patterns of ascospore abortion in *Neurospora crassa* ascus heterozygous for a segmental interchange, following from the events sketched above (*see* Perkins[1]).

interchange will appear to be linked. Recombination between them can only occur through crossing-over between one of their loci and the interchange point. The interchange points themselves will be, in effect, completely linked, and so will any genes which happen to be very close to them. A linkage map constructed on the basis of recombination in an interchange heterozygote will, so far as the two involved linkage groups are concerned, be in the form of a cross, mirroring the cross-shaped pachytene figure. In such a map the regions corresponding to the interstitial chromosome regions will be contracted because of the loss through inviability of a proportion of the products of crossing-over in these segments, as explained in *Fig. 6.2C*. These effects on linkage have been verified in many instances, especially in *Drosophila* and maize. In an interchange homozygote, of course, everything is normal except that segments of two linkage groups are interchanged, with some linkages disappearing and others becoming established.

In general, viable segmental interchanges are reciprocal, if only because unjoined broken chromosome ends are difficult for dividing cells to tolerate (they tend to stick together and form bridges at every anaphase). In practice, however, a very unequal exchange may be difficult to distinguish from a unilateral **transposition**. Apparent transpositions, in which a chromosome has received a segment of another chromosome without giving up anything vital in return, have been found in virtually all species that have been well studied genetically. Transposition heterozygotes, unlike interchange heterozygotes, have the possibility of producing meiotic products with a duplication unaccompanied by a deficiency. Such products may be viable though, if so, they are likely to cause some kind of abnormal phenotype in the next generation. In man, this is one possible origin for Down's syndrome, otherwise caused by trisomy (*see* p. 167).

In Ascomycete fungi, such as *Neurospora* and *Sordaria* species, the effect of different kinds of structural heterozygosity, particularly interchanges and translocations, can be distinguished by the different patterns of ascospore abortion which they bring about in the ordered ascus[1] (*see* foot of *Fig. 6.2* for examples).

True transpositions (as opposed to very unequal interchanges) of very small 'movable' segments are now known to be rather common in many, perhaps all, organisms. This is a separate subject, of great interest and growing importance, and is explored more fully in Chapters 10 and 18.

Chromosome Inversions

When two breaks occur in the same chromosome, three ways of healing the broken ends can be envisaged. First, the three pieces can be rejoined in the original sequence; secondly, the outside pieces can be joined and the piece originally in the middle lost, to give a deletion; finally, the middle piece can be sealed back into the reconstituted chromosome in inverted orientation. We deal with inversions here, and with deletions in the following section.

As a general rule, chromosome inversions have no immediate phenotypic effects. Important exceptions to this statement are found in *Drosophila melanogaster*, where genes close to an inversion break-point, especially if they are brought by the inversion close to heterochromatin, are sometimes inhibited in their activity to an extent that varies between different sectors of tissue. This so-called **variegated position effect**, which can also result from interchanges and transpositions, is further discussed in Chapter 16.

Meiosis in an inversion heterozygote has different effects depending upon whether the inversion includes the centromere or not. When a chromosome with an inversion including the centromere (a **pericentric** inversion) pairs with a structurally normal homologue at meiosis, crossing-over within the inversion leads to segmental duplications and deficiencies as *Fig. 6.3a* explains. The only non-deficient products will be those with either no crossing-over or a two-strand double crossover in the inverted region. The latter class will be relatively rare. Unless it is so short that crossing-over seldom occurs within it, a heterozygous pericentric inversion will reduce sexual fertility. It will always reduce, and may effectively eliminate, recombination of genes within the inverted region.

Inversions not including the centromere (**paracentric inversions**) have more interesting consequences. Crossing-over within the inversion loop will, as *Fig. 6.3b* shows, result in a single chromatid bridge being formed between the two disjoining homologous centromeres at the first anaphase of meiosis. There will also be a fragment not including a centromere at all (an **acentric** fragment) and this will tend

Fig. 6.3

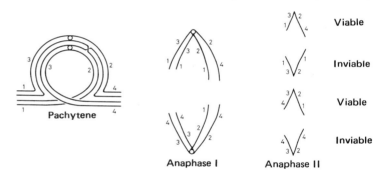

a, Inversion heterozygote with inversion including the centromere (pericentric)

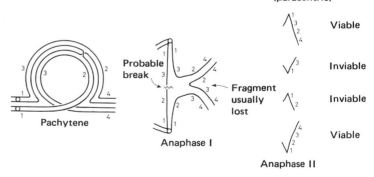

b, Inversion heterozygote with inversion loop not including centromere (paracentric)

Consequences of crossing-over with the inversion loop in an inversion heterozygote. In the case of a pericentric inversion two out of four inviable (duplication-deficiency) products are generated without any mechanical complications at anaphase I. In the case of a paracentric inversion an anaphase bridge and acentric fragment (the former usually broken at some random position, and the latter often lost) leads to the formation of deficient and inviable meiotic products. In *Drosophila* oocytes the bridge has the mechanical consequence of ensuring that the deficient products are excluded from the egg.

to be left behind on the spindle equator and not included in either first-division product. Even if the anaphase I bridge breaks and the rest of meiosis is mechanically normal, the two meiotic products receiving the remnants of the chromatids that took part in the crossover will be deficient and nearly always inviable. As in the pericentric inversion case, crossing-over within the inverted region will be effectively suppressed in the inversion heterozygote.

Special features of gametogenesis in *Drosophila* result in species of this genus largely escaping the deleterious effect of paracentric inversions on fertility. It will be recalled that there is no crossing-over in the *Drosophila* male, and so neither peri- nor paracentric inversions will affect sperm viability. On the female side, only one of the four postmeiotic nuclei is preserved in the egg and the deficient chromatids formed as a result of crossing-over within a paracentric inversion loop are not usually segregated into this nucleus. The reason for this appears to be that the orientation of dyads at metaphase II is affected by whether or not chromatids have been involved in bridge formation at anaphase I; chromatids forming a bridge are evidently pulled to the 'inside' of the first-division spindle and thus tend to orient towards the 'inside' pole of the second-division spindle, as shown in *Fig. 6.3b*. The telophase nucleus in this position is expelled from the oocyte as the second polar body. The overall effect is that paracentric inversions have practically no effect on fertility in *Drosophila* and merely act as crossover suppressors when heterozygous. It is found, indeed, that paracentric inversions are common in *Drosophila* populations, *see* p. 528. In stocks homozygous for inversions the only abnormality (unless there is a position effect—*see* p. 508) will be the inversion of a section of the map of one linkage group.

Paracentric inversions are of great importance in experimental *Drosophila* genetics, especially as the basis for 'balancer' stocks used in the manipulation of whole chromosomes without recombination within them. One application is described in Chapter 8 (p. 222).

Deletions

Deletions often have some dominant effects and, if not dominant lethal, are frequently recessive lethal. This is especially the case in *Drosophila* which has a relatively compact genome with essential functions rather closely packed along the euchromatic parts of the chromosomes. The great usefulness of deletions in genetic analysis lies in the fact that they each remove a set of functions due to a set of contiguous genes, and will 'uncover' any recessive mutation on the homologous structurally complete chromosome which affects any of these functions. In other words, a chromosome with a deletion will fail to complement any defective gene located within the deleted segment. Given a set of point mutations and a set of overlapping deletions in the same chromosome region, it is possible to make a complementation map which will give information on the linear order of the genes. Furthermore, a heterozygote with a point mutation on one homologue and a deletion on the other will yield wild-type recombinant meiotic products only if the mutation falls outside the deleted segment—not if it falls inside it. The complementation approach has been used for the comprehensive cataloguing and mapping of blocks of genes falling within several different *Drosophila* deletions (we see an example in Chapter 15, p. 431), while the recombination analysis is, as we saw on p. 143, the soundest method of mapping sites within genes in those lower eukaryotes where large numbers of meiotic products can be screened.

Three genetic criteria can be used to identify deletion mutations. Firstly, deletion mutants are incapable of mutating back to wild type (apparent exceptions, where the effects of very small deletions involving only one or a few base-pairs can be overcome by a nearby compensating insertion, are explained on p. 362). Where, as in yeast or *Neurospora*, large numbers of mutant cells can be screened for revertants on selective growth medium, this can be a sensitive test. Secondly, deletion mutants characteristically fail to recombine with blocks of closely linked mutant sites which can, albeit with low frequency, recombine with each other. A

Fig. 6.4

a

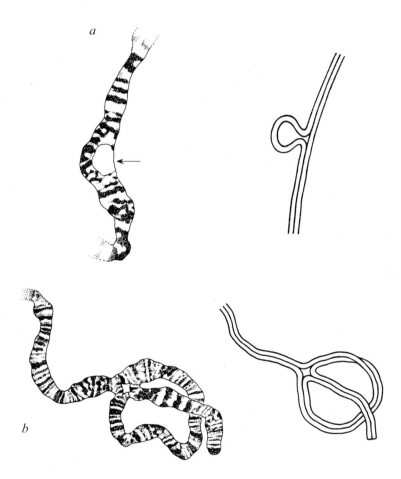

b

Structural aberrations recognized by homologous polytene chromosome pairing in structural heterozygotes of *Drosophila melanogaster*.
a. A deletion in chromosome 2. The arrow points to the stretched deleted homologue bridging between the ends of the non-deleted segment on the normal homologue.
b. An inversion in the distal part of the X-chromosome.
Both drawn from Roberts.[2]
Interpretative diagrams to right.

third criterion, which strengthens the case for a deletion, though not conclusive in itself, is recessive lethality. A combination of these three properties, or even the second one alone if the analysis is thorough, can define a deletion even in the absence of the direct visual evidence referred to in the next section.

Microscopical Analysis of Structural Changes

Under the microscope, structural alterations in chromosomes can be identified in a number of ways. In man, many examples have been found through close study of mitotic karyotypes stained by the highly discriminating banding techniques. *Fig.* 19.4 (p. 571) shows an example.

In *Drosophila*, accurate identification of the chromosome segments involved in interchanges, inversions or deletions is possible through examination with the light microscope of the polytene salivary gland chromosomes in structural heterozygotes. The close somatic pairing of homologues makes it possible to see exactly where the switch from one pairing partner to another takes place (in the case of an interchange or transposition) or exactly where the inversion loop begins and ends. Quite small deletions can also be seen in polytene chromosomes; a structurally normal chromosome in heterozygous combination with a deletion will form a loop-out or bulge in the segment which finds no homologous sequence to pair with. Examples of the use of salivary chromosome analysis to define an inversion and a deletion are shown in *Fig.* 6.4*b* and *a*, respectively.

In other organisms, similar though less detailed information can be obtained through observations of pachytene pairing in structural heterozygotes. *Fig.* 6.5 shows the appearance of an interchange association in *Neurospora*. Structural heterozygosity can be defined even more clearly when pachytene cells are stained to show the synaptonemal complex (*see Fig.* 2.8).

Fig. 6.5

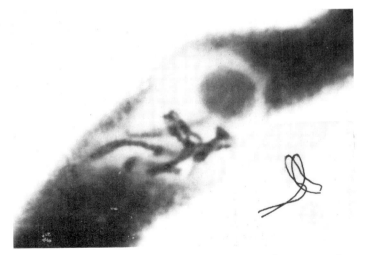

Pachytene stage of meiosis in a *Neurospora crassa* ascus heterozygous for an inversion. Interpretation in diagram to right. Magnification ×4500.
By courtesy of Dr E. G. Barry.

Structural Heterozygosity in Populations

Although, as we saw above, heterozygosity with respect either to inversions or to segmental interchanges normally results in some inviability of meiotic products, both kinds of structural variation can, under special circumstances, exist in wild populations without any deleterious effect on fertility. As was explained on p. 155, paracentric inversions in *Drosophila* are compatible with full fertility and such inversions are, in fact, widespread in wild *Drosophila* populations (*see* Chapter 18, p. 528).

In the case of segmental interchanges, the penalty of infertility can be avoided through the combination of (1) high frequency of chiasma formation in the chromosome arms distal to the exchange points and (2) alternating orientation of the centromeres of the multiple chromosome associations at the first metaphase of meiosis (*Fig. 6.2A*). This is precisely what is seen in several plant species which have been found to harbour high frequencies of segmental interchanges in their populations.

In one section of the genus *Oenothera* (evening primrose) the accumulation of structural interchanges has been carried to the limit. In this group of species the entire diploid chromosome complement of 22 has become involved in a series of interchanges such that a ring of 22 chromosomes, all joined by chiasmata and with strictly alternate centromere orientation, is present at first metaphase of meiosis. Each plant is structurally heterozygous, with one set of chromosomes the arms of which may be numbered 1–2, 3–4, 5–6 ----- 21–22 (the hyphens representing the centromeres) and a completely rearranged set with arm compositions 22–1, 2–3, 4–5 ----- 20–21. The breakpoints of the interchanges are not actually at the centromeres but, since there is no chiasma formation between them and the centromeres, the two sets of chromosomes are each stable and, with rare

Fig. 6.6

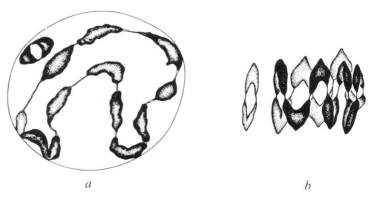

a   *b*

The ring of 12 chromosomes formed in pollen mother cells of the evening primrose *Oenothera lamarckiana, see* text.
a. Diakinesis.
b. Metaphase I. Note that all chiasmata are terminalized. In this cultivated species one chromosome pair remains outside the system of segmental interchanges and forms a separate bivalent. In related species all 14 chromosomes form a ring.
From Cleland.[6]

exceptions, no new combinations of arms are generated. At pachytene the 2 sets of chromosomes pair to form a single figure with 22 pairs of arms radiating from an inner ring consisting of the centromeres and interstitial segments. Chiasma formation within each of the pairs of arms, followed by chiasma terminalization, gives a ring of 22, and this becomes arranged on the spindle equator with all the centromeres of the (arbitrarily designated) 'normal' set directed towards one pole and all the centromeres of the 'interchanged' set directed towards the other (*Fig.* 6.6). The consequence is that two kinds of products are formed, one containing the complete normal set and the other the complete interchanged set. By means of further adaptations which we need not describe, only megaspores receiving one of the two sets develop into embryo sacs, and the only viable pollen grains are those with the other chromosome set. Hence all the fertilized eggs are once again structurally heterozygous and this condition is maintained stably from each generation to the next. All plants are heterozygous with respect to any alleles differentiating the two chromosome sets within these regions.

C. D. Darlington has called species which practice this bizarre breeding system 'permanent hybrids', and the fact that such species exist suggests that stable hybridity can be selectively advantageous. The paracentric inversions which 'float' in *Drosophila* populations are perhaps another example of hybrid advantage, although here the hybridity affects only a proportion of the individuals in a population. We return to these questions of population genetics in Chapter 18.

6.2 Consequences of polyploidy and aneuploidy

Polyploids are organisms with more than two sets of chromosomes. Triploids have three sets, tetraploids four, hexaploids six, and so on. Polyploidy is not widespread in animals and where it occurs, in certain groups of worms and arthropods, especially insects, it is usually associated with parthenogenesis (*see* the review in White's *Animal Cytology and Evolution*). The explanation usually given is that sex determination, which is far more the rule in animals than in plants, depends on a balance between the effects of Xs and Ys, and between sex chromosomes and autosomes, which tends to break down in polyploids.

Polyploidy is probably fairly common in fungi. In the yeast *Saccharomyces*, it occasionally appears in laboratory stocks and the characteristic mode of meiotic segregation which follows from it can be analysed in tetrads (Fincham et al.[9]). It is in vascular plants, especially flowering plants and ferns, however, that polyploidy achieves its greatest importance. In these groups it is demonstrably an important means of evolution and species formation.

Allopolyploidy

The immediate progeny of an interspecific cross are, almost by definition (*see* discussion of the species concept, pp. 549–550), more or less sterile so far as sexual reproduction is concerned. Even where they have some evident homology, chromosomes from different species often pair poorly, so that few bivalents are formed at meiosis and, where pairing occurs, the functional equivalence of the two chromosome sets is not sufficiently close for the mixtures of genes generated by meiosis to function properly together.

If, however, the chromosome number of a sterile diploid hybrid can be doubled, the result will be the presence of a complete diploid set from each of the parent species. In this situation, called **allotetraploidy** (or **amphidiploidy**), there are two diploid chromosome sets present together. If, as is often the case, pairing between the fully homologous chromosomes from the same species occurs to the exclusion of pairing between the partly homologous (**homeologous**) chromosomes from different species, one has, in effect, two independent sets of bivalents in the same cell, and all meiotic products will receive two complete haploid sets, one from each species. This is, in fact, another example of permanent hybridity, and stable amphidiploidy, that is the state of having two different haploid genomes transmitted together in all germ cells, is much more easily attained than is the type of structural hybridity exemplified by *Oenothera*.

Interspecific hybridization is relatively common in plants, and spontaneous chromosome doubling is not a very unlikely event. For example, in tomato (*Lycopersicon esculentum*) tetraploid shoots commonly arise when regeneration occurs from the wound callus tissue formed on a cut stem. In plants in general, any event which results in the abortion of anaphase after the chromosomes have already divided can give rise to tetraploid sectors of tissue, often capable of reproduction. Hence it is not surprising to find tetraploids in wild and cultivated plants. The special advantage of allotetraploids, which gives them greater persistence than autotetraploids, is their regular bivalent formation which enables them to be fully fertile and to breed true to their hybrid character. Though they will usually be capable of crossing with their diploid progenitors, the resulting triploids will be sterile (*see* p. 166 *below*). Thus, a newly formed amphidiploid can fulfil the usual criteria for a separate species: stable transmission of a distinctive phenotype and genetic isolation from related species.

There is good evidence that many natural species of flowering plants and ferns[10] arose in this way. A high proportion of the plants which recolonized northern Europe after the last glaciation were tetraploids, many of them autotetraploids (often reproducing apomictically, *see* p. 520), but many others amphidiploids.

Many important cultivated plants are amphidiploids. One of the best known is the commercial tobacco plant *Nicotiana tabacum*.[3] This species has 48 chromosomes which form 24 bivalents regularly at meiosis. Two related wild species, *N. sylvestris* and *N. tomentosiformis*, each has 24 chromosomes and 12 bivalents. The cross between these diploid species gives a sterile hybrid with 24 chromosomes; the two chromosome sets have too little homology for pachytene pairing. A cross of *N. tabacum* with either of them yields triploid plants ($3n = 36$) which regularly show 12 bivalents and 12 unpaired chromosomes (univalents) at meiosis. This suggests strongly that the *N. tabacum* chromosome complement is amphidiploid, with one set of 12 closely related to, and presumably derived from, *N. tomentosiformis*, and the other set similarly related to *N. sylvestris*. The hybrid origin of *N. tabacum* has been confirmed by synthesizing the species, or something very much like it, artificially. Tetraploidization can be easily induced in plants (and in most other organisms) by treating growing tissues with a drug which will inhibit spindle function by combining with the protein of the spindle fibres (tubulin). The drug most commonly used for this purpose is **colchicine**. Colchicine-induced chromosome doubling in the sterile *N. sylvestris* × *N. tomentosiformis* hybrid produced shoots closely resembling commercial tobacco and forming flowers with regular meiosis.

Allotetraploids, provided that they form only bivalents at meiosis, obey the normal Mendelian rules of inheritance, i.e. 1 : 1 segregation at meiosis of allelic differences in heterozygotes. Their only genetic peculiarity, and even this may not be evident in long-established allotetraploid species, is their tendency to have their gene functions duplicated. The effect of this may be to reduce variability, since if one gene mutates its functional duplicate may still be present in normal form. We might expect that, over long periods of time, some of the duplicated genes would tend to lose their functions or mutate so as to perform somewhat different ones, and there is some evidence that this is the case.

Higher Allopolyploids

The establishment of new true-breeding hybrid forms by adding together of genomes from different species can be extended beyond two species to three or even more. The best-known and most important example is the common bread wheat, *Triticum aestivum*, which is an **allohexaploid** ($6n = 66$). One of its three chromosome sets, the A genome, is homologous with that of the wild diploid species *Triticum monococcum* ($2n = 22$), while a second (D) set is homologous with the genome of a species of a related genus, *Aegilops squarrosa* (also $2n = 22$). The third (B) genome is of unknown origin, but is presumed to be derived from another diploid species of the same taxonomic group; this species has, however, not yet been discovered and may be extinct.

Allopolyploids from Artificial Cell Fusion

In natural evolution and, up to now, in deliberate plant breeding, hybridization has been necessarily restricted to species sufficiently closely related for the pollen tube of one to grow down the style of the other. There may well be pairs of species with genomes capable of combining functionally which cannot be hybridized by the normal sexual process because of pollen-style incompatibility. In recent years it has become possible to overcome such barriers, at least in some cases, by artificially induced fusion of vegetative cells. The procedure is to strip cells (usually obtained by grinding up leaf tissue) of their cell walls by treatment with digestive enzymes which may be obtained, for example, from snail stomach. The resulting **protoplasts**, i.e. cells bounded only by a plasma membrane and needing to be stabilized by a sufficiently high osmotic pressure in their suspending medium, can be induced to stick together and then to fuse by treatment with polyethylene glycol. The nuclei of fused cells may also fuse and a culture of allopolyploid cells may thus be started. In certain plant families, notably the Solanaceae (including tobacco, potato and tomato), it is possible to obtain regeneration of roots and shoots and eventually whole plants from callus tissue formed in such hybrid cultures.

Most of the fertile allotetraploids that have been obtained by this method could have been obtained by normal sexual crossing followed by colchicine-induced tetraploidization (one example is *Nicotiana glauca* × *N. sylvestris*). However, G. Melchers and his colleagues[4] have successfully raised hybrids, obtained by the cell-fusion method, between a diploid line of potato (*Solanum tuberosum*, $2n = 24$) and tomato (*Lycopersicon esculentum*, also $2n = 24$), two species which have not been crossed sexually.[4] It is not yet clear whether the hybrid will set seed. It seems likely that various kinds of functional incompatibility between genomes which are too distantly related, rather than inability to obtain the hybrid cells in culture, will set the limit to the applications of this new technique in plant breeding.

Autotetraploidy

Whereas chromosome doubling in an interspecific diploid hybrid yields an allotetraploid, within a diploid species it results in autotetraploidy with four fully homologous chromosome sets. Autotetraploidy is probably less common in wild plants than allotetraploidy, though the line between the two conditions is not always easy to draw, since there are all gradations from complete

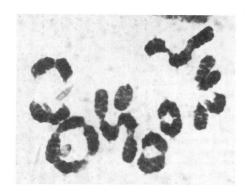
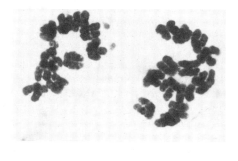

a

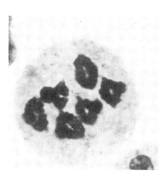
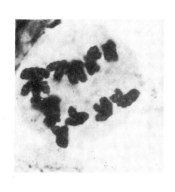

b

Fig. 6.7 — *a*. Metaphase I and anaphase I in the autotetraploid plant *Tradescantia virginiana* (4n = 48). The metaphase shows rings and chains of four (quadrivalents) in various orientations as well as two bivalents (note that the cell is very flattened and some of the terminal chiasmata joining the chromosomes are stretched). The anaphase shows a 13–11 distribution of dyads, a common event due to unequal disjunction of quadrivalents (and occasional trivalents). The faintly stained small bodies are accessory chromosomes (B-chromosomes)—*see* p. 454.
b. Metaphase I and anaphase I in a related diploid species (*Tradescantia paludosa*) for comparison—regular formation of 6 bivalents and 6–6 distribution at anaphase.
Magnification × 750.

synapsis to no synapsis between pairs of related genomes. At all events, autotetraploids showing high frequencies of assocations-of-four (**quadrivalents**) at the first prophase and metaphase of meiosis are not very common in wild species, though they are common enough in cultivated varieties. Their comparative scarcity in nature is probably a consequence of the meiotic irregularities and consequent infertility which is apt to follow from failure to form regular bivalents.

Fig. 6.7 shows an example of a first meiotic metaphase in a pollen mother cell of *Tradescantia virginiana*, a perennial autotetraploid plant much grown in gardens. Attempted two-by-two pairing at pachytene with frequent changes of partner leads, through chiasma formation and terminalization, to the formation of quadrivalents in which all four homologues are joined in chains or rings of four. The quadrivalents usually, but not always, take up orientations on the first metaphase spindle which result in two-and-two disjunction, and to the extent that this happens the products of meiosis have balanced diploid chromosome sets. Frequently, however, the competition for pairing leads to one chromosome of the four failing to form a chiasma with any of its homologues; in this case the result is a **trivalent** plus a lone **univalent**. The trivalents usually disjoin two-and-one while the univalents are distributed to one or other pole at random, or may be left behind at anaphase I and included in neither daughter nucleus. It is obvious that trivalent/univalent formation must lead to meiotic products with chromosome numbers departing from the normal diploid value; even when the number appears to be right this may be due to the presence of three homologues of one type and only one of another.

In animals, the problem of autotetraploid infertility is compounded by the sex chromosome mechanism, which has evolved on the basis of only two sex chromosomes in each cell and a 1 : 1 X : Y balance in the heterogametic sex. Sexual differentiation may fail to work properly with different numbers and ratios (*see* p. 569). Although autotetraploidy in animals does exist, especially in certain groups of worms and arthropods, it is virtually always associated with parthenogenesis, which avoids altogether the hazards of sexual reproduction and the necessity for the male sex.

Genetic Ratios in Autotetraploids

In an autotetraploid sexually reproducing species or variety, Mendel's laws of inheritance are modified, since there are four homologous chromosome sets, instead of two, being segregated into the four products of meiosis. To be more precise there are, after chromosome replication prior to meiosis, eight chromatids and four centromeres of each kind. However, it greatly simplifies the theoretical expectation if we assume that the two chromatids attached to each anaphase I centromere always carry the same allele—in other words, that the allelic difference always segregates at the first meiotic division. *Table* 6.1 explains how, on this assumption, one can make the following predictions:

i. A tetraploid carrying one dominant allele and three recessive (*A a a a*, the so-called **simplex** condition) will produce *Aa* and *aa* meiotic products in equal numbers. This will result in a 3 : 1 ratio on self-fertilization or a 1 : 1 ratio following a backcross to the true-breeding recessive, apparently just like a diploid.

ii. An individual with two dominant and two recessive alleles (*A A a a*, called **duplex**) will give a ratio of 1 *AA* : 4 *Aa* : 1*aa* meiotic products. This ratio is simply

Table 6.1   **Genetic consequences of meiosis in various kinds of heterozygous autotetraploid. Two alleles of one gene shown as *A* and *a***

Type of heterozygote	*A/a/a/a Simplex*		*A/A/a/a Duplex*	
Detectable* crossing-over between *A/a* and centromere	No	Yes	No	Yes†
Anaphase I dyads				
Meiotic products in tetrads	Overall ratio of 1 *a/a* : 1 *a/A*	Overall ratio of 1 *a/a* : 1 *a/A*	Overall ratio of 1 *a/a* : 4 *a/A* : 1 *A/A*	

*Only detectable if the chromosomes involved carry different alleles.
†The case where detectable crossing-over occurs on *both* chromosome pairs is not considered here.
‡This is the only way *A/A* homozygous meiotic products can be formed; otherwise there is a ratio of 1 *a/A* : 1 *a/a*.

derived by consideration of the six possible ways of selecting two from four. A duplex plant will give a 35 : 1 phenotypic ratio on self-fertilization or a 5 : 1 ratio from the test-cross to pure recessive, always provided that A is completely dominant.

iii. The $A A A a$ type (called **triplex**) will give a 1 : 1 ratio of AA and Aa meiotic products, but will show no phenotypic segregation because at least one A allele will be present in all progeny.

All of these predictions are consistent with most of the results obtained in, for example, the breeding of the ordinary cultivated potato, *Solanum tuberosum*. The European cultivated varieties are all tetraploid with $4n = 32$, though many varieties cultivated by the Amerindians of the Andean region, the ancestral home of the potato, are diploid with $2n = 16$.

Even though the assumption of invariable first-division segregation is generally wrong, it is often not so far wrong as to cause a significant departure from the simple prediction. One combination of circumstances, however, gives a result qualitatively different from the predicted ratios. If, due to chiasma formation, a pair of alleles segregate at the second division, and if, furthermore, the dyad crossover derivatives show **adjacent** disjunction at anaphase I, then it is possible for a/a homozygous products to result from meiosis in $A/A/A/a$ (**triplex**) heterozygotes. This result, with two sister alleles getting into the same meiotic product, is called **double reduction**, and it has to be taken into account in the interpretation of tetraploid segregations. Its frequency (often negligible for practical purposes but sometimes considerable) is difficult to predict *a priori* since it depends not only on the frequency of chiasmata between the marker locus and the centromere but also on the relative frequency of adjacent rather than alternative quadrivalent orientation at metaphase I.

Tetrad Analysis in Tetraploids

In the lower eukaryotes, autotetraploidy has been investigated genetically only in yeast (*Saccharomyces cerevisiae*), where it has sometimes been found in laboratory stocks. Yeast tetraploids provide the opportunity of confirming the rules of tetraploid (sometimes called **tetrasomic**) inheritance by tetrad analysis. Several such studies have been undertaken, all tending to confirm the rules set out above, though with indications of double reduction at some loci. An account will be found in Fincham et al.[9]

Triploids and Trisomics

In plants, triploids can be easily produced by crossing tetraploids with diploids or, in yeast, by mating an ordinary haploid with a diploid homozygous for the opposite mating type (obtainable by segregation from a tetraploid ascus). Naturally occurring triploids may occur through the occasional functioning of an unreduced gamete produced by some failure in meiosis. In both yeast and higher plants, triploids are vegetatively normal but sexually sterile, but in animals they are at best poorly viable. In *Drosophila*, triploid flies can be obtained only with difficulty and most triploid larvae fail to pupate. Mammals tolerate triploidy even less well, though it is not uncommon in early embryos. About 6 per cent of aborted fetuses in man are triploid, according to data reviewed by Hassold et al.[5]

The reason for the infertility of triploids, even where they are viable, is clear. A trivalent, or a bivalent plus a univalent, must either segregate two-and-one or lose the odd chromosome at anaphase I. With all the different associations of three behaving independently of one another, the products of meiosis will nearly all have unbalanced sets of chromosomes in numbers varying between haploid and diploid. Precisely haploid or diploid products will be rare; if n chromosomes are being distributed to the anaphase I poles at random, the chance of them all going to the same pole is one in 2^n. The chance of getting only one chromosome in excess of the haploid number will be n times as great $(n/2^n)$ since there are n ways in which it can happen. **Disomic** $(n+1)$ spores or gametes, with just one chromosome extra, are much more likely to be functional than those with larger departures from the haploid state. In plants, the progeny of a cross between a diploid and a haploid will

---

Box 6.1	Trisomic analysis in tomato (data from Rick and Barton[7])

Trisomics of eight different types, carrying no mutant alleles but each with an extra copy of a different identified chromosome, were crossed as female parents to diploids homozygous for recessive mutant alleles of a number of different genes representative of different linkage groups. The minority of trisomic progeny were identified by their distinct phenotypes (leaf shape, growth habit) and backcrossed to the diploid multiple mutant. The **diploid** progeny are recorded in the table as number of wild type : number of mutant.

Results for three trisomics

Trisomic for chromosome	Linkage group and respective genes				
	I $+/d$	II $+/r$	III $+/y$	IV $+/c$	V $+/a$
1	48 : 55	56 : 45	**72 : 29**	56 : 47	54 : 49
7	47 : 56	52 : 45	51 : 46	52 : 51	67 : 63
10	35 : 25	38 : 22	36 : 24	28 : 32	40 : 45

Trisomic for chromosome	VI $+/l$	VII $+H$	VIII $+al$	X $+cot$	
1	51 : 52	53 : 54	No data	53 : 54	
7	47 : 56	58 : 41	52 : 46	**81 : 34**	
10	32 : 28	**95 : 52**	116 : 97	20 : 17	

Interpretation: All the numbers except those in heavy type are consistent (within 0·05 probability limits) with the 1 : 1 ratio expected from a backcross of diploid heterozygote to homozygous recessive (calculate the χ^2 values if you do not believe it). Overall there is certainly some shortage of the mutant phenotypes due, presumably, to somewhat reduced viability of these. The numbers in heavy type are incompatible with a 1 : 1 ratio and compatible with the 2 : 1 ratio expected from a backcross of a trisomic with two copies of the dominant to one of the recessive allele.

    Trisomic 1
e.g. $+/+/y \times y/y$ gives $2 +/y : 1\ y/y$
    but $+/d \times d/d$ gives $1 +/d : 1\ d/d$

Conclusion: Linkage group III corresponds to chromosome 1
    Linkage group X corresponds to chromosome 7
    Linkage group VII corresponds to chromosome 10

be few in number and will usually include a high proportion of **trisomics** $(2n+1)$, resulting from the union of haploid and disomic gamete nuclei.

Trisomic plants are usually viable, but may be less vigorous than the normal diploid. They are nearly always distinguished by some abnormality of morphology or development, the nature of which is determined by the particular chromosome which is present in extra copy. In a number of plant species the complete set of trisomics, one for each chromosome, has been obtained. The first example was *Datura stramonium*, the North American Jimson Weed, which was studied by the pioneer plant and fungal geneticist A. F. Blakeslee. Subsequently complete sets of trisomics have been obtained in several cultivated plants, including tomato and, very notably, wheat.

In animals, trisomy arises, not from triploids (which hardly exist), but from the occasional formation of disomic gametes in normal diploids as a result of non-disjunction of a chromosome pair during meiosis. In man, the data summarized by Hassold et al.[5] show that about 3·5 per cent of all conceptions lead to trisomic embryos (testifying to a rather surprising inaccuracy of human meiosis), and that trisomy accounts for over 20 per cent of spontaneous abortions. Only a few kinds of trisomics in man, those corresponding to the smallest chromosomes, are at all viable. Trisomy with respect to chromosome 21 is the cause of the relatively common Down's syndrome (about 0·1 per cent of all live births and as high as 1 per cent for mothers over 40 years of age), which is fully viable but associated with mental retardation and certain (usually minor) physical abnormalities.

Assignments of Genes to Chromosomes by Trisomic Analysis

The availability of a complete set of trisomics or (in the case of yeast) of disomics provides the means for a relatively easy and unequivocal assignment of new and unmapped mutations to chromosomes. In the case of a flowering plant such as tomato, the general method is to cross a diploid, homozygous with respect to the mutant allele, to each of the trisomics in turn. From the progeny of each cross, trisomics are picked out by virtue of their distinctive phenotype and crossed back to the mutant stock. If the mutation happens to be on the chromosome which is represented three times in the trisomic, the trisomic progeny will have two chromosomes carrying the normal allele and one carrying the mutant. The backcross (assuming the mutant allele is recessive) will give a ratio, among the *diploid* progeny, of two non-mutant to one mutant. If, on the other hand, the unknown mutation is on one of the chromosomes present in only two copies in the trisomic, the backcross will give the normal Mendelian 1 : 1 ratio. Only one trisomic of the whole set will give the 2 : 1 result, and this will define the chromosome on which the known locus mutation is located. *Box* 6.1 sets out an example from tomato.

In recent years, trisomic analysis has become an important tool in yeast genetics.

Nullisomics and Monosomics

An individual with one pair of chromosomes missing completely is called a nullisomic $(2n-2)$ and one deficient in a single chromosome is called a monosomic $(2n-1)$. With the exception of those occasional 'optional extras' the B-chromosomes (*see* p. 454), all chromosomes seem to have unique and essential functions, and nullisomics are generally inviable except in allopolyploid plants

Table 6.2   **System for assignment of genes to chromosomes in wheat using nullisomics**

Nullisomic (20 II) × Hexaploid (20 II + II*)
F_1 monosomic (20 II + I*)

		F_1 pollen grains†	
		96% 20 I + I*	4% 20 I
F_1† eggs	25% 20 I + I*	24% 20 II + II* Hexaploids	1% 20 II + I* Monosomics
	75% 20 I	72% 20 II + I* Monosomics	3% 20 II Nullisomics

II indicates a homologous chromosome pair; I a single chromosome.
* Marks the chromosome pair absent in the nullisomic, carrying a mutation in the hexaploid.
† The different percentages of 20-chromosome and 21-chromosome pollen and eggs are due to differential viability.
After Sears.[8]

where each chromosome set tends to be 'covered' by the more or less equivalent chromosomes of the other genome(s).

Nullisomics in polyploid plants are useful both for assignment of genes to chromosomes and for certain chromosome manipulations of use in plant breeding. In the wheat variety Chinese Spring, E. R. Sears[8] obtained all 21 possible nullisomic lines. All of these lines have distinct phenotypes; from comparisons of each of the nullisomics with the normal full hexaploid, one can obtain some idea of the contribution made by each chromosome to normal development and morphology. Any new mutation can be assigned to one or other of the 21 chromosomes by crossing a fully hexaploid plant homozygous with respect to the new mutation to each of the nullisomics in turn. The F_1 plants will be monosomics and, on self-fertilization, will give a mixture of full hexaploids, monosomics and nullisomics as shown in *Table* 6.2. The hexaploid F_2 progeny must be homozygous for the chromosome which was contributed solely by the hexaploid parent. Thus one cross out of the 21 to the different nullisomics will give *only* homozygous mutants among the hexaploids in the F_2. In the other 20 cases the F_2 will have only 25 per cent homozygous mutant plants in each case.

6.3   Summary and perspectives

The simple rules of Mendelian inheritance, and the normal patterns of recombination within linkage groups, can be completely explained by the chromosome theory of inheritance and our knowledge of the behaviour of chromosomes during meiosis. Numerous kinds of chromosome variation, segmental rearrangements, deletions and duplications, polyploidy and aneuploidy, are associated with more or less drastic departures from the normal genetic rules and patterns.

It is a measure of the strength of the chromosome theory that, given a knowledge of each chromosome variation and its consequences for the mechanics of meiosis, it can be seen that the genetic effect is just what the chromosome theory would have predicted.

The effects of changes in chromosome structure or number can be put to good use in genetic analysis, particularly in assignment of linkage groups to chromosomes and (especially in *Drosophila*) of specific markers to particular parts of chromosomes.

Chapter 6 — Selected Further Reading

Darlington C. D. (1958) *The Evolution of Genetic Systems*, 2nd ed. Edinburgh, Oliver and Boyd.

Fincham J. R. S., Day P. R. and Radford A. (1979) *Fungal Genetics*, 4th ed. Oxford, Blackwell Scientific Publications.

White M. J. D. (1977) *Animal Cytology and Evolution*, 3rd ed. Cambridge, Cambridge University Press.

6.1   In maize the markers B, lg and $gl2$ are linked on chromosome 2 (cf. Problem 4.4, Chapter 4). One plant, heterozygous for all three markers, showed no recombination at all between B and $gl2$ and a reduced level of recombination between lg and the other two. About 40 per cent of the pollen formed by this plant was shrivelled and inviable. Suggest a possible explanation for the behaviour of this plant. How would you attempt to confirm this explanation?

6.2   A male mouse is heterozygous with respect to a segmental interchange between two metacentric chromosomes, the break-points being very close to the centromere of one chromosome and midway along one of the arms of the second chromosome. Assuming that chromosomes attached by a chiasma always disjoin to opposite poles at the first anaphase of meiosis, and that crossing-over occurs between the break-point and the centromere of the second chromosome with a frequency of 30 per cent, what will be the effect on the sperm count of the mouse? What will be the effects on the linkage relationships of markers in the rearranged arm of the second chromosome?

6.3   A strain of *Neurospora crassa*, grown from a spore which had survived X-irradiation, gave, when crossed to a standard strain of opposite mating type, two out of four aborted spore pairs in about half of the asci. The viable ascospores from these asci germinated to give cultures which grew somewhat more slowly than normal. One of these cultures, when crossed to a pyrimidine-requiring mutant, gave numerous asci with 4:0 and 3:1 ratios of pyrimidine-independent and pyrimidine-requiring spore pairs, as well as the expected 2:2 segregations. In all asci there were two spore pairs which were both pyrimidine independent and slow growing. Propose a hypothesis to explain these observations.

6.4   A *Neurospora crassa* strain of mating type A has the peculiarity of giving a proportion of aborted ascospores when crossed to any a strain. Eighty-five viable ascospores from the cross to the double mutant strain *thi-3 met-7 a* were scored as follows:
thi+ met+ A 38, *thi met a* 33, *thi+met a* 1, *thi+met+a*5, *thi met A* 8; * indicates that these isolates resembled the parent A strain in giving aborted ascospores in further crosses. In the standard *Neurospora* map *thi-3* is about 15 centimorgans distant from *met-7* on linkage group 7, unlinked to mating type, which is on linkage group 1. Interpret these results. (Perkins et al. 1962, *Can. J. Genet. Cytol.* **4**, 187.)

6.5   A sporulating culture of *Saccharomyces cerevisiae*, originally derived by crossing two stable strains of opposite mating type, was found to be producing a proportion of ascospores which, when germinated in isolation, gave rise to self-fertile cultures. In some asci all four ascospores were self-fertile, in an approximately equal number of asci two ascospores were a and two α mating type, while in the most common class of ascus one spore was a, one α and two self-fertile. Matings between a and α segregants gave sporulating cultures which again formed asci of all three types. The self-fertile segregants, however, produced asci which all gave normal 2 : 2 segregation for mating type. Suggest an explanation for these results.

6.6   A segregant of mating type *a* from the first culture mentioned in the previous question was mated with a standard α haploid. The resulting sporulating culture produced a considerable proportion of inviable ascospores; one of the viable ones grew to give a culture of mating type *a*, and this was mated to an *ade3* (red-pigmented) strain of mating type α. Asci produced from the resulting sporulating culture showed a regular 2 : 2 segregation of *a* and α in every tetrad, but the white/red (*ade3⁺/ade3*) ratio was 2 : 2, 3 : 1 or 4 : 0, with relative frequencies of about 1 : 2 : 1. What can you infer about the chromosomal constitutions of the primary and secondary *a* segregants?

6.7   The British wild plant *Lythrum salicaria* (purple loosestrife) is a tetra-ploid species. One variety has pink rather than the usual purple flowers. A cross between purple and pink-flowered plants produced only purple-flowered plants in the F_1 generation; when F_1 plants were backcrossed to pink, a ratio of approximately five purple to one pink was obtained. A further backcross of individual purple plants to pink revealed three kinds of purple-flowered plant. The first gave an approximately 1 : 1 ratio of purple to pink, the second an approximately 5 : 1 ratio, and the third almost entirely purple. Several plants of the third type gave, in aggregate, over 2000 purple and one pink in their second backcross progeny. Explain these results and, in particular, how the solitary pink-flowered plant might have arisen. (Based on R. A. Fisher and V. Fyfe, 1955, *Nature*, **176**, 1176.)

Extra-chromosomal Heredity in Eukaryotes

7.1 Criteria for extrachromosomal and extranuclear heredity

The history of extrachromosomal heredity is a long one, but only within the past few decades has this aspect of genetics become at least partly understood following the demonstration of DNA in cell organelles other than the nucleus. There is also the possibility, and we shall discuss one example later in the chapter, of DNA within the nucleus but not integrated into the chromosomes and not following the normal rules of chromosomal segregation. The terms 'extra-chromosomal' and 'extranuclear' need, therefore, to be distinguished, though most of the examples in this chapter can be described in either way.

Chromosomal gene differences segregate at meiosis or occasionally, as we saw in Chapter 5, at the first mitotic division following meiosis. They are transmitted (with the rather rare exceptions reviewed in Chapter 9) without further segregation or recombination through all other mitotic divisions. Extrachromosomal determinants, on the other hand, are distinguished either through their failure to segregate at meiosis or by their propensity to segregate at other times, or by both properties.

In higher organisms, in which the male and female gametes make grossly unequal extranuclear contributions to the zygote, the classical indication of extranuclear heredity is the transmission of a trait through all the female but none of the male meiotic products. Most flowering plants, for example, show predominantly maternal transmission of plastid defects (showing as pale-green, yellow or variegated leaves), although some, including *Pelargonium* species, transmit plastids through the pollen also (*see* Tilney-Bassett[22]).

In the motile unicellular green alga *Chlamydomonas reinhardi*, we find a remarkable example of inequality of extranuclear contributions to zygotes even though the gametes are virtually identical in size and morphology. As was explained in Chapter 2 (p. 54), this habitually haploid organism has two mating types, called plus and minus. It is found that many mutant traits, which can all be plausibly ascribed to chloroplast mutations, are usually transmitted only through the plus mating-type gamete and, when so transmitted, are commonly found in most or all of the meiotic products which are formed on zygote germination. This behaviour contrasts with that of many other unit differences, including mating type itself, which are transmitted equally by the two mating types and segregate in normal 2:2 Mendelian fashion at meiosis. For reasons still not altogether clear (*see* p. 193), the plus mating type acts as a functional female and the minus mating type as a functional male.

Among the non-photosynthetic protozoa, some of the ciliates, notably *Paramecium aurelia* and *Tetrahymena pyriformis*, provide the experimentalist with a unique opportunity for distinguishing nuclear from extranuclear transmission. In these species, sexual fusion of morphologically similar cells is followed immediately by meiosis, reciprocal exchange of haploid nuclei, and karyogamy. In this way a new diploid generation is formed, with completely reconstructed and often genetically recombined haploid contributions from the two parents, but throughout the whole process the extranuclear parts of the conjugating cells usually retain their integrity with little, if any, cytoplasmic exchange. The process of conjugation in *Paramecium* is sketched in *Fig.* 7.1. Note that the germinal

Fig. 7.1

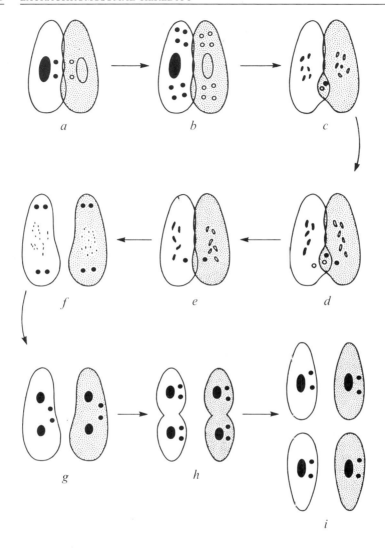

Micronuclear meiosis, exchange and fusion, and macronuclear reconstruction without exchange of cytoplasm at conjugation in *Paramecium aurelia*.
a. Parents each with one macronucleus and two micronuclei.
b. Eight haploid micronuclei formed by meiosis.
c. All meiotic products degenerate except one in each cell; macronucleus begins to disintegrate.
d. Mitotic division of surviving meiotic products, and reciprocal exchange of nuclei between cells.
e. Fusion of immigrant with resident haploid nuclei; resulting diploid nuclei are identical in the two cells.
f. One nucleus gives four by two rounds of mitosis.
g. Two nuclei develop into new macronuclei.
h. Cell division segregates the two new macronuclei, while the micronuclei undergo mitotic division.
i. Normal nuclear constitution, with one macronucleus and two micronuclei per cell, is restored; the cytoplasms and cell cortexes, distinguished by fine stippling, have not been exchanged between the conjugants.
Redrawn after Beale.[25]

nucleus, or **micronucleus**, exerts its influence on the phenotype only by proxy and, in effect, delegates control of the cell during the intervals between sexual reproduction to a massive multiple copy of itself, the **macronucleus**, which divides during mitosis but is replaced by a new macronucleus derived from the micronucleus at each meiosis. Temporary delegation of function to the macronucleus provides an example of one possible kind of cellular differentiation (*see* p. 482). In the present context, the relevance of *Paramecium* lies in the fact that certain cell traits are transmitted as if determined by extranuclear cell components without regard to the reconstruction of the nucleus. Some important examples are described at the end of this chapter.

In baker's yeast, *Saccharomyces cerevisiae*, the haploid cells which undergo sexual fusion are again equal in size, but in this species complete mixing of extranuclear components occurs in the zygote, and the contributions of both parents (with the special exception noted on p. 188 below) survive in the zygote on an equal basis. Extranuclear genetic determinants do not segregate at meiosis, or do so very irregularly, but do segregate with a certain probability at each mitotic budding cycle. Genetic analysis based on this vegetative segregation is described in the following section.

In fungi which can form heterokaryons readily, extranuclear heredity can be diagnosed in the absence of sexual reproduction. The principle is to make a heterokaryon between two strains which differ with respect to a known nuclear marker as well as in the trait, the status of which it is desired to establish. The two types of nuclei are subsequently re-extracted, through isolation of uninucleate conidia or small hyphal tips. If the difference under study is caused by some extranuclear determinant, the reisolates, though pure with respect to their nuclear genotypes, may show persistent 'contamination' of either type of nucleus with the extranuclear determinant originally associated with the other. *Fig.* 7.2 shows an example of the use of this technique—the analysis of the basis of senescence in the fungus *Aspergillus glaucus*.

A variant of this **heterokaryon test** is available in baker's yeast, where it is known as **cytoduction**. Karyogamy following fusion of cells of opposite mating type can be prevented by the presence in each of the parental strains of a mutation *kar*. Subsequent budding of the abortive zygote results in segregation of haploid nuclei of the two unchanged parental types, but each associated with extranuclear components derived from both strains (*see* p. 196 and *Fig.* 7.9).

7.2 Functions of mitochondrial DNA

Since the chromosomal basis of most inherited variation is well established and since, as we shall see in more detail in later chapters, it is the DNA of the chromosomes which carried the hereditary information, our prejudice with regard to extrachromosomal heredity must be that it is due to the DNA of the extranuclear organelles—the mitochondria and, in plants, the chloroplasts as well. We shall see in this section that the evidence for the genetic role of mitochondrial DNA is extremely clear and, in the following section, we shall consider the extent to which there is comparable information with regard to chloroplast DNA.

Fig. 7.2

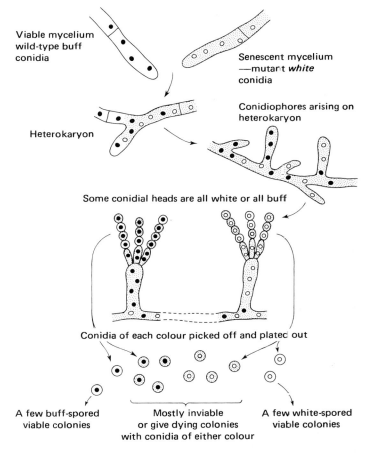

The heterokaryon test used to demonstrate the extranuclear basis of senescence in *Aspergillus glaucus*. Mixing of nuclei in a heterokaryon formed by cytoplasmic fusion of normal and senescent mycelium (the latter carrying a nuclear mutation giving white conidia), followed by reisolation of homokaryons from single uninucleate conidia, shows that nuclei originally in non-senescent mycelium can become contaminated by some extranuclear factor determining senescence. The nature of the factor is not known; it could possibly reside in the mitochondria. The invasive property is comparable to that of the defective mitochondria in suppressive petite mutants of yeast (*see* p. 180). After Jinks.[1]

Two different questions may be asked: 'Which macromolecules (RNA and proteins) have their structures encoded in the mitochondrial DNA?' and 'Which macromolecules are actually made within the mitochondrion?' Logically, the answers to these two questions could be entirely different but, as it turns out, they seem to be very much the same. RNA molecules made within the mitochondrion must be transcribed from mitochondrial DNA (mtDNA) and proteins made within the mitochondrion are made on mitochondrial ribosomes programmed by mRNA transcribed from mtDNA.

A convenient property of mitochondrial (and also chloroplast) protein synthesis is that of being immune to inhibition by certain antibiotic drugs, notably cycloheximide, which potently inhibit cytoplasmic protein synthesis. This resistance is a property of organellar ribosomes, in contrast to cytoplasmic ribosomes. Conversely, the functioning of organellar ribosomes is inhibited by such antibiotics as chloramphenicol, erythromycin and oligomycin, which have little or no effect on cytoplasmic ribosomes. In these properties, organellar ribosomes resemble the ribosomes of prokaryotes and it is these observations which constitute some of the main evidence for the theory, propounded most persuasively by L. Margulis, that the organelles of eukaryotes are vestiges of prokaryotic cells that established symbiotic relationships with primaeval eukaryotic cells.

A consequence of the selective effect of cycloheximide on cytoplasmic protein synthesis is that it allows the investigator to discover just which proteins are made on mitochondrial ribosomes. Thus, if yeast cells are incubated in nutrient medium supplemented, at a given time, with both cycloheximide and a radioactive protein precursor such as $[^{14}C]$leucine, only a limited number of radioactively labelled polypeptide components of proteins is synthesized. These can be separated from one another by electrophoresis in polyacrylamide gel containing the detergent sodium dodecylsulphate and visualized by autoradiography. The technique is explained in *Fig.* 7.3.

Experiments of this kind have led to the identification of a small number of proteins, probably all components of the inner mitochondrial membrane, whose synthesis is maintained in the presence of cycloheximide. These are, of course, by no means all of the mitochondrial protein components, and it seems that most such proteins must be synthesized on ribosomes in the cytoplasm and imported into the mitochondria. A number of proteins with essential functions in respiration are, in fact, of dual origin, with some polypeptide components of mitochondrial and others of cytoplasmic origin. For example, cytochrome oxidase, the essential final link in the chain of electron carriers which leads from respirable substrates to oxygen, contains seven different kinds of polypeptide, four synthesized on cytoplasmic and three on mitochondrial ribosomes. Another component of the inner mitochondrial membrane is the enzyme adenosine triphosphatase (ATPase), which has an essential role in the coupling of respiration to ATP formation. Of the ten polypeptides which enter into the structure of this complex, only two appear to be synthesized on mitochondrial ribosomes and the others are presumably of cytoplasmic origin and nuclear determination.

The mitochondrial ribosomal proteins do not appear to be among the proteins whose synthesis is resistant to cycloheximide, and there is no evidence that their structures are determined by mtDNA. The RNA of the mitochondrial ribosomes, on the other hand, is certainly transcribed from mtDNA, as shown by the fact that both the large (21 S) and the small (15 S) rRNA species form stable base-paired hybrid duplexes with specific regions of the mtDNA.

Fig. 7.3

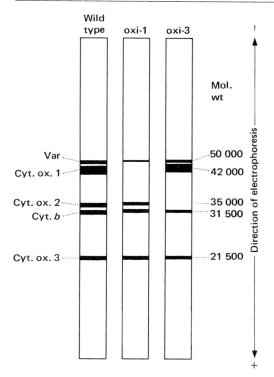

Demonstration of yeast mitochondrial proteins synthesized on mitochondrial (cycloheximide-resistant) ribosomes. Protein synthesis in yeast cells was restricted to the mitochondrial system by the addition of cycloheximide, and the proteins synthesized were radioactively labelled with ^{35}S, present in the growth medium as sulphate. The labelled products were separated on polyacrylamide electrophoretic gels containing sodium dodecylsulphate to dissociate the proteins to their constituent polypeptide chains, which are separated according to their sizes, and the radioactive bands are visualized by exposure of the gel to radiosensitive film. Specific polypeptide deficiencies in two kinds of mit^- mutants are apparent.
From Mahler et al.[2]
Cyt. ox. 1, 2 and 3 are three polypeptide components of cytochrome oxidase; cyt.b is the cytochrome b polypeptide. The 'var' polypeptide is probably a protein component of the small ribosomal subunit.

7.3 | Petite mutants in yeast, and the physical mapping of the mtDNA

Origin and Properties of Petite Mutants

Yeast mitochondrial genetics was initiated by B. Ephrussi's discovery of the **petite** mutants of *Saccharomyces cerevisiae*. These mutants fail to grow at all on carbon sources, such as glycerol, which can only be utilized by respiration, and they form small colonies when grown on sugars, such as glucose, which can be utilized anaerobically by fermentation. The petites occur spontaneously and can be induced in up to 100 per cent of the cells of a culture by treatment with a drug, such as acriflavin or ethidium bromide, which can become intercalated in the stack of base-pairs in double-stranded DNA.

A high proportion of the spontaneously occurring petite mutants, and practically all of those induced by intercalating drugs, show extranuclear inheritance, which takes two different forms in different groups of petite mutants. The **neutral** petites do not transmit their respiratory deficiency to any of the diploid cells stemming from a cross to a strain competent in respiration, nor is the petite character transmitted to any of the haploids formed by meiosis in such diploids. Neutral petites usually lack any detectable mtDNA.

The other class of petite mutants, called **suppressive**, do transmit their deficiency to a proportion of the diploid cells stemming from a cross to non-petite, and also to some of the ascospores formed in such cells. Segregation of petite from non-petite occurs in an irregular way both during meiosis and during vegetative budding. Suppressive petites contain mtDNA, but of a very peculiar kind, as we see below.

Restriction Site Mapping of DNA

The comparison between the structure of normal yeast mtDNA and the altered mtDNAs in the various suppressive petite mutants is greatly facilitated by the technique of **restriction site mapping**. This method, which will feature prominently in later chapters and will, therefore, be described in some detail here, depends on the use of a range of specific endonucleases, isolated from a number of different bacterial species. The useful property of these enzymes is that of cleaving double-stranded DNA at specific base sequences. Their natural significance is probably that of defending their own species against invasion by foreign DNA; each species protects its own DNA, generally by methylation of bases within the otherwise vulnerable sequences, from cleavage by its own endonucleases. *Table 7.1* shows the DNA sequences attacked by each of a number of the more commonly used restriction enzymes and the cutting points in each case. It will be seen that most of the enzymes make staggered cuts at equivalent positions on the two DNA strands within a **palindromic** sequence—that is, to say, a sequence which reads the same both left to right and right to left. A particular restriction enzyme recognizing a specific sequence of six base-pairs will cut double-stranded DNA only at a limited number of sites, spaced, on average, a few kb apart. Any specific piece of DNA will be cut by a restriction enzyme into a characteristic set of fragments with a reproducible array of sizes. The fragments can be separated, and the sizes determined, by electrophoresis through an agarose gel of suitable porosity.

By digesting a given molecule of DNA with two or more different restriction endonucleases, both individually and in combination, it is possible, through deducing the overlaps between the different fragment sets, to work out a map of the restriction sites on the DNA. *Box 7.1* explains the procedure as it applies to one small part of the yeast mtDNA molecule. The whole mitochondrial genome has been mapped in this way; the result is shown in *Fig. 7.4*.

When similar restriction enzyme mapping is carried out on mtDNA isolated from different suppressive petite mtDNAs, it is apparent that much of the sequence and many of the restriction sites are missing. It can be deduced, in fact, that each petite mtDNA molecule includes only a sector, and often only a small sector, of the normal circle. In the majority of cases the DNA sequence retained in the petite

Table 7.1 **Specificities of some restriction endonucleases**

Enzyme	Source	Sequence attacked*
EcoRI	*Escherichia coli* plasmid	5′ G↓A A T T C 3′ 3′ C T T A A↑G 5′
HindII	*Haemophilus influenzae*	5′ G T Y↓R A C 3′ 3′ C A R↑Y T G 5′
HindIII	*Haemophilus influenzae*	5′ A↓A G C T T 3′ 3′ T T C G A↑A 5′
HaeII	*Haemophilus aegypti*	5′ R G C G C↓Y 3′ 3′ Y↑C G C G R 5′
BamHI	*Bacillus amyloliquefaciens*	5′ G↓G A T C C 3′ 3′ C C T A G↑G 5′
PstI	*Providentia stuartii*	5′ C T G C A↓G 3′ 3′ G↑A C G T C 5′
HhaI	*Haemophilus haemolyticus*	5′ G C G↓C 3′ 3′ C↑G C G 5′
HpaI	*Haemophilus parainfluenzae*	5′ G T T↓A A C 3′ 3′ C A A↑T T G 5′
BglII	*Bacillus globigii*	5′ A↓G A T C T 3′ 3′ T C T A G↑A 5′
KpnI	*Klebsiella pneumoniae*	5′ G G T A C↓C 3′ 3′ C↑C A T G G 5′
XhoI	*Xanthomonas holcicolor*	5′ C↓T C G A G 3′ 3′ G A G C T↑C 5′
AluI	*Arthrobacter luteus*	5′ A G↓C T 3′ 3′ T C↑G A 5′

*R means either purine, Y either pyrimidine, provided that the base-pairs are complementary.
Arrows indicate the points where the DNA chains are cleaved, giving 3′-hydroxyl and 5′-phosphate ends.

Box 7.1   **Procedure used for restriction endonuclease mapping of a segment of yeast mitochondrial DNA (data from Morimoto et al.[3])**

HhaI cuts the mtDNA into 11 fragments (H1–H11) which can be separated by size on an agarose electrophoresis gel to give the following pattern:

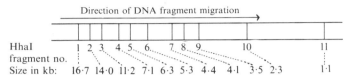

EcoRI and BamHI cut the DNA into 10 and 5 fragments, respectively and all of these fragments have been sized and numbered in order of size (R1–R10 and B1–B5).

Individual fragments from digestion with one enzyme are isolated and digested with a second one, and the sizes of the secondary fragments are determined:

I. Reciprocal digestions with HhaI and EcoRI
(fragment sizes in kb in brackets)

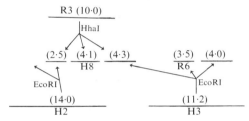

II. Reciprocal digestions with HhaI and BamHI

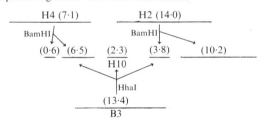

The above relationships establish the sequence of the HhaI fragments.

HhaI	4	10	2	8	3
BamHI		3			
EcoRI				3	6

The exact boundaries of B3 and R3 and R6, and the placing of the fragments indicated by dotted lines, were established in other experiments.

mtDNA represents a continuous sequence from the normal molecule, but sometimes the mutant sequence appears to have been pieced together from two or more separate fragments. *Box 7.2* shows some of the results of petite mtDNA restriction analysis and the reasoning based on these results.

The remarkably altered structure of the mtDNA in suppressive petite mutants

Box 7.2   **Restriction endonuclease analysis of the mtDNA retained in two different petite mutants, retaining some drug-resistance markers (data of Levin et al.[4])**

O_{II}-2 is a petite isolated from an oligomycin-resistant mutant, and it retains the O_{II} marker. F11 was isolated from a strain resistant to several drugs, and it retained the C and E (chloramphenicol and erythromycin resistance) markers. The mtDNA from each strain was digested with restriction endonucleases and the sizes of the resulting fragments were compared with those from wild type by separation in electrophoresis gels. The following patterns were seen:

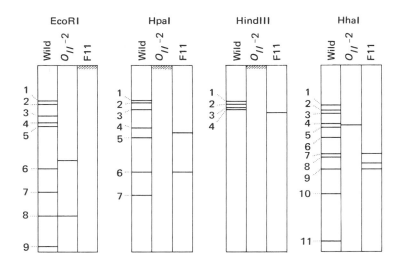

Interpretation

The O_{II} mtDNA retains only EcoRI-8 of the normal restriction fragments. The EcoRI digest also gives a fragment not present in the wild sequence, as does the HhaI digest. O_{II} mtDNA is not digested at all by HindIII or HpaI. The O_{II} marker must fall within the EcoRI-8 fragment or in closely flanking sequences as indicated in *Fig. 7.4*.

The F11 mtDNA retains the HpaI-6 fragment and the HhaI-7 and HhaI-9 fragments; it also retains the HindIII-5 fragment (too small to be seen on the gels diagrammed here). In each case a new fragment not seen in wild-type digests is also produced. F11 mtDNA is not digested at all by EcoRI. Both the C and E markers must fall *within* the EcoRI-1 fragment—*see Fig. 7.4*.

The new fragments found in petite and not in wild-type digests are expected. They presumably represent what is left of the petite mtDNA after the wild-type fragments have been cut out; the ends of a petite mtDNA sequence are not likely to coincide with the normal restriction sites. The undigested petite DNA molecules are much larger than could be accounted for by a single copy of the retained sequence; they evidently contain repeated copies.

provides a strong indication that the extranuclear determinants of the respiratory deficiency are within the mtDNA. This conclusion is supported in detail by investigations of a wide range of mutants which show more restricted alterations of mitochondrial function and which, as we see in the following section, can be mapped to particular regions of the mitochondrial genome.

Fig. 7.4

						9	10				
EcoRI	1		4		2		5	7	8	3	6
HpaI	4	5	6	2		3	7		1			
BamHI	4	1			3		2			5		
HindIII	3	(5)	4	(6)	1			2		3		
HhaI	3	5	7	9	1	6	4	10	2	8	3	

11

← C, E →   ← O_{II} →

Restriction map of mtDNA of a respiration-competent strain of *Saccharomyces cerevisiae*. The map is actually a circle, and the two ends of the bars should be imagined as joined. The regions within which the O_{II}, and C and E markers must fall, as deduced from the data shown in *Box* 7.2, are indicated. The fragments are approximately to scale. Total length 76 kb.
Redrawn from Lewin et al.[4]

Genetic Markers on the Yeast Mitochondrial Genome

Extranuclear mutations in *Saccharomyces* that can be attributed to changes at points on the mtDNA are of two different classes. First, there are those that confer resistance to antibiotics which inhibit mitochondrial function in the wild type. Secondly, there are those that result in deficiencies with respect to specific components of the respiratory machinery, as distinct from the loss of *all* components as seen in typical petites. This second class, known as *mit*⁻, fail to grow on a carbon source that cannot be fermented anaerobically, but form larger colonies than do petites when grown on glucose. *Table* 7.2 summarizes what is known about the mitochondrial functions affected by various kinds of mutation.

In crosses to wild type, both drug-resistant and *mit*⁻ mutants show a mode of inheritance rather like that of suppressive petites, in that the mutant trait is usually transmitted to a proportion of the diploid cells formed by budding from the zygote. However, in contrast to what is found in comparable experiments with suppressive petites, there is no general tendency for either wild or mutant to predominate even though, in any particular mating, one or other parent (which may be either wild or mutant) may make a larger contribution than the other. The general explanation for this sort of inequality is probably that the parental cells happen sometimes to contribute significantly different numbers of mtDNA molecules to the zygote; there is no known mechanism, analogous to chromosomal disjunction, for maintaining a constant amount of mtDNA in each cell. A special case of inequality, affecting a small segment of the mitochondrial genome only, is mentioned on p. 188.

Recombination of Mitochondrial Markers

When a cross is made between yeast strains differing with respect to two or more *mit*⁻ or antibiotic-resistance mutations, the vegetative segregants formed after several cycles of cell budding usually include some in which the parental markers are recombined. This phenomenon, first described by D. Thomas and D. Wilkie, is now exploited routinely for genetic mapping.

Table 7.3 (Cross A) shows the results of an analysis of diploid clones derived from a cross involving three different types of drug resistance. It will be seen that all possible types of recombinants are found together with the parental types, but

Table 7.2   **Mitochondrial genetic markers in *Saccharomyces cerevisiae***

Phenotype	Locus symbol	Function affected
Drug resistance		
Oligomycin	*OLI-1 (O1)* (= *VAR?*)	Mitochondrial ATPase, subunit 9
	OLI-2 (O2)	Mitochondrial ATPase, subunit 6
Antimycin and mikamycin	*ANA-1* (= *COB?*)	Mitochondrial membrane component (cytochrome *b?*)
Erythromycin	*ERY (E)*	Ribosome structure, through primary effect on 21-S rRNA
Chloramphenicol	*CAP (C)*	Ribosome structure, through primary effect on 21-S RNA
Paromomycin	*PAR (P)*	Ribosome structure, through primary effect on 15-S rRNA
Respiratory deficiency (*mit⁻*)		
	OXI-1	Cytochrome oxidase peptide 3
	OXI-2	Cytochrome oxidase peptide 2
	OXI-3	Cytochrome oxidase peptide 1
	COB	Cytochrome *b*/coenzyme Q–cytochrome *c* reductase
	PHO-1,2 (= *OLI?*)	Deficiency in oxidative phosphorylation

For references *see* Bacila et al.[24]

with unequal frequencies. The determinants of chloramphenicol and erythromycin resistance appear to be quite closely linked and both are more distantly linked to paromomycin resistance.

It must be emphasized that one cannot interpret such data as one would the data from meiotic products of a cross between chromosomal markers. It may well be that most of the mitochondrial genomes that segregate from the hybrid diploid have never had the opportunity to recombine. There is no necessity, so far as is known, for pairs of mtDNA molecules of different parental origins to come together at all and, even if they do, a random pairing within a population of two equally frequent types is just as likely to be between like as between unlike molecules. Recombination frequency must depend not only on the distance between the mutant sites (as in classical chromosomal recombination), but also on the numbers of molecules of each kind contributed to the zygote, the number of them that persist and replicate, the number of times they replicate before being sorted out into mitochondrially uniform clones and the probability of 'mating' between pairs of molecules per replication cycle. None of these parameters is known.

Neither is there, as yet, any direct evidence bearing on the mechanism of mitochondrial recombination. Empirically, it is observed that about 25 per cent recombination is the upper limit. Presumably this is because many parental-type mtDNA molecules segregate before they have had opportunity to recombine with the other parental type. Many pairs of markers approach this limiting frequency and only markers known (*see below*) to be relatively close together show much tighter linkage. Thus recombination frequency, while it can indicate close linkage, is not very useful for mapping the mitochondrial genome as a whole.

A definitive map can, however, be obtained using the partially deleted and physically characterized petite mtDNAs. For example, one can start with a respiration-competent strain that carries the determinants of several kinds of antibiotic resistance. One can then isolate a series of petite derivatives of this strain. These derivatives can then be tested, by mating to a wild type, for retention or loss of the resistance markers. If a resistance marker is retained, there will be cells budded from the zygotes that are both capable of respiration and resistance to the antibiotic in question. In the petite derivatives retention of the resistant marker cannot be determined directly because the resistance or sensitivity can only be seen in cells with functional mitochondria. Among petite derivatives of a multiple-resistance strain there will be those which retain only one resistance marker, others which retain two or three, and still others which retain all markers. The part of the physical mtDNA map which is retained in each mutant can then be determined by the kind of restriction fragment analysis outlined above. The result is that the loss of a particular marker can be consistently correlated with the loss of a particular small segment of the mtDNA, and the genetic determinant of the corresponding mitochondrial function can thus be attributed to that segment.

The sites of the various *mit⁻* mutations can be mapped by crossing to a range of different petites representing different and overlapping segments of the normal mtDNA. Whenever the petite mtDNA carries the function that is deficient in the *mit⁻* mutant, recombinants capable of normal respiration will be formed during growth of the colony arising from the zygote. If no such recombinants are formed, the *mit⁻* mutational site is placed within that part of the mtDNA that is missing in the petite tester. By comparing the results obtained from crosses to the whole battery of testers, the *mit⁻* site can be located within a small segment of the mtDNA.

mtDNA Regions from which rRNA and tRNA are Transcribed

E. Southern[6] devised an extremely valuable technique for identifying DNA fragments that have base-sequence homology with specific RNA molecules. This is a blotting procedure whereby a pattern of DNA bands in an agarose gel is transferred to a nitrocellulose sheet after first denaturing the DNA with alkali. If the single-stranded DNA fragments, which are firmly bound to the nitrocellulose, are then exposed to radioactively labelled RNA with a base sequence complementary to one of the DNA fragments, that fragment will bind to the radioactive 'probe' molecule. By drying the nitrocellulose sheet and then exposing it to X-ray film, the restriction fragment(s) from which the RNA was transcribed can be identified. We shall meet other important examples of the application of this technique in later chapters.

The mtDNA map of *Saccharomyces cerevisiae* obtained mainly by restriction site mapping and probing with RNA transcripts is shown in *Fig. 7.5*. It is

Fig. 7.5

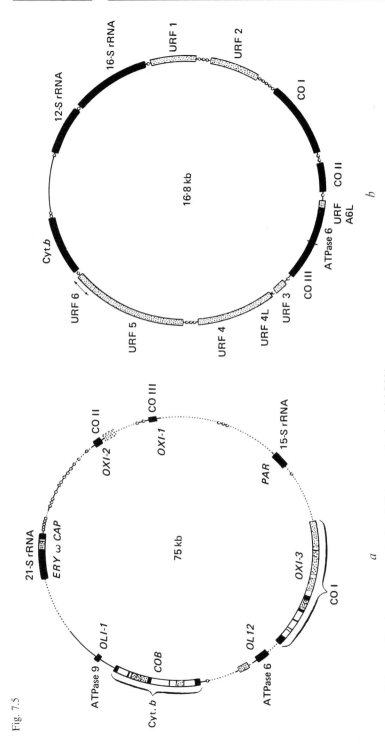

Maps of (a) yeast (Saccharomyces cerevisiae) and (b) human mitochondrial DNA.
On the yeast map both the gene products and the gene symbols are indicated, the latter in italics (see Table 7.2). Black boxes are regions represented in the finally processed transcripts (mRNA or rRNA) and the small circles represent tRNA genes (each less than 100 base-pairs). Open boxes are regions transcribed into RNA segments (intervening sequences) that are lost in processing. Stippled boxes are 'open reading frames', see p. 404, which may code for polypeptide chains of unknown or hypothetical function. In human mtDNA six of these sequences are designated URF 1–6. Dotted lines represent regions which have been identified on restriction site maps but not yet sequenced in the yeast mtDNA. Note the fragmentation of the yeast COB and OXI-3 genes by intervening sequences (see Fig. 15.12, p. 442). The human mtDNA is shown magnified 4 × in proportion to the yeast mtDNA.
Redrawn from Borst and Grivell.[21]

interesting to note that chloramphenicol and erythromycin resistance markers map within the segment hybridizing with 21-S rRNA. In bacteria, mutations which result in resistance to these antibiotics are found within genes coding for ribosomal proteins. In yeast, too, the proteins almost certainly provide the sites through which the antibiotics exert their inhibitory effects. Mutational changes in the 21-S rRNA may affect the structures of ribosomal proteins through conformational interactions within the ribosome.

Unexpected Features of the Yeast Mitochondrial Genome

Two other unexpected findings should be mentioned, even though their detailed description is beyond the scope of this book.

Slonimski, Dujon and their colleagues (Dujon[7]) discovered a strong tendency of certain strains, called omega-plus (ω^+), to transmit markers in the 21-S rRNA region of the genetic map to the exclusion of the corresponding markers in ω^- strains to which they are crossed (*Table* 7.3, Cross B). It is now known that the ω^+

Table 7.3	Analysis of diploid clones following mating between yeast haploids differing in three mitochondrial drug resistance markers

Cross A: $C^sE^sP^s$ (ω^+) × $C^rE^rP^r$ (ω^+) (229 clones analysed from diploid culture grown from cross)

Phenotype	Number	Percentage	Calculations:
			Percentage recombination: / Percentage transmission:
$C^sE^sP^s$	131	57·2	C–E 9·2 C^s 69·4
$C^sE^sP^r$	17	7·4	C–P 21·0 E^s 69·0
$C^sE^rP^s$	9	3·9	E–P 24·0 P^s 73·8
$C^sE^rP^r$	2	0·9	First parent contributes about 70 per
$C^rE^sP^s$	5	2·2	cent of the markers segregated
$C^rE^sP^r$	5	2·2	in the clones
$C^rE^rP^s$	24	10·5	C–E show pronounced linkage; P shows
$C^rE^rP^r$	36	15·7	close to maximal recombination with both

Cross B: $C^rE^sP^r$ (ω^+) × $C^sE^rP^s$ (ω^-) (704 clones analysed)

Phenotype	Number	Percentage	Calculations:
			Percentage recombination: / Percentage transmission:
$C^rE^sP^r$	409	58·1	C–E 21·9 C^r 98·1
$C^rE^sP^s$	129	18·3	C–P 29·9 E^s 76·5
$C^rE^rP^r$	76	10·8	E–P 29·7 P^r 69·6
$C^rE^rP^s$	77	10·9	Strong selective transmission of ω^+
$C^sE^sP^r$	1	0·1	markers C^r and, to lesser extent, E^s
$C^sE^sP^s$	0	0·0	
$C^sE^rP^r$	4	0·6	The selective transmission of C^r tends to
$C^sE^rP^s$	8	1·1	break the linkage with E

Data from Wolf et al.[5]

C, E and P stand for chloramphenicol, erythromycin and paromomycin resistance/sensitivity. Omega (ω) is a segment of DNA present in the CE region of the genome which, when present in one parent and not in the other, confers preferential transmission on markers linked to it.

trait is determined by an extra DNA sequence of about 1 kb in length, inserted into the 21-S rRNA gene and present in ω^+ and not in ω^- strains. The mechanism whereby ω^+ excludes ω^- and markers close to it, when the two are mixed in a zygote, is not yet known. Some process akin to meiotic gene conversion (*see* pp. 130–139) may be involved.

Another striking discovery has been the extreme fragmentation of some of the mtDNA protein-coding sequences.[8] For example the *cob* sequences, coding for the protein of cytochrome *b*, is split by long intervening sequences into at least five segments spread over some 10 kilobases. The coding sequence itself, present in the messenger RNA, is only about one-tenth as long. A rather similar situation is found in *OX13*, which codes for a polypeptide component of cytochrome oxidase[8] (*see Fig.* 7.5). We shall review the evidence for intervening sequences in higher eukaryotes in Chapter 14. The presence of this sort of gene organization in organellar DNA tends to argue against the idea that there is a close affinity between prokaryotes and eukaryote organelles; nothing of the kind has yet been found in prokaryotes.

7.4 Mitochondrial genetics of other eukaryotes

Considerable progress has been made in the identification of mitochondrial genetic markers in fungi other than yeast. The main difficulty, which applies to most fungi and to all higher eukaryotes, is that aerobic respiration and, therefore, mitochondrial function are essential for life, and so the scope for discovering clear modifications of mitochondrial function is very restricted. Thus, though there is good biochemical evidence that the mtDNA of, for example, human cells have much the same range of functions as in yeast, few genetic variants clearly attributable to mtDNA are available.

In various species of mycelial fungi, progressive loss of viability (**senescence**) has been correlated with the proliferation of defective mitochondrial DNA molecules, analogous to those found in yeast petite strains.[26]

Human mtDNA

The relatively small size of mammalian mitochondrial DNA molecules makes it possible, with modern techniques (*see Boxes* 14.1 and 14.2), to determine their entire base sequences. This has now been accomplished for the 16·6 kilobases of human mitochondrial DNA.[9] Merely from inspection of the sequence it has been possible to identify a possible 13 genes encoding polypeptides (five of which have known functions) in addition to the sequences transcribed into the two species of rRNA and a complete set of tRNA molecules. Human mitochondrial DNA seems to serve the same function as that of yeast but is more compact. This is a remarkable example of a complete genetic map obtained without any recourse to conventional genetics in the sense of analysis of inherited variation (*see Fig.* 7.5).

Plant mtDNA

One intriguing case of probable mitochondrial heredity is that of 'cytoplasmically determined' male sterility in maize (*Zea mays*). This type of sterility, which is of commercial importance in connection with the production of hybrid corn, is inherited through the female parent only. The small number of viable pollen grains formed on a 'male-sterile' plant do not transmit the sterility to the next generation, though all the seeds formed on such a plant would do so. In one male-sterile line, the sterility is associated with a maternally transmitted susceptibility to the toxin produced by the fungal pathogen *Helminthosporium maydis*. Recent work suggests that the determinant of both these traits is carried in the mitochondria but is contained, not within the regular mtDNA, but rather in small DNA plasmids (cf. pp. 195–197) that inhabit the mitochondria in affected strains.[27]

The structure of the essential mtDNA of flowering plant is still under investigation. Unlike the situation in yeast or animals there seem to be mtDNA molecules of several different sizes, perhaps incomplete and overlapping versions of the complete mitochondrial genome.[19]

7.5 Chloroplast heredity

The Functions of Chloroplast DNA in Higher Plants

The DNA of chloroplasts (ctDNA) can be physically mapped by restriction fragment analysis in just the same way as mitochondrial DNA, and functions can be attributed to some segments of the molecule by reason of their hybridization with RNA transcripts. Thus both the large and the small rRNA molecules of chloroplast ribosomes (which are different from both cytoplasmic and mitochondrial ribosomes) can be shown to be homologous with, and therefore presumably transcribed from, defined ctDNA segments. In the ctDNA of maize, and other flowering plants which have been investigated, the sequences specifying the rRNA species are duplicated. One such sequence is oriented 'clockwise' in the ctDNA map and the other 'anticlockwise', with a space between them.[10] Sequences corresponding to tRNA species have also been defined.

Several protein-coding sequences have been identified in ctDNA, including one very interesting and important one. The most abundant protein of green leaves is the enzyme ribulose bisphosphate carboxylase, which is responsible for the primary carbon dioxide-fixing reaction in photosynthesis. Each molecule of the enzyme contains sixteen polypeptide chains, eight 'small', about 100 amino acid residues in length, and eight 'large', about 450 residues. The messenger RNA for the large polypeptide, but not that for the small one, can be recovered from chloroplasts and shown to hybridize with a segment of ctDNA (*Fig. 7.7*). Genetic evidence from several flowering plants shows that the small subunit is coded for by a chromosomal gene, but confirms that the large subunit is extranuclear and maternal in its determination.

Fig. 7.6

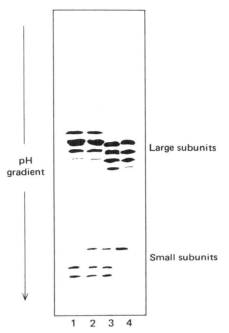

Analysis of the polypeptides of ribulose bisphosphate carboxylase purified from (1) *Nicotiana tabacum*, (2) *N. tabacum* (female) × *N. glutinosa* (male), (3) *N. glutinosa* (female) × *N. tabacum* (male), (4) *N. glutinosa*.
The enzyme samples were dissociated into their component polypeptides by 8 M urea and loaded on to an electrofocusing gel also containing 8 M urea. A gradient of pH was generated in the gel by application of an electrical potential, and the polypeptides migrated to the positions in the gel corresponding to their respective isoionic points (i.e. the pH values at which net surface charge is zero). In the hybrids, the large enzyme subunits are inherited exclusively from the female (seed) parent, while the small subunits are inherited from both parents. The multiple positions of the large subunit are probably due to secondary modification of the polypeptide chain (perhaps successive loss of amide groups) after synthesis; adjacent protein bands differ in one unit charge. The two kinds of small subunit in *N. tabacum* probably reflect its allotetraploid origin (i.e. two different genomes are present in this species—cf. p. 160).
Redrawn from Sakano et al.[11]

S. G. Wildman and his colleagues (Sakano et al.[11]) demonstrated the dual genetic determination of ribulose bisphosphate carboxylase in *Nicotiana* through analysis of the enzyme in the progenies of reciprocal crosses between different species of the genus. Both the large and the small enzyme subunits of the different species can be distinguished from one another by the technique of **isoelectric focusing,** which separates proteins according to their isoelectric points (i.e. the pH values at which their net charges are zero). Reciprocal crosses gave identical results so far as differences in the small subunit were concerned; *both* parental types always appeared in the hybrids. But where there was a difference between the species with respect to the large subunit, the type found in the hybrids was always that of the seed parent, with no sign of a contribution through the pollen. *Fig. 7.6* summarizes results of one such experiment.

The map of the maize chloroplast genome, so far as it is known at the time of writing, is shown in *Fig. 7.7*.

Fig. 7.7

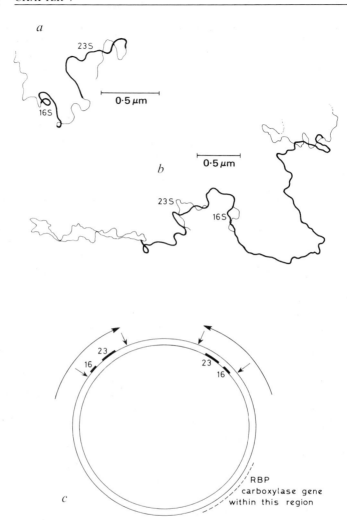

Chloroplast DNA of *Zea mays*.
a. Hybridization of a single strand from an EcoRI restriction fragment with
ribosomal RNA. The DNA was spread in formamide (which prevents
clumping of the single-stranded regions) and coated with cytochrome *c*
(*see* p. 13). The thicker regions are DNA–RNA duplex.
b. A longer section of single-stranded DNA hybridized with rRNA and also
folded back on itself with DNA–DNA duplex formed by inverted repeats,
characteristic of chloroplast DNA in green plants. Where the RNA is
hybridized the complementary DNA sequence is excluded as a single-stranded
loop (R-loop).
c. Map of the circular molecule, showing the positions of the inverted repeats
(arrowed arcs) and the duplicated rRNA genes.
The EcoRI cleavage sites used to generate the fragment shown in (*a*) are
indicated by radial arrows. The entire molecule is about 138 kb; the repeats are
23 kb in length. The region which hybridizes to mRNA for the large subunit of
ribulose bisphosphate (RBP) carboxylase is also indicated.
From Bedbrook et al.[12,13]

Maternally inherited chloroplast defects, showing as yellow or pale-green leaves or sectors of leaves, have been known in a wide variety of flowering plants for a long time. The field has been reviewed by R. Tilney-Bassett.[22] Although the likelihood is that these phenotypes are due to mutations in chloroplast DNA, this has not yet been demonstrated.

Chloroplast Genetics of *Chlamydomonas*

The unicellular motile green alga *Chlamydomonas reinhardi* provides most of our present information about the transmission and recombination of what are presumably ctDNA genes. There are two classes of *Chlamydomonas* mutations which can be plausibly attributed to the ctDNA.

First, as in the mitochondrial system of yeast, there are mutations which confer resistance to various antibiotics that inhibit protein synthesis in organelles but not in cytoplasm. For example, mutants can be selected that are specifically resistant to streptomycin; some of these are actually *dependent* on the antibiotic for growth. Other mutants resist other drugs, including erythromycin and spectinomycin.

The second class of maternally inherited mutations includes those which result in a growth requirement for acetate as carbon source. The basis of this phenotype is a failure to fix carbon dioxide by photosynthesis. A number of specific defects in the photosynthetic apparatus are now known; they can be positively selected because they confer resistance to arsenate which is toxic for photosynthesizing cells.

All of these mutations, both the drug resistances and the photosynthesis defects, show preferential transmission through the plus mating type. So far as these markers are concerned all products of meiosis normally resemble the mat^+ parent and the mat^- parent makes no contribution. The explanation for this uniparental mode of inheritance is still in some doubt. R. Sager, the main pioneer of *Chlamydomonas* chloroplast genetics, obtained evidence that the chloroplast DNA contributed to the zygote by the minus mating-type parent was selectively degraded. This appeared to provide the most direct indication that the uniparentally inherited traits were determined by ctDNA. This conclusion seems highly probable, even though there has been some debate about the evidence on DNA degradation in mat^+ (*see* discussion by Gillham[23]).

Whatever the basis for uniparental transmission, it certainly places an obstacle in the way of genetic analysis. Sager overcame this difficulty with her demonstration that exceptional zygotes, with chloroplast markers from *both* parents, do occur and that their frequency could be greatly increased by irradiating the plus mating-type parent with ultraviolet light prior to mating. By the use of this technique, segregation of chloroplast markers contributed by both parents can be followed in the products of meiosis and their immediate mitotic descendants. Detailed analysis of such segregation by Sager and Ramanis, and also by Gillham and his colleagues (*see* review by Gillham[23]), has shown that recombination occurs and that different pairs of markers showed different degrees of linkage.

Several methods of analysis have been used. Perhaps the simplest is that of **zygote analysis**. Zygotes formed in a cross between two strains differing in several chloroplast markers are induced to germinate (and undergo meiosis) on agar plates, and allowed to form colonies. A sample of cells is taken from the zygote colony (usually by replica plating, *see* p. 346) and tested for the presence of recombinant types (e.g. doubly resistant cells where the parents were each resistant to a single antibiotic). A colony, shown by this means to contain markers from both

Table 7.4   **Zygote clone analysis of recombination of extranuclear (presumed chloroplast) determinants in *Chlamydomonas reinhardi* (data from Harris et al.[14])**

Cross: streptomycin-resistant (S^r) mating-type$^+$ × erythromycin-resistant (E^r) neomycin-resistant (N^r) spectinomycin-resistant (Sp^r) mating-type$^-$

Zygotes formed in the cross were induced to undergo meiosis and germinate and were allowed to form colonies on plates without added drugs. Colonies (each from a single zygote) containing cells with drug-resistant markers from the *mat*$^-$ as well as from the *mat*$^+$ parent (7 per cent of zygotes in this cross) were identified by testing small samples of their cells on drug-containing media. Such colonies were picked off the plate, suspended in water, suitably diluted and plated on drug-free medium to allow single cells from the zygote colony to grow into secondary colonies. These were tested for growth on media containing the drugs, singly and in combination.
3746 colonies, stemming from 59 biparental zygotes, were tested.

Pattern of drug resistance*	Number of colonies	Percentage total	Interpretation in terms of exchanges in a linear sequence†
S^r + + + (*mat*$^+$ parent type)	2089	56	No exchange
+ E^r N^r Sp^r (*mat*$^-$ parent type)	1058	28	
S^r E^r + +	82	2·2	Single exchange E—N
+ + N^r Sp^r	108	2·9	
+ E^r + +	46	1·2	Double exchange E—N and N—S
S^r + N^r Sp^r	0	0	
+ + + +	360	9·6	Single exchange N—S
S^r E^r N^r Sp^r	2	0·005	
+ + + Sp^r	0	0	Double exchange N—Sp and ?
+ E^r N^r +	1	0·003	
Total	3746		

* + stands for the wild-type (drug-sensitive) trait in each case. Other possible combinations, not listed, did not occur in this sample.
† The most plausible linear sequence, allowing the commonest recombinant types to be single exchanges, is E—N—S. Sp is very close to N and could be on either side of it.
S^r N^r combinations appear to be poorly viable.

parents, can then be suspended in water and plated on fresh agar medium so as to allow each single cell to grow in its turn into a colony which can then be scored for the relevant markers. The numerous rounds of mitotic division which will have occurred since meiosis in the zygote are sufficient for chloroplast marker segregation to have been completed in most cells. A minority of cells may remain hybrid (i.e. capable of further segregation) with respect to one or more markers. The frequency of recombination can be expressed as—

$$\frac{\text{pure recombinants}}{\text{total pure clones}} \quad \text{or, alternatively, as} \quad \frac{\text{pure recombinants}}{\text{total clones}},$$

including hybrid clones in the total; the former measure is more in accord with the usual principles of recombinational analysis.

Table 7.4 shows the results of zygote colony analysis applied to a cross segregating with respect to four antibiotic-resistance markers. The whole procedure shows a marked resemblance to the diploid clone analysis employed in yeast mitochondrial genetics (*Table* 7.3). The fact that meiosis occurs in the zygotes of *Chlamydomonas*, but not of yeast (unless deliberately induced), is not a fundamental difference when one is dealing with organelle markers which are not affected by meiosis.

Another method, analogous to coconversion or cotransduction analysis (*see* pp. 132 and 295), depends on sampling from cell pedigrees stemming from germinated zygotes and noting the frequency with which a pair of markers undergo segregation simultaneously. Those markers which are close together in the chloroplast genome will **cosegregate** frequently, while those far apart will more easily become separated—one segregating while the other remains hybrid.

For all the effort, much of it successful, which has gone into the analysis of chloroplast determinants of *Chlamydomonas*, the progress made does not yet compare with that in the parallel field of yeast mitochondrial genetics. The reason for this is clear. In the yeast, but not in the *Chlamydomonas* system, there is available an array of viable deletion mutants by reference to which point mutants can be mapped and which can be correlated with physical changes in the DNA. There is undoubtedly a special extranuclear genome controlling a number of chloroplast properties, but the identification of this with chloroplast DNA is still not formally established, highly probable though it appears to be.

7.6   Other determinants of extranuclear heredity: plasmids and symbionts

We should not assume that the DNA molecules of mitochondria and chloroplasts are the only vehicles of extranuclear heredity. There are other organelles, present in certain kinds of cells, that contain DNA. One example is provided by the **kinetoplasts**, which occur in close association with mitochondria at the bases of the flagella in certain motile protozoa. These structures contain large amounts of highly repetitive DNA in the form of a multitude of small circles.[28] It is not known whether this DNA is transcribed and translated into protein. No inherited differences have yet been attributed to it.

Plasmids

Some cases of extranuclear inheritance may be due to DNA which is not part of the essential apparatus of the cell at all, but is rather of external infective origin. **Plasmids**, relatively small autonomously replicating DNA molecules that are sometimes present and sometimes not, are best known in prokaryotes, but they are also present in some eukaryotes. The plasmids present in the mitochondria of certain strains of maize were mentioned on p. 190. A much more thoroughly studied example is the **2-μm plasmid** which is present in most laboratory strains of *Saccharomyces cerevisiae*. This is a closed-loop molecule about 6 kilobases (2 μm) in length which is transmitted in crosses independently of

Fig. 7.8

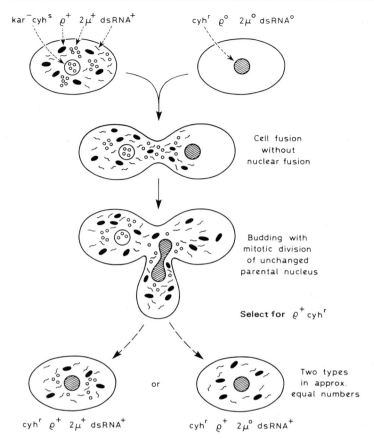

'Cytoduction' analysis of the inheritance of the *Saccharomyces* '2-µm' plasmid. The nuclei of the haploid parent strains were differentiated by the marker cycloheximide resistant/sensitive, and one of them carried the nuclear mutation *kar*⁻ which prevents karyogamy. The *cyh*ʳ strain lacked mitochondrial DNA (ρ°), the 2-µm plasmid (2µ°) and the double-stranded RNA which determines the 'killer' trait. Budded cells from the abortive zygotes were selected for cycloheximide resistance and respiratory competence, and their content of 2-µm plasmid DNA and dsRNA was determined. Described by Livingston.[17]

both the nuclei and the mitochondria. The demonstration of this is a good example of the use of cytoduction (*see* p. 176) as an analytical technique (*Fig.* 7.8).

The status of this plasmid as a genetic determinant in its own right is in some doubt; it is not clear that it determines anything but itself. It has not been associated with any clear component of the phenotype, unless the appearance of a DNA band in an electrophoresis gel can be considered as such. However it has been put to extensive use by molecular geneticists as a 'cloning vehicle' for DNA fragments of various origins. By manipulation *in vitro* (*see* pp. 271–274), the replication origin of the 2-µm plasmid has been fused with a part of the plasmid cloning vector pBR322 (*see* p. 274) to give a hybrid DNA circle which is capable of multiplying both in yeast and in *E. coli*.[20] This useful hybrid, called ScpI, has been used for the molecular cloning of the yeast mating-type loci (*see* p. 488).

Viruses and Virus-like RNA

Another kind of extranuclear, non-organellar, self-replicating nucleic acid in *Saccharomyces cerevisiae* is the linear (i.e. two-ended non-circular) double-stranded ribonucleic acid molecules (dsRNA) discovered by E. A. Bevan and his colleagues in London.[16] These *are* associated with a phenotype, conferring on the cells harbouring them the ability to kill yeast cells that do not contain this kind of RNA. They also make their host cells immune to the extracellular toxic protein that they themselves produce. The dsRNA molecules of yeast are clearly derived from an RNA-containing virus, since they are associated within the cell with virus-like protein structures which resemble closely the coats (capsids) of known infective RNA viruses. However, those kinds of 'killer' dsRNA which have been investigated so far do not make infective viruses, and their associated protein particles must be regarded as traces of an earlier existence as a cell-free infectious particle, superseded by the possibly less risky option of direct cell-to-cell transmission of the RNA by mating. It is easy to imagine that a self-replicating RNA species of this kind might lose all vestiges of its viral origin and become recruited as a normal, and possibly advantageous, part of the equipment of the cell.

An example from *Drosophila melanogaster* of a trait determined by a virus is that of **carbon dioxide sensitivity**. The lethal effect of CO_2 for certain strains, discovered through the use of the gas as an anaesthetic during the experimental manipulation of the flies, is a classical case of maternal inheritance. Its viral basis was indicated by experiments in which the sensitivity was shown to be transferable by injection of one fly with tissue extracts from another. The virus involved, called **sigma**, appears to have RNA as its genetic material.

Bacteria

In the ciliated protozoan, *Paramecium aurelia*, various **killer** strains, recognized by their ability to kill sensitive *Paramecium* strains in mixed culture, are associated with small bacteria, clearly visible under the light microscope in the cytoplasm of the killer cells. As must be the case with all successful killer determinants, these parasites (or symbionts as they are better called, since they do not harm their hosts) confer resistance to their own killer substance. The killer property was one of the first to be shown to exhibit extranuclear inheritance in *Paramecium*.

Dependence on the Nucleus

It is common to find that adventitious extranuclear self-replicating entities, whether they are plasmids, viruses or bacteria, are critically dependent on the nuclear genotype. Both in killer yeast and in killer *Paramecium*, the extranuclear determinant is lost whenever nuclear alleles, essential for maintenance, are removed by meiotic segregation.

7.7   Direct transmission of cellular characteristics

There are still many examples of extrachromosomal heredity lacking clear molecular explanations. As we see, there is no shortage of possibilities involving extrachromosomal nucleic acid of one kind or another. There is also the possibility that some cases of transmission of cell traits through cell division may not depend on nucleic acid at all. This is the subject of the final section of this chapter.

Wherever there is constancy of cell type over many generations there is the possibility that structures, or patterns of metabolic reactions, may be capable of self-perpetuation without having to be encoded in nucleic acid. There are a few examples of this kind of direct, as opposed to coded, transmission and the clearest ones come from *Paramecium aurelia*.

The *Paramecium* Cell Cortex

In *Paramecium* there is no cell differentiation and the same cell structure is largely preserved not only through mitosis but also through meiosis when the cell nuclei are reconstructed (*Fig.* 7.1). Especially notable for its constancy of pattern is the exterior, or cortex, of the cell, which consists of orderly rows of repeating units, each one with a pair of cilia and their associated basal bodies. The posterior basal body connects to a fibre which extends from it in an anterior direction and to the left. Each unit is asymmetrical and shows a difference between front and back and between left and right. All, in the normal cortex, are in the same orientation. T. M. Sonneborn[18] showed that it was possible to cut out a small section of the cortex and to replace it back to front. The remarkable finding was that all descendants of a cell which had undergone this operation reproduced the inverted arrangement in a segment of their cortex. If the original change was made in the middle of a row of units, the inverted pattern was copied into new units as they were formed in adjacent positions, until eventually it extended in one row along the whole length of the animal. It was never, however, extended into flanking rows.

The abnormal pattern obviously did not depend on any change of the micro- or macronuclei; in fact it survived sexual conjugation during which the micronucleus was reconstructed by karyogamy and meiosis and the macronucleus was entirely regenerated from the micronucleus. Two cells emerging from conjugation are always identical in nuclear genotype, yet in Sonneborn's experiments each exconjugant retained the cortical pattern of the parent from which its cortex had been derived. There is no question in this case of DNA being involved. What we see here is the laying down of new structure on the pattern of pre-existing structure.

This kind of direct non-coded inheritance of structure must be taken into account as a possibility whenever cell type is constant. In any organism which reproduces itself by means of specialized cells and has to regenerate all the structures of its other cell types afresh in each generation, there is obviously much greater dependence upon coded information.

Surface Antigens of *Paramecium*

The unique genetic system of *Paramecium* has allowed the demonstration of self-perpetuating states of the cell that are not detectable at the morphological level but can be distinguished by serological methods. G. H. Beale showed that *P. aurelia* cultured for long periods at different temperatures came to display different reactions to specific antibodies raised in rabbits by infection of *Paramecium* cells of different types. At 29 °C one major surface antigen, type G, was produced; anti-G antibodies caused agglutination of G cells, but not of cells of the same strain which had been cultured at 25 °C. These latter were of a different antigenic type, D, which was agglutinated by anti-D serum. At a still lower temperature, say 18 °C, yet another antigen, S, was produced.

The different antigens were mutually exclusive and each reacted only with antibodies formed in response to its own type. A switch in temperature, for example from 29 °C to 25 °C, brought about, after a number of cell fissions, a shift in antigenic type, in this case from G to D. At an intermediate temperature, say 23 °C, both antigenic types remain stable for long periods without switching and so, at this temperature, crosses between cells in the different states, G and D, could be made. When this experiment was done it was found that each exconjugant cell retained the antigenic type of the parental cell whose cytoplasm and cell cortex it had inherited, in spite of the fact that any nuclear genetic differences between the two cell types must have been equalized by the reciprocal fertilization following meiosis.

Thus the inheritance of antigenic type is formally similar to that of cortex pattern, the difference being that the origin of the different heritable states is more easily achieved (by a mere shift in temperature rather than by surgery) and more easily reversible. The basis for the self-perpetuation, within temperature limits, of distinct antigenic states is not understood, but there is a clear indication that patterns of macromolecular synthesis can sometimes resist change just as morphological pattern evidently can. The phenomenon parallels antigen switching in *Salmonella* (*see* p. 485), although there is no indication that the mechanisms are similar in the two cases.

7.8   Summary and perspectives

Although the bulk of the genetic information in eukaryotes is encoded in the chromosomes, certain essential items, directly connected with organelle function, are carried in the DNA of the mitochondria and, in green plants, the chloroplasts.

The genes of organelles can, to a certain extent, be mapped by the frequencies of recombination which they show. Such recombination occurs in cells in which different organelle genomes are present together and is not especially associated with meiosis.

Much more detailed mapping is possible through physical methods, making use of the relatively small size of organelle genomes and the consequent possibility of comprehensive restriction site mapping. The location of mutations on restriction site maps can be accomplished by deletion analysis and specific gene transcripts (mRNA, tRNA, rRNA) can be assigned to particular DNA segments by molecular hybridization. The determination of the complete DNA base sequence of an organelle DNA is now possible and has been accomplished in the case of the human mitochondrial genome.

Extrachromosomal heredity is not always attributable to organelle DNA; it may, in special cases, be due to DNA or RNA plasmids, viruses or even symbiotic bacteria. There is also the possibility, exemplified by *Paramecium,* of direct transmission of variations in morphological pattern or antigenic type without these variations being encoded in nucleic acid at all.

Chapter 7 | # Selected Further Reading

Bacila M., Horecker B. L. and Stoppani A. O. M. (eds) (1978) *Biochemistry and Genetics of Yeasts,* 594 pp. New York, Academic Press.

Beale G. H. (1954) *The Genetics of* Paramecium aurelia, 177 pp. Cambridge, Cambridge University Press.

Beale G. H. and Knowles J. (1978) *Extranuclear Genetics,* 142 pp. London, Edward Arnold.

Birky C. W. Jr (1978) Transmission genetics of mitochondria and chloroplasts. *Annu. Rev. Genet.* **12**, 471–512.

Gillham N. W. (1978) *Organelle Heredity,* 602 pp. New York, Raven Press.

Margulis L. (1981) *Symbiosis in Cell Evolution,* 420 pp. San Francisco, Freeman.

Tilney-Bassett R. A. E. (1978) In: Tilney-Bassett R. A. E., ed., *Inheritance and Genetic Behaviour of Plastids,* 2nd ed., Part 2, pp. 251–521. Amsterdam, Elsevier.

Problems

7.1 In the flowering plants *Mirabilis jalapa* and *Pelargonium zonale*, variegated green-and-white ornamental varieties are known as well as completely green varieties. In *Mirabilis* the green × variegated cross (seed parent written first) gives entirely green progeny while variegated × green gives predominantly variegated progeny. In *Pelargonium* the progeny tend to be variegated whichever way round such a cross is made. Metzlaff et al. (*Theoret. Appl. Genet.* 1981, **60**, 37) found a restriction-site difference between the chloroplast DNA of a predominantly white *Pelargonium* variety and that of a green variety. The cross between the two varieties yielded variegated plants. Chloroplast DNA was isolated from green and white parts of one such plant; that from the green tissue showed the restriction-site pattern characteristic of the green parent while that from the white tissue followed the pattern of the white parent. What can you conclude about the inheritance of green–white variation in *Pelargonium*? How may *Mirabilis* differ?

7.2 *Saccharomyces cerevisiae* strains can be classified as *killer* (*K*), *neutral* (*N*) or *sensitive* (*S*) depending on whether they produce and are resistant to a toxin, are resistant to the toxin but do not produce it, or neither produce nor are resistant to it. *K* and *N* strains contain double-stranded RNA while *S* strains do not. Somers and Bevan (*Genet. Res.* 1969, **13**, 71) mated a neutral strain *N*1 to two different sensitive strains *S*1 and *S*4. Tetrads formed by the *S*4 × *N*1 diploid showed no segregation; all the ascospores gave neutral colonies. Tetrads from the *S*1 × *N*1 diploid showed segregation, with two spores from each tetrad giving stably neutral colonies and the other two giving colonies which, after a period of growth, behaved like typical sensitives. However, when cells were isolated from very young colonies of the latter type, after five cycles of budding or less from the ascospore, and mated to cells of strain *S*4, some of the resulting diploid cells gave *neutral* colonies which, on sporulation, showed 2:2 segregation of *sensitive* and *neutral*. Is this nuclear or extranuclear inheritance, or a combination of both? Explain.

7.3 Jinks (*J. Gen. Microbiol.* 1959, **21**, 397) studied an ageing phenomenon in the fungus *Aspergillus glaucus*. With successive transfers, senescent cultures produced increasing proportions of inviable conidia until, eventually, all the cells were inviable. When growing hyphae of two strains, one normally vigorous and one senescent, were made to come into contact so that anastomosis of hyphal tips occurred, the originally vigorous hyphae grew to give senescent mycelium even when they were cut away from the originally senescent hyphae. *A priori*, this apparently infectious senescence could have been due to migration of either nuclei or of extranuclear elements. Given that nuclear genetic markers (e.g. *buff* rather than green conidia) exist in *A. glaucus*, explain how you could distinguish between these two alternative explanations.

7.4 Certain strains of *Drosophila melanogaster* are abnormally sensitive to carbon dioxide. F_1 flies from a cross of sensitive female × normal male were all sensitive, and F_1 females backcrossed to normal males again gave entirely sensitive progeny. However, F_1 flies from the cross normal female × sensitive male were mostly normal with only a few sensitives, and this normality was transmitted, again with few exceptions, to the progeny of a backcross to sensitive male. Normal females, after injection with extracts of sensitive flies, transmitted sensitivity to some of their progeny. Such transmission did not occur when normal males were injected and mated to normal females. What do you think might be the cause of the CO_2 sensitivity? What other kinds of evidence might be sought?

7.5 Lewin et al. (*Mol. Gen. Genet.*, 1978, **163**, 257) characterized the HpaI fragments of *Saccharomyces cerevisiae* both in wild type and in a series of petite mutants which had been isolated from a strain resistant to chloramphenicol and paromomycin (*C* and *P*). The petite mutants were tested by complementation to show whether they retained the mitochondrial genes responsible for these resistances. The table summarizes the results from several of the strains.

Petite no.	Genes retained P	C	HpaI restriction fragments (kb) 23·7	20·6	14·7	7·1	5·6	3·2	2·3	Other
1	−	+	−	−	−	−	−	+	−	5·5
2	+	+	−	+	−	−	−	+	−	10·8
3	+	+	−	+	+	−	−	+	+	23
4	−	+	−	−	−	+	+	+	−	17
5	−	+	−	−	−	−	+	+	−	19
6	+	−	−	−	+	−	−	−	+	17·3
7	+	−	+	+	+	−	−	−	+	10
Parent strain	+	+	+	+	+	+	+	+	+	None

Assuming that each petite mutation is the deletion of a continuous segment of a circular mitochondrial genome, followed by recircularization, deduce the wild-type restriction site map and, so far as you can, the limits of the deleted and retained portions in each petite mutant. (You may assume that the extra fragments in the mutants are the result of rejoining of broken ends.) What can you conclude about the locations of the *C* and *P* resistance determinants?

7.6 In yeast, *Saccharomyces cerevisiae,* oligomycin resistance (*OLI*) and erythromycin resistance (*ERY*) are dominant mitochondrial markers, and leucine-requirement (*leu⁻*) and canavanine resistance (*canʳ*) are recessive nuclear markers. The nuclear mutation *kar* prevents karyogamy (nuclear fusion) of those haploid nuclei into which it is segregated. The symbols cir⁺ and cir⁰ have been used to designate strains with and without 2-µm plasmid circles, and cir⁺-l and cir⁺-s to designate strains with larger and smaller forms of the plasmid. Two crosses were analysed with a view to characterizing the mode of transmission of the plasmid:

a. MATα leu⁻ kar OLI cir⁺ × *MATa leu⁺ ERY* cir⁰
Mated cells were plated on medium containing both oligomycin and erythromycin and *OLI ERY* colonies were picked; a sample of those that proved to be haploid (non-sporulating) were tested for the *leu* marker

and for presence or absence of the plasmid, with the following results:

MATα leu⁻		MATa leu⁺	
cir⁺	cir⁰	cir⁺	cir⁰
8	0	3	3

b. *MATα leu⁻ kar canˢ OLI* cir⁺-l × *MATa leu⁺ canʳ ERY* cir⁺-s
Again *OLI ERY* colonies were selected; *leu⁺ canʳ* haploids were tested for plasmid type (no *leu⁻ canˢ* haploids were recovered):

MATa leu+ canʳ	
cir⁺-l	cir⁺-s
0	7

What do these results suggest concerning the mode of transmission of the plasmid? (Data of Kielland-Brandt et al., *Carlsberg Res Commun.* 1980 **45**, 119–124.)

The Analysis of Continuous Variation

8.1 Continuous versus discontinuous variation

Our account of genetics so far has been based on mutations with clear-cut effects. These are almost all of laboratory origin and most have been induced by mutagens (*see* Chapter 12). Only through the use of mutations with clear individual effects is it possible to study the structures and functions of individual genes. However, most inherited variation that exists in non-laboratory populations, whether these are wild or domesticated, is of a less clearly defined kind, taking the form of a **continuous** distribution of values over a range, without any sharp breaks to allow the demarcation of distinct phenotypic classes. This ubiquitous continuous variation is in one sense rather uninteresting, since it does not provide good material for the study of individual genetic loci, which is the main preoccupation of most geneticists.

From other points of view, however, continuous variation is of great importance. It is the main resource used in the selective breeding of plants and animals, and it may also (though here our knowledge is very uncertain) be the basis upon which, in past ages, major evolutionary developments have been built by natural selection. We must, therefore, seek ways of analysing continuous variation, even though the information to be gained is bound to be incomplete. In order to understand the methods which can be used it is first necessary to be familiar with some elementary statistics.

8.2 Some statistical measures

Normal Distribution, Standard Deviation and Variance

If a population showing variation with respect to any measurable characteristic is divided into classes representing equal small steps covering the range of values, and the number in each class is represented in the form of a histogram, the outline of the histogram can usually be approximated by a symmetrical bell-shaped curve (*Fig.* 8.1). Such a curve can be fitted to a mathematical form known as the **normal distribution**.

In order to see how a normal distribution might arise, imagine that each member of the population is targeted on the population mean (the value represented by the centre and summit of the symmetrical curve) and that it is being pushed off target by a number of random and independent disturbances, acting positively or negatively with equal likelihood. Most individuals will experience a mixture of positive and negative influences, probably not quite balancing but nevertheless not taking them far from the mean. Individuals on which the perturbing influences are nearly all positive or nearly all negative will occur, but with very much lower probability; these will be at the fringes of the distribution. Such a situation can be

Fig. 8.1

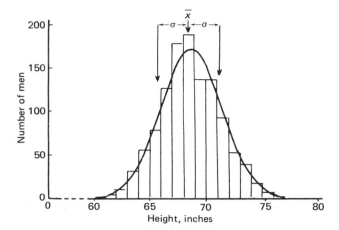

An example of continuous variation in a population approximating to the
normal distribution. The histogram represents the frequency distribution of
stature in a population sample of 1164 men, divided into one-inch classes. The
curve is the normal distribution, calculated from the formula given in the text
on the basis of the mean height ($\bar{X} = 68.6$) and taking the value 2·70 for the
standard deviation (σ). It will be seen that the calculated curve fits closely to
the outline of the histogram, and that σ is the distance from the mean to the
point of maximum slope on each side of the mean.
Redrawn from Darlington and Mather.[9]

shown to generate a normal bell-shaped curve. To take a concrete example, the
normal distribution is a two-dimensional representation of the distribution of
shots across the bull's-eye of a target, where the aim of the marksman is subject to
the cumulative effect of numerous independent small errors. Pursuing this analogy,
we can make it easier for the marksman by defining the bull's-eye as the centre of
his cluster of shots.

The general formula for the normal curve is

$$f = \frac{1}{\sigma\sqrt{2\pi}}e^{-(x-\bar{x})^2/2\sigma^2}$$

where x is the value of the continuous variable, f is the frequency of the class falling
within a certain narrow range of mean value x, $\bar{x}$ is the mean of the distribution,
and e and π are the standard mathematical constants. The quantity σ (sigma) is a
characteristic of the particular distribution, and it is, in fact, the distance from the
mean to the two points on either side of the mean where the curve has its maximum
slope. Sigma is the measure of the spread of values about the mean, and is called
the standard deviation of the distribution.

The proportion of the population falling within a particular distance from the
mean is a function of the ratio of that distance to the standard deviation. A most
useful rule of thumb is that approximately 95 per cent of a normally distributed
population falls within two standard deviations from its mean. Plus and minus two
standard deviations from the mean are sometimes called the 95 per cent **confidence
limits** of the distribution, within which one can be 95 per cent sure that any
individual can be found.

The square of the standard deviation is called the **variance** of the distribution, and it can be calculated in the following way. If the n members of the population have values x_1, x_2, x_3, ..., x_n, so that the population mean is

$$\frac{x_1 + x_2 + x_3 + \dots + x_n}{n} = \bar{x}$$

then the individual deviations from the mean are $x_1 - \bar{x}$, $x_2 - \bar{x}$ etc. and the variance is the sum of the squares of all these deviations divided by $n-1$.

$$V_x = \frac{\Sigma(x - \bar{x})^2}{n-1} \quad \text{and} \quad \sigma_x = \sqrt{V_x}$$

Variance has the useful property of being **additive**. That is to say, if fluctuations from the mean are due to two or more independently acting disturbances, the total resulting variance will be the sum of the variances that these would produce individually. Conversely, starting with the overall variance, it may be possible to **partition** it into additive components attributable to different perturbing factors, provided that the latter act independently.

A feature of the variance formula which needs some explanation is the use of $n-1$ as the denominator rather than n, which would give a simple average mean square deviation. What variance assesses is the amount of deviation from the mean in proportion to *opportunities* for deviation. A single observation cannot deviate from the mean; with a sample of only one, the value of that one must *be* the mean. With a sample of n there are $n-1$, not n, opportunities for deviation from the mean, and it is thus in relation to $n-1$ that the sum of squares must be assessed. It is said that the data have $n-1$ **degrees of freedom**.

Covariance

Covariance is a useful derivative of the variance concept. It assesses the extent to which the variation in one set of values can be related to (correlated with, perhaps attributed to) variation in the other. Consider the example of the correlation between fathers and sons with respect to some measurable trait such as height. Suppose that the heights of a population of adult males are x_1, x_2, ..., x_n, and that their fathers were of heights x_{f1}, x_{f2}, ..., x_{fn}. The variance (V_s) of the population of sons is

$$\frac{\Sigma(x - \bar{x})^2}{n-1}$$

and that of the fathers (V_f) is

$$\frac{\Sigma(x_f - \bar{x}_f)^2}{n-1}$$

The covariance of the two populations is given by the formula:

$$W_{fs} = \frac{\Sigma(x - \bar{x})(x_f - \bar{x}_f)}{n-1}$$

The analogy with the formula for variance is easily seen; instead of *squared* deviations the covariance formula has *products* of deviations—the deviation of each son multiplied by that of his father.

If the two populations were entirely uncorrelated, fathers and sons would vary from the means of their respective generations in opposite directions as frequently as in the same direction—in other words, the product of their deviations would be as often negative as positive, and the terms comprising the covariance would add up to approximately zero. If they were completely correlated, on the other hand, their covariance would be equal to their individual variances, if the latter were identical, or, more generally, to $\sqrt{V_f V_s}$, the square root of their product. The **correlation coefficient** is defined as the ratio $W_{fs}/\sqrt{V_f V_s}$. It can take any value from zero, for no correlation, to plus or minus one for complete positive or negative correlation.

Testing the Significance of Differences between Means

When the difference between two populations is a quantitative rather than a qualitative one (i.e. one of degree rather than of kind), one needs to be convinced that the difference is significant before looking for the reason for it. Neither population will be absolutely uniform; each will be characterized by a mean and variance. The estimate of each mean $\bar{x}$ will itself have a variance $V_{\bar{x}}$ and standard deviation (called the **standard error**, s.e.) depending on the size of the sample on which it is calculated. Specifically:

$$V_{\bar{x}} = \frac{V_x}{n} \quad \text{(where } n \text{ is the sample size)}$$

and

$$\text{s.e. (mean)} = \sqrt{\frac{V_x}{n}}$$

The variance of the *difference* between two means $(\bar{x} - \bar{y})$ is the sum of their individual variances, and the standard error of the difference is therefore given by the formula:

$$\text{s.e. (difference of means)} = \sqrt{\frac{V_x}{n_x} + \frac{V_y}{n_y}}$$

One assesses the significance of the difference between two means $\bar{x}$ and $\bar{y}$ by calculating the difference as a multiple of its standard error, i.e.

$$(\bar{x} - \bar{y}) \bigg/ \sqrt{\frac{V_x}{n_x} + \frac{V_y}{n_y}}$$

a ratio known to statisticians as *t*. Here again we can apply the useful rule of thumb that, if the difference is more than twice its standard error, the probability that such a large difference would have been obtained through an accident of sampling from populations which really had identical means is less than one in twenty. Lower values of the ratio imply higher probabilities of the observed difference being due merely to sampling error. The precise probability represented by a particular value of *t* can be looked up in standard statistical tables, and one can be correspondingly confident or cautious about the significance which one gives to the observed difference.

8.3 | Components of variance

Analysis Based on Inbred Strains

Continuous variation was a considerable puzzle to early geneticists until it was realized that the results of crosses could well be explained by the assumption that the variation observed was due to the segregation of a large number of gene differences with individually small but cumulative effects, together with non-genetic variation attributable to the environment.

Fig. 8.2 shows an early example of the typical result obtained by measuring F_1, F_2 and backcross progenies from an original cross between two inbred strains differing with respect to a quantitative characteristic. In this case the experimental organism was a plant, but broadly similar results are obtained with animals, such as mice, in which highly inbred strains are available. The parental strains, following many generations of propagation through self-fertilized single plants, must have been homozygous at virtually all loci, and may be presumed to have had very little genetically caused variance. They were, indeed, relatively uniform and the variation which they did show can be attributed to environmental fluctuations. The F_1 generation from the cross between them was similarly uniform, distributed narrowly about a mean, intermediate between the parental means. The F_2, on the

Fig. 8.2

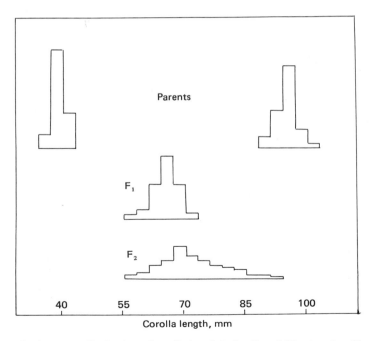

Parents

F_1

F_2

40   55   70   85   100

Corolla length, mm

The frequency distribution of corolla length in families of *Nicotiana longiflora*. The two parental strains (P) were highly inbred and effectively homozygous for virtually all genes.
Data of East;[1] diagram redrawn from Darlington and Mather.[9]

other hand, while showing almost the same mean as the F_1, had a greatly increased variance. This is just what would be expected if the difference between the parental lines were due to the cumulative effect of a number of gene differences each having a small effect. The F_1 plants would then be uniformly heterozygous with respect to every locus at which the parental strains differed, while in the F_2 there would be a number of largely independent $1:2:1$ segregations with superimposed effects.

Fig. 8.3 shows the result that would be expected if the two inbred strains differed with respect to only three genes, assumed arbitrarily to contribute equally and additively to a quantitative difference, with the heterozygotes exactly intermediate between the homozygotes in their effects (i.e. no dominance). If there were absolutely no environmentally related variance, the result of this simple situation would be the stepped distributions shown, but the broad outlines of the distributions mimic quite closely the type of experimental result exemplified by *Fig.* 8.2, and one can see that only a little rounding of the corners would be required to give approximately normal distributions. Increasing the number of gene differences would make the detection of steps in the distributions even more difficult. Very much the same kind of result would follow even without the simple arbitrary assumptions of equal contributions of all gene differences and absence of dominance, provided that the number of gene differences was fairly large, the contribution of each small and the dominance not all in one direction.

The working model, then, for the inheritance of continuous variation invokes the existence of numerous gene differences with *small, similar* and *supplementary* effects. In as much as reciprocal crosses usually give the same F_1 mean (within sampling error), the contribution of extrachromosomal differences, though doubtless present, is usually discounted. Mather has called the genes involved **polygenes**; we return below (p. 224) to the question of what, in material terms, polygenes may be.

A general type of analysis which allowed certain conclusions to be drawn concerning polygenic inheritance, even though the individual genes could not be identified, was developed by K. Mather with a number of collaborators, notably J. L. Jinks.[10] We cannot go very far into the analysis in this book, but a description of its first few steps will at least serve to indicate its general nature.

Imagine first a single-gene difference (G/g) between inbred strains contributing a quantity $2a$ to the difference between the two strains. We can move our measuring scale arbitrarily so that the mean of the contributions of the two homozygotes is zero; GG will then contribute $+a$ and gg $-a$. Allowing for the possibility of dominance, let the contribution of Gg be $+d$, where d can take any value between zero, for no dominance, and $+a$, for complete dominance of G.

Providing that the parental strains are homozygous at all loci they will have no genetically based variance, and neither will the F_1 generation, for the latter will be uniformly heterozygous at all loci with respect to which the parents differ. All three populations will exhibit some variance due to the impossibility of eliminating all environmental variation, and this we may symbolize by E. If environmental fluctuation affects all three genotypes equally, which is implausible as a general proposition but which may not be very far wrong as an approximation, E will be the same for all three.

The F_2 generation, in contrast, will show both genetically and environmentally based variation. We can calculate the contribution of the single-gene difference G/g to the genetic part of the variance. The calculation is summarized in *Box* 8.1, and it includes two terms, one, $a^2/2$, due to the difference between the homozygotes, and the other, $d^2/4$, due to the dominance exhibited in the heterozygote.

Fig. 8.3

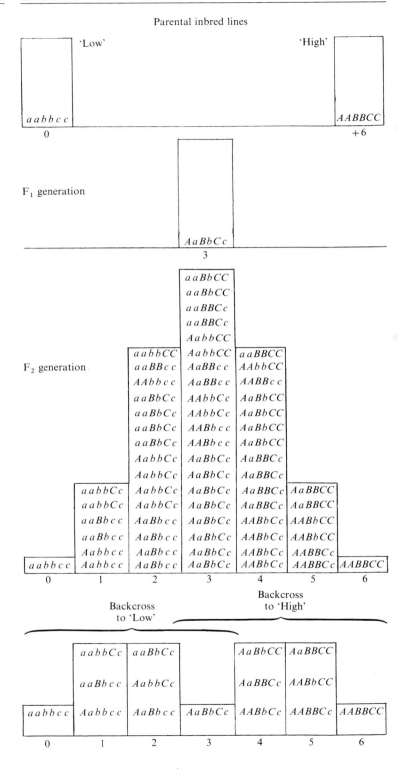

Fig. 8.3   A simple model of polygenic inheritance based on only three gene differences, A/a, B/b, C/c, with the capital letter alleles each supposed to contribute $+1$ to the character being measured, and the lower-case alleles zero. The three effective alleles are supposed to act additively without dominance or interaction (epistasis).

Box 8.1	Partitioning of the variance due to a single-gene difference: F_2 and backcross progenies

Phenotypic values: $GG + a$

$Gg \; +d$ (Departure from mid-parent value due to dominance)

$gg \; -a$

Initial cross: $GG \times gg$; F_1 all Gg

F_2 (from self-fertilization or sib-mating in F_1) will be composed as follows:

$\frac{1}{4}GG$, phenotype $+a$
$\frac{1}{2}Gg$, phenotype $+d$ F_2 mean $= \frac{1}{4}a + \frac{1}{2}d - \frac{1}{4}a = \frac{1}{2}d$
$\frac{1}{4}gg$, phenotype $-a$

Deviations from F_2 (mean): $a - \frac{1}{2}d$ for GG individuals
$d - \frac{1}{2}d = \frac{1}{2}d$ for Gg individuals
$-a - \frac{1}{2}d$ for gg individuals

Sum of squares of deviations from F_2 (mean):

$\frac{1}{4}(a - \frac{1}{2}d)^2 + \frac{1}{2}(\frac{1}{2}d)^2 + \frac{1}{4}(a + \frac{1}{2}d)^2$

$= \frac{1}{4}a^2 - \frac{1}{4}ad + \frac{1}{16}d^2 + \frac{1}{8}d^2 + \frac{1}{4}a^2 + \frac{1}{4}ad + \frac{1}{16}d^2$

$= \frac{1}{2}a^2 + \frac{1}{4}d^2$

If A and D are the variance components corresponding to a^2 and d^2

$$V_{F2} = \tfrac{1}{2}A + \tfrac{1}{4}D \qquad (1)$$

Backcross to GG parent (B_1) will give progeny as follows:

$\frac{1}{2}GG$, phenotype $+a$
$\frac{1}{2}Gg$, phenotype $+d$ Mean $\frac{1}{2}a + \frac{1}{2}d$ Deviations from mean: $\frac{1}{2}a - \frac{1}{2}d$
$\frac{1}{2}d - \frac{1}{2}a$

Sum of squares of deviations from mean:

$\frac{1}{2}(\frac{1}{2}a - \frac{1}{2}d)^2 + \frac{1}{2}(\frac{1}{2}d - \frac{1}{2}a)^2 = (\frac{1}{2}a - \frac{1}{2}d)^2$

$= \frac{1}{4}a^2 - \frac{1}{2}ad + \frac{1}{4}d^2$

Giving the same meanings to A and D as in F_2,

$$V_{B1} = \tfrac{1}{4}A + \tfrac{1}{4}D - \tfrac{1}{2}(ad)$$

For the backcross (B_2) to the gg parent the progeny will be:

$\frac{1}{2}Gg$, Phenotype $+d$
$\frac{1}{2}gg$, phenotype $-a$ Mean $\frac{1}{2}d - \frac{1}{2}a$ Deviations from mean: $\frac{1}{2}d + \frac{1}{2}a$
$-\frac{1}{2}a - \frac{1}{2}d$

Sum of squares of deviations $= \frac{1}{2}(\frac{1}{2}d + \frac{1}{2}a)^2 + \frac{1}{2}(\frac{1}{2}d + \frac{1}{2}a)^2 = \frac{1}{4}a^2 + \frac{1}{2}ad + \frac{1}{4}d^2$

$$V_{B2} = \tfrac{1}{4}A + \tfrac{1}{4}D + \tfrac{1}{2}(ad)$$

$$V_{B1} + V_{B2} = \tfrac{1}{2}A + \tfrac{1}{2}D \qquad (2)$$

Using Eqns (1) and (2) as simultaneous equations, both A and D can be evaluated.

There will, of course, also be contributions from other gene differences. If, in conformity with the simplest polygene model, all their effects are additive (i.e. no epistasis), then the total genetic variance will be the simple summation of all the $a^2/2$ components and all the $d^2/4$ components, giving sums which we will represent by A/2 and D/4. These are the **additive** and **dominance** components of the total F_2 genetic variance. The total F_2 phenotypic variance (V_{F2}) is given by the formula:

$$V_{F2} = \frac{A}{2} + \frac{D}{4} + E$$

The corresponding formulae for the two backcross progenies can also be calculated (*Box* 8.1), and adding them together gives the convenient formula:

$$V_{B1} + V_{B2} = \frac{A}{2} + \frac{D}{2} + 2E$$

Doubling the F_2 variance and subtracting it from the sum of the backcross variances gives a direct estimate of the additive genetic variance:

$$V_{B1} + V_{B2} - 2V_{F2} = \frac{A}{2}$$

Substituting the value thus obtained into either of the first two formulae, and relying on the parental and F_1 variances for an estimate of E, a value can be obtained for D.

A point of interest is whether D has any significant magnitude—in other words, does dominance exist for polygenes? Without describing how one can assess the error in the calculation of D and the other variance components, which would take more space than would be reasonable in a general book, the reader must be content with the statement that dominance, though often significant in its effect, usually contributes much less than the additive component to the overall genetic variance.

The calculations which have just been outlined use up all the available information in the calculation of A and D, leaving none for testing the assumptions on which the analysis is based—namely that the effects of epistasis and linkage are negligible. In order to show whether these assumptions are approximately valid it is necessary to analyse further generations. For example, if one is dealing with self-fertile plants, one can take a sample of individuals from the F_2 generation and self-pollinate them to produce a series of F_3 progenies. Further algebra shows that the covariance of F_2 individuals and F_3 progeny ($W_{F2, F3}$) means is given by the formula:

$$W_{F2, F3} = \frac{A}{2} + \frac{D}{8}$$

Estimates of A and D derived from the F_2 and backcross families can be fed into this formula, and the covariance so predicted can be compared with that observed. Agreement supports the simple assumptions; disagreement can be attributed to epistasis or linkage or a combination of the two.

It comes as no surprise that quantitative analysis often confirms that the well-known and ubiquitous phenomena of linkage and epistasis must exist. What is perhaps more notable is how close the simple model, in which they are neglected, comes to fitting most bodies of data. This apparent paradox can be explained in a rough way by saying that, so far as continuous variation is concerned, these components of variance are usually small compared with the additive component and that epistasis consists of a large number of gene interactions which individually have small effects, act in different directions on the phenotype and tend to cancel each other out.

8.4 Outbred populations and heritability

Estimates of Heritability

We have seen how, through measurements on inbred lines and their F_1's, F_2's and backcross progenies, one can distinguish between genetic and environmental variance and, within genetic variance, between additive and dominance components. This three-way partition of phenotypic variance (V_P) can be applied, given certain assumptions, to outbred as well as inbred populations. In general:

$$V_P = V_A \text{ (additive)} + V_D \text{(dominance)} + V_E \text{ (environment)}$$

In its precise sense, **heritability** is that fraction of the total phenotypic variance accounted for by the additive genetic component; for reasons which need not worry us here it is symbolized by h^2:

$$h^2 = \frac{V_A}{V_P}$$

It is, in fact, that fraction of the total variance which can be fixed genetically by selection—the contribution of dominance to the phenotype disappears as selection makes more and more genes homozygous.

The term heritability is also sometimes used in a broad sense, meaning the fraction of the total variance which is accounted for by genetic variation, *including* the dominance component:

$$\text{Heritability (broad sense)} = \frac{V_A + V_D}{V_P}$$

Heritabilities in populations can be estimated on the basis of correlations between relatives without the availability of inbred lines and even without controlled crosses, so long as the relationships between individuals are known. The calculations depend on ignoring the possible effects of linkage and epistasis, the justification for which was discussed briefly above. It also depends on certain questionable assumptions about the effects of environment. The first assumption is that every kind of environmental fluctuation will produce the same variance in one genotype as in another—in other words that there is no **genotype–environment interaction**. This cannot be strictly true, but animal breeders do not seem to be led too far astray by supposing that it is, at least where environmental differences are not extreme. The second assumption about environment is that the environments of relatives are no more similar on average than those of non-relatives. When one is dealing with domesticated or experimental plants or animals, this can be made to be true or nearly so (with the important exception of sibs in animals where maternal effects are often significant); but it is obviously false in human populations. We shall return to the vexed question of the effects of environmental correlation on human resemblances on p. 220 and in Chapter 19.

Related individuals will necessarily have a certain proportion of their genes in common because of their relatedness. For example, parents must share one-half of their genes with each of their offspring. Indeed, because many alleles are the common property of much or even all of the population, parents have even more than 50 per cent genetic identity with their offspring, but this additional genetic similarity will be no more than they have, on average, with essentially unrelated individuals. What we are concerned with is the *extra* similarity due to relatedness, and this can be expressed as the covariance of parents with offspring:

$$W_{(\text{parent–offspring})} = \tfrac{1}{2} V_A$$

Note that the dominance component of the genetic variance does not contribute to the covariance since, so far as dominance has an effect, it will be as likely to reduce parent–offspring resemblance as to enhance it. Putting the point another way, similarity with regard to dominance relationships depends on similarity with respect to *both* alleles of a given gene, and only one is inherited from a single parent.

The situation is rather different in the case of **full sibs**, i.e. offspring of the same two parents. These, on average, have half of their genes in common by virtue of their common parentage; from each pair of alleles in each parent, two sibs are as likely to inherit the same one as different ones. Thus the additive contribution to the covariance is again $\tfrac{1}{2}V_A$. There is, however, also a dominance contribution in this case, because there is a one-in-four chance of a pair of sibs inheriting a particular allele from one parent along with a particular allele from the other. Thus:

$$W_{(\text{full-sibs})} = \tfrac{1}{2}V_A + \tfrac{1}{4}V_D$$

Because of the presence of a dominance component which cannot be independently measured, full-sib covariances are not as useful as parent–offspring covariances for estimating V_A, which is what one needs for heritability estimations. From this point of view, **half-sibs**, i.e. individuals with only one parent in common, provide more usable data. Half-sibs share one-quarter of their genes by virtue of

Table 8.1	Heritability of abdominal bristle number in *Drosophila melanogaster* estimated from different relationships (data from Clayton et al.[2])
Relationship	*Estimated heritability*
Parent–offspring	0.51 ± 0.07
Half-sibs	0.48 ± 0.11
Full-sibs	0.53 ± 0.07

Notes: The method of calculating the standard errors of the estimates is given in Falconer.[11] The difference between the estimates is not significant, taking into account the errors. As noted in the text, heritability based on full-sib covariance may be inflated because of the contribution of dominance to the covariance; here the effect of dominance, even though it may be present, is too small to be significant.

their common parent, but never more than one allele at a single locus. There is, therefore, no special similarity due to dominance and

$$W_{\text{(half-sibs)}} = \tfrac{1}{4}V_{\text{A}}$$

Another advantage of half-sib over full-sib data is that, when the common parent is the father, they are free of the environmental effects of common maternal parentage, which inflates the covariance to an extent which cannot be easily assessed and allowed for.

Table 8.1 gives some examples of heritabilities measured on the basis of different degrees of relationship. It will be seen that there is a very good agreement between the different estimates, even including the one based on full-sib data—another indication that the effect of dominance on genetic variance is sometimes small enough to ignore.

Heritability and Parent–Offspring Regression

The **regression** of offspring on parents is the slope of the line obtained when individual offspring values are plotted against the corresponding parental values. The regression, symbolized as b_{OP}, can also be calculated from the formula

$$b_{\text{OP}} = \frac{W_{\text{(parent–offspring)}}}{V_{\text{P}}}$$

where V_{P} is the phenotypic variance of the population. We saw above that $W_{\text{(parent–offspring)}}$ is equal to half the additive genetic component of the variance V_{A}, from which it follows that:

$$b_{\text{OP}} = \tfrac{1}{2}\frac{V_{\text{A}}}{V_{\text{P}}}$$

That is to say, the regression is an estimate of half the narrow-sense heritability.

A graphical plot of offspring against parents will often give a scatter of points with no obvious straight line relationship among them. One can always calculate a best-fit straight line (i.e. one which minimizes the squares of the deviations of the points from the line) and also the statistical error in the estimate of its slope. The greater the scatter, the greater the error. The numerical example given as an exercise at the end of this chapter is one in which something close to the best fit can be obtained by eye.

The Connection between Heritability and Response to Selection

The main relevance of h^2, or narrow-sense heritability, is that it bears a direct relation to the response which can be expected to artificial selection. Suppose we have a population showing a normal distribution with respect to some quantitative aspect of the phenotype and we select, as parents of the next generation, individuals differing by a quantity S (which may be either positive or negative) from the population mean M. S is then a measure of the selective pressure which one is applying to the population. If heritability were complete (i.e. if h^2 were equal to 1), the response to selection would be complete—the mean of the next generation would be $M + S$, the same as that of the selected parents. If heritability were zero there would be no response—the mean of the next generation would be M, the same as that of the starting population.

Suppose the mean of the progeny population is in fact $M + R$, with R taking a value somewhere between S and zero. Then the ratio R/S, which can be regarded as a sort of gearing ratio relating output in the form of response to selection to input in the form of selection pressure, is equal to the heritability h^2. A formal proof of this simple relationship will not be offered here, but the reader will probably feel able to accept it as reasonable or even intuitively obvious.

At first sight it would seem that heritability must be a very useful indicator of the response to be expected to selection in practical breeding situations. On further thought, however, it is not so obvious why it should be useful at all. The reason for this uncertainty is that heritability is not a fixed quantity for a given character, but depends on the amount of genetic variation left in the stock. In so far as selection increases the frequency of certain alleles at the expense of others it must lead to progressively higher frequencies of homozygosity at the relevant loci, and hence to a reduced genetic variance. Heritability ought constantly to decrease if selection in one direction is continued over several generations. Hence the only heritability value which one really knows is that of the starting population. This will indeed enable us to predict as far as the next generation, but not necessarily any further, and we may already have looked at the next generation in any case in order to estimate the heritability.

The situation is, however, not usually as bad as this argument might make it seem. In practice heritability may not decline appreciably for many generations, and the response to selection will remain approximately constant over this time (e.g. *Fig.* 8.4). The general explanation for this is that the pool of genetic variation which one starts with is usually too large to be substantially depleted by a single generation of selection. Whether this is the case depends, of course, on the intensity of selection, and this in turn is limited by the size of the population which one is selecting from and by the proportion of that population which one needs as parents of the next generation in order to keep up the numbers. Given a population of very large size and very high fecundity, one might hope to pick out the most extreme selectable phenotype in the first generation (though even so a further generation would be required to establish that the selected individuals did not owe their qualities to environmental accident). But, in practice, it is not usually possible to screen enough individuals to progress more than a short distance towards the ultimate goal of total homozygosity for all 'desirable' alleles. Furthermore there is the possibility that more genetic variation may actually be added during the selection programme by recombination of linked loci. Not all combinations will necessarily be present in the finite starting population and, if some of the relevant loci are closely linked, it may be many generations before rare crossovers generate some of the favoured recombinant chromosomes.

Even so, response to selection can obviously not continue for ever. Eventually, if homozygosity with respect to all genes affecting the trait being selected provides the best phenotype from the point of view of the selection programme, then such homozygosity will be attained. At that point genetic variance will have been eliminated, all variation will be due to environment, and heritability will be zero. Not only will further selection in the original direction be ineffective, it will also be impossible to go into reverse and select back towards the mean of the starting population.

In practice it has often turned out to be extraordinarily difficult to achieve this

Fig. 8.4   *a*

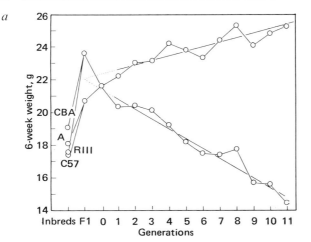

b

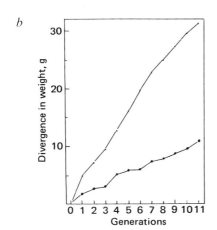

a. The results of selection for increased and decreased body weight in mice, starting with a mixed hybrid population. The starting population (generation 0) was derived from four inbred lines, CBA, RIII, A and C57, by intercrossing the F_1 generations from the two crosses CBA × RIII and C57 × A. Each generation consisted of six families, each consisting of two litters (a total of about 16 offspring) from a single pair of mice. The two 'best' mice from each family were chosen as parents of the next generation and matings were made between mice from different families. The points are generation means, and the best straight-line fits to these points are drawn.
From Falconer.[3]

b. The response to selection compared with the intensity of selection in the same experiment as shown in (*a*). The larger points are the differences between the means of the large and small selected lines in each generation, and the small points show the cumulative selection differentials. The selection differential in each generation is the increased difference between the means of the selected large and small mice as compared with the difference between the means of the populations from which they were selected. The ratio of the slopes of the two lines gives the heritability and it will be seen that it does not change much over the eleven generations—the overall ratio is 0·35. With further generations of selection the heritability gradually declined and eventually approached zero.
From Falconer.[3]

predicted state of genetic fixity. Back-selection is often effective long after any response to forward selection has ceased, even though it may not be possible to go all the way back to the starting mean. The probable reason for this persistent genetic variance is that certain alleles cannot be fixed because they are lethal or at least very poorly viable in homozygotes, especially, perhaps, when they are in company with many homozygous alleles at other loci, all pulling the phenotype in the same direction. In such circumstances artificial selection may become ineffective because of natural selection operating against it rather than because genetic variation has been exhausted.

Heritability in Human Populations

As we noted above, the calculation of heritability values from correlations between relatives in human populations is beset with the complication of environmental correlation. When families have different environments, a part of the covariance of relatives may be due to environmental circumstances shared by members of the same family but not by different families. For example, for the full-sib relationships we should write:

$$W_{fs} = \tfrac{1}{2}V_A + \tfrac{1}{4}V_D + W_{E(fs)}$$

where $W_{E(fs)}$ is the component of full-sib covariance due to environmental correlation.

Each kind of relationship will have its own W_E component. The magnitude of this component of covariance is expected to be greatest for twins reared together, somewhat less for non-twin full-sibs (since, although they may be brought up in the same home, they will not coincide in time), very considerably less again for the parent–offspring relationship (since the circumstances of successive generations, even within the same family, may be very different) and perhaps least of all for half-sibs reared apart. So there is no general correction which one can apply to eliminate the effect of environmental correlation. The best one can do is to compare relationships for which W_E may be the same but the genetically based covariance is different.

The most plausible attempts at estimating heritabilities in man have been based on twin studies. For monozygotic (mz) and dizygotic (dz) twins, respectively, the following formulae should hold:

$$W_{mz} = V_A + V_D + W_{E(mz)} \quad \text{and} \quad W_{dz} = \tfrac{1}{2}V_A + \tfrac{1}{4}V_D + W_{E(dz)}$$

Provided we assume that $W_{E(mz)} = W_{E(dz)}$, then

$$2(W_{mz} - W_{dz}) = V_A + \tfrac{3}{2}V_D$$

This expression, divided by V_P, the total phenotypic variance, is equal to the narrow-sense heritability, if V_D is negligible, and not very different from the broad-sense heritability. In either case it gives an estimate of the fraction of the total variance which is due to genetic causes.

The assumption that dizygotic twins' environments are matched as closely as those of monozygotic twins must be erroneous in some degree, since each twin is an important part of the environment of the other. The extent of the error can only be guessed at; an obvious guess is that it is negligible for physical characteristics but sometimes important where psychological differences are concerned.

It will be seen that heritability estimates in man may be overestimated if environmental correlation is neglected; yet we lack the data which would enable us to take it into account in a quantitative way. The extent of the error, however, would be expected to vary depending on the relationship being used to calculate heritability. In at least some instances, heritability estimates obtained from different relationships have been in fair agreement, encouraging the conclusion that, in these cases at least, the effects of environmental correlation are small compared with those of the genetic correlation.

8.5 Location and number of genes involved

Chromosome Assay

While the biometrical analysis of the combined effects of gene differences that cannot be individually distinguished provides a fascinating challenge for those of mathematical inclination, and can provide reassurance that the data are at least consistent with the paradigm of chromosomal heredity, we cannot feel altogether satisfied with a picture that shows only a cloud of undifferentiated genes. Efforts to give more concreteness to the polygene model have achieved some success in *Drosophila melanogaster*, where the small number of chromosomes, and the detail in which they have been mapped, provide an exceptionally favourable opportunity for assaying the effects of particular chromosomes or even chromosome segments. In this species it can fairly easily be shown how the determinants of a quantitative difference are distributed over different chromosomes.

An early example of chromosome assay in *Drosophila* was provided by K. Mather and B. J. Harrison,[4] who studied the genetic basis of a difference in average number of abdominal hairs between two wild-type strains collected from two widely separate geographical regions. The method that they developed depended on measuring the effect of each of the three major chromosomes (X, II and III) in each of the wild strains against the effect of a standard chromosome carrying a dominant visible marker. The standard chromosomes carried long inversions to prevent them from recombining with the wild chromosomes when in heterozygous combination with them (cf. p. 154, Chapter 6).

Each wild type was crossed to a stock carrying all three marked chromosomes, and F_1 flies showing all three dominant markers were backcrossed to the parental wild type. Because of the inversion heterozygosity, the chromosome pairs present in the F_1 segregated as integral units, and the backcross progenies consisted, with rare exceptions, of only eight genotypes, in approximately equal numbers and all phenotypically distinguishable. Among these eight classes, four pairs could be used to assess the effect of substituting the wild-type X-chromosome for the marked X-chromosome. Another four pairs showed the effect of substituting the wild-type chromosome II for the marked chromosome II, and another four the effect of the chromosome III substitution. Thus, there were four independent measures of the effect of each wild-type chromosome, measured in each case against the marked standard. Only female flies were used since only this sex provided information about the X-chromosome. If the effects of the chromosomes are

Box 8.2	Assaying the contributions of different chromosomes to variation in number of abdominal hairs in *Drosophila melanogaster* (data of Mather and Harrison[4])

Two wild-type inbred strains of *Drosophila,* Oregon (O) and Samarkand (S) had means of 43·5 and 59·2, respectively. The difference was analysed by comparison of both with a standard set of chromosomes carrying viable dominant markers and long inversions to prevent recombination between them and their wild-type homologues. The standard chromosome set carried *Bar* (*B*) on the X-chromosome, *Plum* (*Pm*) on chromosomes II and *Stubble* (*Sb*) on chromosomes III.

Each wild type was crossed to a strain carrying the three standard chromosomes, and the *B*, *Pm*, *Sb* triply heterozygous progeny were backcrossed to the wild type. The mean abdominal hair number was determined for each of the eight classes of female progeny:

	Mean no. of hairs in class			
Source of wild-type chromosomes	B Pm Sb + + +	B Pm + + + +	B + Sb + + +	B + + + + +
S	51·9	52·4	55·6	56·7
O	48·1	43·6	48·0	46·7

	Mean no. of hairs in class			
Source of wild-type chromosomes	+ Pm Sb + + +	+ Pm + + + +	+ + Sb + + +	+ + + + + +
S	54·3	55·9	56·4	58·9
O	49·0	45·2	49·0	46·0

Estimates of effects of substituting wild-type chromosome for standard:

Chromosome				
X(S):	B Pm Sb — + Pm Sb	2·4		
	B + Sb — + + Sb	0·8	Mean 2·2	
	B Pm + — + Pm +	3·5		Difference S–O due to X estimated as +1·5
	B + + — + + +	2·2		
X(O):	B Pm Sb — + Pm Sb	0·9		
	B + Sb — + + Sb	1·0	Mean 0·7	
	B Pm + — + Pm +	1·6		
	B + + — + + +	−0·7		
Chromosome				
II (S):	B Pm Sb — B + Sb	3·7		
	B Pm + — B + +	4·3	Mean 3·3	
	+ Pm Sb — + + Sb	2·1		Difference S–O due to II estimated as +2·3
	+ Pm + — + + +	3·0		
II (O):	B Pm Sb — B + Sb	−0·1		
	B Pm + — B + +	3·1		
	+ Pm Sb — + + Sb	0·0	Mean 0·95	
	+ Pm + — + + +	0·8		
Chromosome				
III (S):	B Pm Sb — B Pm +	0·5		
	B + Sb — B + +	1·1	Mean 1·4	
	+ Pm Sb — + Pm +	1·6		Difference S–O due to III estimated as +4·5
	+ + Sb — + + +	2·5		
III (O):	B Pm Sb — B Pm +	−4·5		
	B + Sb — B + +	−1·3		
	+ Pm Sb — + Pm +	−3·8	Mean −3·1	
	+ + Sb — + + +	−3·0		

Total difference due to the three major chromosomes estimated as +8·3.

Box 8.2 continued

The estimated effect of substituting one X, one II and one III chromosome from Oregon with X, II and III from Samarkand is $1·5 + 2·3 + 4·5 = 8·3$. The two wild-type strains in fact differ from each other with respect to both the chromosomes of each kind present in the diploid set and so, if there is no dominance, the total effect of the three large chromosome pairs should be $2 \times 8·3 = 16·6$, which is in good agreement with the observed difference between the strains of 15·7 hairs. Thus, in this case, the quantitative difference can be wholly accounted for by additive effects of the major chromosomes acting without dominance. Chromosome IV is neglected in this analysis; it is so small and carries so few known genes that it is not expected to contribute to the difference to any appreciable extent.

additive, the difference between the effects of two homologous wild-type chromosomes should be given by the difference between their respective differences from the marked standard. The total chromosomal effect (neglecting the possible contribution of the relatively tiny chromosome IV) should then be obtained by adding together the differences attributed to the three chromosomes individually, doubling the answer to allow for the fact that the two wild types differ with respect to chromosome *pairs* and the effects measured by reference to the marked standards were due to only *single* chromosome substitutions. The results obtained by Mather and Harrison are summarized in *Box* 8.2. It will be seen that they are in very reasonable agreement with the simple hypothesis of additive effects without dominance. It seems that the genes responsible for the very marked difference between the Oregon and Samarkand wild strains were distributed over all three major chromosomes, with the effect of chromosome II particularly strong.

The Number of Effective Loci

It is possible to carry the analysis further and to determine at least the approximate distribution *within* a chromosome of the loci contributing to a quantitative phenotypic difference. The chromosome to be dissected is brought into heterozygous combination with a structurally normal and multiply marked standard chromosome and allowed to cross over with it at meiosis. The various phenotypically distinguishable crossover chromosome classes are transmitted to different flies of otherwise similar chromosomal constitutions and the average effect of a wild-for-standard substitution in each intermarker interval is assessed by making the appropriate pairwise comparisons. This kind of experiment has sometimes suggested that the quantitative difference under investigation may be attributed to rather few loci.[5] As we saw above, the additive effects of as few as three or four allelic differences of individually small effect can easily, given a modest amount of environmentally caused variation, account for the appearance of continuous variation. One should not conclude, however, that this is the typical case in natural populations, which may well contain more heterozygosity at more loci than could be generated by a cross between any pair of inbred laboratory strains.

Some attention has been given to the possibility of estimating the number of loci effectively contributing to a quantitative difference on the basis of the rate at which variation within families disappears during successive generations of inbreeding (self-pollination in the case of self-fertile plants). In each generation, the proportion of families which no longer show significant genetic variance indicates the proportion of individuals in the previous generation which had become homozygous for all the relevant loci and this, in turn, provides a basis for calculating how many loci were effectively heterozygous in the starting population. This method, the possibilities of which have been explored mainly by Jinks and his colleagues,[6] has the limitation that it has to make an arbitrary distinction between allelic differences whose effects are respectively just below and just above the chosen level of statistical significance; in other words, one can increase or reduce the estimate by relaxing or tightening one's statistical criteria. There is also a problem arising from linkage—one will count as single units closely linked pairs or clusters of loci that do not happen to recombine during the experiment.

In summary, the problem of counting the loci contributing to 'polygenic' inheritance is a difficult and intractable one, but in favourable circumstances it is possible to learn a good deal about the broad distribution of such loci over the chromosomes and it may sometimes be possible to identify individually some of those with the largest effects.

8.6   What are polygenes?

When Mather first coined the term 'polygene' he supposed that the entities to which he referred might well be a special class of genes with relatively unspecific effects. They were contrasted to the 'major' genes, mutations in which had sharp and unique effects on the phenotype, providing the unmistakable markers used in classical chromosome mapping.

During the subsequent decades, this concept of polygenes as distinct from major genes has been largely discredited as more has been discovered about the range of allelic variation of which all genes are capable. As we shall see in Chapter 18, it is now clear that for many genes there is no such thing as a standard wild-type allele. Rather, there exists a variety of forms of the gene, all adequately functional but showing subtle differences. The same genes which show this insignificant, or barely significant, variation can also, through other kinds of mutation, give rise to alleles with extreme or even lethal effects on the phenotype. Whether a gene is associated with a major or a small effect depends on the variants of that gene that are present in the population under study. It seems easy to account for polygenic inheritance as due to the cumulative effects of 'small' differences in 'major' genes.

Yet, though some modern developments might lead us to discount the polygene concept, others have shown that genes with small, similar and supplementary effects definitely exist. These substantiated polygenes include the highly repetitious genes coding for ribosomal RNA (cf. Chapter 1, p. 39), transfer RNA and histones and perhaps also some of the moderately or highly repeated sequences of other kinds which abound in the genomes of higher eukaryotes but for which no definite functions have yet been shown.[7] Could repetitious DNA sequences of any of these kinds account for at least some of the genetic variation seen in outbred populations?

So far, quantitative phenotypic variation has been attributed to only one of the families of repeated genes referred to in the last paragraph. In *Drosophila melanogaster* the mutant *bobbed* phenotype of shortened bristles and reduced growth rate can range in intensity from barely discernable to lethal. It correlates closely with the quantity of ribosomal RNA produced, which itself is proportional to the number of rRNA-coding genes. The *bobbed* phenotype becomes apparent when, as a result of deletion, the gene number falls below about 50 per cent of the normal number of around 200. A reduction to below about 15 per cent of normal is lethal.

However, in several respects the *Drosophila* rRNA genes have quite different properties from those attributed to typical polygenes. Instead of being scattered in location and for the most part unlinked, they occur in tandem array in tight clusters at the ends of the X and Y-chromosomes—in the nucleolar organizer regions, as we noted in Chapter 1 (p. 6). A cluster depleted by deletion behaves as a single-locus mutation; *bobbed* was known as a sex-linked locus, virtually the only one common to X and Y, long before the molecular explanation of its phenotype was known. The rRNA genes are indeed, in a sense, good examples of polygenes, but their tight clustering enables them to masquerade collectively as a single gene capable of mutation to a graded series of alleles.

Another feature of the *Drosophila* rRNA genes which certainly does not accord with the original polygene concept is their capacity for **magnification**.[8] This phenomenon is seen when an X-chromosome, carrying a *bobbed* 'allele' with a rRNA gene number less than 50 per cent of normal, is transmitted through a number of generations in males in which the Y-chromosome has a deletion covering the nucleolus organizer region. This severely deficient chromosome combination initially confers an extreme *bobbed* phenotype. Surprisingly, in the light of our usual assumptions about the accuracy of chromosome replication, the phenotypic abnormality becomes less severe with each succeeding generation. After four or five generations the chromosome, which has descended directly and without ordinary recombination (since there is no crossover in the male) from the original *bobbed* chromosome, is found to have almost the normal quota of rRNA genes.

The mechanism underlying this striking magnification of gene number has been much disputed; what appears to be the best-supported hypothesis invokes unequal sister-strand exchange (*see* Chapter 12, p. 323). Where there are many identical sequences in regularly spaced tandem array, as is the case with the rRNA gene cluster, it is easy to envisage that pairing could occur between chromatids one or several copies out of register, and we know from cytological evidence that sister-strand exchange is a fairly frequent event in those eukaryotic species in which it has been looked for. The sister-strand exchange mechanism should generate chromo-somes with reduced numbers of rRNA genes as well as ones with magnified numbers, but the former will tend to be lost because of reduced viability while the latter will be favoured by natural selection in a *bobbed* strain.

There is no obvious reason why the possibility of magnification and diminution of rRNA genes should be confined to *Drosophila*. The arrangement of these genes is similar in all eukaryotes so far as is known, and several other kinds of repetitive genes, notably those coding for 5-S RNA and for histones, are also often arranged in tandem clusters. There is also some clustering of genes for species of tRNA, though here the arrangement is less regular in the examples so far analysed. Magnification and diminution in the numbers of these types of polygenes may well account for a significant amount of continuous variation in populations.

Two other components of the genome which are certainly subject to quantitative, and also positional, variation are the middle-repetitive movable sequences and the highly repetitive simple-sequence 'satellite' sequences, both of which are discussed in Chapter 14. These can certainly influence the phenotype in special ways; movable sequences are mutagenic (*see* p. 325) and heterochromatin, consisting largely of satellite DNA, exerts position effects in *Drosophila* (*see*

p. 508). It is not clear, however, that they play any necessary or normal role in determining the phenotype. Some of the dispersed movable sequences are transcribed (sometimes abundantly) into RNA but are not known to be translated into protein; satellite DNA is not even transcribed. Both classes of sequence seem much more labile in quantity and distribution over the chromosomes than is the single-copy DNA which comprises the essential and specific protein encoding genes.

8.7   Summary and perspectives

A great deal of the variation found in actual populations takes the form of a continuous gradation of phenotypes without any clear division into classes that could be attributed to segregation of gene differences with major effects. Statistical methods have been developed for the analysis of this continuous variation. Some of it is usually due to environmental fluctuation but a proportion can often be shown to be genetically determined. The fraction of the variation due to genetic differences of additive effect is called the heritability, and this parameter is of use in predicting the effect of selection in plant or animal breeding. A part of the genetic variance is attributable to gene interaction and to dominance and does not contribute to heritability in the strict sense.

The genetic differences that contribute to continuous variation show the property of segregation and can sometimes be located in defined chromosomes or chromosome segments. They behave as one would expect of multiple gene differences distributed over many loci and individually of small effect. Continuous genetic variation does not need to be attributed to a special class of genes ('polygenes'). Although variations in the numbers of repetitious genes may contribute to continuous variation, the presence in populations of functionally almost equivalent multiple alleles of unique genes could, by itself, explain the observations. The same gene can, by allelic variation of minor effect, contribute to continuous variation and also, by occasional mutation to a severely defective allele, make an obvious individual mark on the phenotype.

Chapter 8   Selected Further Reading

Falconer D. S. (1981) *An Introduction to Quantitative Genetics*, 2nd ed. p. 165. London, Longman.

Mather K. and Jinks J. L. (1977) *Introduction to Biometrical Genetics*, 231 pp. London, Chapman and Hall.

Chapter 8	**Problems**

8.1 In his experiment on the inheritance of corolla length in *Nicotiana*, East (*see Fig.* 8.1) obtained the following values for the means and variances of the parental strains and the F_1 and F_2 generations. First parent: mean 21·4, variance 1·39; second parent: mean 81·8, variance 25·8; F_1: mean 40·8, variance 4·8; F_2 mean 38·3, variance 36·0. Estimate the heritability (in the broad sense) of the F_2 variation. What further information would you need in order to estimate the narrow-sense heritability (i.e. the additive, fixable variation)?

8.2 Two different inbred lines obtained from the same plant species were tested for a possible heritable difference in plant height at first flowering. Ten plants of each line, grown in the same controlled environment, gave means and variances of 139·4 and 30·7 for the first line, and 134·1 and 29·6 for the second. How strong is the evidence that the two lines are genetically different with respect to determinants of plant height? Supposing that the means and variances were based on twenty plants of each line rather than ten, what effect does this have on your answer?

8.3 In an experiment in which he selected for large and small body size in mice (not the same experiment as shown in *Fig.* 8.4), Falconer obtained the results shown.

	Mean weight at 6 weeks				
Generation no.	*All mice*	*Selected*	*All mice*	*Selected*	*Control (unselected)*
0	22·2	24·1	22·2	20·4	22·2
1	23·6	25·2	22·6	20·9	23·1
2	24·3	26·2	21·4	19·7	22·6
3	24·8	26·9	20·7	19·8	22·8
4	26·2	28·2	21·0	19·7	23·4
5	27·0	29·2	20·5	19·2	23·1
6	28·1		19·6		22·7

Construct a graph relating selection differential to response and estimate the heritability of body weight.

8.4 In a mouse selection experiment the following 10-week weights were determined for ten father–son pairs:

Father: 28·0 30·5 31·0 37·9 34·9 32·3 38·0 28·0 25·0 30·9

Son: 30·0 31·5 32·0 32·5 32·0 31·6 33·6 31·0 29·3 30·9

By means of a graph, estimate the regression of sons on fathers, and hence the heritability of 10-week body weight.

8.5 Recombinant inbred (RI) lines of plants or animals are fixed (homozygous) for an assortment of genes assembled at random from different inbred strains. Each RI line is initiated by self-fertilizing or sib-mating individuals from the F_2 generation of an interstrain cross and continuing such inbreeding in each subsequent generation. Supposing that two varieties of a cultivated plant differ with respect to 20 unlinked genes, how many different recombinant inbred lines are possible? How many generations of self-pollination would be necessary to make each RI line homozygous for all 20 genes with 95 per cent probability?

Parasexual Analysis

9.1   Introduction

A severe limitation of the classical methodology of genetics is that it can be applied only to the analysis of organisms that can be crossed sexually. This excludes from investigation not only differences between sexually reproducing organisms that are too distantly related to form fertile hybrids, but also those species with no known sexual reproduction. There is also no possibility, by orthodox genetical methods, of analysing the causes of differences that may arise between somatic cells, either during the development of the organisms or in culture on artificial media.

The scope of genetics has been greatly expanded by the development of methods for bringing about genetic mixing and segregation without the normal sexual cycle. The pioneers of such **parasexual analysis** were G. Pontecorvo and his colleagues in Glasgow, who demonstrated its feasibility in the fungus *Aspergillus nidulans* and related species. Essentially the same principles have now been applied to animal and human cells in culture. Briefly, what is involved is the fusion (either natural or contrived) of cells of two different genotypes, the selection of more or less stable hybrids in which nuclear fusion has occurred, and the subsequent observation of segregation (either spontaneous or induced) of parental markers during hybrid growth.

This chapter will first summarize the method which was worked out for *Aspergillus*, and since extended to several other fungi, and will then go on to review some of the ways in which parasexual methods have been applied to mammalian cells.

9.2   Parasexual analysis in *Aspergillus nidulans*

In fungi the initial step of the parasexual cycle, the bringing together of two genetically different nuclei into the same cell, can often be accomplished by making use of the widespread natural process of hyphal fusion (anastomosis). This does not always occur readily and never occurs between hyphae of different species; even within a species, isolates from different geographical areas are more often than not incompatible in this respect. However, if one starts with a single genetically homogeneous strain and obtains mutations in that strain, there is usually no difficulty. Heterokaryotic mycelium can be obtained simply by growing two different mutant strains together and selecting for mycelium combining the wild-type features of both phenotypes. Having obtained a heterokaryon, the next step is to select a diploid derivative in which the two parental genomes are present within the same nucleus. One can then look for genetic segregation during subsequent growth.

As an example, one of the classical early experiments may be cited.[1] Of two mutant strains of *Aspergillus nidulans*, one carried mutations giving *white* (*w*) conidia and nutritional requirement for proline (*pro*), and the other mutations giving *yellow* conidia (*y*) together with requirements for adenine (*ad*) and *p*-aminobenzoate (*paba*). From mixed cultures of these two strains, heterokaryons able to grow on unsupplemented medium were isolated; the *white* component provided the capacity for adenine and *p*-aminobenzoate synthesis while the *yellow* component supplied the capacity for proline synthesis.

Aspergillus conidia, unlike those of *Neurospora*, cannot be heterokaryotic since each contains only a single nucleus. Conidia formed on a *white/yellow* heterokaryon are therefore, in general, a mixture of white and yellow (the pigment development of each conidium being dependent on its own nucleus), and have the nutritional requirements associated with these colours in the parental strains. Occasionally, however, one finds patches of mycelium, or single conidial heads or parts of heads, with *green* conidia. In the experiment under discussion these conidia differed from the parental strains in being able to grow on minimal medium. The behaviour of cultures established from these exceptional conidia, which were about double the volume of normal conidia, showed that they must have been diploid, with the two originally separate genomes combined in a single nucleus and complementing one another with respect both to pigmentation and to nutritional status.

The most obvious evidence for diploidy lay in the tendency of the green-spored colonies to produce sectors of growth with either yellow or white conidia. Such sectors were not abundant, but one or two were present on most plate cultures. It was later found that their frequency could be greatly increased by irradiation of the growing colony with ultraviolet light. Further analysis of cultures established from spontaneously arising sectors of different colour showed that they had usually segregated with respect to some or all of their nutritional markers.

Table 9.1 shows the classification of colour sectors in one experiment. The white and yellow segregants could be divided into two groups, diploid and haploid, on the basis both of conidial size and of capacity for further segregation. The haploid segregants gave no further sectors. The *yellow* haploids always required adenine and *p*-aminobenzoate but not proline. Thus the three nutritional markers, *ad, pro*$^+$ and *paba*, appear to be completely linked to yellow during the vegetative segregation of a haploid genome. The *white* haploid segregants were equally frequently *ad*$^+$ *pro paba*$^+$ and *ad pro*$^+$ *paba*, and further testing showed that the latter genotype included *yellow*, masked by the epistatic *white*. Thus *white* segregated independently of the other four markers. Everything was consistent with the idea that haploidization involved random segregation of different chromosome pairs as in meiosis but, unlike meiosis, without any crossing-over *within* chromosome pairs. This absence of crossing-over, with the resulting absolutely clear distinction between free reassortment and complete linkage, makes mitotic haploidization the ideal system for sorting out markers into linkage groups.

Later analysis, especially by E. Käfer,[2] showed that haploidization in diploid *Aspergillus* was a step-wise process. First one or a few chromosomes were lost at random, giving slow-growing aneuploid mycelia which tended to be rapidly overgrown by sectors which had lost further chromosomes and more nearly approach the haploid condition. Normal growth depends on a balanced chromosome complement, and it is only the final haploid derivatives which were fully vigorous and stable. The sequential loss of chromosomes, with accompanying loss of genetic markers, is, as we see below, the key to the assignment of markers to specific chromosomes in animal cell cultures.

Table 9.1   **Mitotic segregation in a diploid of *Aspergillus nidulans***

Original diploid

```
  w      +      pro    +      +
 ──O────────────O──────────────────
   I          II   III   IV
  +      ad     +      paba   y
 ──O────────────O──────────────────
```

Segregants selected by colour	Ploidy	Nutritional phenotype	Genotype	No. of occurrences	Interpretation
Yellow	$2n$	Prototroph	$\dfrac{w\ +\ pro\ +\ \ \ y}{+\ \ ad\ +\ \ paba\ \ y}$	32	Crossover region IV
	$2n$	p-Amino-benzoate requiring	$\dfrac{w\ +\ pro\ paba\ \ y}{+\ \ ad\ +\ \ paba\ \ y}$	78	Crossover region III
	n	p-Amino-benzoate + adenine requiring	$w^{+}\ ad\ +\ \ paba\ \ y$	19	Haploidization
White	$2n$	Prototroph	$\dfrac{w\ +\ pro\ +\ \ \ +}{w\ \ ad\ +\ \ paba\ \ y}$	56	Crossover region I
	n	Proline-requiring	$w\ +\ pro\ +\ \ \ +$	25	Haploidization
	n	Adenine + p-amino-benzoate requiring	$w\ \ ad\ +\ \ paba\ (y)^{*}$	25	Haploidization

Data from Pontecorvo and Käfer[1]. The data are somewhat simplified by the omission of a *biotin* marker which contributes little information, and also by the omission of several rarer segregant classes which require more complex explanations (non-disjunction, double crossing-over or haploidization accompanied by crossing-over). The mechanism accounting for homozygosity of markers distal to crossovers is the same as explained in *Fig.* 9.1 for the *Drosophila* example.

* Note that the *yellow* marker, though presumed to be present, cannot be seen in this class because of the epistatic *white*.

Adenine-requiring diploid segregants also occur as the result of crossing-over in region II but, since they show no colour segregation, were not detected in this experiment.

For the explanation of how crossing-over during the G_2 phase of mitosis can make markers distal to the crossover homozygous, *see Fig.* 9.1.

Turning now to the diploid segregants listed in *Table* 9.1, these evidently arise through a more complex process which turns out to be less useful as a model for the genetic analysis of higher cells but, nevertheless, provides an important method of genetic mapping in *Aspergillus*. Within the group of markers *ad pro paba y*, which all segregate together in haploids and hence may be assumed to be on the same chromosome, *ad* segregates in diploids independently of the other three. Within the subgroup *pro paba y*, *pro* and *paba* both remain heterozygous in a proportion of the homozygous *y* sectors. Other yellow sectors are homozygous for *paba* as well as for *y*, with *pro* still heterozygous. Others again are homozygous for all three markers.

Fig. 9.1

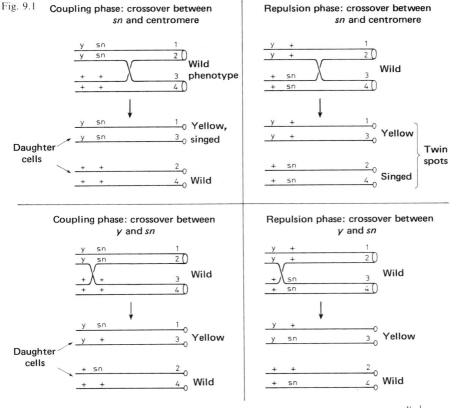

Coupling phase: crossover between *sn* and centromere

Repulsion phase: crossover between *sn* and centromere

Coupling phase: crossover between *y* and *sn*

Repulsion phase: crossover between *y* and *sn*

The origin of yellow/singed twin spots by mitotic crossing-over in $\dfrac{y\ +}{+\ sn}$ *Drosophila melanogaster* (after Stern[3]). The figure also shows the consequences of single crossovers between *y* and *sn* (yellow spots only) and the corresponding results in the 'coupling' heterozygote $\dfrac{y\ sn}{+\ +}$ (either yellow or yellow-singed spots depending on the position of crossing-over but not twins). This figure simplifies the situation since it does not include double crossovers, which do occur (though with lesser frequency) in this system. Only the disjunction mode which segregates daughter centromeres 1 with 3 and 2 with 4 is considered here; the equally likely alternative, 1 with 4 and 2 with 3, does not lead to homozygosity in the daughter cells following single crossovers (work it out). Exactly the same explanation applies to the homozygous diploid segregants in the *Aspergillus* example of *Table* 7.1.

The explanation of these regularities in the data had already been provided in essence by the work of Curt Stern on *Drosophilia*, nearly 20 years earlier.[3] He showed that females heterozygous with respect to the X-linked markers yellow-body (*y*) and 'singed' bristles (*sn*) showed occasional small patches of yellow cuticle which mostly had singed bristles when the markers were in 'coupling' (i.e. $\dfrac{y\ sn}{+\ +}$).

When the two recessive mutations were in the 'repulsion' phase (i.e. $\dfrac{y\ +}{+\ sn}$), the yellow patches themselves had normal bristles but many of them were 'twinned' with adjacent and similarly sized patches with normal colour and singed bristles. The two markers were known to be fairly closely linked and relatively distant from the centromere. The interpretation, which all subsequent studies have confirmed, was that, rather rarely, crossing-over occurs between the chromatids of homologous chromosomes leading, in 50 per cent of cases, to homozygous segregation of all originally heterozygous markers distal to the crossover. The twinned patches were thus interpreted as the two reciprocal products of crossing-over (*Fig. 9.1*).

It follows from this interpretation of the mitotic segregation of homozygous from heterozygous diploids that, where a mitotic crossover occurs between two markers linked in the same chromosome arm, the more distal of the two will (with 50 per cent probability) become homozygous in the daughter cells while the proximal marker will remain heterozygous ('proximal' and 'distal' being used with reference to the centromere). It is evident that this model provides a criterion for ordering markers within a chromosome arm. The order indicated by the data of *Table 9.1* is (centromere, *pro*)–*paba*–*y*; *ad* must be in the other arm of the same chromosome. Note that this mitotic crossover analysis cannot show linkage across a centromere; by its very nature it destroys such linkage by its requirement for a crossover between marker(s) and centromere. It is only the haploid analysis which shows the linkage of *ad* and *paba*.

Similar methods of analysis have been applied to a number of other fungal species. These include several in which, as in *Aspergillus*, diploid mycelia can be obtained by artificial selection though they are absent or rare in nature, and the yeast *Saccharomyces cerevisiae*, in which diploidy is the dominant phase of the life history. Attempts to obtain diploids in *Neurospora* have not been successful.

9.3 Application to animal and human cells

Cell Fusion

Starting in the 1940s, techniques have been developed for the culture of animal and human cells *in vitro*. On appropriate media, usually incorporating fetal calf serum, skin cells (fibroblasts) will grow to form coherent sheets of cells, while various kinds of cancer cells are capable of multiplying in suspension, almost like free-living unicellular micro-organisms. Sexual reproduction in animals depends on the differentiation of sex organs and specialized germ cells, and there would seem, at first sight, to be no way in which recombinational genetics could be applied to vegetative cells growing in culture. The means for carrying out genetic analysis in cell cultures was provided by the discovery that it was possible to induce the fusion of dissimilar cell types with the formation, in the first instance, of binucleate heterokaryons.

An early method for the induction of cell fusion involved the use of a chemically inactivated virus (Sindbis virus). Even when this virus has been killed by the chemical (usually β-priopiolactone) its protein coat still binds to the outer cell membrane of cultured mammalian cells, causing them to clump together and frequently to fuse completely. It has since been found that, provided that one has some selective procedure favouring growth of the hybrid cells, it is possible to obtain hybrids from mixed cell cultures even without the virus treatment. Once cell fusion has occurred to give a heterokaryon, nuclear fusion may follow. Thus a hybrid cell line is formed with nuclei containing, at least initially, all of the chromosomes from the two kinds of (generally diploid) cells which were fused.

The karyotypes of animal cells in culture are usually rather unstable, with a tendency towards aneuploidy. In the fused nucleus of a hybrid cell there will, assuming the two diploid genomes to be more or less equivalent functionally, be a four-fold redundancy in most gene functions, and so it is not surprising to find that chromosomes can be lost without any marked effect on growth rate. In some cases of interspecific cell hybrids, chromosome loss is non-random and occurs with a high frequency. The importance of the preferential loss of human chromosomes from man–mouse or man–hamster hybrids is explained below. It has been suggested that the basis for this effect is the failure of human chromosomes to replicate fast enough to keep up with the rodent chromosomes, whose replication cycle is geared to a shorter cell generation time.

The Use of Selectable Markers

Like all genetic analysis, that of cultured cells depends on the availability of genetic markers. Auxotrophic mutants, on which so much reliance is placed in fungal genetics, are difficult to come by in mammalian cells, though mutant cell lines requiring glycine were isolated early in the history of cell genetics. It was at first something of a puzzle that such mutants, which were expected to be recessive (and, as we see on p. 240, actually are so), should be obtainable in diploid cells at all, since a mutation in one chromosome should be concealed by the continuing functioning of the homologue. The answer to the problem is believed to lie in the frequency of chromosome loss in cell culture.[8] Within a population of cells there will usually be some with only one copy of the chromosome in which mutation is being sought, and in such cells dominance will not be a problem.

The most useful markers are those which enable cells carrying them to be efficiently selected from a mixed population, and it is an added convenience if selection can be exercised either for or against the mutant type, depending on the cultural conditions. These advantages are offered by a number of types of mutation which affect the utilization of nucleosides or free bases for nucleotide synthesis. Free bases and nucleosides are not intermediates in *de novo* nucleotide synthesis, but, in so far as they are products of nucleic acid degradation, it is an obvious economy for the cell if these ready-made building blocks can be reused. A number of 'salvage' enzymes are able to perform this recycling function. For example, thymidine can be converted to thymidylic acid through the action of **thymidine kinase**:

$$\text{Thymidine} + \text{ATP} \longrightarrow \text{Thymidine 3'-monophosphate} + \text{ADP}$$
$$\text{(i.e. thymidylic acid)}$$

Free guanine and free hypoxanthine can be converted to guanylic acid and inosinic acid, respectively (the latter compound being the normal immediate precursor of adenylic acid); the enzyme functioning in both conversions is **hypo-xanthine–guanine phosphoribosyltransferase** (HGPRT):

Guanine + phosphoribosyl pyrophosphate $\longrightarrow$ Guanidine 3'-
                                                               monophosphate
                                                               (i.e. guanylic acid) +
                                                               pyrophosphate
Hypoxanthine + phosphoribosyl pyrophosphate $\longrightarrow$ Inosinic acid +
                                                               pyrophosphate

There is an analogous enzyme, adenine phosphoribosyltransferase (APRT), which functions in the conversion of adenine to adenylic acid.

A significant technical advance in the genetics of mammalian cells was the invention of a growth medium on which both thymidine kinase (TK) and HGPRT are essential for growth. This medium, called **HAT** medium, contains the drug **aminopterin**, which blocks the *de novo* synthesis of both thymidylic acid and adenylic acid, but compensates by also providing **thymidine** and **hypoxanthine** for use via the salvage pathways. Any mutant which is unable to form either one of the relevant salvage enzymes is unable to grow on HAT. It is also possible to provide conditions under which one of the salvage enzymes is *dis*advantageous to the point of being lethal. The general principle is to include in the growth medium a base or nucleoside analogue which is highly toxic if incorporated into nucleotide and hence nucleic acid, but is more or less harmless if it cannot be so incorporated for want of the appropriate salvage enzyme. Thus, bromodeoxyuridine, a thymidine analogue, is toxic only in the presence of thymidine kinase, while the toxicity of thioguanine is dependent on HGPRT and that of diaminopurine (an adenine analogue) on APRT. Mutants resistant to one of these drugs are generally deficient in the activity of the corresponding salvage enzyme.

Assignments of Markers to Chromosomes

Let us now consider how cell and nuclear fusion, and subsequent chromosome loss, can lead to the assignment of human genes to microscopically identifiable chromosomes.

A good simple example is provided by the work of O. J. Miller and his colleagues[4] on the genetic determination of thymidine kinase in human cells. They made hybrids between normal human fibroblasts and mouse cells of a mutant line selected as resistant to bromodeoxyuridine and deficient in thymidine kinase. The hybrid cells were propagated for many cell cycles on HAT medium. The need for thymidylic acid guaranteed that the human chromosome carrying the determinant of thymidine kinase was retained in the growing cell culture even if all the other human chromosomes were lost. In fact, after prolonged culture, most of them *were* lost, and a number of clones, grown from single cells, proved on cytological analysis of the karyotype to have only one remaining human chromosome (*see Fig.* 9.2). The surviving chromosome was always the same one, and was identified on the basis of size, centromere position and Q-banding pattern (*see* p. 35, Chapter 1) as chromosome 17. Thus it was concluded that this chromosome carried information which was both necessary and sufficient (in the background provided by the thymidine kinase-deficient mouse genome) for the formation of the enzyme. It should be added that, by all available criteria, it is the *human* form of thymidine kinase which is produced in this situation and not the mouse one. Many other

Fig. 9.2

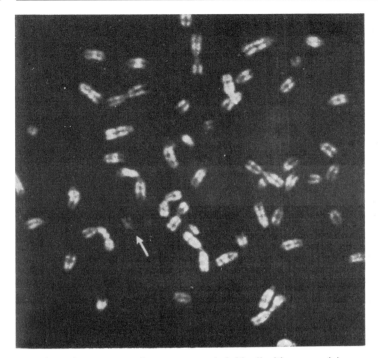

Metaphase chromosomes of a man–mouse hybrid cell with one surviving human chromosome fragment, identified as a fragment of chromosome 17. The staining is by quinacrine fluorescence. Human chromosomes in general stain much less intensely with this reagent (as also with Giemsa) than those of mouse. *See text.*
From Miller et al.[4] reproduced by permission.

human enzymes have been attributed to specific chromosomes by analogous experiments (*see* Chapter 19, p. 563).

In the experiment just described only one enzyme marker was used. Where several are available in the same hybrid cell line, decisive evidence can be obtained on linkage or independence of markers. F. H. Ruddle and his colleagues fused human fibroblasts with cells of a mouse tumour cell line, the latter being unable to grow on HAT medium because of a deficiency in HGPRT. Following cell fusion, culture on HAT eliminated the mouse cells, leaving the advantage with the hybrid cells which were able to outgrow the fibroblasts. Repeated plating of the hybrid culture after periods of growth, with isolation of clones grown from single cells, led to the isolation of many clones which differed from one another in the number of human traits retained. Ruddle et al.[5] observed seventeen enzymes in which the human and mouse cells differed, the distinguishing criteria not being presence/absence but rather the electric charge on the enzyme proteins as shown by rate of migration in electrophoretic gels. Newly selected man–mouse hybrid cultures produced both the human and the murine form of each enzyme. As cloning and subcloning proceeded, human enzymes were lost, presumably as the result of the loss of human chromosomes. Comparisons of the arrays of enzyme in a large number of subclones showed that fourteen of the seventeen human enzymes were lost independently of each other and of the other three, implying that they were all determined by genes on different chromosomes. The other three enzymes, HGPRT, glucose 6-phosphate dehydrogenase (G6PD) and phosphoglycerate kinase (PGK), were always lost together. Other studies have shown that the presence or absence of these three markers correlates completely with the presence or absence of the human X-chromosome.

That HGPRT and G6PD are determined by sex-linked genes was already known from human pedigree analysis. Rare recessive alleles present in human populations cause deficiencies in these enzymes and, when homozygous, have clinical manifestations. HGPRT deficiency is associated with a severe behavioural disorder (Lesch–Nyhan syndrome), and G6PD deficiency with extreme sensitivity to the antimalarial drug quinacrine. Both traits show classical sex-linked inheritance (p. 84).

Fig. 9.3

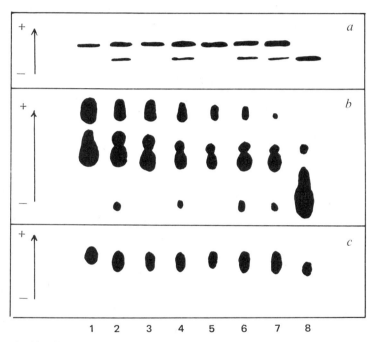

The identification by electrophoresis and specific enzyme staining of the mouse and human forms of three enzymes in man–mouse hybrid cell clones.
(a), (b) and (c) are different starch gels in which the proteins extracted from mouse cells (1), human cells (8), hybrid cells (2 and 7) and subclones isolated from such cells after culture (3–6) were separated by electrical charge in a voltage gradient. The gels were stained, after the electrophoretic separation, with specific reagents which produce an insoluble dye in the presence of the activity of the enzyme being analysed: (a) 6-phosphogluconate dehydrogenase, (b) phosphoglucomutase and (c) peptidase C.
Note that the phosphoglucomutase in each species consists of a family of molecules with different charges, but that the whole set is more negatively charged (i.e. moves farther towards the positive electrode) in mouse than in man. The two species-specific peptidases are not well separated, but the mouse variety is consistently more negative. In subclones 3 and 5 the human forms of 6PGD and PGM have been lost together (they are both determined by chromosome 1, *see text*). The human peptidase C is present in all the subclones.
Redrawn from Douglas et al.[6]

A further refinement of cell fusion and segregation analysis can be illustrated by a study explained in *Fig.* 9.3. The investigators fused a HGPRT-negative hamster cell line with human fibroblasts, and looked for the human versions of phosphoglucomutase (PGM) and 6-phosphogluconate dehydrogenase (6PGD) in subclones isolated from the hybrids. The two enzymes were always lost together, with the correlated loss of chromosome 1, the largest chromosome of the human complement. In one subclone in which both of the enzymes were missing it was found that the greater part of chromosome 1 was still present—only the distal quarter of the short arm was deficient, presumably as the result of a spontaneous chromosome break. Another enzyme, peptidase C, known also to be determined by chromosome 1 from its absence in clones in which the whole chromosome had been lost, was retained in the subclone with the small terminal deletion. This enzyme must, therefore, be specified by a gene located in the non-deleted part of chromosome 1. This kind of analysis, combined with observations of the effects of whole chromosome loss, and with a comparatively minor contribution from classic pedigree analysis, has led to the location of a large number of human enzyme determinants in more or less defined regions of the human karyotype. V. McKusick[7] has summarized the progress made in such mapping up to 1980.

Does Mitotic Crossing-over Occur in Mammalian Cell Culture?

The examples just considered are akin to haploidization analysis in *Aspergillus* in that they depend on chromosome loss rather than on the type of

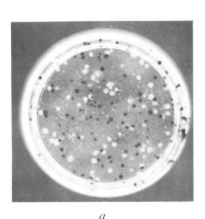

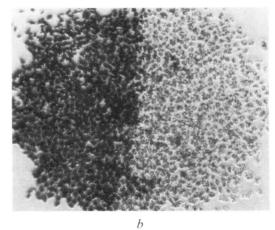

a   *b*

Fig. 9.4

a. A mixture of G6PD$^+$ and G6PD$^-$ hamster cell colonies stained to show presence or absence of glucose 6-phosphate dehydrogenase (G6PD). The staining mixture contains glucose 6-phosphate, NADP (the natural coenzyme which is reduced as the glucose 6-phosphate is oxidized by the action of the dehydrogenase) and nitroblue tetrazolium, a synthetic dye which is chemically reduced by the reduced NADP to form an insoluble deep purple dye. The G6PD$^+$ colonies, possessing the enzyme, thus stain dark while G6PD$^-$ colonies, lacking the enzyme, remain unstained.
b. A close-up of contiguous G6PD$^+$ and G6PD$^-$ colonies.
From Rosenkraus and Chasin.[8] Photographs by courtesy of the authors.

Aspergillus segregation due to mitotic crossing-over between homologous chromosomes. Indeed, one would not expect to find such crossing-over between the contrasted chromosome sets in interspecific cell hybrids since they will almost certainly lack the necessary degree of homology. It is interesting, therefore, to consider what happens following fusions of differentially marked cells of the *same* species.

In one experiment two mutant derivatives of a hamster cell line were fused.[8] Both required glycine for growth as a result of different and complementary mutations attributed to different genes *glyA* and *glyB*. In addition, the *glyA* cell line carried three further mutations causing deficiencies in HGPRT (selected by azaguanine resistance), APRT (diaminopurine resistance) and G6PD (requiring a special procedure for its isolation—*see Fig.* 9.4). The hybrid cells were independent of glycine because of the complementary action of the two *gly* genes, one of which was still functional in each cell line. They were not initially resistant to either azaguanine of diaminopurine, since the *glyA* genome contributed the ability to form both of the relevant salvage enzymes.

Selection for resistance to one or other analogue led to the isolation of HGPRT⁻ and APRT⁻ subclones. It should be noted that the possibility of stringent selection for the resistance phenotypes in this example compensates for the lower rate of spontaneous chromosome loss as compared with the man–rodent hybrid cells. It was found that the diaminopurine-resistant (APRT⁻) subclones were nearly always still G6PD⁺, as would be expected if the two enzymes were determined by genes on different chromosomes. In contrast, subclones exhibiting the azaguanine-resistant (HGPRT⁻) phenotype were G6PD⁻ in over 90 per cent of cases. Conversely when, by a more laborious selective procedure, G6PD⁻ segregants were obtained and screened for presence of the other two enzymes, they were found in more than 90 per cent of cases to be APRT⁺ HGPRT⁻.

The near-complete association of the HGPRT⁻ and G6PD⁻ traits confirmed that, as in man, the genes determining these two enzymes are on the same chromosome (again the X, as it turns out). The interesting question was whether the occasional exceptional segregants which had lost the one enzyme without the other could be due to mitotic crossing-over, making one locus homozygous for the negative allele and leaving the other locus heterozygous. If this were the case one should be able to recover subclones carrying structurally normal but genetically recombinant chromosomes. Such recombinant chromosomes have, however, never been demonstrated, either in the case under discussion or in other experiments which should have revealed their presence had they existed. It appears, rather, that the infrequent segregants in which two originally linked markers have been separated are due to chromosome breakage and loss of chromosome fragments, unconnected with any process of reciprocal crossing-over.

It may be that mitotic crossing-over is a special case. It occurs in filamentous fungi, perhaps because they are habitually haploid and are not adapted to diploidy without meiosis. In *Drosophila* there seems to be an unusual tendency for somatic chromosomes to be associated in homologous pairs during mitosis, and this may be conducive to mitotic crossing-over.

9.4 | Chromosome-mediated gene transfer in mammalian cells

In the type of analysis described in the preceding pages, different sets of chromosomes are mixed together by means of spontaneous or induced fusion of whole cultured cells. Over the past few years means have been developed for the transfer of *isolated chromosomes* from one cell type into another. This bizarre deviation from normal genetic practice results from the ability of animal cells to ingest chromosomes added to their culture medium. The isolated chromosomes are obtained by differential centrifugation of carefully broken cells whose mitotic cycle has been arrested in metaphase by the use of the drug colchicine. Metaphase chromosomes are comparatively robust and easy to isolate.

Cell line		Human chromosome or fragment retained (Giemsa stain)*	TK	GK	PC1
TK⁻ mouse clone transformed to TK⁺ by human chromosome transfer—unstable			+	+	+
Different subclones: (all selected as still TK⁺)	a		+	+	+
	b	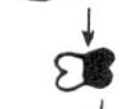	+	−	+
	c		+	−	+
	d		+	−	+
	e		+	+	+
	f		+	+	+
	g	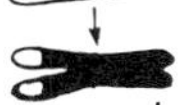	+	+	+
	h		+	+	+

*Drawing based on the original photographs with light/dark (human/mouse) contrast accentuated.

Fig. 9.5 — The effects of a fragment of human chromosome 17, and of a series of smaller derivatives of it transposed to mouse chromosomes, when present in an otherwise wholly mouse genome. The left-hand side of the figure shows the appearance of the fragment or transposition product in a series of cell lines, and the right-hand columns show the presence or absence of the human versions of thymidine kinase (TK), galactokinase (GK) and procollagen 1 (PC1) in each cell line. The chromosomes were stained with Giemsa reagent, which stains mouse chromosomes, in general, much more intensely than human chromosomes. Drawings of chromosomes are based on the original photographs, with the contrast between darkly stained mouse and lightly stained human segments exaggerated. Arrows indicate centromere positions.
From Klobutcher and Ruddle.[9]

They can even be fractionated, to a certain degree, according to their size, although it is as yet hardly possible to obtain pure suspensions of a single chromosome type.

The uptake of a chromosome by a cell is a rather rare event, and can be detected only by the use of selectable genetic markers. Again, considerable use has been made of HAT medium and the enzymes which are required for growth on it. In one instructive study,[9] chromosomes were isolated from human tumour cells to mouse cells of a line deficient in thymidine kinase. One mouse cell in between 10^6 and 10^7 was transformed to the TK^+ phenotype and was able to form a colony on HAT medium. Two clones thus established could be shown each to contain a human chromosome fragment—in one case translocated to the end of a mouse chromosome and in the other as an independent small chromosome which seemed, from its G-banding pattern (see Chapter 1, p. 34), to be a piece (including the centromere) of human chromosome 17 (Fig. 9.5). This chromosome, as we have already seen, carries the gene determining thymidine kinase.

Further study of the clone possessing the independent chromosome 17 fragment showed that it was unstable in its TK^+ character, but yielded stable subclones on further selection. In some of these subclones, the chromosome fragment could be seen to have been transposed to the end of one of the mouse chromosomes, and the human centromere had been lost. In three cases, in which the transposed piece was particularly small, a second human chromosome 17 enzyme marker, galactokinase, had been lost along with the human centromere, while a third one, the non-enzymic human protein procollagen 1, had been retained. This strongly suggested that the genes determining the formation of the three marker human proteins were all in the same arm of chromosome 17 (since the longer transposed fragments of this arm are associated with the retention of all three) and that the galactokinase determinant was closer to the centromere than the other two.

This study illustrates once again the difficulty which human chromosomes have in maintaining themselves in cells of predominantly mouse genotype. The only way in which human genetic markers can become stabilized in such cells is by 'escaping' on to mouse chromosomes. The process is parallel to the bacteriophage-mediated transduction in bacteria discussed in Chapter 11. In both cases a small fraction of the genome is segregated from the donor cell before transfer (prezygotic rather than postzygotic segregation). However, interspecies chromosome transfer differs from bacterial transduction in that the donated fragment is added to the recipient genome rather than substituted for a homologous recipient sequence by recombination. Indeed, in a two-species system such as we have been discussing, the homology between recipient and donor chromosomes is probably too weak for substitutional recombination to occur.

9.5   Transformation of cells with free DNA

At almost the same time as chromosome-mediated gene transfer was demonstrated in mammalian cells, convincing evidence was presented for the transfer of genes to such cells from solutions of free DNA. The process, called **transformation**, had long been known to be possible in bacteria. Transformation was first studied in *Diplococcus pneumoniae*. The identification by Avery and his colleagues of the transforming principle as DNA was the first direct evidence of the chemical nature of the genetic material. The phenomenon was later intensively studied in *Haemophilus* spp., in *Bacillus subtilis* and (once the requirement for calcium was discovered, *see below*) in *Escherichia coli*. We shall return to bacterial transformation in the following chapter (p. 275). The more novel extension of DNA-mediated transformation to mammalian cells needs to be considered here.

As in the case of chromosome-mediated gene transfer, the demonstration of DNA-mediated transfer depends on screening large numbers of treated cells. Markers subject to selection on HAT medium have again been extremely useful. The DNA works best when added as a precipitate with calcium phosphate.

A good example is the transformation of TK$^-$ mouse cells to the TK$^+$ phenotype, able to grow on HAT medium, by treatment with the DNA of the human virus, **herpes simplex** (HSV).[10] This virus includes among its genes one coding for thymidine kinase. The virus enzyme is clearly distinguishable by electrophoretic mobility from the thymidine kinases of mouse or man. The TK$^+$ mouse transformant colonies produced an enzyme which was demonstrably of the HSV type, with no sign of the native mouse enzyme. HSV DNA was favourable material for the experiment since it contains relatively few genes, so those that are present are in much higher concentration than are single-copy genes in a mammalian genome. It was, in fact, possible to increase the transforming potency of the DNA preparation still further by digestion with a restriction endonuclease, BamHI (*see Table* 7.1, p. 181), and fractionating the resulting fragments by size. The TK gene in the HSV genome is contained with a BamHI fragment of 3·4 kb, and a fraction enriched with respect to fragments of about this size was more effective than the unfractionated DNA. Even so, only of the order of one in a million cells was transformed to ability to make thymidine kinase with the concentrations of fractionated DNA which it was practicable to supply. Those cells that were transformed each appeared to have incorporated just one copy of the TK$^+$ viral gene at some chromosomal locus—probably a different locus in each transformed cell.

R. Axel's group also showed that the mouse TK$^-$ cells could be transformed to TK$^+$ by DNA isolated from other cells, though the efficiency in terms of number of transformants per μg of DNA applied was lower than when HSV DNA was used. When human cells were used as the DNA source, the thymidine kinase found in the transformed cells was of the human type, distinguishable from the mouse enzyme by electrophoretic mobility. The end result is quite similar to that of the chromosome-mediated transfer experiment outlined in the preceding section. The important difference is that DNA fragments, unlike chromosomes and chromosome fragments, are usually too small to carry more than one genetic determinant—*only* TK$^+$ rather than both TK$^+$ and the linked procollagen-determining gene.

A most important discovery, which arose from the work just described, was that cells which had taken up one selectable DNA marker were very likely to have taken up other DNA fragments as well, where these had been supplied along with the marker DNA. A particularly dramatic experiment, again by Axel's group,[11] resulted in the acquisition by mouse cells of the capacity to produce the *rabbit* β-globin. The gene for rabbit β-globin was supplied in the form of a DNA fragment cloned in a bacterial plasmid (*see* pp. 274 and 394). The cloned DNA was mixed in 100-fold excess with HSV DNA and used to treat TK$^-$ mouse cells, which were then selected on HAT medium for acquisition of the HSV TK$^+$ gene. Eight of the selected transformants were examined with a DNA 'probe' (*see* pp. 386 and 390) for the presence of the rabbit DNA sequence. Six of these clones, three-quarters of those tested, were shown to have acquired the rabbit sequence. Most of them carried it in more than one copy and one had about twenty copies. Moreover, the rabbit gene was to some degree expressed in some of the mouse cell clones, which were found to be producing detectable amounts of the rabbit β-globin polypeptide. It was evident that the selection for TK$^+$ transformants had automatically selected for cells which were competent to take up DNA in general, and that the uptake was very efficient in this small minority of the cell population. Other experiments showed that cells selected by this technique can take up DNA of virtually any kind—for example, genomes of the bacteriophage φX174 (*see* pp. 284, 425) were in one experiment taken up by mouse cells in many copies and transmitted in stable fashion to daughter cells through mitosis.

Foreign DNA, once it has gained entry into the cell, seems to find its way into the cell nucleus and then becomes integrated into the chromosomes. Proof of chromosomal integration comes from the demonstration that long DNA restriction fragments, which must from their size be predominantly chromosomal in origin, include sequences (identified as described on p. 390) derived from the transforming DNA.

Until very recently, transformation of mammalian cells with DNA was possible only in cell culture, without any possibility of introducing a new gene into a whole animal for transmission to future generations. The prospect has been markedly changed by reports from two different groups[12,13] of the results of injecting rabbit β-globin gene sequences directly into the pronuclei (i.e. the haploid nuclei before fusion) of recently fertilized mouse eggs. On reimplantation into the oviducts of female mice (made pseudopregnant by hormone treatment), some of the treated eggs developed into young mice. About one-half of these progeny were shown to carry rabbit β-globin gene sequences. Several of the males were mated to normal females to see whether the rabbit gene could be transmitted to a further generation. In all cases the result was positive; a proportion of the progeny again carried the

Fig. 9.6

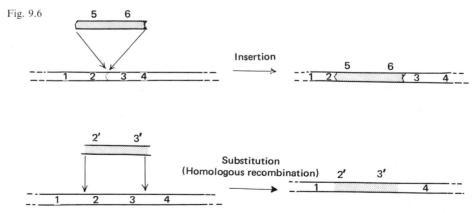

The difference between insertion and substitution in DNA-mediated transformation. In yeast, both modes of integration occur but in experiments on mammalian cells insertion appears to predominate. Where single-copy DNA is concerned there will be only one site at which a given fragment can be substituted by homologous recombination, but there are probably many sites at which insertion can occur.

rabbit gene. In one case[12] the location of the foreign gene was investigated by the technique of *in situ* hybridization (cf. p. 32) and was shown to be on chromosome 1, the normal mouse β-globin gene being on chromosome 7. The expectation, based on the results of transformation in cell culture, is that integration will have occurred at many different loci.

Integration of transforming DNA seems likely to be a much less specific process in mammals than in *E. coli* (or, so far as present information goes, in yeast). In bacteria the donor DNA fragment is integrated into the donor genome only at the site with which it has homology. Some kind of *recA*[+]-dependent homologous recombination is involved, and the homologous sequence previously in the recipient cell is displaced. Homologous substitution may also occur when eukaryotic cells are transformed by DNA (indeed, in yeast, there is good evidence that it does), but in animal cell experiments donor DNA fragments may be inserted by some unknown mechanism at many different sites, presumably without displacement of resident genetic information. The distinction is illustrated in *Fig.* 9.6.

Transformability: a General Attribute of Cells

It seems probable that transformability with foreign DNA is a general attribute of eukaryotic as well as prokaryotic cells. The main problem is that of obtaining cells in a suitably receptive state. In baker's yeast (*Saccharomyces cerevisiae*) transformation is carried out using **protoplasts**, i.e. cells which have been deprived of their cell walls by enzyme treatment and are left with only their plasma membrane as a barrier to the uptake of macromolecules. A similar technique can be applied to other fungi and probably also to cells of green plants. In yeast, especially, the possibility of introducing genetic DNA directly into cells has opened up exciting possibilities for the purification and analysis of genes.[14]

9.6 | Summary and perspectives

We have in this chapter surveyed a variety of parasexual mechanisms—ways of cell-to-cell transfer and subsequent segregation of genetic material outside the normal sexual cycle. The first examples which we considered were of interest in the first instance as providing methods of genetic *analysis* in organisms in which the absence of normal sexuality made conventional genetics impossible. But all modes of parasexual genetic transfer imply the possibility of *synthesizing* new genotypes which, in some cases, could not have been obtained in any other way. DNA-mediated, or chromosome-mediated, gene transfer can indeed work between distantly related species. Any suggestion that this opens up the possibility of making monstrous hybrids between, say, man and animals would be wide of the mark, since the cells successfully transformed by foreign genes always retain a vast majority of genes from their original species. Nevertheless, DNA-mediated transformation may well come to have some applications, especially perhaps in plants where regeneration of whole organisms from single cells is much more practicable than it is in animals.

One can only speculate whether and, if so, to what extent interspecific transfer of DNA has been significant in evolution. Natural selection, as we understand it, works by amplifying the effects of rare events. These events have generally been assumed to be mutation and recombination occurring *within* each species, but rare transfer of DNA sequences *between* species may also have played some part. The recent evidence that cells of virtually all kinds can not only take up alien DNA but even admit it, in relatively undegraded form, into their nuclei and then integrate it into their chromosomes, is surprising and thought provoking.

Whatever the significance of irregular modes of gene transfer in eukaryotes, there is no doubt of their importance in prokaryotes. Bacteria have no regular system of sexual reproduction and rely entirely on parasexual processes for genetic recombination. The substitutes for sex found in bacteria are the main theme of the following two chapters.

Chapter 9

Problems

9.1 A heterokaryon was made from *white* (*w*) and *yellow acriflavin*-resistant (*y Acr*) strains of *Aspergillus nidulans*, and a diploid derivative was purified from a green (w^+ y^+) sector which appeared in a heterokaryotic colony. This diploid proved to have a level of resistance to acriflavin less than that of the yellow parent but greater than that of the wild type. Segregation from the diploid was looked for (i) by searching plate cultures for white and yellow sectors and (ii) by selection for full acriflavin resistance. Segregants were classified as haploid or diploid on the basis of conidial size, and the colour segregants were tested for their acriflavin resistance.

All of the white segregants, both haploid and diploid, were as sensitive as wild type to acriflavin. All of the haploid yellow segregants were fully resistant; all of the diploid yellow segregants had the same intermediate level of resistance as the parental diploid. Haploid fully resistant segregants were yellow or green with approximately equal frequency; diploid fully resistant segregants were mostly stable green but with a substantial minority with one or more white sectors. From this information deduce as much as you can about the chromosomal locations of the three markers.

9.2 A *Saccharomyces cerevisiae* diploid strain was obtained by mating triple mutant *ade2 ade3 mal* (*adenine*-requiring, *maltose* non-utilizing) and double mutant *ade2 ade6* cells of opposite mating type. Note that the *ade2* single mutant is red because of the accumulation of a metabolic intermediate; the synthesis of this intermediate is blocked by *ade3* or *ade6* and so the parent strains were both white. The derived diploid was red because of $ade3^+–ade6^+$ complementation. Roman (*Cold Spring Harbor Symp. Quant. Biol.* 1956, **21**, 175) found on plating the diploid cells that a small minority of them gave white colonies. Analysis of 27 of these revealed the following genotypes (all were $\dfrac{ade2}{ade2}$):

$\dfrac{+\ \ ade3\ \ mal}{+\ \ ade3\ \ mal}$	$\dfrac{+\ \ ade3\ \ mal}{ade6\ \ ade3\ \ mal}$	$\dfrac{ade6\ \ ade3\ \ +}{+\ \ ade3\ \ mal}$	$\dfrac{ade6\ \ +\ \ +}{ade6\ \ +\ \ +}$
2	9	5	11

Deduce from these data the linkage relationships of the three markers. Explain the origin of each of the white colony types.

9.3 Ruddle et al. (*Nat. New Biol.* 1971, **232**, 69) fused human and mouse cells in culture and characterized a large number of hybrid subclones for their retention of human enzymes which could be distinguished electrophoretically from their mouse equivalents. The table, in which + indicates the presence and − the loss of the human enzyme in question, presents a small part of the data.

What can you conclude about the chromosomal locations of the determinants of the different human enzymes, given that G6PD is known to be sex linked from human pedigree analysis?

Clone no.	1	2	3	4	5	6	7	8	9	10
Enzyme:										
Lactate dehydrogenase (LDH)	−	+	+	−	+	−	+	+	−	+
Peptidase A (PEPA)	+	+	+	+	+	−	+	+	+	−
Glutamate–oxaloacetate transaminase (GOT)	−	−	−	−	+	−	+	+	−	−
Glucose phosphate isomerase (GPI)	+	+	+	−	+	−	−	+	+	+
Glucose 6-phosphate dehydrogenase (G6PD)	+	+	+	+	+	+	+	+	+	+
Phosphoglucomutase (PGM)	−	−	−	−	−	+	−	−	−	−
Phosphoglucokinase (PGK)	+	+	+	+	+	+	+	+	+	+

9.4 *Neurospora* conidia can, under certain conditions (removal of cell walls with enzyme, stabilization by high osmotic pressure, presence of Ca^{2+} ions) be genetically transformed by purified DNA. In one experiment, conidia of an *am* mutant (deficient in glutamate dehydrogenase, growth requirement for an amino acid) treated with wild-type DNA under transforming conditions and a frequency of about 10^{-6} grew to give nutritionally independent and glutamate dehydrogenase-producing colonies. How might this experiment be performed so as to rule out the possibility that the apparently transformed colonies are merely reverse mutations? (It is relevant here to note that some *am* mutants are deletions and also that certain *am* alleles produce varieties of the enzyme which are active but distinguishable from the standard wild type by various *in vitro* tests.) Given that the colonies really are transformants, how would one determine whether the transforming DNA had been integrated at the normal *am* locus or elsewhere in the genome?

9.5 The methods of parasexual analysis described in this chapter for *Aspergillus nidulans* are applicable in principle to *Penicillium chrysogenum*, a fungus without a sexual cycle and of great importance as a producer of penicillin. Considerable improvements have been obtained in penicillin yield by induction of mutations within *P. chrysogenum* clones. Outline a parasexual procedure whereby progess towards greater penicillin yield might be accelerated. Note that, like *Aspergillus*, *Penicillium* species have uninucleate conidia and are capable of mutation to conidial colour differences.

Bacteria and Bacterial Plasmids

10.1   Introduction

Until 1949, when genetic recombination in *Escherichia coli* was demonstrated by J. Lederberg and E. L. Tatum, bacteria were thought to reproduce purely clonally, subject to genetic variation only by mutation. Bacterial cells are haploid, and have no regular systems of sexual reproduction. We now know, however, that bacteria can undergo genetic recombination whenever homologous DNA molecules with different combinations of genetic markers are brought together into the same cell, and a vast development of bacterial genetics, mostly with *E. coli*, has been based upon this fact.

The opportunities for recombination between homologous but different genetic sequences in bacteria are limited, and only exist at all because of sporadic and generally rare transfer from cell to cell of fragments of DNA. Such transfer is brought about not by any process essential to bacterial reproduction, but by various kinds of invading DNA molecules which can establish themselves in the bacterial cell. These extraneous molecules are of two kinds: **plasmids**, which are autonomously replicating closed-loop DNA molecules transmitted only by cell division or cell-to-cell contact, and virus genomes, which have the ability to get themselves packed into protein coats (capsids) to form infective virus particles which are released into the extracellular environment. The bacterial viruses (bacteriophages) require a chapter to themselves. In this chapter we shall review the role of plasmids in bacterial genetics.

10.2   Transmissible plasmids

Col-, R- and F-plasmids

Plasmids affect the phenotypes of their host cells in various ways. The first plasmid-transmissible trait to be studied was the ability to produce **bacteriocins**—antibiotic proteins which are excreted into the medium by certain bacterial strains and which kill other strains. When produced by *E. coli*, they are called **colicins**. A colicinogenic plasmid always confers resistance to its own colicin. P. Fredericq, who was responsible for most of the earliest information about colicins, showed that the ability to produce (and be resistant to) a given colicin was transmitted from cell to cell in infectious fashion under conditions which permitted cell-to-cell contact. The infective principle was not released into the culture medium. The factors determining the colicins were called Col, with a distinguishing letter in each case. For example, ColI and ColV determine colicins I and V, respectively.

The property of infective transmission by cell contact, which is characteristic of Col-plasmids, is also displayed by another and even more widespread class of elements—the resistance-transfer or R-plasmids. These were discovered by T. Watanabe in Japan as infectious determinants of resistance to antibiotics and other antibacterial substances. It is common for one plasmid to confer resistance to a number of different drugs. For example R1, which was isolated from *Salmonella paratyphi*, but is transmissible to *E. coli* as well as to a wide range of other intestinal (enteric) bacteria, carries determinants of resistance to ampicillin, chloramphenicol, kanamycin, streptomycin, spectinomycin and sulphonamide. R6, which has a similar host range, lacks the ampicillin and spectinomycin determinants but includes paromomycin, neomycin and tetracycline resistances in its repertoire. Such resistance-transfer factors have now been found in samples of enteric bacteria from many parts of the world. Presumably they have existed for a long time at low frequences in bacterial populations, but the modern widespread use of antibiotics has greatly increased their prevalence.

One of the most intensively studied transmissible plasmids, the F (fertility) factor of *E. coli* strain K12, does not determine any colicin or drug resistance and was recognized initially only as a factor conferring on its host the ability to transfer genetic markers to other *E. coli* cells. The uses of F in genetic analysis are dealt with below (pp. 258–264).

In addition to the transmissible plasmids which have just been mentioned, there are others which are not infectious by themselves but can be assisted to pass from cell to cell through the presence, in the same host, of another plasmid which is self-transmissible. We discuss the mechanism of transfer below (p. 262). Especially notable among non-self-transferable plasmids is ColE1 which determines the colicin after which it is named and which, as we shall see, has become of great importance as a starting point for the construction of 'cloning vehicles' (pp. 271, 274).

10.3   Plasmid DNA

All bacterial strains harbouring infectious determinants of the kinds just reviewed can be shown to possess, in addition to the basic complement of DNA contained in their main 'chromosome' (p. 260), a minor DNA component consisting of relatively small, covalently closed, supercoiled molecules. These can be equated with the infectious determinants on several grounds. Most simply, acquisition or loss by a bacterial strain of colicinogenicity. drug resistance or fertility, as the case may be, always follows the acquisition or loss of the plasmid DNA. More elaborately, structural changes in the plasmid DNA, either occurring spontaneously or deliberately brought about by 'engineering' procedures (pp. 271–275), can be monitored and correlated with changes in the plasmid-borne genetic markers. In what follows, the role of plasmid DNA will be assumed and justified as we go on.

The plasmid DNA can be separated from the major DNA of the bacterial cell by virtue of the physical and chemical properties associated with the compact supercoiled circular structure. One method involves sedimentation of the DNA to equilibrium in a caesium chloride density gradient (cf. p. 12) containing ethidium bromide, a drug capable of intercalating into the stack of base-pairs which runs down the core of the duplex DNA molecule. A molecule with intercalated ethidium has its base-pairs forced apart and is consequently less dense, forming a band higher up the density gradient than would otherwise be the case. The closed supercoiled DNA circles are much less easily penetrated by the ethidium and thus remain closer to the normal density. Thus a DNA preparation from a plasmid-containing bacterium will show, in a caesium chloride–ethidium bromide density gradient, a minor band of supercoiled plasmid DNA further down the tube than the main band of broken relaxed DNA fragments from the bacterial 'chromosome'. The positions of the different components can be clearly seen under ultraviolet light, which induces brilliant fluorescence of the bound ethidium bromide.

A second method, which gives information about the size of the plasmid, is to run the DNA from a lysed bacterial culture through an agarose gel by electrophoresis. The relatively small and compact plasmid molecules will be able to penetrate a gel of a concentration sufficient to hold the much larger fragments of the chromosomal DNA almost immobile. The plasmid DNA in the gel can be seen by fluorescence following staining with ethidium, and its size can be assessed by reference to a suitable series of standards. A separation of this type, but used for bacteriophage rather than plasmid DNA, is shown in *Fig.* 11.10 (p. 306).

Plasmid molecules, separated by one of the methods just outlined, can be visualized under the electron microscope by the Kleinschmidt procedure already mentioned in Chapter 1 (p. 13), and their sizes measured directly. This is not so easy if both strands of the closed circular duplex are intact throughout, so that the whole molecule is supercoiled, but is straightforward when one of the strands has been 'nicked', as often happens during the preparation. In the latter case, one strand of the duplex is freed to spin around the other until the whole circle achieves its most relaxed open form. *Fig.* 10.1 shows the comparison between the supercoiled and relaxed circular forms of the DNA of resistance plasmid R6K).

Different plasmid DNAs differ greatly in size. The transmissible plasmids are usually of the order of 100 kilobases (kb), but non-transmissible plasmids may be much smaller. ColE1, for example, contains only 4·2 kb. The number of copies per cell also varies over a wide range. Whereas F, and most of the other larger plasmids, are restricted to one or a few copies, others can attain much larger numbers. ColE1 has the convenient characteristic of accumulating to high concentration in cells whose protein synthesis is inhibited by chloramphenicol.

Fig. 10.1

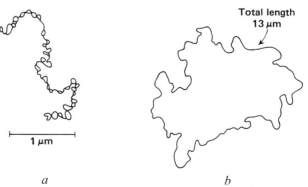

Total length
13 μm

1 μm

a   *b*

Electron microscope comparison of (*a*) covalently closed circular supercoiled double-stranded DNA of the ampicillin-resistance plasmid R6K, and (*b*) the same molecule nicked in one strand and hence relaxed into the open circular form. Note that this plasmid is not the same as R6 (*Fig.* 10.3) which is different in origin, is different in incompatibility group and resistance determinants, and is nearly twice as large.
Drawn from the photograph of Kontomichalou et al.[1]

The essential feature of plasmid DNA molecules is their ability to replicate independently of the bacterial chromosome. This depends on the presence of a replication origin (*see* p. 15): a special sequence where separation of the component strands of the duplex can begin. The ability to replicate probably also depends on the availability of a site on the cell membrane to which the replication origin can attach.

10.4 | The property of transmissibility

A characteristic of *E. coli* cells harbouring transmissible plasmids is their possession of **pili** (singular: pilus), which are surface filaments, distinct in thickness from other filamentous appendages on the cell surface and sometimes as long as the cell itself. The pili have an essential role in cell-to-cell plasmid transmission, but its nature has yet to be fully explained. Evidently the pilus tips can stick firmly to other cells within the specific host range of the plasmid, and pairs of cells can thus be anchored together as a preliminary to plasmid transfer. According to one idea, the pilus, the protein of which is known to be encoded by the plasmid, acts as a conducting tube through which a single strand of the plasmid DNA can be threaded. This is an intriguing notion, since a pilus is very similar in structure and dimensions to the ensheathing protein coat of a filamentous bacteriophage (*Fig.* 10.2). A pilus containing plasmid DNA could be regarded as a filamentous bacteriophage adapted for transmission by cell-to-cell contact rather than through free infective particles. Proof that transmission of the plasmid DNA

Fig. 10.2

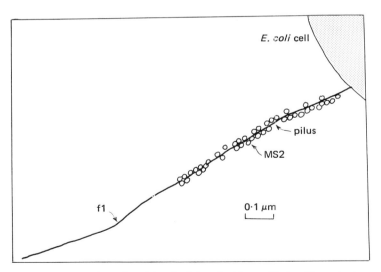

A pilus on the surface of an F⁺ cell of *E. coli* with two kinds of bacteriophage adsorbed to it: a filamentous phage (f1) attached end-on to the tip and numerous spherical phages (MS2) attached laterally.
Drawn from the electron micrograph of E. Boy de la Tour and L. Caro, reproduced in Broda.[13]

is indeed via the pilus rather than through some separate connection between the donor and recipient cells has, however, not yet been obtained.

The protein of which the pilus is composed is coded for by a gene in the plasmid DNA, but (in the F-plasmid which is the most intensively studied) this is only one of some ten identified plasmid genes, all of which are necessary for plasmid transmission. The functions of the other nine genes remain largely unexplained at the time of writing. References to the relevant original papers can be found in the recent book on plasmids by Paul Broda.

The process of plasmid transfer involves DNA synthesis which starts at a special origin on the plasmid DNA which is *not* the same as the origin used for normal plasmid replication within one cell. Only a single DNA strand is transferred; it is always the same strand, and its leading end comes from the transfer origin. It is likely that the mechanism is similar to 'rolling-circle' replication (*see Fig.* 1.7, p. 14), with a single strand being displaced from the plasmid duplex and 'pushed' into the recipient cell while, at the same time, the circle itself is repaired by new synthesis. Once transferred to the recipient cell, the single DNA strand acts as template for the synthesis of a complementary strand, and the resulting double-stranded molecule is made circular presumably by recombination between repeated ends of a complete linear sequence.

One other aspect of plasmid transfer must be mentioned. Most transmissible plasmids transfer most readily from cells into which they have only recently arrived. It seems that, with prolonged residence of the plasmid DNA, the cells accumulate a repressor of the plasmid transfer functions, so that very few pili are produced and further transfer becomes very inefficient. The F-factor is an exception in being permanently derepressed and always capable of efficient transfer—this is how it came to be recognized as a fertility factor. Derepressed mutants of normally repressed plasmids can also be obtained and it is evident that the repressor is the product of a plasmid gene. In F, this gene is inactive, but its function can be supplied, and the transfer of F inhibited, by the presence of another compatible plasmid of a related repressed type in the same cell. This phenomenon, which is an example of complementation between plasmid genomes, is known as **fertility inhibition** (*fi*). For example, R1 is fi^+ with respect to F, but derepressed (*drd*) mutants of R1 are fi^-.

10.5   Relationships among plasmids

Pilus Type and Plasmid Interactions

Plasmids can be classified according to a number of different criteria. The transmissible plasmids may, for example, produce pili like those associated with the *E. coli* F-factor or pili of a different kind, the best known of which is the type associated with ColI. F-type and I-type pili provide the specific adsorption sites through which several different kinds of bacteriophages attack the *E. coli* cell (*Fig.* 10.2), and the phages which attach to F-pili will not attach to I-pili or vice versa. A second criterion of relatedness to F, mentioned in the preceding paragraph, is the ability to repress F transfer by supplying the repressor function missing in the latter plasmid. A third property which permits rather finer

Fig. 10.3

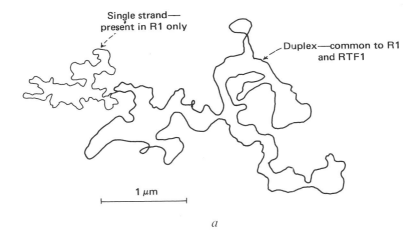

Single strand—
present in R1 only

Duplex—common to R1
and RTF1

1 μm

a

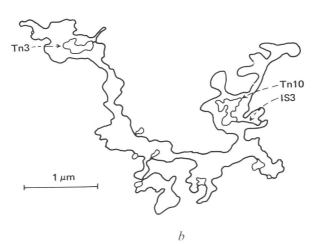

Tn3 --

Tn10
IS3

1 μm

b

a. Electron micrograph of heteroduplex DNA formed by R1 (95 kb) and RTF1 (65 kb). RTF1 (resistance transfer factor) was derived from R1 and has lost all the drug-resistance determinants carried by the latter plasmid. It retains all the functions necessary for replication and transfer and is largely homologous with the F-factor with respect to these.
Drawn from the photograph of Sharp et al.[2]

b. Heteroduplex formed by R1 and R100–1. The latter plasmid is very similar to R6 except that it lacks the determinant of resistance to kanamycin/neomycin. The heteroduplex shows a number of small differences and two major ones. One 9·2-kb sequence (transposon 10—Tn10) is present in R100–1 and not in R1; it consists of a 6·4-kb segment bounded by inverted repeats of the 1·4-kb IS3. Tn10 is present in a number of different plasmids and carries a gene for tetracycline resistance. A 4·8-kb segment (Tn3) is present in R1 but not in R100–1; it is flanked by only short inverted repeats and includes a gene for ampicillin resistance.
Drawn from the photograph of Sharp et al.[2]

discrimination between groups of plasmids is that of **incompatibility**, that is the inability of two different plasmids to replicate in the same host cell. This may be a consequence of competition for attachment to a specific cell membrane site, which is probably an essential prerequisite for replication. Plasmids which are mutually incompatible in this way are said to be in the same incompatibility group, and this is an indication that they are quite closely related, at least so far as their replication origin is concerned. Plasmids which resemble one another in pilus type, and also agree in being fi^+ or fi^- with respect to F, may still belong to different incompatibility groups.

Heteroduplex Analysis

The permutations of different characteristics shown by a range of plasmids suggest that reshuffling of plasmid genotypes has occurred in their evolution, and this is strikingly confirmed by studies of plasmid sequence

Fig. 10.4

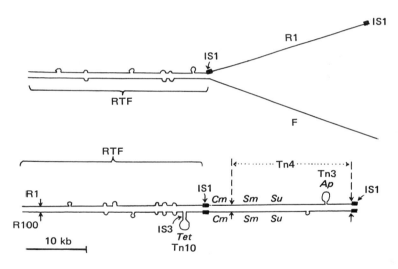

Diagrams summarizing heteroduplex comparisons of pairs of plasmids: F and R1 (above) and R1 and R100–1 (below). Complete lack of homology is indicated by divergent lines; locally non-homologous segments within regions of overall homology are shown as small loops (cf. *Fig.* 10.3*b*). The largely homologous region in the R1/F comparison is the resistance transfer factor (RTF) which carries all the functions needed for replication and transfer (cf. *Fig.* 10.3*a*).

Note how different plasmids differ with respect to blocks of DNA bracketed by repeated sequences—one block often inserted within another. The entire resistance-determining part of R1 is between *direct* repeats of IS1; the tetracycline-resistance determinant (*Tet*) between inverted repeats of IS3 in the transposon Tn10; the ampicillin-resistance determinant (*Ap*) between short inverted repeats in the transposon Tn3 (*see text*); the block containing the streptomycin and sulphonamide-resistance determinants (*Sm, Su*), but excluding the chloramphenicol-resistance determinant (*Cm*), between short inverted repeats indicated by arrows—this transposon (Tn4) is present in R1 and R100–1 but missing from some other related plasmids. For simplicity of drawing the circular DNA molecules are shown as if arbitrarily broken at one end of the RTF.
From Hu et al.[3]

relationships using the technique of **heteroduplex mapping**. If a mixture of two plasmid DNAs, fragmented and heated to give linear single strands, is allowed to reanneal under appropriate conditions of temperature and salt concentration, a proportion of hybrid duplex molecules will be formed if the two plasmids have sequence homology. If the sequences are only partly homologous, the hybrid duplexes (heteroduplexes) will show double-stranded regions where the single strands are fully complementary, and single-stranded loops and tails where they are not. The principle is illustrated by *Fig.* 10.3*a*, which shows the heteroduplex formed between a single strand of R1 and a partly complementary single strand of a derivative of it (RTF1) which has lost the drug-resistance determinants but retains the functions necessary for plasmid replication and transfer. This is a simple relationship, with one strand related to the other by a single continous deletion, but much more complex differences may be revealed by the technique, as shown in *Fig.* 10.3*b* and summarized in the diagram of *Fig.* 10.4. Reannealing of single strands may also reveal sequences within the *same* strand which can pair with each other in inverted orientation to give stem-and-loop structures. This, as is explained in *Fig.* 10.5, is liable to occur wherever inverted repeats occur close together in a DNA duplex.

When two different plasmid DNAs are compared by the heteroduplexing method, it is often found that some sequences are shared while others are present in one plasmid but not in the other. *Fig.* 10.4 shows the relationship between F and R1, both with DNA amounting to about 95 kb. About 45 kb of the sequence of

Fig. 10.5

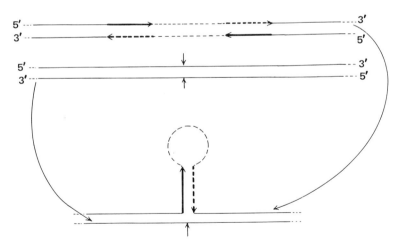

Explanation of the origin of the stem-and-loop structure formed by a single strand from a DNA duplex in which a unique sequence (thin broken line) is flanked by inverted repeats (thick lines) with the two complementary strands distinguished (3'→5' polarity indicated by arrowheads). The diagram shows the result of forming a heteroduplex from two DNA duplexes, one possessing a transposon which the other lacks. The site of insertion of the transposon is indicated in the duplex lacking it by arrows. An example of the situation shown is the Tn10 loop, with stem formed from inverted repeats of IS3, seen in the R1/R100 heteroduplex (*see Figs* 10.3*b* and 10.4). Note that the same stem-and-loop structure would be formed by one single strand by itself when inverted repeats are within a few kilobases of each other; this is the explanation for the extremely rapidly reannealing 'fold-back' fraction of DNA found in most experiments on kinetics of reannealing (*see* Chapter 1, p. 19).

each plasmid is largely shared with the other. This common sequence is the same as that retained in the RTF1 derivative of R1 (*Fig. 10.4*), and it evidently contains everything that is necessary for replication and transfer. The remainder of the R1 DNA (that missing in RTF1) contains all the R1 drug-resistance determinants. It is not known what functions, if any, are encoded in the substantial part of F which is not represented in R1.

A significant observation, made by very detailed heteroduplex mapping, is that the resistance-determining part of R1 is bounded by repeated copies (in the same orientation) of an approximately 800-base sequence. This sequence is found in scattered locations in both chromosomal and plasmid DNA in *E. coli* and must therefore be able, in some way, to move around the genome; it is called IS1 (**insertion sequence** 1) and its entire sequence of 769 base-pairs has now been determined (*see Box* 14.1). It, and other IS and analogous repeated sequences, are thought to play a key role in plasmid recombination and evolution.

Other important features are revealed by heteroduplex comparisons between different R-plasmids; *Fig.* 10.3*b* and 10.4 show the comparison of R1 and R100. Present in R100 but not in R1 is a sequence of about 5 kb bounded by 1·2-kb inverted repeats. This sequence is *within* the RTF region (though it is absent from that region in F and R1), is found in a number of R-plasmids including R6, and is associated with tetracycline resistance (*Tet^r*). It is called transposon 10 (Tn10) and is an example of a class of DNA segments which are bounded by repeated sequences and are capable of transposition from one plasmid or chromosomal site to another. Other examples of transposons are explained in the legend for *Fig.* 10.4. The 1·2-kb inverted repeats are, like IS1, present in multiple scattered copies in *E. coli* DNA, and are called IS3.

The structures and relationships of numerous different R-plasmids have now been determined in some detail. The picture which has emerged is of a number of different plasmid vehicles, belonging to different incompatibility groups, to which have been added further blocks of DNA in the form of transposons. We return to the question of the mechanism of transposition at the end of the chapter.

10.6   The role of the F-plasmid in *E. coli* recombination

F$^+$, F$^-$ and Hfr Strains: Time-of-entry Mapping

The F-factor renders *E. coli* 'sexually' fertile in the sense that an F$^+$ strain can transfer any of its genetic markers, though the efficiency of transfer is low for chromosomal markers compared with that of the plasmid itself. The original observation, made in 1949 by J. Lederberg and E. L. Tatum, was that a mixture of two strains carrying different auxotrophic markers yielded a small number of recombinants, usually selected as prototrophs, combining markers from each of the original strains. It was later found that at least one of the strains had to be F$^+$ in order for recombinants to be formed and that much, if not all, the recombination activity could be attributed to a minority of cells in the F$^+$ parent strains. These cells could be grown to give clones capable of **high-frequency recombination** (Hfr). The main structure of *E. coli* genetics has been built on the basis of Hfr strains crossed to strains lacking F (F⁻).

A striking feature of the recombinants recovered after mixing Hfr and F⁻ cells is that they inherit a limited set of markers (the composition of which depends on the particular Hfr strain being used) from the Hfr parent with high frequency, but that the bulk of their genome is inherited from the F⁻ parent. W. Hayes made a major advance with his demonstration that the Hfr parent always acted as a donor and the F⁻ as a recipient. He showed that the Hfr cells could be rendered inviable by antibiotic treatment prior to mixing and still be capable of donating markers to recombinants; only the continued survival of the F⁻ cells was necessary. A second key to the understanding of the Hfr × F⁻ cross, worked out by Hayes in London and by F. Jacob and E. Wollman in Paris, was the time dependence of the marker transfer. This was demonstrated by the classical **interrupted mating** experiment, an early example of which will now be described.

Cells of Hayes' first Hfr strain, called HfrH, were mixed with cells of an F⁻ strain carrying a number of mutations: *thr* and *leu*, threonine and leucine requirement; *azi*ʳ, resistance to azide; *ton*ʳ, resistance to bacteriophage T1; *lac*⁻ and *gal*⁻, inability to use lactose or galactose as carbon source; *str*ʳ, resistance to streptomycin. The Hfr strain was wild type with respect to all these markers. On

Fig. 10.6

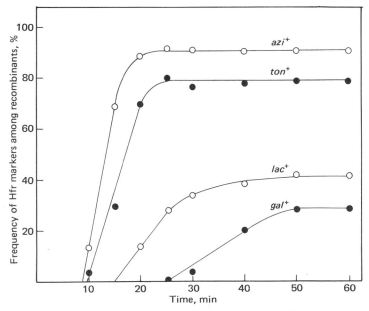

The sequential entry of HfrH genetic markers into the F⁻ cell during conjugation. After different periods of mating between Hfr and F⁻ cells, cell contact was broken by mechanical agitation and samples of cells were plated on minimal medium plus streptomycin. This selected for recombinants with *thr*⁺ and *leu*⁺ from the Hfr parent (these markers happen to enter very early from HfrH, with *thr*⁺ as the leading marker and *leu*⁺ following within a minute) and *str*ʳ from the F⁻ parent. Such recombinants were tested for the presence of *azi*ʳ (azide resistance), *ton*ʳ (bacteriophage T1 resistance), *lac*⁺ (lactose fermentation) and *gal*⁺ (galactose fermentation) from the Hfr parent. After Jacob and Wollman.[12]

Table 10.1	The sequence of transfer of markers from Hfr to F⁻ cells shown by different Hfr strains

Hfr strain	
HfrH	*thr-leu-azi-ton-pro-lac-pur-gal-trp-his-gly-str-mal-xyl-mtl-ile-met-thi*
Hfr1	*leu-thr-thi-met-ile-mtl-xyl-mal-str-gly-his-trp-gal-pur-lac-pro-ton-azi*
Hfr2	*pro-ton-azi-leu-thr-thi-met-ile-mtl-xyl-mal-str-gly-his-trp-gal-pur-lac*
Hfr3	*pur-lac-pro-ton-azi-leu-thr-thi-met-ile-mtl-xyl-mal-str-gly-his-trp-gal*
Hfr4	*thi-met-ile-mtl-xyl-mal-str-gly-his-trp-gal-pur-lac-pro-ton-axi-leu-thr*
Hfr5	*met-thi-thr-leu-azi-ton-pro-lac-pur-gal-trp-his-gly-str-mal-xyl-mtl-ile*
Hfr6	*ile-met-thi-thr-leu-azi-ton-pro-lac-pur-gal-trp-his-gly-str-mal-xyl-mtl*
Hfr7	*ton-azi-leu-thr-thi-met-ile-mtl-xyl-mal-str-gly-his-trp-gal-pur-lac-pro*
AB311	*his-trp-gal-pur-lac-pro-ton-azi-leu-thr-thi-met-ile-mtl-xyl-mal-str-gly*
AB312	*str-mal-xyl-mtl-ile-met-thi-thr-leu-azi-ton-pro-lac-pur-gal-trp-his-gly*
AB313	*mtl-xyl-mal-str-gly-his-trp-gal-pur-lac-pro-ton-azi-leu-thr-thi-met-ile*

For each Hfr strain the marker showing the earliest and highest frequency transfer is on the left, and the marker entering last and with the lowest frequency on the right of the series. The ensemble of sequences can be put together to form a circular map. Note that, in terms of this map (*Fig.* 10.7), transfer can occur clockwise or anticlockwise depending on the Hfr strain.
From Jacob and Wollman.[12]

Fig. 10.7

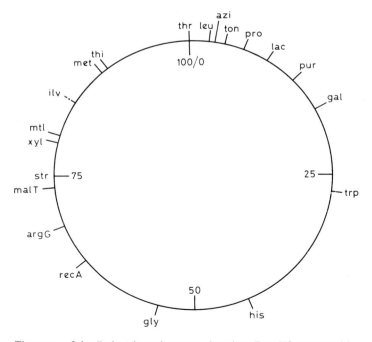

Time map of the *Escherichia coli* genome, based on F⁻ × Hfr crosses with interrupted mating. The map is calibrated from 0 to 100 minutes, this being close to the time required for transfer of the entire chromosome under standard conditions. Compare with *Table* 10.1 above. The markers which feature in the table are shown here, together with some others of special interest.

mixing, Hfr and F⁻ cells rapidly became joined by virtue of the F-pili produced by the Hfr strain. At different intervals following mixing the cell suspension was subjected to vigorous mechanical agitation, the effect of which was to break apart the joined cells, and plated, after appropriate dilution, on minimal medium containing streptomycin. In this way selection was imposed for $thr^+ leu^+ str^r$ recombinants. No such recombinants were found when cell conjugation was interrupted less than 8 minutes after mixing. Beginning at about 8 minutes, a few $thr^+ leu^+ str^r$ colonies began to appear on the plates, but up to 9 minutes these contained no Hfr markers other than the $thr^+ leu^+$ which had been selected for— all other markers came from the F⁻ parent. At 10 minutes the ton^s Hfr marker had begun to make its appearance among the selected recombinants, and lac^+ and gal^+ followed at 16 and 25 minutes, respectively. Each of the Hfr markers appeared to enter the F⁻ cell beginning at a fixed time after the onset of mating and, as *Fig.* 10.6 shows, thereafter increased in frequency among the recombinants, ultimately attaining a plateau frequency which was highest for the earliest-entering markers and progressively lower with later times of entry. In fact, after gal^+ had entered, further HfrH markers were transferred only at low frequencies.

That there were, in fact, markers beyond *gal* was readily shown by experiments with Hfr strains isolated independently of HfrH. Each of these was found to transfer a block of markers in a fixed time sequence and with decreasing frequency with increasing time. The composition of the block, and the time sequence within it, varied from one Hfr strain to another. *Table* 10.1 shows the marker sequences derived from eleven independently isolated Hfr strains. It will be seen that some of the sequences overlap, and that, with the aid of the overlaps, it is possible to position all the markers in a circular map encompassing all known *E. coli* marker mutations. Such a map, which is calibrated in units of time, is the standard way of representing the *E. coli* genome (*Fig.* 10.7).

Mechanism of DNA Transfer

The F-factor evidently determines the pili and other functions which are necessary for the cell conjugation and marker transfer, but what is the difference between an ordinary F⁺ strain, with its low frequency of transfer, and a derived Hfr strain with its thousand-fold higher frequency? Two pieces of evidence helped to answer this question. First, Hfr strains, in contrast to ordinary F⁺, did not transfer the F-factor or did so very rarely. In acquiring the capacity to transfer chromosomal genes they seemed to have lost F, or at least its infectivity. In fact, F is not lost in Hfr strains since such strains generally revert to the ordinary F⁺ condition with appreciable frequency. The second key observation was that the vast majority of recombinants formed in F⁻ × Hfr crosses were F⁻ and not Hfr. Only after a full 90 minutes of mating, corresponding to the time of transfer of the entire *E. coli* chromosome, was a small number of recombinants found which had inherited the Hfr property; these could often be shown to have extreme 'tail-end' markers from the Hfr parent.

These observations suggested that the formation of an Hfr from an F⁺ cell involved the **integration** of the plasmid into the *E. coli* chromosome. This integration could then be supposed not to destroy the normal ability of F to get itself transferred, but to ensure that in order to do so it had to take the whole chromosome along too. In fact, ordinary F transfer and chromosomal transfer mediated by an integrated F can be explained by the same mechanism. In both cases a free single-stranded end (probably 5′) is thought to be peeled off the

Fig. 10.8

Diagrammatic representation of a hypothesis to explain the efficient transfer of the F-plasmid from an F$^+$ to an F$^-$ *E. coli* cell without mobilization of the chromosome (*A, B, C*), and the efficient transfer of a chromosomal segment from an Hfr to an F$^-$ cell without (usually) complete transfer to the F-plasmid itself, which is integrated into the chromosome in Hfr strains (*D, E, F*). The plasmid DNA is represented by a thicker line. Its length is exaggerated here; in reality it is only about one-fortieth of the length of the chromosome.

Broken lines indicate newly synthesized DNA filling the place of the transferred strand. It has been demonstrated that it is always the same single strand of F which is transferred; in this hypothesis the transferred strand is the one which is nicked to initiate transfer.

In (*C*) and (*F*) the contacts between the conjugating cells are shown as broken after a time sufficient for complete transfer of F from F$^+$, or of the leading markers a, b, c but not the 'tail-end' markers x, y, z from Hfr. The direction of transfer, i.e. clockwise or anticlockwise, a→z or z→a, depends on which way round the plasmid sequence is inserted into the chromosome sequence.

plasmid sequence from a single-strand break at a replication origin. This free end then enters the recipient cell followed, in the case of the integrated plasmid, by the attached chromosomal DNA sequence. The transfer is believed to be driven by new DNA synthesis as pictured in the 'rolling-circle' model (*Fig.* 1.8, p. 14). When the F-plasmid is free the transfer of its DNA is complete within a minute or so, but in Hfr strains the F sequence has become part of the much larger DNA loop of the *E. coli* chromosome, which all has to be transferred before the trailing part of the plasmid sequence can enter the F⁻ cell to become reunited with its leading part. The transfer process may be terminated by accidental breaking apart of the conjugating cells at any time, and the odds are heavily against its persisting long enough for complete transfer to take place. *Fig.* 10.8 summarizes this interpretation.

Recombination

The transfer of a segment of chromosome from the donor Hfr cell is only the first step in the formation of stable recombinants. The transferred segment is normally unable to replicate in its own right, and the markers which it carries are only recovered in recombinant colonies if they are integrated into the recipient chromosome. Such integration requires two exchange events—one on each side of the marker(s) being integrated (*Fig.* 10.9). The frequency of exchange per unit length of DNA is high in *E. coli*. A length of chromosome equivalent to

Fig. 10.9

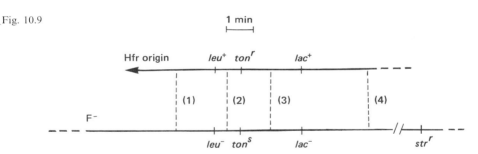

Recombination analysis of the same F⁻ × Hfr cross as gave the results shown in *Fig.* 10.6. In this experiment selection was made for *lac⁺* from the Hfr parent and *str^r* from the F⁻ (i.e. on medium containing streptomycin and with lactose as sole carbon source, supplemented with threonine and leucine). The selected recombinants, all of which must have arisen from F⁻ cells which received the whole Hfr segment between the Hfr origin and *lac⁺*, were scored with respect to the *leu* and *ton* markers. The integration of the Hfr *lac⁺* marker into the F⁻ chromosome requires *at least* two exchanges, one between *lac* and *str* (region 4) and one between the origin and *lac* (region 1, 2 or 3). The percentage frequencies of the recombinant genotypes were as follows:

> *leu⁺ ton^r lac⁺ str^r* 62 (exchange in 1),
> *leu⁻ ton^r lac⁺ str^r* 8 (exchange in 2),
> *leu⁻ ton^s lac⁺ str^r* 30 (exchange in 3),
> *leu⁺ ton^s lac⁺ str^r* 1 (triple exchange in 1, 2 and 3).

The distances between the markers are shown approximately to scale in time units; because no marker is known at the extreme leading end of the Hfr segment the distance from the origin to *leu* is known only roughly.
A probability of exchange of about 10 per cent for each minute on the time map is indicated.
Data from Jacob and Wollman.[12]

only 1 minute of transfer time, about one-ninetieth of the whole genome, has an exchange formed within it with about 10 per cent probability. Thus among the progeny clones arising after transfer by HfrH of the whole *thr–gal* segment, amounting to about 15 minutes of transfer time, all possible recombinations of the markers *azi, ton, gal* and *lac* are found in frequencies not far short of those expected from random reassortment. Only when markers are within 5 minutes of each other on the time map do they show significant linkage by the criterion of frequency of recombination. It is only time-of-entry experiments which give a clear picture of the overall structure of the genome. Recombination analysis is, however, useful in fine-structure mapping of sites within genes or of clusters of contiguous genes.

10.7   F-prime (F′) factors: plasmids carrying chromosomal segments

It has been mentioned that Hfr strains can revert to ordinary F^+, which implies, if our ideas about the nature of the Hfr state are correct, that F can not only be integrated into the chromosome but can also be excised from it. The integration hypothesis is supported, and some insight is obtained into the nature of the integration and excision processes, by examples of **inexact excision**, as a result of which the F DNA sequence becomes joined in a larger plasmid with a segment of chromosomal DNA coming from the region adjacent to the point at which F had been integrated. Such augmented F-plasmids are known as F-primes (F′) and are recognized in the first instance by the fact that they carry chromosomal markers with them whenever they are transferred from cell to cell. They can be isolated from $F^- \times$ Hfr crosses by selecting after short times of mating for Hfr markers which would not normally appear in recombinants until almost the whole chromosome had been transferred. The recombinants carrying precociously transferred markers turn out to possess F′-plasmids which have 'captured' the genes in question.

The detailed structures of several independently isolated F′-plasmids have been investigated by heteroduplex mapping. One structure (F′13) is represented in *Fig. 10.10c*. The plasmid circle includes the whole of the F sequence together with a larger segment of chromosomal DNA. Bracketing the F sequence, and forming the boundary between it and the chromosome fragment, are repeated copies, both in the same orientation, of a 1·3-kb sequence called IS2—a member of the class of **insertion sequences** (IS) which occur at scattered locations in bacterial and plasmid genomes. The IS sequences are thought to play an essential role in the integration of the F-plasmid. The proposed mechanism is sketched in *Fig. 10.10a, b*. It involves crossing-over between homologous IS sequences present both in the plasmid and the chromosome. The effect is the integration of the small plasmid circle into the large chromosomal one. Excision of the plasmid, with reversion of the Hfr to the F^+ condition, is supposed to occur through the same mechanism acting in reverse.

The plasmid F′13 was derived from an Hfr strain, 13, which transfers a chromosomal segment with *proC* at the leading end followed after about 2 minutes by *purA*. The *lac* locus, which in the circular *E. coli* map is within 1 minute

Fig. 10.10

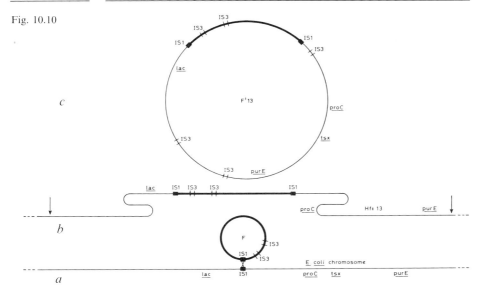

Hypothesis to account for the origin of Hfr13 from F⁺, and for F'13 from Hfr13.

a. Exchange between the IS1 sequence known to be present in F and a homologous IS1 sequence in the *E. coli* chromosome. Note that there are also two IS3 sequences in F, and that other Hfrs could arise by exchange between either of these and one of the IS3 sequences present in the chromosome.

b. Inferred structure of Hfr13 in the region of the integrated F-plasmid. Hfr13 injects *proC*, *tsx* and *purE* with high efficiency into F⁻ cells, and *lac* with low efficiency as the 'tail-end' marker. The duplicated IS1 sequences flanking the integrated F are the reciprocal products of crossing-over between the chromosomal and F IS1 sequences shown in (a).

c. Structure of F'13 as demonstrated by heteroduplex mapping and inferred from the fact that it carries the *proC, tsx, purE* and *lac* markers, all of which it transfers with high efficiency.

Note the IS3 sequences, which were presumably present in the chromosome prior to plasmid integration and excision. The F'13-plasmid was evidently formed by cutting and rejoining of the chromosome sequence at the arrowed positions. The much more common excision of the standard F-plasmid would occur by a reversal of the crossover event shown in (a). Other F'-plasmids have been shown to arise through one cut outside the F sequence and one cut within it, so that the F' contains only a part of the F DNA (a part which must include the replication and transfer origins if the excised circle is to act as a transmissible plasmid).

Structure of F'13 from Hu et al.[3]

of *proC* but on the opposite side to *purA*, is transferred by Hfr13 only very rarely and at the extreme end of the chromosome sequence. It is thus evident that F must have integrated between *proC* and *lac*. F'13 carries the wild-type alleles of *proC*, *purA* and *lac*, and so it looks as though it arose through excision of a DNA segment including F and substantial chromosomal sequences on each side of it. How this anomalous type of excision occurred is not yet clear, but it is likely to be an event of the same kind as leads to normal F excision, but utilizing sequences with some degree of homology with each other some distance away from F on each side, rather than the immediately flanking IS sequences. Other F'-plasmids appear to be the result of exchange between some site *within* F and another site outside it. These F-primes carry chromosomal sequences from only one side of the site of integration and only part of the F sequence. Since their recovery depends on their ability to replicate they always include the F replication origin.

F'-plasmids are important in *E. coli* genetics as a means of introducing a second copy of a chromosomal fragment into a cell already possessing a full genome. Strains harbouring F'-elements are partial diploids (or **merodiploids**) and can be used for studies of *cis* and *trans* interactions between homologous chromosome segments. They have been much used in complementation studies, which are essential for establishing functional allelism between mutants, and for tests for dominance and recessivity, which are important for the understanding of regulation of gene action (pp. 371–378).

10.8   Mechanism of general recombination in bacteria

It is necessary to distinguish between two types of recombination of DNA molecules. The first is **general recombination** which can operate between any pair of homologous DNA segments. The second is **site-specific** recombination, depending on special sequences, not necessarily homologous, in the molecules being recombined.

The Role of *recA*

The most definite lead we have to the mechanism of general recombination is the identification of an *E. coli* gene, called *recA*, which is absolutely essential for the process. Mutants of the *recA⁻* class form no recombinants at all when used as the F⁻ parent in an F⁻ × Hfr cross, and are also deficient in other homology-dependent recombination processes, including transformation (p. 275), transduction (p. 295) and integration of the F-plasmid.

The protein which is lost as a result of *recA⁻* mutation has been identified. As 'protein X' it had been known for some time as a molecule produced in unusually high concentration following ultraviolet (UV) irradiation, and thus presumed to be involved in repair of UV-damaged DNA (*see* p. 330). This protein has the property of forming a complex with single-stranded DNA and ATP and, when so complexed, of binding to and unwinding double-stranded DNA. When the single strand complexed with the protein is homologous with the double-stranded DNA which is being unwound, a three-stranded structure ('D-loop') is formed. The invading free single strand pairs with one strand of a duplex, displacing the complementary resident strand as a loop (*Fig.* 10.11).

Thus it is plausible that an early step in general recombination is the *recA*-mediated insinuation of a single DNA strand into a homologous duplex molecule. The following steps are more obscure and in *E. coli* there appear to be at least two different pathways each involving some limited digestion of DNA strands. Two genes, *recB* and *recC*, are necessary for the formation of an ATP-dependent nuclease which will degrade one strand of a DNA duplex starting from a local

Fig. 10.11

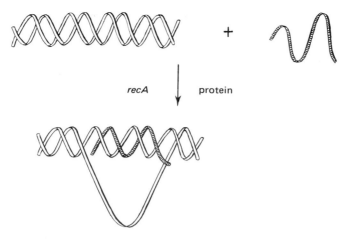

recA ↓ protein

The process catalysed by the protein product of the *recA* gene of *Escherichia coli*. The protein can catalyse the insertion of a DNA single strand into a homologous duplex, with displacement of a 'D-loop'.
After Cunningham et al.[4]

region of unwinding (a D-loop for example), but elimination of the function of either of these genes merely reduces the frequency of recombination rather than abolishing it entirely. The residual recombination in *recB/C* mutants is dependent on other genes, *sbcA, sbcB* and *recF*, the functions of which we will not dwell upon except to note that they involve other kinds of nucleases.

The absolutely essential function of *recA* in both pathways is taken to be the formation of an obligatory recombination intermediate—perhaps a D-loop. A possible sequence of events is that postulated in the hypothesis of Meselson and Radding (*Fig.* 5.3*B*, p. 138), which may apply both to eukaryotes and to prokaryotes.

10.9 | Site-specific recombination: transposons

Inverted Repeats and IS Elements

Recombination at specific sites or within specific sequences has been studied most extensively in prokaryotes, though there are now important examples of site or sequence-specific recombination in eukaryotes also (*see* Chapter 17). In *E. coli*, dependence on *recA* function is diagnostic of general recombination, whereas site-specific recombination is independent of it. In the following chapter we shall encounter the special recombination system concerned in the integration into the chromosome of the genome of bacteriophage lambda. In this chapter the relevant examples concern the 'wandering' DNA sequences which are found at variable locations within bacterial and plasmid DNA.

So far as they have been investigated, wandering sequences in *E. coli* are flanked

Fig. 10.12

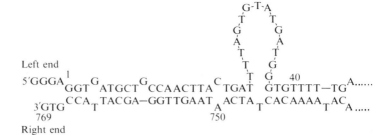

Alignment of the left and right ends of one strand of IS1. Note that the two ends, apart from a few single-base mismatches and gaps and one major insertion in the left-end sequence, are mutually complementary over a length of 30 bases. That is to say, one end of the double-stranded sequence is an approximate inverted copy of the other.
Data of Ohtsubo and Ohtsubo[8] obtained by the method outlined in *Box* 14.1.

by inverted repeats, so that the two ends of each such sequence look the same. This obviously fits with the idea that there is a special enzyme for cutting the ends of the sequence, excising it from the chromosome.

The most freely movable of prokaryotic DNA sequences are the transposons (cf. *Fig.* 10.3), which are often responsible for the drug resistance conferred by R-plasmids. In most of the transposons that have been closely studied, the terminal inverted repeats are provided by various IS sequences which are themselves movable (though at a much lower frequency) and are found at scattered and variable locations in *E. coli*, both in the chromosome and in plasmids. For example Tn10, referred to above as carrying a determinant of tetracycline resistance, is bounded by two copies of the 1·3-kb sequence IS3 in mutually inverted orientations. In apparent contrast Tn9, which confers chloramphenicol resistance, is flanked by two copies of IS1 (850 bases) in the *same* orientation at each end. However, analysis of the complete base sequence of IS1 has shown that the two ends of IS1 are themselves, with a few gaps and mismatches, inverted copies of one another (*Fig.* 10.12); the same applies, therefore, to the two ends of Tn9. A third transposon Tn3 (referred to in *Fig.* 10.3*b*), which confers ampicillin resistance, has what at first appeared to be its own unique pair of 39-base inverted terminal repeats. These repeats turn out to be rather closely related to those found in an IS sequence called, with some inconsistency of nomenclature, γδ.

Genetic Control and Mechanism of Transposition

Where resides the genetic information for the specific enzymes that cut at the ends of transposons so as to permit their transposition? Are they encoded in the transposons themselves or are they part of the general equipment of the cell, encoded in the bacterial chromosome? The latter alternative might seem unlikely, in as much as transposition is a function which the cell does not obviously need. It would fit better with the view of transposons as largely autonomous and ultimately 'selfish' sequences if they were to carry their own transposition apparatus. Nevertheless, at least some transposons may be dependent for their movement on host cell genes. The sequence of Tn9 has been completely deter-

Fig. 10.13

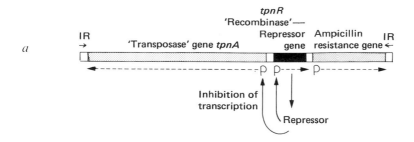

a

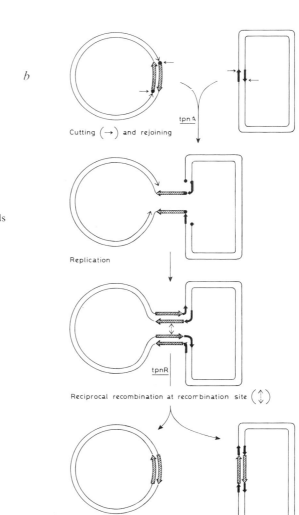

b

a. The structure of Tn3.
b. Postulated mechanism of transposition of Tn3 from one plasmid to another. The two plasmids are distinguished for diagrammatic purposes by shape—circle and square, respectively. The DNA double helix of each is simplified to two parallel lines. The Tn3 single strands are shown as cross-hatched arrows and the 9-base sequences which become duplicated in the recipient plasmid as solid arrows. Arrowheads indicate 3′ ends and round dots 5′ ends.
After Sherratt et al.[10]

mined[7] and almost the whole of it is occupied by the gene coding for chloramphenicol transacetylase, the enzyme responsible for the drug resistance. There is clearly no room for any further gene concerned with transposition unless it is within IS1, as is still possible.

The situation is very different in the case of Tn3, the sequence of which is also completely defined. In addition to the gene coding for β-lactamase, which accounts for the ampicillin resistance, this transposon contains *two* other genes, *tpnA* and *tpnR*, concerned with transposition.[5, 6, 9] The complementary nature of these genes is shown by the fact that two different mutant Tn3 transposons in the same cell, one with *tpnA* and the other with *tpnR* inactivated by deletion, are able to transpose normally although neither could do so alone. The two gene products can evidently act on their DNA targets from a distance and in *trans*.

A model summarizing our understanding of *tpnA* and *tpnR* function is shown in *Fig.* 10.13. There are three well-established facts about Tn3 transposition which are accounted for by this model. First, when Tn3 is transposed to the new site it is retained at the old site—in other words it is replicated in the process of transposition. Second, wherever it occurs it is flanked by nine base-pair direct repeats with a different sequence for each occurrence of the transposon. Whenever it is transposed to a new site it generates a new pair of flanking sequences by duplication of a nine base-pair sequence previously present at that site in only one copy. It thus looks as if, as well as getting itself replicated during transposition, the transposon also induces the replication of nine base-pairs at the new site of integration. The third line of evidence is based on the identification of **cointegrates**—fusions of the donor and recipient plasmids with Tn3 copies bridging between them at each of the two junction points. These structures are accumulated when the Tn3 is defective either in the *tpnR* gene or in a centrally located recombination sequence. They are evidently transposition intermediates, formed as a result of *tpnA* activity and requiring *tpnR* activity, acting at the recombination sequence, for their further conversion (the **resolution** step of *Fig.* 10.13*b*).

The mechanism postulated may be briefly summarized as follows:

i. The product of *tpnA* ('transposase') cuts in the donor plasmid at the 5′ ends of the Tn3 inverted terminal repeats and, at the same time, cuts the recipient plasmid at staggered sites nine base-pairs apart; the cut 5′ ends of the transposon are then joined to the 3′-OH ends generated by the cuts in the recipient plasmid. While the first cuts are specific for the Tn3 terminal sequences, the second cuts are apparently randomly placed subject only to the nine base-pair spacing which, it is conjectured, is determined by the geometry of the transposase protein.

ii. New synthesis initiated at the cut 3′-OH ends in the donor plasmid replicates both the transposon and the nine base-pair recipient sequence and creates a fully double-stranded cointegrate.

iii. Reciprocal recombination, specifically catalysed in the recombination sequence of the transposon by the *tpnR* product ('recombinase' or 'resolvase'), resolves the cointegrate into two separate plasmids, each now with one copy of the transposon. The function attributed here to *tpnR* is akin to that of *int* in lambda bacteriophage (*see* Chapter 11, p. 300).

The *tpnR* product has a second remarkable property, that of repressing transcription both of its own gene and of *tpnA*; *tpnR*⁻ mutant transposons produce unusually large amounts of *tpnA* protein.

At least some features of this scheme are likely to apply to the movement of other transposons and isolated IS elements, and perhaps also to freely movable sequences in the genomes of eukaryotes (*see* Chapter 15, p. 445). Inverted terminal repeats, replication during transposition and generation of short direct repeats in the recipient site seem to be general properties of freely movable DNA sequences wherever they are found. Note, however, that the *controlled* movements and rearrangements associated with cases of cellular differentiation considered in Chapter 17 are governed by different rules.

10.10 | Plasmids as cloning vehicles

Plasmids are of great importance in modern experimental genetics because they provide the means of amplifying and purifying DNA sequences that one may wish to study in detail. We have already seen how F'-plasmids can act as vehicles for the selective amplification of the segments of *E. coli* chromosomal DNA which they incorporate. This is an example of naturally occurring **cloning** of DNA sequences, and the formation of hybrid bacteriophage genomes, which is described in the next chapter, is another such example. The development of techniques for cutting and rejoining DNA sequences has led to the possibility of using plasmids for artificial cloning of any sequence which one is able to recognize.

To serve as a cloning vehicle a plasmid should possess a single site for cutting by a restriction endonuclease. For example, ColE1 was used in some of the earlier experiments because, as well as its other advantages of high copy number and amplification in the presence of chloramphenicol, it had a single EcoRI (*see Table 7.1*) site. This restriction enzyme will open the plasmid circle to give a linear molecule with complementary single-strand tails—5'-AATT....— at its ends. Under conditions appropriate for reannealing, these tails will stick together by complementary base-pairing to reform the original circle, but they will also stick to the identical termini of any other EcoRI restriction fragment, whatever the origin of the DNA. It is thus possible, by reannealing a mixture of EcoRI-cut ColE1 molecules with EcoRI fragments of some other DNA, to reconstitute circles in which the exotic DNA fragments are inserted between the ColE1 cut ends. The association can be stabilized by treatment of the annealed mixture with DNA ligase, which joins the juxtaposed 3'-OH and 5'-phosphate single-strand termini thus:

Fig. 10.14

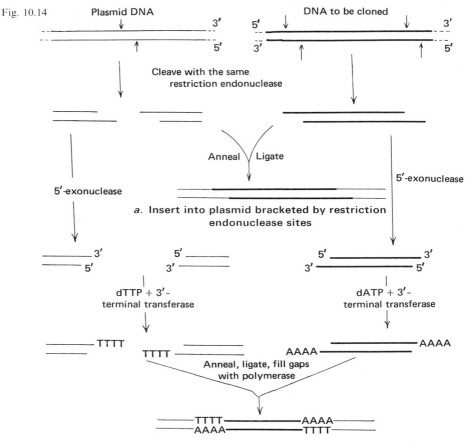

a. Insert into plasmid bracketed by restriction endonuclease sites

b. Insert into plasmid bracketed by oligo (dT)/oligo(dA) linkers

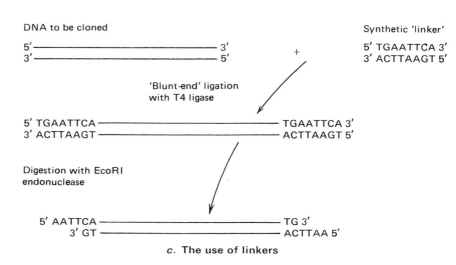

c. The use of linkers

Fig. 10.14

Methods for insertion of an extra piece of DNA from any source into a plasmid.
a. Cleavage with a restriction endonuclease such as EcoRI or HindIII, generating complementary single-stranded 'tails', which can be annealed and then sealed with DNA ligase.
b. Generation of fragments either by restriction endonuclease or by mechanical shearing, followed by 5'-exonuclease digestion to expose the 3'-OH termini. Sequential addition of A or T, with dATP or dTTP and 3'-terminal transferase gives mutually complementary oligo(dA) and oligo(dT) 'tails'. Only a section of the plasmid DNA is shown—the ends have to be imagined as joined in a closed loop.
c. The principle of using synthetic 'linkers' for joining of DNA fragments. The fragment to be cloned may be generated by mechanical shearing or by a restriction endonuclease. In the latter case the single-stranded tails, which will be present if the endonuclease is one which makes staggered cuts, must be converted to double strand with DNA polymerase and a deoxynucleoside triphosphate mixture. The linker is a self-complementary synthetic sequence of 10 bases, containing the EcoRI recognition sequence. Both fragment and linker are treated with polynucleotide phosphorylase and ATP to ensure that the 5'-terminal ends are phosphorylated; they are then mixed and ligated with the ligase isolated from *E. coli* cells infected with T4 bacteriophage ('T4 ligase') which catalyses end-to-end joining of fully double-stranded ('blunt-ended') duplex DNA. If the linker is present in the mixture in excess, most of the fragments will become joined to a linker sequence at each end. Treatment of the reaction mixture with EcoRI endonuclease generates EcoRI-specific 'sticky ends' which enable the fragment to be inserted into a EcoRI plasmid site. The EcoRI treatment also degrades to small easily removable fragments any end-to-end linker polymers which are formed in the ligase reaction.
Based on Heynecker et al.[11]

When introduced into an *E. coli* cell (*see* next section), the artificial hybrid will replicate just like normal ColE1 and the inserted segment will be amplified along with the ColE1 DNA.

A different technique permits the insertion of a DNA fragment into a plasmid even when the fragment has not been generated by the same restriction enzyme as was used to open the plasmid circle. Indeed the fragments need not be obtained by endonuclease digestion at all, but can be generated by mechanical shearing, which may be preferred if one wishes to fragment the DNA in a random way. The method is to add synthetic mutually complementary 'tails' to the 3' ends of the sequences to be joined. A necessary preliminary to this addition is the limited trimming back of the 5' termini by brief treatment with an exonuclease, so that the DNA duplexes are left with short 3'-OH-terminated tails. Then a string of adenylate or thymidylate residues can be added to these 3'-OH ends by incubation with dATP or dTTP and the enzyme **terminal transferase**. For example, oligo(dA) tails may be added to the DNA of the opened plasmid, and oligo(dT) to the ends of the fragments which it is desired to clone. Reannealing of the two differently tailed components will yield hybrid circles exclusively, through the A–T pairing. The tails will not all be of the same length, and the synthetic circles will thus tend to have single-strand gaps at their junctions, but these can easily be filled in by repair synthesis catalysed by DNA polymerase.

An even more sophisticated application of the artificial tailing principle (shown in *Fig*. 10.14) involves the addition to the exposed 3'-OH termini of prefabricated sequences corresponding to restriction enzyme cutting sites—for example 5'-GAATTC-3'-OH, corresponding to EcoRI. Reannealing and ligation of these 'linker' sequences will create an EcoRI restriction site at each plasmid-insert junction; this is very convenient if, as is often the case, one wishes to cut out the inserted sequence after it has been amplified in the hybrid plasmid.

Rather than using a simple naturally occurring plasmid, such as ColE1, as a cloning vehicle, most workers in the field now use more or less elaborately constructed compound plasmids for the purpose. It is desirable that the vehicle should carry markers to signal both its presence in the host cell and the fact that it is carrying a DNA insert. The artificial plasmid pBR322 is a good example. It derives its replication origin from ColE1, and shares with this plasmid the ability to replicate to very high copy numbers in the presence of chloramphenicol. It also carries two drug-resistance determinants, for ampicillin and tetracycline. Sequences for cloning may be inserted in any one of four unique restriction sites (*see Table* 7.1), for EcoRI, HindIII, BamHI and PstI, respectively (*Fig.* 10.15). It happens that the second and third of these are within the tetracycline resistance (Tc^r) gene, while the PstI site is within the ampicillin-resistance (Ap^r) gene. An extra segment of DNA inserted at any of these three sites inactivates the gene within which it falls. Thus a pBR322 derivative with an insert at the HindIII or at the BamHI site can be recognized through its determination of ampicillin but *not* tetracycline resistance. Likewise, an insert at the PstI site will eliminate the ampicillin resistance without interfering with the tetracycline resistance.

By making a sufficiently large collection of *E. coli* strains harbouring hybrid plasmids, it is possible to clone an entire genome in fragments. Such a collection is known as a **genome library**. To make use of such a library one needs efficient methods for identifying the items which it contains. There are two general methods which can be used in different cases. First, one can sometimes select clones for the *function* of a particular genomic fragment. This is only possible when the sequence in question is capable of expression (i.e. both transcription and translation into a recognizable protein or enzyme) in the *E. coli* cell, and there are many reasons,

Fig. 10.15

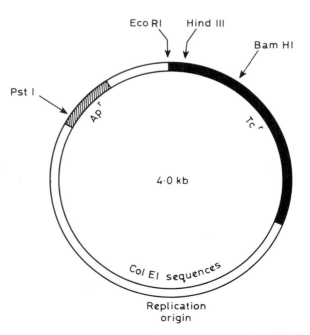

The structure of the artificially constructed cloning plasmid pBR322. *See text.* After Bukhari et al.[14]

including the presence of intervening sequences in the genomic fragments (*see* pp. 436–437), why such expression may fail. The recent development of cloning in yeast (p. 197) opens up more possibilities for detecting expression of eukaryotic genomic fragments. The second general method of recognizing specific sequences is to 'probe' for them using radioactively labelled complementary DNA or RNA sequences. The availability of a probe usually depends, in the first instance, on the isolation of the messenger RNA or other RNA transcript of the gene one is looking for. We shall meet several examples of the identification and characterization of specific sequences from genomic libraries in the later chapters of this book (*see* particularly pp. 392–394).

10.11 Transformation of bacteria with DNA

In its special genetic sense the word 'transformation' means the changing of the genotype of a cell by supplying it with free DNA from the outside. It was first demonstrated in the pathogenic bacterium *Diplococcus pneumoniae*. Heat-killed bacteria of one serological type (the effective antigen being the specific capsular polysaccharide on the cell surface) were shown to be capable of conferring their type of capsule on live cells which had lost their own capsular antigen after prolonged culture in the laboratory. W. H. Avery later showed that it was the DNA of the killed cells that was the active principle, and conditions were worked out for transformation of *Diplococcus* in the test tube using purified DNA preparations. Transformation was later shown to be possible for any heritable difference generated by mutation and not only for capsule type.

Following the early work on *Diplococcus*, attention turned to other bacteria, and it was shown that purified DNA would also cause transformation in species of *Haemophilus* and in the non-pathogenic saprophyte *Bacillus subtilis*. Certain characteristics of the transformation process seemed to be the same in all the bacteria in which it could be made to work. With rare exceptions, transformation occurred with respect to only one genetic trait at a time, which was taken to mean that the piece of DNA taken into the cell and integrated into the bacterial chromosome was generally too small to carry more than one gene or, at most, a few immediately adjacent genes. Studies with isotopically labelled DNA, with isolation of DNA from the recipient cells at early times after treatment, provided evidence that, although duplex DNA might be taken up by the cell, only a single strand was integrated into the chromosome at any one point. This is consistent with general models for recombination which involve single-strand assimilation with initial formation of a heteroduplex (cf. pp. 138 and 267).

In all the bacteria susceptible to transformation, care had to be taken to ensure that the recipient cells were in a receptive, or **competent**, state. This was a matter of determining the correct cultural conditions or, in the case of *Diplococcus*, the most receptive phase of the cell division cycle. For many years all attempts to find conditions under which *E. coli* could be transformed failed. It was eventually found that treatment with 0·05 M calcium chloride was extremely effective in enabling *E. coli* cells to take up DNA. This discovery has been of importance mainly because it has made it easy to introduce isolated plasmid DNA into host cells. Recombinant plasmids can now be fabricated in cell-free systems and then readily introduced into *E. coli* by the calcium chloride technique. Once in the cell they can replicate indefinitely.

10.12   Summary and perspectives

Plasmids are separately replicating DNA molecules which can confer a number of occasionally useful properties on bacterial cells. These properties include resistance to antibiotics and the ability to produce and be resistant to bacteriocins. Many plasmids also possess the means for bringing about their own transfer from cell to cell. In some cases, the best known of which is F, the *E. coli* fertility plasmid, the plasmid can integrate into the bacterial chromosome and bring about the cell-to-cell transfer of chromosome fragments. The F fertility system provides the means of mapping the *E. coli* genome in a single closed-loop linkage group.

Drug-resistance plasmids are often of modular construction with extensive blocks of DNA, called transposons, which are able to transpose from one plasmid to another. The sequence-specific recombination involved in these transpositions depends on gene products, presumed to be enzymes, which are different from those involved in general recombination. Some transposons encode their own transposition functions. Others, bracketed by the rather widely distributed IS elements, probably rely on transposition enzymes specific for these elements.

By getting into transmissible plasmids with different bacterial host ranges, a transposon may carry its drug-resistance determinant into bacteria of many different kinds The movement of transposons between plasmids and the movement of plasmids between cells have had dire consequences for medical practice; the usefulness of many antibiotics has been greatly reduced by the spread of resistant strains of pathogenic bacteria.

Plasmids can occasionally incorporate segments of chromosomal DNA and thus bring about natural 'cloning' and transfer of these segments between different cells and possibly even different species. This may well have important implications for bacterial evolution.

Through the use of enzymes that will specifically cut and rejoin segments of double-stranded DNA, it is now a relatively simple matter to use plasmids as artificial cloning vehicles, enabling one to purify and propagate DNA segments from virtually any source. The technique of transformation permits any gene cloned in and purified from a plasmid to be reintroduced and established in the bacterial cell. Such techniques have brought about a revolution in experimental genetics, not only of bacteria but of higher organisms as well.

| Chapter 10 | Selected Further Reading |

Broda P. W. (1979) *Plasmids*, 197 pp. Oxford, Freeman.

Bukhari A. I., Shapiro J. A. and Adhya S. L. (eds) (1977) *DNA Insertion Elements, Plasmids and Episomes*. New York, Cold Spring Harbor Laboratory.

Hayes W. (1968) *The Genetics of Bacteria and their Viruses*, 2nd ed. Oxford, Blackwell Scientific Publications.

Jacob F. and Wollman E. (1961) *Sexuality and the Genetics of Bacteria*, 374 pp. New York, Academic Press.

Kleckner N. (1981) Transposable elements in prokaryotes. *Annu. Rev. Genet.* **15**, 341–404.

Willetts N. and Skurray R. (1980) The conjugation system of F-like plasmids. *Annu. Rev. Genet.* **14**, 41–76.

| Chapter 10 | Problems |

See Problems at end of Chapter 11.

11

Bacteria and Bacteriophages

11.1 Introduction: a survey of *Escherichia coli* bacteriophages

The second aspect of bacterial genetic analysis, that which does not involve plasmids, is dependent on the bacterial viruses or bacteriophages (phages for short). Several of these have become important objects of genetic analysis in their own right. Some of them have a role in promoting the cell-to-cell transfer of fragments of bacterial DNA. This latter process of phage-mediated exchange of bacterial genes is called **transduction**, and a great deal of genetic analysis has been based upon it. What follows is mainly concerned with phages which attack *Escherichia coli*, but it should be realized that all groups of bacteria have their own viral parasites.

Structure

E. coli phages show a great range of size and complexity. They all, so far as they are known, consist of a single molecule of nucleic acid (either RNA or DNA, and either single stranded or double stranded) packaged in a protein coat or **capsid**.

A summary of the most important features of some of the most famous *E. coli* phages is given in *Table* 11.1. In some of the smallest, including the single-stranded DNA phages fd and φX174 and the single-stranded RNA phages such as MS2 and Qβ, there is only a single type of protein molecule in the virus particle—the one of which the capsid is built. For several of these simple phages the entire nucleotide sequence has been determined, and each of the very limited number of genes has been defined. We shall consider MS2 and φX174 in Chapter 15 as examples of gene organization in very small genomes (pp. 425–427).

The other extreme is represented by T4 and the other closely related 'T-even' phages T2 and T6, which have some thirty times as much DNA per particle as φX174. T4 resembles simpler phages in having a single capsid ('head') protein, but in addition is equipped with a complex tail, including six fibres which have the function of attaching specifically to the *E. coli* cell wall, a base plate which becomes fastened to the wall by six pins, a tubular core which serves as an injection syringe for the DNA, and a contractile sheath which, attached to the head at one end and the base plate at the other, drives the core into the cell (*Fig.* 11.1*a*). All of these tail components have their own distinctive proteins, each of which is encoded in the viral DNA. The complex phages, such as T4, also specify in their own genomes many of the proteins and enzymes involved in their DNA replication.

Replication

The replication of bacteriophage nucleic acid is a complex and, in some cases, controversial subject. Some of the simplest systems are the best understood. The RNA of MS2 and Qβ is replicated in linear form through the activity of an RNA replicase which is encoded in one of the three genes which constitute the total genome of these simplest of viruses (*see* p. 427). The 'plus' strand, which is present in the infecting particle, is used as template for the formation of complementary 'minus' strands, which are then repeatedly transcribed into plus strands for a new generation of phage particles. The relatively simple DNA phage, φX174, has single-stranded circular DNA in its infective particles, and this is made double stranded through the action of DNA polymerase after entering the cell. The double-stranded circles act as intermediates in replication; only one of the two strands, the newly synthesized 'minus' strand, acts as a template for further synthesis, which results in the formation of multiple copies of the 'plus' circle which are encapsidated to form new phage particles.

In **lambda** phage (*see below*) there is good evidence, most directly from electron microscope studies, that the DNA forms a closed loop ('circle') before it is replicated in the host cell. Circularization takes place through the annealing of the complementary single-stranded tails ('sticky ends') which are a feature of the structure of the DNA of the lambda infective particle (*Fig. 11.2c, d*). After the circle has been formed, and the whole structure covalently sealed by DNA ligase action, the lambda DNA is replicated by divergent replication forks, just like a bacterial chromosome but on a smaller scale (*see* pp. 12–15, Chapter 1). At a late stage of infection the mode of replication changes, and concatenates of repeated end-to-end copies of the genome are formed by the 'rolling-circle' mechanism. Phage-sized pieces of double-stranded DNA for packaging in the lambda capsid are formed from concatenates through the action of endonuclease making staggered cuts in a specific sequence.

The rolling-circle model may apply to the replication of the DNA of the T-even phages of *E. coli*. The DNA of the infective particles has a feature which seems well suited to circularization—that of **terminal redundancy**, in which a considerable sequence (about 2 per cent of the genome) is repeated, in the same orientation, at both ends of the linear DNA of the phage particle. Recombination between the homologous terminal sequences could yield a circle. Once formed, the circle might replicate in the rolling-circle mode to generate a concatamer, which could then be cut at random into lengths equal to 102 per cent of the unit genome. This view accounts nicely for the fact that the DNA molecules in the phage particles do not have fixed ends but are rather a family of **cyclically permuted** sequences (in the same sense as the marker sequences injected by different Hfr *E. coli* strains are cyclically permuted—*see Table* 10.1). An alternative view of the T4 life cycle supposes that concatamers are produced by recombination between cyclically permuted monomers (themselves perhaps the products of linear replication) rather than by rolling-circle replication.[1]

The DNA which is packaged in the T4 head is of fixed length but has random ends. In T4, and also in phage P1, a minority of the phage particles have unusually small heads; this is possible because of an alternative three-dimensional geometry of packing of the head protein subunits. These smaller heads contain correspondingly smaller DNA molecules—still with cyclically permuted sequences but terminally deficient rather than terminally redundant. Thus it seems likely that the size of the DNA molecule which is cleaved out of the concatamer is governed by the size of the head available for packaging it; this idea is known as the 'head-full' hypothesis.

Other kinds of phage have yet other modes of replication. T7, which has terminal redundancy with fixed ends, appears to replicate as a linear molecule. Phage *Mu*, which is notable for its induction of host cell mutations by virtue of its ability to integrate its DNA anywhere in the host chromosome, replicates in the integrated state, transposing as it does so.

Fig. 11.1

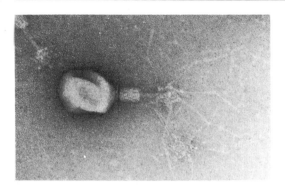

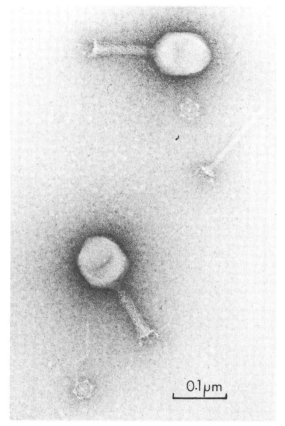

a

Bacteriophages.
a. T4, showing contracted tail sheath and tail fibres (upper picture) and uncontracted sheath, base plate with six spikes but stripped of tail fibres (lower picture). Two detached hexagonal base plates (one with a tail fibre still attached) are seen end-on. Magnification × 140 000.
b. Lambda.
c. Lambda phage attached to an *Escherichia coli* cell.
Electron micrographs by courtesy of Dr P. J. Highton.

Fig. 11.1 cont.

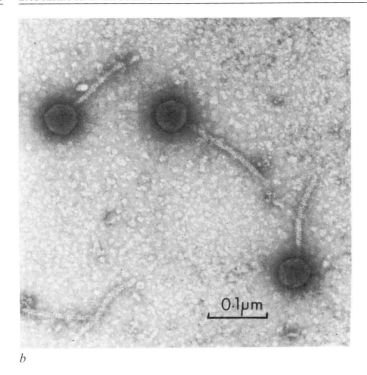

b

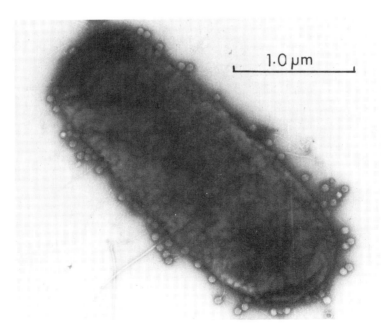

c

Table 11.1   **Summary of the main properties of some important bacteriophages**

Name	Form of particle	Dimensions
fd (also f1 M13, F12)	Filament	circa 800×6 nm
f2, MS2, R17 (Qβ a distant relative)	Spherical particle	25 nm
φX174	Spherical particle	25 nm
T4, T2, T6 ('T-even')	Head + rigid tail with contractile sheath	Head 120×80 nm Tail 95 nm
T7	Head + vestigial tail	Head 63 nm Tail 15 nm
P1	Head + rigid tail with contractile sheath	Head 85 nm Tail 220 nm
P22	Head + vestigial tail	Head 60 nm Tail 7 nm
Lambda	Head + flexible tail, no contractile sheath	Head 55 nm Tail 150 nm
Mu-1	Head + rigid tail with contractile sheath	Head 54×61 nm Tail 100 nm
M13	Spherical particle	15 nm

*ss, ds = single-stranded, double-stranded.
All the phages listed are specific for *E. coli* except for the P22, which is a *Salmonella typhimurium* phage.

Size and form of nucleic acid*	Capable of lysogeny	Properties of prophage
ss DNA circle 5·7 kb	No (but only mildly virulent)	—
ss RNA linear 3·6 kb (MS2 3569 b)	No	—
ss DNA circle 5375 b	No	—
ds DNA, linear, terminally redundant, cyclically permuted 170 kb	No	—
ds DNA, linear, terminally redundant, not cyclically permuted 37 kb	No	—
ds DNA, linear, terminally redundant, cyclically permuted 85 kb	Yes	Free circular plasmid
ds DNA linear, terminally redundant, cyclically permuted 40 kb	Yes	Integrated near *pro* in *Salmonella* genome
ds DNA, linear, with ss cohesive ends 48 kb	Yes	Integrated between *gal* and *bio* in *E. coli* genome
ds DNA, linear, short inverted repeats at termini flanked by host sequences 38 kb	Yes	Integrated anywhere in *E. coli* genome
ss RNA, closed loop	No	—

'male-specific' phages fd. f2 etc.. which will attack any bacterium with a plasmid determining an F-type pilus, and

Fig. 11.2

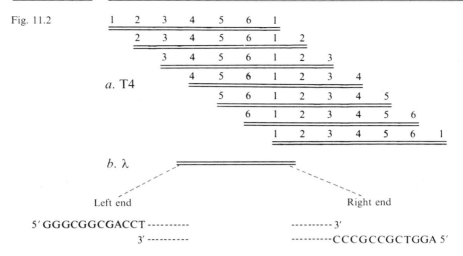

The reasons (*a*) why bacteriophage T4 has a circular genetic map even though the DNA in the infective virus is not circular, and (*b*) why bacteriophage lambda is able to form circular DNA in the infected host cell even though it has linear DNA and a linear two-ended genetic map. DNA molecules in the infective T4 virus are a family of cyclically permuted, terminally redundant sequences. In lambda, where there is only one type of DNA molecule, the ends have mutually cohesive single-stranded 'tails'; the nature of the lambda cohesive ends was worked out by Wu and Taylor.[2]

Lysogeny

Perhaps the most important distinction to be made between phages is that between exclusively **virulent** types, whose interaction with their host cell is confined to killing it, and those **temperate** phages which are capable of establishing a balanced relationship with their host in which the viral DNA is present in a latent state, known as a **prophage**. The state of harbouring a prophage is known as **lysogeny** because, under certain conditions brought about experimentally or by chance, the prophage can be unleashed, with lysis of the host cell and liberation of infective particles (a process called **induction**).

Some prophages, such as that of P1, undergo controlled replication as free DNA circles, just like a plasmid of low copy number. A more usual situation, however, is for the prophage to become integrated into the host DNA, to replicate in this state for an indefinite number of cell cycles, and to become excised again when induction occurs.

11.2 T4: a model virulent bacteriophage

The T4 Life History

Bacteriophage T4, and its close relatives T2 and T6, are exclusively virulent in their interaction with *E. coli* cells, a fact which limits their usefulness in the study of the genetics of the host cell. But as models for virus genetics the T-even phages have probably had more impact than any other virus.

The basis for T-even phage genetics was laid by the early 1940s by M. Delbrück with his demonstration of the basic features of the life cycle and his working out of simple procedures for titrating the phage particles released after a single cycle of infection. When a suspension of particles is added to an *E. coli* culture, most of the phages adsorb to bacterial cells and inject their DNA within a few minutes. About 22 minutes later all the infected bacterial cells undergo lysis and each one liberates of the order of 100 T4 particles. The standard way of following this process is to take samples of the infected culture at intervals and to plate them after suitable dilution on growth medium (solidified with agar) in mixture with a large excess of T4-sensitive *E. coli* cells. The latter will grow to form a continuous 'lawn' over the surface of the plate, with circular clearings or **plaques** marking the centres of infection by T4. Each centre of infection may be due either to an infected cell which had not yet lysed at the time of plating or to a single phage particle released by lysis. Thus, in an experiment of this kind, the plaque count increases about 100-fold at the end of the latent period of infection.

Mutants

The classical genetics of T-even phages, like that of any other virus or organism, depends on mutations for identifying the functions of the genetic determinants and marking their positions in the genome.

In bacteriophage, the scope for easily visible morphological change is obviously limited, but mutations affecting plaque size or appearance are easy to find (*Fig.* 11.3). In T-even phages there are three classes of **rapid lysis** (*r*) mutants which give larger-than-normal plaques on *E. coli* strain B. These, on the basis of recombination experiments (*see below*), map at three distinct loci; mutants at the *rII* locus have the additional characteristics of not growing at all on *E. coli* strain K12. The inhibition of *rII* mutants in K12 is due to the presence in that strain of the lambda prophage. Another readily scorable type of mutation (*tu*) results in 'turbid' plaques with fuzzy edges. Others extend the host range—for example mutants of the *h* series can form plaques on strains of *E. coli* that carry a mutation conferring resistance to the wild-type phage (h^+).

The greater part of T-even (especially T4) phage genetics, however, is based on **conditional mutants**—that is mutants whose growth is restricted to certain conditions. These include **temperature-conditional** mutants, which form plaques at 25 °C but not at 42 °C, and **suppressor-sensitive** mutants, which form plaques only on strains of *E. coli* that carry one of a series of genetic suppressors which can overcome the effects of a whole class of mutations called *amber*. As we see in Chapter 13, amber mutations are changes from amino acid coding to chain-terminating codons (cf. p. 358), and the suppressor mutations bring about changes in the anticodons of specific species of tRNA, enabling amber codons to be translated as if they coded for amino acids. Virtually any gene which codes for a protein can give rise to both temperature-sensitive and amber mutants, and the whole array of such mutants can be sorted out into groups corresponding to functional genes or cistrons by the complementation criterion.

In T4, the simple test of complementation is mixed infection. If two mutants individually are unable to form plaques at 42 °C, or unable to do so at any temperature in a host not carrying a suppressor mutation, but grow under the

Fig. 11.3

a. Plaques formed by bacteriophage T2 particles released from *E. coli* cells infected by a mixture of two mutant phages. One mutant, *r*, is distinguished from the wild type by forming larger plaques; the other, *h*, by its ability to grow on the T2-resistant *E. coli* strain B/2. The mixedly infected *E. coli* cells were of strain B which permits growth of both *h* and h^+ phages. The liberated progeny phages were plated with a mixture of B and B/2 cells, so that *h* phages made clear plaques, being able to lyse both cell types, and h^+ phages turbid plaques in which B cells had been lysed but B/2 cells remained. The plaques formed were of four types: two parental (hr^+, small clear and h^+r, large turbid) and two recombinant ($h^+ r^+$, small turbid and *hr*, large clear). The plaques shown occurred on a section of a plate which showed, overall, about 40 per cent recombinants.
Drawn from the photograph in Hayes.[13]
Magnification ×2.

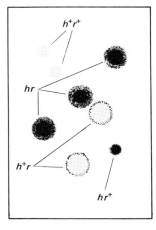

a

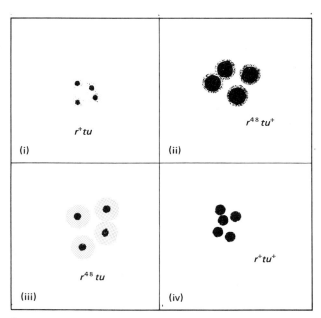

b

b. The appearance of plaques formed by the two parental and two recombinant phage T4 types generated by one of the crosses referred to in *Fig.* 11.4.
Drawn from the photographs of Doermann and Hill.[3]

restrictive condition in mixed infection, then it is inferred that the two mutations are affecting different functions, so that each mutant can supply the function missing in the other. Conversely, failure of the mixed infection under the restrictive condition indicates that the two mutants cannot complement each other in *trans* (*see* p. 129) and are thus, by definition, affected in the same cistron.

Recombinational Analysis in T4

If two different T4 mutants are allowed to multiply together in mixedly infected cells, a proportion of the phage particles released are of recombinant genetic types. For example, two different conditional lethals may yield anything up to 20 per cent wild-type recombinant phage particles capable of forming plaques at the restrictive temperature or on the non-permissive host. A mixed infection with two mutants characterized by visually distinguishable plaques may likewise produce up to 20 per cent wild-type recombinants and another 20 per cent of progeny phage carrying both mutations. *Fig.* 11.3*a* shows an example from the genetics of bacteriophage T2.

Forty per cent is about the maximum recombination frequency found from mixed infections with T4 (or T2). We should not expect to find the 50 per cent recombination characteristic of unlinked or distantly linked markers in classical eukaryote genetics since, in a phage cross, we are not observing the outcome of a single round of recombination in which we can be sure that all the input molecules have participated. Rather we are mixing two phage genomes in a ratio which may be known as an average, but which must be indeterminate for any one cell; we then leave them to replicate together for a number of generations which again may be known as an average but is certainly variable between one molecular lineage and another, and we do not know *a priori* how frequently during replication the phage genomes pair and recombine.

The highest frequencies of recombination are found when the numbers of the two phage genotypes put into the cross are equal, and it is easy to see why this should be so. If the proportions are p and $1-p$, random pairwise associations will involve two of one kind with a frequency p^2, two of the other kind with frequency $(1-p)^2$, and one of each kind (which is the only case which can lead to genetic recombination) with frequency $2p(1-p)$. Clearly this latter frequency achieves a maximum value of 0·5 when $p = 0·5$; any departure from this equality will increase the proportion of encounters between like molecules at the expense of those between unlike molecules.

N. Visconti and M. Delbrück, who first worked out the theory of random mating of phage genomes within the mixedly infected *E. coli* cell, calculated that the upper limit of about 40 per cent recombination is what one would expect if, with equal numbers of the two mutants, there were five rounds of replication with pairing and genetic exchange at each one. There is evidence for multiple rounds of recombination in that, when the mixed infection is by *three* different strains (e.g. $a^+ b c$, $a b^+ c$ and $a b c^+$), recombinants carrying markers from all three (e.g. $a^+ b^+ c^+$) can be recovered. However, this result does not prove the occurrence of repeated pairings since it could as well be a consequence of *simultaneous* interaction between three or more molecules, analogous to multivalent formation at meiosis in a polyploid (cf. p. 162).

Probable recombination intermediates, in the form of branched DNA molecules with segments derived from different parental phages, have been detected in electron microscope preparations.[4] These appear to arise through the association,

Fig. 11.4

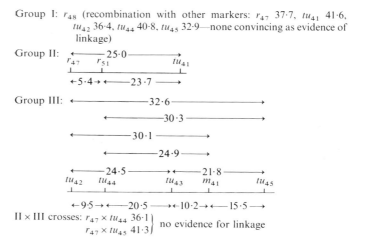

Group I: r_{48} (recombination with other markers: r_{47} 37·7, tu_{41} 41·6, tu_{42} 36·4, tu_{44} 40·8, tu_{45} 32·9—none convincing as evidence of linkage)

Group II:

Group III:

II × III crosses: $r_{47} \times tu_{44}$ 36·1 } no evidence for linkage
$r_{47} \times tu_{45}$ 41·3

Genetic mapping of bacteriophage T4 using two-point crosses. The distances between markers are shown in terms of percentage recombinants released from *E. coli* strain B cells mixedly infected with two single mutants in approximately equal numbers. The phenotypes of the phage plaques are described by the symbols: *m* = minute, *r* = rapid lysis (large plaque), *tu* = turbid (*see Fig.* 11.3). As explained in the text, only recombination frequencies below about 40 per cent can be taken as evidence for linkage.
Data of Doermann and Hill.[3]

through complementary base-pairing, of single-stranded regions of the parental duplexes. As in other recombination systems which we have considered (Chapter 5), heteroduplex formation is probably a key step.

Although linkage was detected quite early in the study of T-even bacteriophages, attempts to construct an overall linkage map were hampered by the high proportion of two-point crosses in which no linkage was detectable. The earlier studies, performed mostly with mutants giving visibly distinct plaques, defined three linkage groups, with a clear linear sequence within each group but no, or only doubtful, linkage between them. Some of the data are shown in *Fig. 11.4*. Later work by G. Streisinger and others[5] resulted in the establishment of connections between the groups and the construction of a circular map. The completion of the circle was made possible by the use of three and four-point crosses. The principle relied upon was one well established in classical genetic mapping—namely, that the most probable marker sequence was the one that required the smallest number of crossovers to account for the commonest recombinant classes (cf. p. 107). Thus, in a cross $a + + \times + bc$, with a and b closely linked and c farther away, recombinants selected as $+ +$ with respect to the a and b markers could be scored with respect to c. Without linkage there is no reason why $+ + +$ and $+ + c$ recombinants should not be equally frequent. If, on the other hand, c is significantly linked to either of the other two markers we would expect a departure from this 1 : 1 ratio. If the c allele occurs among the selected recombinants with a frequency significantly greater than 50 per cent, the linkage sequence c–a–b is indicated, since this allows the commonest recombinant class to be explained in terms of a single exchange thus:

$$
\begin{array}{ccc}
+ & a & + \\
\hline
c & + & b
\end{array}
$$

A frequency of c of less than 50 per cent indicates the sequence a–b–c, consistent with the majority class arising thus:

$$\underset{+\ \ b\ \ c}{\overset{a\ +\ +}{\diagup}}.$$

Table 11.2 shows some of the data which led to the closing of the circle.

When the circularity of the linkage map was first demonstrated it was in apparent contradiction to the finding that the DNA in the phage head was a linear molecule. One way of resolving the paradox was to assume that the molecules present in phage particles were a family of cyclically permuted sequences, as if derived through single random cuts in circular molecules. On the basis of this model, the *average* distance between two markers in the population of linear derivatives would increase as their distance apart on the hypothetical circular progenitor increased. If the markers are separated by a shortest distance d in the

Table 11.2	Three-point crosses with bacteriophage T4 mutants, establishing the circularity of the linkage map (data from Streisinger et al.[5])		
Cross	Scoring	Progeny ratio*	Inferred marker sequence
$+\ +\ r_{47} \times$ $r_{67}\ r_{73}\ +$	Only $r_{47}{}^+\ r_{73}{}^+$ form plaques on K12(λ)†; $r_{67}/+$ scored by eye	$\dfrac{r_{67}\ +\ +}{\bullet\ \ +\ \ +} = 0\cdot33$	r_{67}–r_{73}–r_{47}
$tu_{44}\ +\ + \times$ $+\ tu_{42}\ r_{67}$	$tu^+\ r$ and $tu^+\ r^+$ scored by eye	$\dfrac{+\ +\ +}{+\ +\ \bullet} = 0\cdot74$	tu_{44}–tu_{42}–r_{67}
$+\ +\ tu_{44} \times$ $r_{48}\ tu_{45}\ +$	$tu^+\ r$ and $tu^+\ r^+$ scored by eye	$\dfrac{+\ +\ +}{\bullet\ \ +\ \ +} = 0\cdot72$	r_{48}–tu_{45}–tu_{44}
$r_{73}\ +\ + \times$ $+\ r_{47}\ r_{48}$	Only $r_{73}{}^+\ r_{47}{}^+$ form plaques on K12(λ)†; $r_{48}/+$ scored by eye	$\dfrac{+\ +\ +}{+\ +\ \bullet} = 0\cdot29$	r_{73}–r_{47}–r_{48}

* • indicates either wild type or mutant at the locus indicated.
† r_{47}, r_{51}, r_{73} are *rII* mutants which grow on *E. coli* B but not on K12(λ) (*see text*, p. 292); the relative positions of r_{51} and r_{73} are not defined.

Derived circular map:

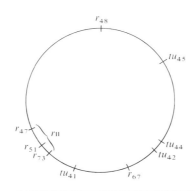

circular genome, d being expressed as a fraction of the total genome length, they will be d units apart in a proportion $(1-d)$ of phage particles and $(1-d)$ units apart in a proportion d. Their average separation $2d(1-d)$ will be close to $2d$ when d is small and will reach a maximum value of 0·5 when the markers are at diametrically opposite points on the circle. We now know from physical evidence that linear T4 DNA is indeed cyclically permuted, though the terminal redundancy which it shows is not consistent with the linear molecules being cut from circles of unit genome length, but demands rather that they be derived from a concatameric intermediate as discussed on p. 281 (*Fig. 11.2a*).

Fine Structure of the T4 Genome

The most extensive collections of T4 mutants consist of conditional lethals, and these are particularly convenient for the detection of very low frequencies of recombination. The progeny of a cross between two such mutants can be assayed under restrictive conditions, at the non-permissive temperature or on a bacterial host not carrying an amber suppressor, and a small number of recombinants capable of forming plaques can easily be detected among a large excess of inviable parental types. The resolving power of the method is limited only by the spontaneous mutation frequency, which is several orders of magnitude lower than the smallest recombination frequencies (about 10^{-4}) actually found. Thus any two conditional lethals which can recombine with each other at all can easily be shown to do so.

An extremely important and pioneering study of the fine genetic structure of T4 was carried out in the early 1950s by S. Benzer. He used the *rII* mutants, which are conditional lethals in that they will form plaques on *E. coli* strain B but not on strain K12(λ). All these mutants are relatively closely linked; the most recombination that any two of them show is about 10 per cent. Two subclasses, representing different cistrons, were distinguished by complementation tests; any *rIIA* mutant complements any *rIIB* mutant to lyse K12(λ) in mixed infection, but two different A mutants or two different B mutants are ineffective, except for some instances of relatively weak complementation between B mutants (*see* allelic complementation, p. 365).

Benzer[6, 7] devised a rapid and sensitive method for mapping sites of mutation within the *rII* region based on an extensive series of deletion mutants, identified as such by their inability to revert to *rII*⁺ by mutation and by their failure to recombine with sets of other mutants which would recombine with each other. His procedure was to seed plates of nutrient agar with a mixture of B and K12(λ) cells in a ratio of about 1 : 100, together with sufficient T4 particles of one of the deletion mutants to infect about 10 per cent of the cells. Then, at marked positions on the plate, drops of suspensions of other *rII* mutant phages were applied. Three kinds of result were observed in different cases:

i. Firstly, one may find scattered plaques within the area of the drop, due to *rII*⁺ recombinants released from the mixedly infected cells of strain B, on which both mutants can grow. This result shows that the two mutants do not complement (since the whole background of K12(λ) cells is not lysed), but they *will* recombine and must therefore have non-overlapping mutational damage.

ii. Secondly, there may be no lysis at all, showing that the mutants are neither complementary nor capable of recombining. This indicates that the mutation in the phage added in the drop must overlap or fall within the deletion present in the phage already present in the plate.

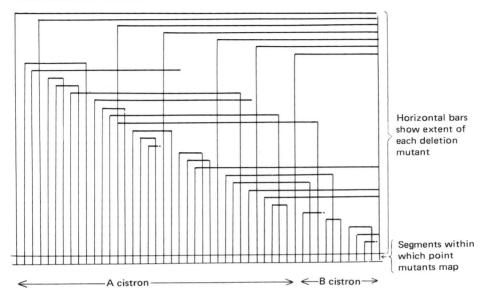

Horizontal bars show extent of each deletion mutant

Segments within which point mutants map

←———————— A cistron ————————→ ←—B cistron—→

Fig. 11.5

a. Deletion map of the *rII* region of the T4 map. Non-overlapping deletions show recombination in mixed infections. Deletions confined to region A complement those confined to region B; deletions overlapping both regions do not complement any other *rII* mutant. Point mutations can all be mapped into one or other of the numbered and lettered segments on the basis of presence or absence of recombination with each number of the set of deletion mutants. Redrawn from Benzer.[6]

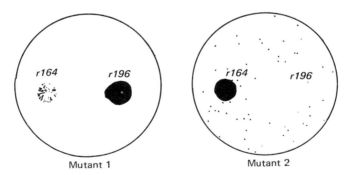

Mutant 1   Mutant 2

b. Spot tests used in the classification of *rII* mutants. Each plate was seeded with bacteria about 10 per cent of which were infected with phage carrying an *rII* mutation. The plate was spotted with one drop of a suspension of phage particles with deletion *r164* in region A (left) and one drop of phage with deletion *r196* in region B (right). The mutant 1 present all over the left-hand plate is a deletion mutant in region A since it complements with the B deletion to give complete clearing in the right-hand drop and does not revert to *rII*⁺ (no plaques formed from the 10⁸ phages present in the plate outside the area of the drops). It must be non-overlapping with the *r164* deletion since *rII*⁺ recombinants giving plaques are formed within the *r164* drop. The mutant 2 in the right-hand plate has its mutation in region B since it complements to give complete clearing with the A deletion; it must itself be a point mutant since it reverts to *rII*⁺ (small plaques present scattered over the plate). It must fall within the *r196* deletion since it forms no recombinants when mixed with *r196*. Adapted from the photograph of Benzer.[7]

iii. The third possibility is clearing of the whole area covered by the drop, showing that the K12(λ) cells were all lysed by the mixed infection and hence that the two mutants were complementary. This means that the effect of the deletion is confined to one cistron, with the second mutation located in the other. Deletion mutants overlapping both cistrons will not complement any *rII* mutant.

On the basis of presence or absence of recombination in many such tests, it proved possible to arrange many different deletion mutants in an unambiguous linear sequence on the basis of the pattern of overlaps. Most were confined to either the A or the B cistron, though a few overlapped both. The two cistrons occupied adjacent but non-overlapping regions of the map.

The many mutants which, because they were capable of reverse mutation to rII^+, were evidently not deletion mutants, were with rare exceptions able to recombine with each other in all pair-wise combinations. These **point mutants** could easily be placed without ambiguity into one or another of a large number of linearly ordered segments defined by the deletions and their overlaps. Their almost universal recombinability showed that they nearly all represented mutations at unique sites. Benzer mapped several hundreds of such sites, and the rate of discovery of new ones was consistent with the idea that there were as many mutable and mutually recombinable sites in the two cistrons as there were base-pairs in the DNA (the latter number was not known, but on the plausible assumption that *rIIA* and *rIIB* code for polypeptide chains it seemed likely to be of the order of a thousand). *Fig. 11.5* shows some of the results. The *rII* system became a model for fine-structure genetic analysis both in bacteriophage and in cellular organisms.

The Complete T4 Map

Many conditional lethal mutants (temperature-sensitives and 'amber' suppressible) have now been classified by complementation and mapped by recombinational analysis. The circular map now shows some 65 end-to-end cistrons. The functions of many of these have been defined either by biochemical tests or by electron microscopy of the defective phages or phage components made by conditional lethal mutants under non-permissive conditions. In general the results are fully in accord with the generalization (*see* Chapter 13) that each cistron codes for one kind of polypeptide chain. We shall return to consider the significance of the gene arrangement, or 'genetic architecture', of the T4 genome in Chapter 15 (p. 423).

11.3 | Temperate phages: general transducing agents

Bacterial DNA in Defective Phage Particles

Two of the temperate phages listed in *Table 11.1*, P1 of *E. coli* and P22 of *Salmonella typhimurium*, have been of great importance, less as objects of study for their own sake than as aids for the genetic analysis of their host bacteria.

In this connection the fact that they are temperate, i.e. able to lysogenize cells as an alternative to lysing them, is important, since it means that many cells survive infection. Among such survivors, it is easy, by appropriate selective screening, to detect the presence of genetic markers that could only have come from the bacterial strain on which the infecting bacteriophage was last grown. This phenomenon, discovered in the *S. typhimurium*–P22 system by N. Zinder and J. Lederberg, is called **transduction**, meaning transfer of genetic markers between bacterial strains through the agency of bacteriophage grown on one and used to infect another.

It was later shown that the transducing capacity was confined to a special class of **defective** phage particle, in which the normal phage DNA had been wholly replaced by a DNA fragment of equivalent size apparently taken at random from the bacterial genome. Packaging of bacterial DNA occurs in the formation of less than one in a thousand phage particles and, since each particle can contain only about one to two per cent of the bacterial genome, any particular bacterial gene is transduced by only about one particle in 10^5. Nevertheless, transduction can be easily shown for any selectable mutation, e.g. one conferring prototrophy on an auxotroph, or resistance to phage or a drug on a sensitive strain. All this applies equally to P22 in *S. typhimurium* and P1 in *E. coli*, although in many other respects particularly in their lysogenic states (cf. *Table* 11.1), these phages are very different.

Linkage and Recombination Investigated by General Transduction

Since each transducing particle can package only a little more than 1 per cent of the bacterial genome, simultaneous transduction of two different donor markers, at anything more than the extremely low frequency expected for the coincidence of two independent events, is evidence of rather close genetic linkage. To take a simple example, if P1 is grown on wild-type *E. coli* and used to infect a *thr⁻ leu⁻* double mutant, one can select, on medium supplemented with threonine but not with leucine or vice versa, for transduction of just one of the two markers and then ask whether the selected transductants have also received the other marker. In this case the number of double transductants (**cotransductants**) is only 1 or 2 per cent of the number of single transductants. Even this low frequency is some orders of magnitude greater than could be accounted for by simultaneous transduction by two different transducing particles, and constitutes strong evidence that the two genes are relatively close. At the same time, they are only just close enough for **cotransduction**. Time-of-entry mapping shows them to be separated by about 1·5 minutes on the time map (cf. *Fig.* 11.9), or about 1·5 per cent of the genome.

Though very different in mechanism, transduction shares important features with F-mediated conjugation (pp. 258–264). In both systems only a fragment of the donor genome is transferred to the recipient, the fragment being about ten times larger on average in conjugation than in transduction. In both cases the donor DNA cannot replicate in the recipient cell until it has been integrated into the chromosome by heteroduplex formation or (so far as longer tracts of DNA are concerned) by double or even-numbered multiple crossovers. Since transduced fragments are much smaller than most of those transferred by conjugation, multiple crossovers are much less important in the former case. Both in transduction and in conjugation the fragmentation of the donor genome prior to integration into the recipient occurs in two stages—**prezygotic**, or before transfer, and **postzygotic**, that is by recombination after transfer. Prezygotic separation of donor markers is relatively more important in transduction than in conjugation because of the smaller size of the transferred donor fragments.

Transduction involves too fine a fragmentation to be very useful for working out the large-scale architecture of the genome; for the latter purpose, F-mediated conjugation with time-of-entry analysis is more suitable. But for fine-structure analysis, transduction is an extremely valuable technique and is often the method of choice. Both deletion mapping and the selective analysis of recombinants from three or multipoint crosses can be readily carried out through the use of transducing phage. Some examples are given on pp. 298 and 310.

11.4 Lambda

Functions and Genome Structure

Bacteriophage lambda, which was discovered as a latent phage in *E. coli* K12, can either lyse the cells that it infects or make them lysogenic. These two possible outcomes of lambda infection represent two alternative developmental pathways, the choice between which is governed by a delicately balanced switch mechanism which will be discussed in outline in Chapter 17. The two pathways depend on two different sets of genes, most of which have been mapped by procedures that are essentially the same as those already described for bacteriophage T4. The lambda map obtained by analyses of the progenies of mixed lytic infections is linear rather than circular. This result is expected since λ DNA is packaged in lambda particles as linear molecules with fixed ends, rather than as a set of cyclically permuted sequences as in T4 (*Fig.* 11.2). The arrangement of the genes is significant both from the point of view of understanding how their activities are regulated (*see* p. 490, Chapter 17) and in relation to the use of lambda for the construction of cloning vehicles (Section 11.6). In brief, the central section of the map contains genes essential for lysogenization but not for lytic growth, while the flanking arms contain all the genes involved in the replication of the viral DNA prior to cell lysis, in coding for the head and tail proteins of the infective virus particles, and in the lysis of the host cell.

An essential feature of the DNA of lambda particles is the presence, at the two ends of each linear molecule, of mutually complementary single-stranded 'tails'. The sequence at one end is 5'-GGGCGGCGACCT..... (the dots indicating the commencement of double-stranded structure) and, at the other end, ...CCCGCCGCTGGA-3'. These single-stranded termini, which are 'sticky' by virtue of their ability to anneal by complementary base-pairing, provide the means for circularization of the DNA after it has entered the host cell. They are generated by the endonuclease that cleaves the concatenated DNA (formed in the last stage) of λ DNA replication—*see* p. 15) with cuts staggered twelve base-pairs apart.

The Nature of the Lysogenic State

E. coli cells that have survived infection by lambda harbour the phage genome in latent form and are **lysogenic**, in that the latent phage (**prophage**) can be **induced** to grow and cause lysis. The simplest inducing treatment, which works with high efficiency, is irradiation with low doses of ultraviolet light such as would be insufficient to kill a non-lysogenic bacterial strain. The second distinguishing feature of lysogenic cultures is their immunity to lytic infection by free lambda particles.

Wild-type lambda makes turbid plaques, since, although a majority of the infected bacteria are lysed, there is always a background of growth due to the cells which have survived the infection and become immune. Mutants of lambda that are unable to establish the lysogenic state are recognized through their giving clear plaques. Of three classes of clear-plaque mutants (c_I, c_{II} and c_{III}), the first is deficient in the formation of a **repressor** protein which both maintains lysogeny, by preventing induction of the prophage, and confers immunity by similarly inhibiting the growth of any secondarily infecting lambda virus. We shall give further consideration to the functions of the c_I suppressor and the other c gene products when we deal with the genetic control of differentiation in Chapter 17.

What is the nature of the prophage in the lysogenic cell? A partial answer to this question came with the discovery, totally unexpected at the time, that in a conjugational cross between a lysogenic (and hence lambda-immune) F⁻ strain and a lambda-sensitive Hfr strain, the sensitivity/immunity to the bacteriophage was inherited as if associated with a chromosomal locus between *gal* (galactose fermentation) and *bio* (biotin synthesis)—respectively at 16·7 and 17·2 minutes on the time map. Sensitivity to lambda was transferred from the Hfr to the F⁻ parent starting at 17 minutes. When the cross was made the other way round, with the lysogenic strain as the Hfr parent and the lambda-sensitive strain as the F⁻, a strikingly different result was obtained. Starting at 17 minutes from the beginning of conjugation, the F⁻ population initiated phage production with the ultimate release of free infective lambda particles. It was evident that the latent phage was being transferred at this time from the Hfr to the F⁻ cells which, lacking immunity, were unable to prevent the prophage from starting immediate uncontrolled replication.

The presence or absence of the prophage, then, appeared to be a genetic marker associated with a specific short segment of *E. coli* DNA. What was not immediately clear was the mode of association—whether the phage genome was integrated into the linear sequence of the *E. coli* DNA or whether it was in some way attached to the host DNA as a lateral appendage. Detailed studies on the relation of the prophage genetic map to that of the host chromosome proved that the former alternative was correct. *Table* 11.3 summarizes the results of a transduction cross between two *E. coli* strains, one lysogenic with wild-type lambda prophage, and the other with a prophage carrying four suppressible (amber) mutations. The first strain carried a *gal⁻* mutation (unable to use galactose as carbon source), while the second was *gal⁺*; both carried an amber–suppressor mutation, permitting growth of amber phage mutants. P1 bacteriophage was grown on the *gal⁺* parent and used to transduce the *gal⁻* parent. Selection was made for *gal⁺* transductants and these were induced by ultraviolet light in order to reveal what kind of lambda they now contained. The lambda phage released from each transductant was scored by complementation tests—mixed infection into suppressor-negative cells with reference lambda strains

Table 11.3   *A.* **The lambda prophage map is a cyclical permutation of the vegetative map**

gal⁺ transductant class	Number found	Simplest interpretation
gal^+ + + + +	2147	gal^+ transduced alone
gal^+ sus_7 sus_5 sus_{10} sus_{96}	344	All donor markers cotransduced—all markers linked to gal
gal^+ sus_7 + + +	77	gal^+, sus_7 cotransduced—sus_7 closest to gal.
gal^+ sus_7 sus_5 + +	41	gal^+, sus_7 sus_5 cotransduced—contiguous markers
gal^+ sus_7 sus_5 sus_{10} +	24	gal^+, sus_7, sus_5, sus_{10} contiguous—sus_{96} at the end of row

gal^+ sus_7 sus_5 + sus_{96}	8	
gal^+ sus_7 + + sus_{96}	3	
gal^+ + sus_5 + sus_{96}	2	
gal^+ + sus_5 + +	2	Less frequent types resulting from multiple exchanges between the transducing fragment and the *E. coli* chromosome
gal^+ + sus_5 sus_{10} sus_{96}	2	
gal^+ + + + sus_{96}	3	
gal^+ sus_7 + sus_{10} +	1	
gal^+ sus_7 + sus_{10} sus_{96}	1	
gal^+ + sus_5 sus_{10} +	1	

Deduced prophage sequence:
gal–sus_7–sus_5–sus_{10}–sus_{96}

Transduction cross: P1 phage grown on wild-type *E. coli* lysogenic with lambda carrying the suppressor-sensitive mutants sus_{10}, sus_{96}, sus_7 and sus_5, used to transduce gal^- *E. coli* lysogenic with wild-type lambda. The sequence of the *sus* markers in the 'vegetative' map obtained from recombination frequencies in mixed lytic infection (cf. *Fig.* 11.4 for analogous mapping in T4) is $sus_{10}sus_{96}sus_7sus_5$.

B. **The *gal–bio* distance is increased in lambda lysogens**

Cross (P1 transduction)	No. bio⁺ among gal⁺ transductants	No. gal⁺ among bio⁺ transductants	Average percentage cotransduction
$gal^+ bio^+$ donor × $gal^- bio^-$ recipient	416/939	233/470	47
$gal^+(\lambda)bio^+$ donor × $gal^-(\lambda)bio^-$ recipient	128/3280	18/633	3·5

Data from J. Rothman.[8]

carrying each of the amber mutations singly. In this way it was possible to show that either one, two, three or all four of the prophage mutations could be cotransduced with gal^+, and that they showed a clear gradient of cotransduction strongly suggestive of a linkage group extending linearly from gal at one end (*Table* 11.3*A*). Similar studies, in which bio^+ was used as the selective marker in transduction, showed the same linkage order of the prophage markers connecting to *bio* at the other end. This genetic evidence showed that the prophage map was indeed linearly inserted into the sequence of chromosomal markers between *gal* and *bio*. Also consistent with this conclusion was the demonstration that, by the criterion of frequency of cotransduction, *gal* and *bio* were considerably more widely separated in strains lysogenic for lambda than in lambda-sensitive strains (*Table* 11.3B).

The prophage map, obtained by methods such as those just described, can be compared with the map based on the frequencies of recombinants generated during lytic growth following mixed infection, i.e. the vegetative map. Remarkably enough, the one map is a cyclic permutation of the other. It is as if one map were derived from the other by joining the two ends to form a circle and then cutting the circle again at a point almost diametrically opposite the join. In fact, this is precisely what happens between infection and lysogenization.

Biochemical studies, including the complete determination of the relevant part of the lambda DNA sequence and the sequence of the lambda attachment site on

Fig. 11.6

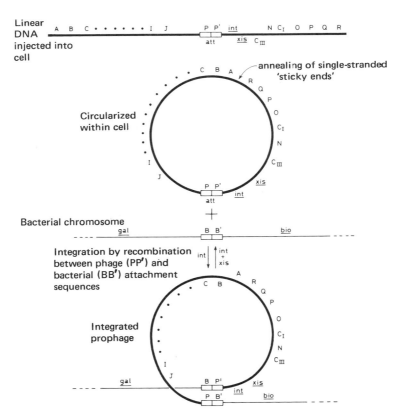

Integration of the lambda genome into the *E. coli* chromosome to give a prophage with a genetic sequence which is cyclically permuted in comparison with the sequence in the infective particle. The 'left-arm' genes (A–J, not all shown in the diagram) code for head and tail proteins. Those in the 'right arm' are concerned with lysogeny (*int*, *xis*, c_I, c_{II}, c_{III}), DNA synthesis (*O*, *P*) or cell lysis (*R*) —*see also Fig.* 17.4.

the *E. coli* chromosome (*Fig.* 11.6), have shown that the integration of the phage genome into the chromosome depends on a site-specific crossover between the chromosomal attachment region and a region in that central part of the phage genome which is essential for lysogenization. The two regions are not, in fact, homologous except for a short sequence of 15 base-pairs in the centre of each (*Fig.* 11.8), but they are both recognized by a special enzyme, coded for by the *int* (integration) gene of the lambda central region. This enzyme makes double-stranded cuts in both sequences and then rejoins them in crossed-over con-figuration. The effect, as *Fig.* 11.6 shows, is to integrate the phage genome linearly into the chromosome. The excision of the phage genome, which occurs on induction of a lysogenic strain, is also dependent on the *int* function and in addition requires another lambda gene (*xis*—excision) not required for integration. The activity of the *int* enzyme in catalysing recombination of the specific DNA sequences has been demonstrated in a cell-free system and the enzyme has been purified. There may soon be available a description of the mechanism through which it acts so precisely. Paradoxically, the action of the *int* product fits much better the classical idea of reciprocal crossing-over than do the currently favoured models of general recombination in eukaryotes, for which reciprocal recombination was first proposed. Little is known as yet about the *xis* function.

Lambda is but one of a large family of closely related phages which have been shown to have their genomes integrated into the host DNA in the lysogenic state. Two others deserve mention. Phage 434 integrates at a locus distant from the one used by lambda and confers its own specific immunity—cells lysogenic for 434 can be lysed by lambda but not by 434, and vice versa. Lambda–434 hybrid recombinants are formed freely during mixed infections, and derivatives which are largely lambda in origin but with the immunity characteristics of 434 have been useful in various experimental studies. A second lambda-like phage, called φ80, is notable for its site of integration—usefully located adjacent to the *trp* operon of *E. coli* (cf. pp. 376–377).

11.5   Specialized transduction

The Origin and Structures of λd*gal* and λd*bio*

Lambda is not a general transducing phage—the specificity of its DNA packaging mechanism, in which the encapsidated molecules are not only of standard size but also have specific terminal sequences, precludes the accidental encapsidation of random pieces of host DNA. However, lambda *will* transduce the *E. coli gal* and *bio* genes which, as we saw above, flank the lambda integration site in the *E. coli* chromosome.

When lambda phage grown on *gal*[+] *E. coli* is used to transduce a *gal*[−] strain, and *gal*[+] transductants are selected on medium containing galactose as carbon source, the frequency of such transductants is only about 1 per 10^6 infecting lambda particles. Nearly all the transductant colonies are lysogenic and, on induction with ultraviolet light, they yield high-frequency transducing (HFT) lysates, in which close to a half of the phage particles are transducing with respect to *gal*[+].

Infection of *gal*⁻ cells with HFT lysates at high dilution, so that few cells are infected by more than one phage particle, reveals that the gal^+-transducing phages are defective—they can integrate into the *E. coli* chromosome but, having done so, they do not yield phage particles upon induction by ultraviolet light. In other words, the gal^+ transductants are **defective** lysogens. They can be made to yield another generation of phage particles only when their missing lysis functions are supplied by superinfection with a competent 'helper' phage. Note (*Fig.* 11.6) that the defective transducing phages retain the lambda immunity function (c_I) and so superinfection by λ requires a low dose of UV to destroy the immunity of the defective lysogen.

The nature of the deficiencies in defective *gal*-transducing lambda phages (called λd*gal*) can be determined by complementation analysis—that is by superinfecting with a range of mutant helper phages with different functions eliminated, so as to determine which functions of the helper are really needed. It turns out that the λd*gal* phages are deleted with respect to a block of genes varying somewhat in extent but always extending from the right (*bio*) end of the prophage. The quantity of DNA needed for packaging in a lambda particle is made up by the inclusion at the other end of the prophage of a segment of bacterial chromosome including part or all of the *gal* gene cluster (**operon**—*see* p. 317).

The λd*gal* genome is inserted into the chromosome of a transduced cell by reciprocal recombination between the transducing gal^+ and the chromosomal gal^- regions. This generates a gal^+-gal^- tandem duplication flanking the prophage. That the transductants still harbour *gal*⁻ is shown by the fact that they segregate this mutation with high frequency (about 2×10^{-3} per cell division) during ordinary growth. A bacterial strain carrying two different alleles of the same gene (e.g. gal^+/gal^-) is called a **heterogenote**, and the segregation from it of homogenotes appears to be due to pairing and recombination (presumably of the gene conversion type, cf. Chapter 5) between the two gene copies.

The model for the formation of specialized transducing phage genomes which is implied by their genetic structure is illustrated in *Fig.* 11.6. It is an extension of the Campbell hypothesis, which we have already seen applied to the integration and excision of the F-plasmid (p. 265, *Fig.* 10.10) and to the normal integration and excision of lambda (*Fig.* 11.6); it is closely analogous to the model proposed for the origin of F'-plasmids. The essence of the idea is that the looping-out of the prophage prior to excision is sometimes inexact, in that, although the length of DNA in the loop is roughly correct for encapsidation in a lambda head, its ends are displaced; a block of phage DNA is excluded from the loop and a similar-length stretch of *E. coli* DNA is included in compensation. The crossover event which frees the loop from the chromosome does not involve the normal attachment sequence but some other pair of sequences which may happen to have some limited homology, sufficient for low-efficiency general recombination to take place. It is known that the lambda genes *int* and *xis*, whose functions are essential for normal prophage excision, are not required for the abnormal excision involved in the origin of the λd*gal* and other transducing sequences.

The defective lambda derivatives responsible for transduction of bio^+, called λd*bio*, appear to be similar in their origin to λd*gal*, but with the looped-out DNA segment displaced in the opposite direction. The properties of λd*bio* phages are different from those of λd*gal* since they have deletions at the left end (as it is conventionally drawn) of the prophage rather than at the right end. The effect of this is that defective phages carrying *bio* have all the genes essential for lysis and infective particle formation, but need to be complemented by helper phage in order

Fig. 11.7

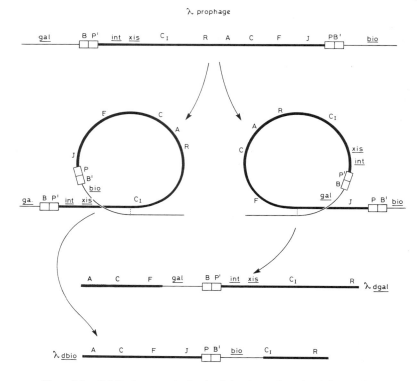

The origin of defective transducing lambda phages from lambda prophage.
Thick and thin lines indicate lambda and *E. coli* DNA, respectively. Looping
out and aberrant pairing between (supposedly) partially and accidentally
similar lambda and *E. coli* sequences, followed by recombination, releases a
circular defective phage genome which then becomes linear by normal cleavage
in the 'sticky end' sequence between A and R. *See Fig.* 11.6 for the
corresponding normal excision of prophage.
Redrawn after Gottesman and Weisberg.[12]

to integrate. The structures and modes of origin of the two classes of transducing
phage are compared in *Fig.* 11.7.

Although lambda has a single regular site of integration, and thus usually acts as
a specialized transducing agent with respect to the *gal–bio* region only, it is capable
of rare integration at other sites and thus of extending its transducing range. Such
'illegitimate' insertions can be isolated from among the survivors of lambda
infection in strains of *E. coli* in which the normal lambda attachment site has been
removed by a deletion mutation. The secondary attachment sites, which can be
used, though at a very low frequency, in the absence of the normal one, have had
their base sequences analysed in a few cases and they all show a good deal of
similarity to the fifteen-base 'core' sequence which is common to the normal
lambda and bacterial interacting regions. Integration of lambda prophage at a
secondary site resembles normal integration in depending on the *int*-coded enzyme,
which has a strong preference for the normal 'core' sequence but does not

Fig. 11.8

P–P′   5′ --- GTTC**AGCTTTTTTATACTAA**GTTGG ---- 3′

B–B′   5′ --- AGCCT**GCTTTTTTATACTAA**CTTGA ---- 3′

Δ–Δ′   5′ --- CGCGC**CTTTGTTTTCAAAAA**CCTGC ---- 3′
                                           ↓
P–Δ′   5′ --- GTTC**AGCTTTTTTATAC**A**AA**CCTGC ---- 3′
                                        ↓
Δ–P′   5′ --- CGCGC**CTTTGTTTTCA**A**TAA**GTTGG ---- 3′

Sequences of *E. coli* and lambda phage attachment regions. P–P′ is the lambda attachment sequence present in normal lambda DNA. B–B′ is the bacterial attachment sequence present in a *gal–bio* transducing phage obtained by recombination between λd*gal* and λd*bio* (cf. *Fig.* 11.7). Δ–Δ′ is an abnormal low efficiency bacterial attachment site *within* one of the *gal* genes; P–Δ′ and Δ–P′ are combinations found in transducing phages excised from this site in a manner analogous to the origin of λd*gal* and λd*bio* from the normal site of integration. The 'core' sequences are in bold type—the flanking sequences show little or no homology between B, P and Δ. Only one strand of the DNA is shown.
From Bidwell and Landy.[9] The arrow shows the apparent position of the crossover in integration/excision.

absolutely require it. Some of the DNA sequences that have been shown to be involved are compared in *Fig.* 11.8.

Once integrated at an unusual site, transducing lambda phage carrying nearby genes can be formed in just the same way as already postulated for λd*gal* and λd*bio* (*Fig.* 11.7). For example, lambda derivatives carrying both *lac* and *trp* have been isolated. Very much the same tricks can be played with the lambdoid phage ɸ80 which, as mentioned above, normally integrates close to *trp*.

Once isolated and characterized, a specialized transducing lambda can be propagated indefinitely with helper phage, and the segment of *E. coli* chromosome which it carries can be multiplied up to any desired quantity—in other words the bacterial sequence can be **cloned**.

Lambda transducing phages have proved extremely useful for the fine analysis of the bacterial chromosome segments which they carry. A λd*gal* genome, for example, carrying a block of genes of the *gal* operon, can acquire, by recombination between the duplicated segments in a heterogenote, any mutation available in any of these genes. It can then be used to carry the mutation into a cell of another *gal* mutant type; the resulting heterogenote then provides a test for complementation between the two mutations and for the occurrence of recombination between them. Deletions with a λd*gal* genome can be characterized by complementation and recombination and their various extents can also be defined physically by heteroduplex mapping (cf. p. 255).

11.6   Artificial cloning of DNA in phage vectors

Construction of Lambda Vectors

In the last chapter we saw how drug-resistance or colicinogenic plasmids, or artificially constructed hybrids of the two, can be adapted for use as vehicles for cloning DNA sequences from various sources. Using the same kinds of techniques for cutting and joining DNA sequences, it is equally possible to use lambda as a cloning vector.

Cloning of foreign DNA sequences in lambda is subject to one limitation, which can in fact be turned into an advantage. In order to be packaged in the phage head the DNA has to be within a certain size range—too large a molecule could obviously not be fitted into the fixed geometry of the protein shell and, more surprisingly, too small a molecule cannot be encapsidated either. Fortunately, the problem of making room for an insertion can be overcome by making a compensating deletion of non-essential material in the lambda genome. The upper diagram in *Fig.* 11.9 shows the positions in the linear infective DNA of the sites cut by EcoRI. The two centrally located EcoRI fragments, B and C, contain, respectively, no known genes and genes essential only for lysogenization and induction from the lysogenic state. R. W. Davis and his colleagues, by digestion of lambda DNA with EcoRI, allowing the fragments to reanneal end to end (through their mutually complementary four-base single-stranded tails, *see Box* 7.1) and treatment with DNA ligase, obtained a number of types of reconstituted genome with one or more EcoRI fragments deleted. Such derivatives can be introduced into *E. coli* cells by the $CaCl_2$ transformation procedure (*see* p. 275) and will be replicated and packaged in infective particles if they are big enough. Davis chose a phage lacking the B segment and also deficient, because of a deletion in the phage used as starting material, in a substantial part of the E segment (the *nin* deletion, *Fig.* 11.9). This phage DNA was just large enough for packaging. It was further modified by two mutations which, by single base-pair changes, eliminated the EcoRI sites on the D/E and E/F boundaries. The result was a substantially shortened phage genome with only two remaining EcoRI sites—those flanking the C segment. This derivative is called λgtλC, gt standing for *general transduction* of any DNA segment carried between the EcoRI sites.

Replacment of the C segment by another EcoRI-generated DNA fragment can be brought about by the procedure outlined in *Fig.* 11.9. EcoRI fragments of λgtλC DNA are allowed to anneal with ECoRI fragments of DNA from any other source and, after ligation, the reconstituted DNA molecules are used to transform *E. coli* cells. Only two kinds of products are capable of replication and packaging. One is the reconstituted λgtλC and the other is a class of recombinant with the C segment replaced by a piece of the alien DNA. The derivative with the C segment merely left out is too small to be packaged, and so the minimum size requirement provides automatic selection against this unwanted type. The hybrid recombinants can be readily distinguished from the original λgtλC type because they give clear plaques rather than turbid ones, having lost the functions (including *int*) which are present in the C segment and are necessary for the establishment of lysogeny. The

Fig. 11.9

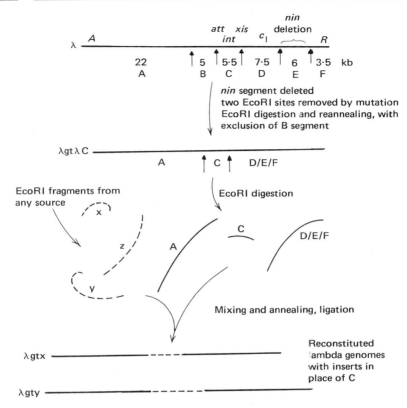

The derivation of a general cloning vector λgtλC from lambda bacteriophage, and of derivatives carrying DNA segments from any source. The lambda segment B does not include any genes essential either for lytic growth or for lysogenization, but its removal, together with that of the equally dispensable *nin* segment, makes the genome too small for packaging in the lambda head when the C segment is also removed. This provides a means of selecting derivatives with foreign inserts substituting for segment C over those which simply have segment C deleted.
After Thomas et al.[10]
The analogous vector λgtλB, used by Leder et al.,[11] has certain advantages for biological containment in that it cannot lysogenize. Numerous other variations on the theme of foreign insert carried between more or less normal lambda arms have been devised.

hybrids are fully capable of lytic growth and can, with their cloned inserts, be grown through an indefinite number of cycles of infection and cell lysis.

λgtλC was only the first of many especially engineered lambda derivates to be used for DNA cloning. Some of the further developments were made in the interests of biological containment, it being feared that some of the cloned sequences might present some hazard if carried in a vector which could possibly survive in natural populations of bacteria. One derivative, λgtλB, has the B segment instead of the C segment between the EcoRI sites. This, since it lacks the C segment genes necessary for lysogenization, is unable to establish itself as a prophage, even before the replacement of the lambda middle piece by the alien

Fig. 11.10

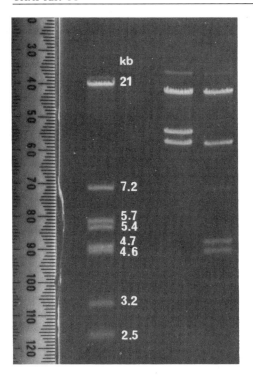

Size analysis of segments of *Neurospora crassa* DNA cloned by *in vitro* insertion between the EcoRI sites of the lambda vector Charon 4A. The *Neurospora* DNA was only partially digested with EcoRI, and so most of the cloned fragments contained internal EcoRI sites. Two of the hybrid phages were digested with EcoRI and run on an agarose electrophoresis gel in parallel with size markers (sizes in kb are indicated). The two Charon 4A 'arms' are common to each clone, but the two inserted segments differ in size and in the number and spacing of the EcoRI sites that they contain. The gels were stained with ethidium bromide and photographed under ultraviolet light. The very faint slow-running band seen in each of the two clone digests is probably due to terminal adhesion of the lambda arms through their single-stranded 'sticky ends'—*see* p. 286.

Photograph by courtesy of Dr Jane Kinnaird. Charon 4A *Neurospora* 'library' by courtesy of Dr John Bishop.

segment. It was further modified by three 'amber' mutations in the genes W, E and S, which make the phage totally dependent on the presence of an amber suppressor mutation in the host cell—a rarity in nature. A further extensive series of vectors, many of them with built-in bacterial markers to permit the easy recognition of derivatives with inserts displacing these bacterial functions, have been named Charon phages after the sinister ferryman of Greek mythology.

Lambda-based cloning vectors have certain definite advantages over the plasmid-based ones discussed in the last chapter. Plaques containing highly amplified cloned sequences are somewhat easier to screen for specific sequences than bacterial colonies harbouring the same sequences cloned in plasmids. Furthermore, a sequence packaged in a lambda phage is relatively stable and inert, and can be stored for long periods without the bacterial host. As against these advantages there is the disadvantage, which does not apply to the plasmid system, of an upper limit to the size of the DNA fragment which can be cloned. In the case of λgtλC, an insert of 12 kb restores the full lambda genome length and nothing much greater can be encapsidated. The Charon series on the other hand can accommodate up to 20 kb.

Cloning in M13

The filamentous DNA phage M13 has a special advantage as a cloning vector since, owing to the tubular structure of the capsid which can be indefinitely extended to accommodate enlarged phage genomes, there is no upper limit on the size of the DNA fragment which can be cloned. The fact that the DNA in the M13 particle is single stranded is not a serious disadvantage since it can easily be made double stranded *in vivo* or *in vitro* and then cut with restriction enzymes to create insertion sites for cloning. Double-stranded M13 DNA, carrying inserts of a range of lengths, can be introduced into *E. coli* by the usual transformation procedure and subsequently amplified, packaged and released in virus filaments.

Cloning in M13 has become a method of choice for amplification of specific DNA sequences prior to sequence analysis. The procedure is outlined in *Box* 14.2.

11.7 Summary and perspectives

Bacteriophages are at the same time relatively simple and extremely suitable for genetic analysis. Bacteriophage T4, in particular, has been valuable in the investigation of gene structure and function and the nature of mutation. Bacteriophage lambda is the classical and most completely investigated example of a virus capable of integrating into the genome of its host.

As well as being important objects of genetic investigation in themselves, many bacteriophages (the temperate phages) provide, as agents of transduction, means for the detailed analysis of the genetic structure of the host bacterium.

The ability of temperate phages, such as lambda, to 'pick up' fragments of host cell DNA provides a 'natural' means of cloning bacterial genes. Artificial methods, involving *in vitro* cutting and rejoining of isolated DNA sequences, make it possible to use phage derivatives as vehicles for the cloning of virtually any DNA sequence from any source. Lambda vectors complement plasmids in providing a wide choice of cloning methods.

Chapter 11

Selected Further Reading

Hayes W. (1968) *The Genetics of Bacteria and their Viruses*, 2nd ed., 925 pp. Oxford, Blackwell Scientific Publications.

Hershey A. D. (ed.) (1971) *The Bacteriophage Lambda*, 792 pp. New York, Cold Spring Harbor Laboratory.

Nash H. A. (1981) Integration and excision of bacteriophage lambda: the mechanism of conservative site-specific recombination. *Annu. Rev. Genet.* **14**, 143–168.

Stent G. S. (1963) *Molecular Biology of Bacterial Viruses*, 474 pp. San Francisco, Freeman.

Problems for Chapters 10 and 11

1 Cells of two strains of *E. coli*, one *leu⁻ his⁺* sulphonamide-resistant (*sulʳ*) and the other *leu⁺ his⁻* colicin-producing (*col⁺*) were mixed and, after incubation for 1 hour, diluted and plated on three types of medium: (*a*) minimal, with no leucine or histidine, (*b*) minimal supplemented with leucine and (*c*) minimal supplemented with histidine. No colonies grew on (*a*), but numerous colonies grew on (*b*) and (*c*). All those growing on (*b*) were *leu⁻ sulʳ* but none was *col⁺*. All those growing on (*c*) were *his⁻ col⁺* and about 10 per cent were *sulʳ*. Cells derived from one of the *col⁻ sulʳ* colonies from (*c*) were mixed with cells of a wild-type *sulˢ col⁻* strain and, after incubation for 1 hour, the mixture was diluted and plated on minimal medium. Of the many colonies which grew, about 10 per cent were *sulʳ* and, of these, about a half were *col⁺*. None were *sulˢ col⁺*. Propose explanations for the modes of transmission of the *sulʳ* and *col⁺* determinants.

2 *E. coli* Hfr and F⁻ cells were mixed and, 10 minutes later, the mixture was diluted and plated. A colony was recovered which had all the markers of the original F⁻ strain but which, by the criterion of susceptibility to male-specific bacteriophage, had acquired F-pili. Cells of the strain established from this colony, which was *thyA⁺, argA⁺, gal⁺* and *strˢ*, were mixed with cells of an F⁻ *thyA⁻ argA⁻ gal⁻ strʳ* strain and after 10 minutes the mixture was diluted and plated on medium containing streptomycin (to select for *strʳ*), arginine and thymine and with glucose as carbon source. Among 1090 *strʳ* colonies, 886 were F⁻ *thyA⁻ argA⁻ gal⁻* and 204 were F⁺ *thyA⁺ argA⁺ gal⁻*. One of the latter type was used to establish a culture which was grown in the presence of acridine orange. Plating following the acridine treatment gave 390 colonies of the following types (all *strʳ gal⁻*):

$$371 \text{ F}^- \ thyA^- \ argA^- : 13 \text{ F}^+ \ thyA^+ \ argA^+$$

$$3 \text{ F}^- \ thyA^- \ argA^+ : 2 \text{ F}^- \ thyA^+ \ argA^-$$

A control plating of a culture not exposed to acridine orange gave 809 colonies of type F⁺ *thyA⁺ argA⁺* and one each of types F⁻ *thyA⁺ argA⁻* and F⁺ *thyA⁺ argA⁻*. On the standard *E. coli* time map, *thyA, argA, gal* and *str* are at 55, 54, 24 and 64 min, respectively.
What was the nature of the sex factor in these experiments? Explain the results so far as possible. (Ishibashi et al. *J. Bacteriol.* 1964, **87**, 554).

3 Hfr strains have been isolated in two species of *Salmonella, S. typhosa* and *S. typhimurium*, by transfer of the F-plasmid from *E. coli* and selection for transferability of chromosomal markers. A comparison was made between three Hfr strains, one from each *Salmonella* species and one from *E. coli*. They were found to transfer chromosomal markers to

F^- strains beginning at the following times (min) after mixing:

	ile	met	pro	arg
Hfr 1	28	20	5·5	22
(*E. coli*)				
Hfr 2	4	18	47	22
(*S. typhimurium*)				
Hfr 3	51	33	8	36
(*S. typhosa*)				

What do these data say about (*a*) the comparative chromosome maps of the three species, (*b*) the position and orientation of the integrated F factor in each of the three Hfr strains and (*c*) the kinetics of transfer mediated by each Hfr? (Johnson et al., 1964, *J. Bacteriol.* **88**, 395.)

4 Mutants of *E. coli* deficient in alkaline phosphatase (*pho⁻*) map in a tight cluster close to *lac*. HfrC transfers *pho* after 9 minutes, *lac* after 10 minutes and *thr* and *leu* after 20 minutes. Garen (*Symp. Soc. Gen. Microbiol.* 1960, **10**, 239) investigated the linkage relationships of a series of *pho⁻* mutants by making crosses of the form HfrC *phoᵃ thr⁺ leu⁺ strˢ* × F⁻ *phoᵇ thr⁻ leu⁻ strʳ*, selecting *thr⁺ leu⁺ strʳ* recombinants and determining the proportion of *pho⁺* colonies. Explain the rationale of this experimental design. The following matrix shows the recombination frequencies (in terms of percentage *pho⁺* among *thr⁺ leu⁺ strʳ* colonies) among five of the *pho⁻* mutants investigated:

U7	U3	U12	E1	
0·065	0·13	0·18	0·045	U18
	0·063	0·09	0·008	U7
		0·022	0·033	U3
			0·036	U12

Can you deduce a linear order of *pho* sites from these data? How might the use of a *lac* marker help towards a more certain ordering? Taking recombination frequency as a true measure of distance, and using the approximation that 1 minute on the *E. coli* time map corresponds to 10 per cent recombination, calculate the distance in base-pairs between U18 and U12.

5 If P1 bacteriophages grown on wild-type (*leu⁺ thr⁺*) *E. coli* are used to infect *leu⁻ thr⁻* cells, and the infected cells are plated on minimal medium supplemented with leucine only, about 10^{-3} of the surviving (lysogenic) cells grow to form *thr⁺* colonies about 3 per cent of which are *leu⁺* and about 97 per cent *leu⁻*. Conversely, when plating is made on minimal supplemented with histidine only, about 10^{-3} of the survivors grow to form *leu⁺* colonies of which about 3 per cent are *thr⁺* and about 97 per cent *thr⁻*. Explain these results. Assuming (which is approximately true) that *leu* and *thr* are separated by 2 per cent of the time map of the *E. coli* genome, what can you conclude about the capacity of bacteriophage P1 in transduction of fragments of *E. coli* DNA?

6 *E. coli trpA* and *trpB* mutants have a specific nutritional requirement for tryptophan; *trpE* mutants will grow on anthranilic acid as well as on tryptophan. In transduction experiments using phage P1, the following frequencies of cotransduction were found: *trpA* and *trpB* 98 per cent, *trpA* and *trpE* or *trpB* and *trpE* 80–90 per cent. In one experiment the double mutant *trpB⁻ trpE⁻* was used as donor and *trpA⁻* as recipient, and selection was made for transductants that would grow on medium with anthranilic acid as sole supplement. On testing for anthranilic acid requirement, 90 per cent of these transductants proved to be *trpE⁺* and 10 per cent *trpE⁻*. From this information deduce the order of the three genes. (From Yanofsky and Lennox, 1959, *Virology*, **8**, 425.)

7 Jacob and Wollman (*Sexuality and the Genetics of Bacteria*, 1961) crossed *E. coli* F⁻ *thr⁻ leu⁻ lac⁻ gal⁻ trp⁻ his⁻ str*ʳ to three HfrH strains which were, respectively, non-lysogenic, lysogenic with lambda phage and lysogenic with the lambda-like phage 424. The crosses were made over a prolonged period with minimal mechanical disturbance so as to maximize the entry of Hfr markers. In each case *thr⁺ leu⁺ str*ʳ recombinants were selected and scored for the presence of *lac⁺*, *gal⁺*, *trp⁺* and *his⁺*. The following frequencies (%) were observed:

Donor: HfrH	*lac⁺* 58	*gal⁺* 45	*trp⁺* 35	*his⁺* 14
HfrH(λ)	*lac⁺* 26	*gal⁺* 3,	*trp⁺* 6,	*his⁺* 3
HfrH(424)	*lac⁺* 72,	*gal⁺* 46,	*trp⁺* 42,	*his⁺* 2

How can the broad pattern of these data be explained? Are there any details which remain unexplained?

8 When *gal⁻ E. coli* is infected with lambda phage from an HFT (*gal⁺*) lysate at a ratio of five phage particles per bacterium, virtually all of the colonies selected as having been transduced to *gal⁺* are lysogenic, releasing a mixture of λ and λd*gal* phages. When, on the other hand, the ratio is 10^{-3} phage particles per bacterium, virtually all of the transductants are defective lysogens, capable of lysis after induction with UV but not releasing any phage particles. When one of these defective lysogens is induced and then superinfected with wild-type lambda, lysis occurs with release of a mixture of normal λ and λd*gal*. Explain these findings. Given the availability of a full range of conditional (temperature sensitive or *amber*) lambda mutants, explain how one could determine the extent of the deficiency in the prophage of each defective lysogen.

9 Six *rII* mutants of phage T4 were selected as non-revertible. They were tested in all pairwise combinations for complementation [ability in mixture to grow on *E. coli* K(λ)] and for recombination. The results were as follows (○ indicates recombination, ● recombination and complementation and — neither):

	1205	164	187	1299	638	
	—	—	●	●	●	H88
		—	○	—	—	1205
			●	●	●	164
				○	—	187
					—	1299

Ten classes of revertible mutants were distinguished by their different interactions with the six above mutants:

Class no.:

1	2	3	4	5	6	7	8	9	10	
—	○	○	●	●	—	—	●	●	●	H88
○	—	—	—	—	—	—	○	○	○	1205
○	○	—	●	●	○	—	●	●	●	164
●	●	●	○	○	●	●	—	○	○	187
●	●	●	○	—	●	●	○	—	○	1299
●	●	●	—	—	●	●	—	—	—	638

Mutants in the ten revertible classes recombine with one another in all combinations between classes and in the great majority of combinations within classes.

Make a map of the section of the *rII* region covered by these mutations, showing the segments within which point mutations can be localized and the limits of the overlapping deletions. (Benzer and Freese, 1958, *Proc. Natl Acad. Sci, USA* **44**, 112.)

10 Certain artificially fabricated lambda-cloning vectors have the central part of the phage genome deleted and partly replaced by a DNA sequence from *E. coli* containing the *lac* gene, coding for β-galactosidase. These artificial lambda phages make clear blue plaques when plated on *lac⁻* bacteria on medium containing the complex galactoside known for short as X-gal. The blue colour is due to the hydrolysis of X-gal by the β-galactosidase present in the plaques but not in the uninfected *lac⁻* cells. One such vector (Charon 16) has a single EcoRI restriction site located within the *lac* gene, and is about 15 kb pairs short of the maximum amount of DNA which can be packaged in a lambda particle. Explain why these features make Charon 16 a convenient cloning vector. (Blattner et al. 1977, *Science*, **196**, 161.)

11 Plasmid R6·5 contains a gene determining kanamycin resistance, flanked by EcoRI sites. ColE1 has a single EcoRI site. From these components a 5 kb *kanʳ* cloning plasmid was constructed with a single EcoRI site. How do you think this was done?

12 A derivative of bacteriophage lambda was obtained with a deletion removing its *att* site and with an inserted copy of Tn5 which determines resistance to the antibiotics kanamycin and neomycin. *E. coli* cells which had been infected with this phage (λ:Tn5) were plated on medium containing neomycin and broth to provide most possible nutritional requirements. About one per cent of the neomycin-resistant colonies which appeared were auxotrophic mutants of many different kinds. Very few of these were lysogenic for lambda. Suggest a likely origin for the mutants. How might you confirm your suggestion? (Shaw and Berg, 1979, *Genetics* **92**, 740.)

13 Standard curves relating DNA fragments size and electrophoretic mobility in agarose gels can be conveniently obtained by plotting size against the reciprocal of the distance travelled. Construct such a curve by measurements on *Fig*. 11.10. From the curve, estimate the sizes of the two Charon 4A 'arms' and of the segments of *Neurospora* DNA included in each clone.

Mechanisms of Mutation

12.1 | Kinds of mutations and their effects

Classification

We saw in previous chapters how mutations can be mapped genetically and following chapters will show how they can be characterized in terms of biochemical effect and DNA structure. In this chapter we consider the question of how they arise.

To recapitulate briefly, mutations can be classified into three broad categories. The simplest are **base-pair substitutions**, which may in turn be subdivided into **transitions** and **transversions**. In transition mutants, a purine is replaced by a purine and a pyrimidine by a pyrimidine, i.e.

$$A \rightleftharpoons G$$
$$T \rightleftharpoons C$$

In transversion mutants a purine is replaced by a pyrimidine and vice versa, i.e.

$$A \rightleftharpoons T \quad \text{or} \quad G \rightleftharpoons C \quad \text{or} \quad A \rightleftharpoons C$$
$$T \rightleftharpoons A \qquad\qquad C \rightleftharpoons G \qquad\qquad T \rightleftharpoons G$$

The second class of mutations are **small deletions** or **insertions**, one or a few base-pairs in extent. When they occur within sequences coding for polypeptide chains, such mutations can often be recognized as **frameshifts**.

Thirdly, there is the very broad category of relatively large-scale **structural changes** in chromosomes, including larger deletions and duplications of chromosome segments, transpositions, inversions and reciprocal segmental interchanges. The consequences of several of these kinds of structural changes were reviewed in Chapter 6. Many of them, in eukaryotes, are large enough to be analysed with the light microscope.

It is not really possible to draw a line between very small deletions or tandem duplications of material, such as might be recognized as frameshift mutations, and very large ones which might attain microscopic visibility. There is, in fact, a continuous gradation of size. Deletions of the order of hundreds or thousands of base-pairs, occurring within a gene or spanning a few genes, are of great importance in the functional analysis of the genome (*see*, for example, p. 431).

Phenotypic Effects and Revertibility

Only a few limited generalizations can be made relating the molecular nature of mutations at the DNA level to their phenotypic effects. Mutants which are readily revertible to wild type are likely to have base-pair substitutions, although some frameshift mutations will also revert with fair frequency. Deletions are incapable of reversion and are apt to have lethal effects in haploids or when homozygous in diploids, especially when they span more than one gene. Mutants with base-pair substitutions usually produce (when the mutation is within a gene coding for protein structure) a recognizable altered gene product, whereas deletion and frameshift mutants produce, at most, shortened polypeptide chains which are often unstable or otherwise difficult to recognize.

Rearrangements involving no overall gain or loss of material frequently have no phenotypic effect, especially in higher eukaryotes where, because of the high proportion of apparently functionless or redundant DNA, there is a good chance that the break-points will not intersect any functionally important gene. In *Drosophila*, however, there is an important phenomenon known as *position–effect variegation*, in which gene action is inhibited in certain cell clones when a gene, normally located in euchromatin, is brought close to heterochromatin through structural rearrangement. We shall return to this effect in Chapter 17.

'Silent' Mutations

Although phenotypic effect is normally essential for the recognition of a mutation, it should be borne in mind that many, almost certainly the great majority, of the possible changes in DNA structure will be phenotypically 'silent'. Especially in higher eukaryotes, much of the DNA is not transcribed, and this part of the genome is (at least within broad limits) free to 'drift' through random and inconsequential mutation without constraint by natural selection. Some regions are transcribed but not translated, and here only a small proportion of the bases have specific functions, for example in signal sequences for RNA processing or for transcription or translation starts and stops. Even within sequences coding for polypeptides, the majority of base changes in the third position of codons make no difference to the amino acid coded for and, even when an amino acid replacement does result from a base-pair substitution, there is, in most proteins, a good chance that the function of the protein will not be critically affected. In summary, one can say that mutations with marked phenotypic effects direct us to points in the genome which are particularly sensitive—specific nucleotides which are involved in control of transcription, RNA processing or translation, or which are parts of codons for critical amino acid residues. There is indeed an enormous number of such points, but there is a 'silent majority' of nucleotides which are subject to the same mutational processes but whose mutations we can only know about through direct analysis of DNA sequence.

12.2 Spontaneous mutation

Frequency and Proof of Spontaneity

In calling a mutation spontaneous we mean that it was not deliberately induced by mutagenic treatment. Spontaneous mutations occur at unpredictable times in all genes and all organisms. As we see below, they are more frequent in some genotypes than in others, but there is no known way of preventing them from happening. Their frequency is low; Drake has estimated a frequency of 10^{-8} per base-pair per round of DNA replication in bacteriophage T4 and at least a thousand-fold less than that in eukaryotes. The frequency per gene (in the sense of polypeptide-coding sequence) will be of the order of 10^3 times higher.

When one is attempting to measure such low frequencies it is essential to have some selective system that will pick out the rare mutants from the vast majority of non-mutants. At one time it was a very real question whether the 'spontaneous' mutations may not actually have been induced by the selective regime. The simplest proof that spontaneous mutations arise independently of the means used to select them makes use of the technique of replica plating, described in Chapter 13 (p. 346). This technique, applicable to bacteriophage, bacteria and yeast-like fungi, identifies the positions of mutants on a master plate from the positions of the selected plaques or colonies on replica plates. From the fact that repeated replicas from the same master plate produce mutant colonies in the same positions, and that cells taken from the corresponding positions on the master plate (which has never been subject to the selective conditions) give cultures rich in mutants, it is clear that mutant cells were already present in the original culture before any

Fig. 12.1

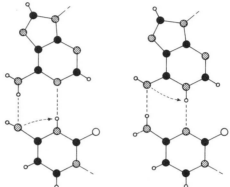

Common amino form   ADENINE   Rare imino form

Rare imino form   CYTOSINE   Common amino form

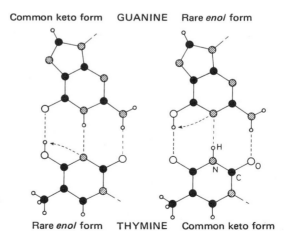

Common keto form   GUANINE   Rare *enol* form

Rare *enol* form   THYMINE   Common keto form

Tautomeric forms of DNA bases and their modes of hydrogen-bonded pairing. Broken lines indicate hydrogen bonds and dotted arrows indicate shifts of positions of protons in the tautomeric transitions.
Drawing after J. W. Drake.[13]

selective procedure was applied. This kind of proof was first obtained in the case of bacteriophage-resistant mutants in *E. coli*, and was subsequently used in many other cases both in bacteria and yeasts.

Mechanisms and Control of Base-pair Substitutions

Tautomerism

Watson and Crick, at the same time as they put forward their double helical model of DNA structure, proposed that errors in base-pairing might be a source of spontaneous mutation. They invoked the known **tautomerism** of the DNA bases, in which each base can exist in two alternative forms interconvertible by the shifting of a double bond and a redistribution of hydrogen atoms within the molecule. The equilibrium in each case greatly favours one tautomeric form, but the existence of the minority form at the instant of replication is a possibility and may be expected to result in pairing with the 'wrong' partner base. The idea, illustrated in *Fig.* 12.1, appears very plausible in the light of what is known about the tautomeric equilibria. Indeed, it is a problem to explain why spontaneous mutations from this source are not more frequent than they are. The answer probably lies in the various mutation avoidance devices possessed by living cells; these are discussed below.

Transitions Versus Transversions

The Watson–Crick tautomerism hypothesis predicts that spontaneous mutations should in general be **transition** base-pair substitutions, with purine substituted for purine and pyrimidine by pyrimidine. Some indication of the kinds of proportions of base-pair substitutions that have occurred spontaneously in the DNA of our own species is given by the extensive information available on variant haemoglobins. *Table* 12.1 summarizes a part of the data. There is ample confirmation of the importance of single base-pair substitutions, as opposed to more complex changes, and transitions do indeed seem to be more frequent than one would expect if all base-pair substitutions were equally probable. But clearly transversions are also an important source of genetic variation in the globin genes. No specific mechanism for the origin of transversion has been proposed.

Proof-reading Function of DNA Polymerase

Rather convincing, though indirect, evidence for the importance of the incorporation of 'wrong' bases as a cause of spontaneous mutation is provided by studies on the effects of changes in the properties of the replicating enzyme, DNA polymerase. DNA polymerases can act on the 3'-hydroxyl ends of growing DNA chains in two ways. They can add further deoxynucleotide residues, using the deoxynucleoside triphosphate as substrate and an already-synthesized single DNA strand as template. Or they can hydrolyse off the terminal nucleotide, reversing the previous synthetic step. The 'reverse' hydrolytic reaction is particularly apt to occur when the 3'-terminal base is mismatched with its partner base on the template strand. The hydrolytic activity thus acts as a 'proof-reading' mechanism—if a wrong base is inserted at the growing end of the chain the polymerase is likely to remove the deoxynucleotide containing it, and repeat the

Table 12.1   **Nature of mutations in human globin genes, deduced from amino acid replacements in rare variants**

Amino acid change From	To	Code change in mRNA From	To	Probable nature of mutation	No. of occurrences
Glutamic acid	Lysine	GAR	AAR	Transition*	7
	Alanine		GCR	Transversion*	2
	Glycine		GGR	Transition*	1
	Valine		GUR	Transversion*	2
	Glutamine		CAR	Transversion*	3
Aspartic acid	Histidine	GAY	CAY	Transversion*	6
	Glycine		GGY	Transition*	1
	Valine		GUY	Transversion*	1
	Tyrosine		UAY	Transversion*	1
	Asparagine		AAY	Transition*	4
Lysine	Glutamic acid	AAR	GAR	Transition*	4
	Threonine		ACR	Transversion*	1
	Asparagine		AAY	Transversion*	3
Arginine	Serine	CGY	AGY	Transversion*	1
	Leucine	CGN	CUN	Transversion*	1
	Glutamine	CGR	CAR	Transition*	1
Histidine	Glutamic acid	CAY	GAR	Complex*	1
	Aspartic acid		GAY	Transversion*	2
	Tyrosine		UAY	Transition*	4
	Glutamine		CAR	Transversion*	1
	Arginine		CGY	Transition*	1
Glycine	Aspartic acid	GGY	GAY	Transition*	7
	Arginine	GGR	AGR	Transition*	5
Alanine	Glutamic acid	GCR	GAR	Transversion*	1
	Aspartic acid	GCY	GAY	Transversion*	3
Valine	Glutamic acid	GUR	GAR	Transversion*	1
	Methionine	GUG	AUG	Transition	2
Leucine	Arginine	CUN	CGN	Transversion*	2
	Proline	CUN	CCN	Transition	4
Phenyl-alanine	Valine	UUY	GUY	Transversion	1
	Leucine		CUY	Transition	1
	Serine		UCY	Transition	1
Serine	Lysine	AGY	AAR	Complex*	1
	Arginine		AGR	Transversion*	2

Table 12.1
(*cont.*)

| Amino acid change | | Code change in mRNA | | Probable nature | No. of |
From	To	From	To	of mutation	occurrences
Tyrosine	Histidine	UAY	CAY	Transition*	1
	Phenyl-alanine		UUY	Transversion	1
Glutamine	Glutamic acid	CAR	GAR	Transversion*	1
	Arginine		CGR	Transition*	1
Asparagine	Lysine	AAY	AAR	Transversion*	1
	Threonine		ACY	Transversion	1

Amino acid replacements from the table of Rucknaegel and Winter.[1] Both alpha and beta chain variants are included. The variations detected are not a random sample of all possible variants since in many cases detection depended on changes in electrical charge, but the ratio of transitions to transversions should not be greatly biased.

* Indicates a change resulting in a charge difference in the haemoglcbin. N indicates any base, R a purine and Y a pyrimidine. The codons cited are the ones consistent with the mutations being single base-pair substitutions. The variants designated 'Complex' cannot be explained by single base-pair substitutions and must be due to at least two base-pair changes in the DNA.

In total, the data indicate, making the simplest interpretation of each variant: 45 transitions, 38 transversions and 2 complex events.

Among those not bringing about charge changes, a sample which may be subject to a different bias in selection, the numbers are: 8 transitions, 3 transversions and no complex events.

If all base-pair substitutions were equally probable, there should be approximately twice as many transversions as transitions.

The data suggest that transitions occur more readily than transversions, but the latter clearly occur with an important frequency.

synthetic step, with a high probability of the correct base being inserted at the second try.

In *Escherichia coli*, certain mutants producing temperature-sensitive DNA polymerase III, and unable to grow at all at the (higher) restrictive temperature, have about a ten times higher than normal frequency of base-pair substitution mutations at temperatures permitting growth. One explanation would be that the mutant polymerase was relatively deficient in the 'proof-reading' function. Such an explanation is supported by studies of temperature-sensitive mutants of bacteriophage T4 with modifications in the structure of the T4-encoded DNA polymerase. Several of these mutant polymerases were found to have, in different cases, either higher or lower 3'-exonuclease activities relative to their polymerase activities. The reduced exonuclease activities were associated with enhanced spontaneous mutation frequencies in all genes tested, while the increased exonuclease activities were associated with *reduced* mutation frequencies. Some of the data are summarized in *Table* 12.2. The obvious interpretation of these observations is that the exonuclease activity of DNA polymerase is antimutagenic because it corrects errors of misincorporation.

Effects of DNA Methylation

The proof-reading function of DNA polymerase is probably not the only error–correction mechanism. One possibility is that mismatches of base-

Table 12.2   **Correlation between the 'proof-reading' function of bacteriophage T4 DNA polymerase and mutation frequency**

The polymerase is capable of catalysing the two sequential reactions:

$$DNA + ndNTP \xrightarrow[\substack{\text{polymerase}\\\text{function}}]{} DNA\text{-}(dNMP)_n \xrightarrow[\substack{\text{3'-exonuclease}\\\text{function}\\\text{(proof-reading)}}]{} DNA\text{-}(dNMP)_{n-1} + dNMP$$

where dN is any deoxyribonucleoside

The polymerase activity is measured by incorporation of radioactively labelled dATP or the 2-aminopurine analogue (dAPTP) into DNA, and the exonuclease function by the release of the label in the form of dAMP or dAPMP. The table shows the results of such assays on the DNA polymerase produced by wild-type T4 and two temperature-sensitive mutants in gene 43, which codes for the polymerase structure, the two mutants showing altered spontaneous mutation frequencies.

Phage strain	Spontaneous mutation frequency	Activities with dATP			Activities with dAPTP		
		Polymerase (a)	Exonuclease (b)	$\left[\dfrac{a}{a+b}\right]$*	Polymerase (c)	Exonuclease (d)	$\left[\dfrac{c}{c+d}\right]$*
Wild type	Normal	1·31	0·20	0·87	0·098	0·052	0·65
tsL141	Lower than normal	0·40	1·58	0·20	0·009	0·093	0·09
tsL98	Higher than normal	1·24	0·16	0·89	0·128	0·049	0·73

*These ratios provide a measure of the probability of the incorporated nucleoside staying incorporated (i.e. not removed by 'proof-reading'). Note that this probability is in all cases lower when the unnatural adenine analogue is incorporated instead of adenine. The higher the ratio the greater the spontaneous mutation frequency, and the greater also the susceptibility of the phage to the mutagenic effect of 2-aminopurine.
Data from Schaar et al.[2]

pairs may be corrected by excision and replacement after, rather than during, synthesis. The mechanism could be similar to that which we invoked to explain gene conversion (*see* Chapter 5).

One apparent difficulty with mismatch correction as a mutation avoidance mechanism is that, without some means of distinguishing the 'wrong' from the 'correct' base, it is as likely to correct the mutation 'in' as to correct it 'out'. It seems that a means exists whereby *E. coli* can discriminate between a pre-existing, presumably non-mutant DNA strand and a newly synthesized strand containing an erroneous base. It depends on the fact that in *E. coli*, as well as in many other species, some of the adenine in DNA becomes methylated in the 6-amino position. Mutants called *dam* were found to be deficient in their ability to carry out this methylation, and to have about a ten-fold increase in spontaneous mutation frequency as well as over a hundred-fold increase in susceptibility to the chemical mutagen 2-aminopurine[3] (*see* p. 336). One hypothesis to explain this finding is that many spontaneous mutations in *E. coli* are avoided by an excision–repair mechanism in which newly synthesized (still unmethylated) DNA strands are corrected by matching with the pre-existing (methylated) complementary strands.

Yet another error–correction mechanism may come into play in *E. coli* when cytosine is accidentally deaminated to give uracil, as apparently occasionally happens. Uracil has the hydrogen bonding properties of thymine, and if allowed to remain in the DNA, will pair with adenine at the next round of replication and thus lead, after one more round, to a G-C→A-T transition. In *E. coli* strains possessing the restriction endonuclease EcoRII, the second base of the sequence recognized by this enzyme (5'-CCAGG-3') is methylated to 5-methylcytosine, making the sequence resistant to cleavage. It has been shown[4] that these methylated cytosines are mutational 'hot-spots' in strains producing the EcoRII nuclease and its accompanying protective methylase. The mutations occurring at these sites are of the G-C→A-T transition type. The explanation seems to lie in the existence of an enzyme, uracil DNA glycosidase, which removes uracil residues from DNA and permits their replacement by cytosine. Methylcytosine is converted on deamination to thymine which is not removed. On this interpretation, the bacterium, by methylating certain of its cytosines, deprives itself of one means of avoiding mutation at those particular sites. This hypothesis is supported by the properties of mutants which lack uracil DNA glycosidase (*ung*⁻); these turn out to be mutators, showing a high frequency of transition mutations at *all* G-C sites.

Mutator Mutations

Apart from *ung*⁺, the best-known example of an antimutator *E. coli* gene is *mutT*⁺ (the so-called Treffers gene, named after its discoverer). Mutations in this gene result in a greatly enhanced frequency of base-pair substitution mutations, the factor of increase being at least 100-fold. Surprisingly, the extra mutations are all of one **transversion** type, namely A-T→C-G. One interpretation of this observation is that such mutations occur by some kind of spontaneous error or chemical instability which it is the normal function of the *mutT*⁺ gene to avoid or correct. A number of other genes affecting mutation frequency are known in *E. coli* and yeast; their modes of action are not yet understood, but presumably they effect error–correction of one kind or another.

'Spontaneous' Mutation Frequency is Genetically Controlled

It will be evident from the foregoing that, at least so far as base-pair substitutions are concerned, 'spontaneous' mutation frequencies are by no means unalterable properties of DNA, as frequencies of radioactive disintegration are unalterable properties of certain atoms. They rather depend on the properties of DNA polymerases and the efficacy, which can be varied by mutation, of a variety of 'proof-reading' and other repair and mutation-avoidance mechanisms available to the cell. There is reason to think that these mechanisms are themselves subject to mutation and natural selection, and that the spontaneous mutation frequencies which we find in natural populations of organisms of all kinds are adaptive. They presumably represent a compromise between the short-term need to replicate already well-adapted DNA sequences with sufficient precision and the longer-term advantage of variability.

Frameshifts and other Consequences of Sequence Reiteration

Frameshifts

Although good data on relative frequencies are difficult to obtain, it is probable that frameshifts account for a considerable proportion of spontaneous mutations; they may be even more frequent than base-pair substitutions. Analysis of mutations in genes for which the protein products are known and base sequences in messenger RNA can be deduced (cf. *Table* 12.2), shows that frameshifts are particularly apt to occur in sequences containing runs of similar bases. G. Streisinger proposed that, in sequences of this kind, occasional 'slippage' of a free end of a DNA strand could occur either forwards or backwards along the template strand, resulting in failure to replicate, or repeated replication of, one or a few bases. This model (*Fig.* 12.2) seems very plausible, and is also used to explain the mutagenic action of intercalating chemicals (p. 337), which are believed to increase the frequency of replication slippage.

Fig. 12.2

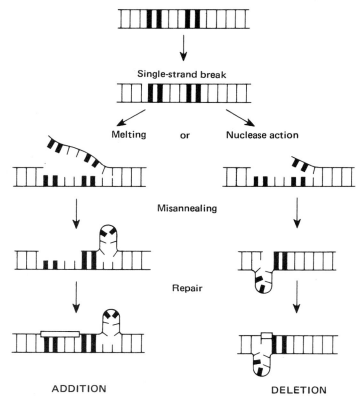

Single-strand break

Melting   or   Nuclease action

Misannealing

Repair

ADDITION   DELETION

The hypothetical mechanism for the origin of frameshift mutations proposed by Streisinger et al.[6] The heavy bars in the schematic double-stranded DNA molecules represent repeated bases or short repeated sequences; the open bars indicate regions of repair synthesis. In this model, the strand 'slippage' is shown as following single-strand breakage in already synthesized double-stranded DNA, but it could also occur at the free end of a growing strand during regular DNA synthesis.
Redrawn after Drake.[13]

Instability of Tandem Repeats

DNA sequences containing longer and more complex repeats, either in tandem or in inverted orientation, are a source of larger scale changes. Directly repeated sequences are prone to diminution or further increase in the number of repeats. An important example is the genetic region carrying numerous tandemly repeated rRNA genes which, at least in *Drosophila*, can be shown to be capable of either increasing or decreasing in number in response to selection (cf. p. 225). The mechanism is commonly supposed to be unequal sister-strand crossing-over following out-of-phase pairing, thus:

$$\frac{\text{X X X X X X X}}{\text{X X X X X X X}} \longrightarrow \frac{\text{X X X X X X X X X}}{+}$$
$$\overline{\text{X X X X}}$$

Unequal crossing-over within tandemly repeated gene clusters is expected to occur at meiosis, but, in an organism such as *Drosophila* in which mitotic crossing-over is known, it could also occur during mitotic multiplication of germ-line cells.

Where the number of repeats is large, pairing errors of the kind just discussed will in general have only small effects on the phenotype, but unequal crossing-over between tandem sequences present in only two copies can have a striking phenotypic effect. A classical case is that of the dominant X-linked *Bar* mutation in *Drosophila melanogaster*, which is visible under the microscope as the tandem duplication of a polytene chromosome band. *Bar* is somewhat unstable, giving rise occasionally to wild type, with the extra copy eliminated, and to a more extreme allele called *Ultrabar*, which turns out to have the chromosome region triplicated.

The instability of tandemly duplicated segments is also a well-known phenomenon in bacterial genetics and, whenever one finds a mutant which reverts with high frequency to wild type, the involvement of a tandem repeat is one of the first possibilities to be considered.

It should be noted that sister-strand crossing-over is not the only possible means of eliminating a tandem duplication. One can also envisage crossing-over within a chromatid or DNA duplex, paired on itself by a loop, thus:

This mechanism could account for loss of repeats and also for their extra-chromosomal amplification by replication of the excised circle, as is known to occur for rRNA genes in certain oocytes.[7]

Instability of Inverted Repeats

Other kinds of accident, either in regular DNA synthesis or in repair synthesis, may occur in sequences containing extensive inverted repeats. This has been most clearly shown by studies of nucleotide sequences of the *Escherichia coli* IS2 element (*see* Chapter 10, p. 258) and some of its derivatives. It seems that errors in the replication of IS2 are associated with the presence of inverted repeats, two of which, eight and six base-pairs long, respectively, occur

Box 12.1   The sequence of events postulated to account for the origin of an expanded
DNA element IS2–6 from the *Escherichia coli* inserted sequence IS2

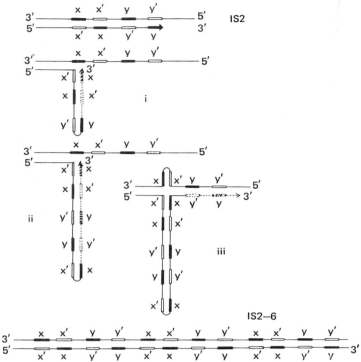

The diagram shows a part of the IS2 sequence which includes two inverted
repeated (palindromic) sequences, called here x and y, which are respectively eight
and six base-pairs long; the complementary sequences on the two strands of the
DNA (x and x′ for one repeat, y and y′ for the other) are distinguished as filled and
open boxes. Newly synthesized strands are shown as broken lines or boxes; arrows
indicate the direction of synthesis. In (i) a strand of IS2 being synthesized (in the
usual 5′→3′ direction) is shown as peeling ('slipping') off its template strand and
folding back on itself with pairing of y and y′; it then continues synthesis using
itself as a template (broken line). In (ii) 'slippage' again occurs, with folding-back
of the detached strand at the second x–x′ repeat; this is followed by a second round
of reverse-self-copying. In (iii) synthesis continues along the proper template
strand, but with the omission of one x–x′ repeat because of self-annealing (hairpin
formation) in the template. Through a further round of DNA replication the
miscopied strand acquires a complementary strand, and IS2–6 is established as a
variant double-stranded element. The structures of IS2 and IS2–6, the latter
differing from the former in possessing an inverted insertion of two extra copies of
the x–x′—y–y′ region, have been fully established by DNA base-sequence
determination. The postulated mechanism is not proved beyond doubt, but is a
most plausible interpretation, given the structures of the two elements. The whole
DNA region is very A-T rich, making slippage relatively likely. Note that, if valid,
the events shown provide a general model for the generation of repeats in DNA
regions containing closely spaced palindromic sequences, and also (*see* Stage iii)
for the origin of deletions when the template strand loops out at a palindrome.
The IS2–6 element is itself quite unstable, being prone to spontaneous deletion of
repeated sequences (Ghosal and Saedler[8]).

close together. A feature of the whole DNA region containing these repeats is that it is unusually rich in A-T base-pairs, which are more prone to come apart than G-C pairs. The diagram in *Box* 12.1 shows how this feature of the IS2 sequence is thought to have led to the formation, by a sequence of replication errors, of the longer derivative IS2–6. The essence of the idea is that the relatively weak pairing of bases in the AT-rich sequences provides the opportunity for the growing end of a DNA strand to peel off its template and fold back on itself, using the base-pairings which are provided by its own inverted repeat. The folded-back strand can then continue synthesis using itself as a template, generating a sequence which, following another round of replication, will become a more extensive inverted repeat.

As *Box* 12.1 shows, *two* successive slippages of this kind are needed to account for the origin of the IS2–6 sequence and, in addition, a further abberation, this time concerning the template strand rather than the newly synthesized strand, has to be postulated. The template, left in single-stranded condition during the self-copying of its partner, is itself thought to loop out to form a hairpin through one of its inverted repeats, and this results in a failure to replicate the looped-out region. If this part of the model is valid it obviously provides a general mechanism for the generation of deletions.

While the analysis of the molecular nature of sequence rearrangements in IS2 is so far unique, it may well be of general relevance. We may expect that closely spaced inverted repeats will in general be sources of genomic instability. Since such instability in essential coding regions will be difficult for an organism to tolerate, they are more likely to be found in non-coding or redundant regions of the genome.

Movable Genetic Elements as Sources of Mutation

A new dimension has been added to our understanding of genomic variability by the description of insertion sequences (IS elements) and transposons in bacteria (*see* Chapter 10, pp. 267–271) and the subsequent discovery of similar elements in higher organisms (*see* Chapter 15). We shall not deal with them in detail in this chapter, but it is appropriate here to list some of their known effects:

i. They can become inserted into genes, usually eliminating gene activity.

ii. Relatively frequently, they generate breaks at one or other of their ends which, in conjunction with other more or less nearby breaks, lead to chromosome deletions of varying extents or, occasionally, to inversions of chromosome segments.

iii. They undergo occasional transposition to other chromosome loci, carrying with them segments of chromosome from the loci at which they were previously inserted. In this way they lead to transpositions of blocks or portions of genes.

In both *E. coli,* and *Drosophila melanogaster* (*see* pp. 253, 448), a high proportion of 'spontaneous' mutations that abolish gene activity are due to the activities of apparently anarchic movable elements. For example, it has recently been shown by W. Bender that most of the mutations in the *Drosophila* **bithorax** region are so attributable. We discuss the origin and significance of this source of genetic variation in Chapter 15.

12.3 Induction of mutation and the action of mutagenic agents

Mutations can be very greatly increased in frequency by the use of any one of a variety of mutagenic agents or **mutagens**. These include radiation (ionizing radiations or ultraviolet light) and many chemicals of various types. The remainder of this chapter will be concerned with mutagenic treatments and their mechanisms of action, so far as these are known.

Ionizing Radiation

Direct Effects: Proportionality to Dose

A number of different types of high-energy radiation produce ionizations—that is the generation of charged atomic and molecular particles by

Fig. 12.3

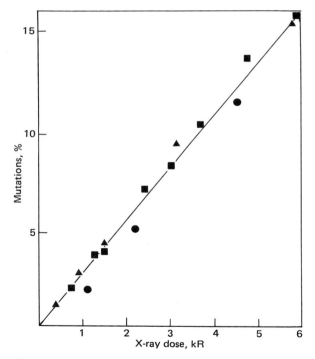

The direct proportionality between yield of recessive lethal sex-linked lethal mutations in *Drosophila melanogaster* and dose of X-rays.
The *ClB* technique for determining the frequency of sex-linked recessive mutation was used (for explanation, *see Box* 12.2).
The data, from three different authors, were compiled by Timoféeff-Ressofsky and Zimmer (1947) and previously reproduced in Auerbach.[14]
■, Timoféeff-Ressofsky; ▲, Oliver; ●, Efraimson and Schechtmann.

displacement of electrons. This property is shared by short wavelength electromagnetic radiation (including X-rays and the still shorter wavelength gamma-rays) and by beams of high-velocity atomic particles such as neutrons and helium nuclei (alpha particles).

All ionizing radiation induces chromosome breaks and the chromosomal

Box 12.2	The *ClB* procedure for detecting X-ray induced sex-linked mutations in *Drosophila melanogaster*

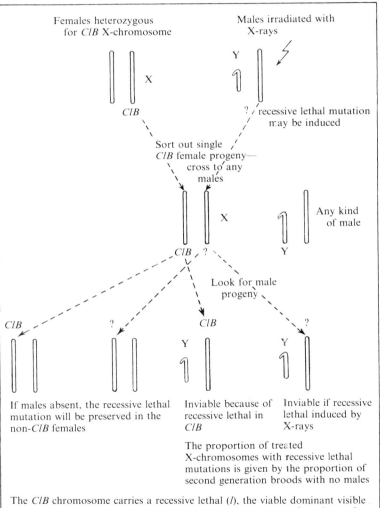

Females heterozygous for *ClB* X-chromosome

Males irradiated with X-rays

ClB

?/ recessive lethal mutation may be induced

Sort out single *ClB* female progeny— cross to any males

Any kind of male

ClB / ?

Look for male progeny

ClB   ?   *ClB*   ?

If males absent, the recessive lethal mutation will be preserved in the non-*ClB* females

Inviable because of recessive lethal in *ClB*

Inviable if recessive lethal induced by X-rays

The proportion of treated X-chromosomes with recessive lethal mutations is given by the proportion of second generation broods with no males

The *ClB* chromosome carries a recessive lethal (*l*), the viable dominant visible marker *Bar* (*B*) and a long inversion to prevent the recovery of products of crossing over between *l* and any newly induced lethal mutation. *ClB* has been largely superseded by the Müller-5 X-chromosome, which has a complex (more recombination-proof) inversion, and the w^a (apricot) marker as well as *B*, but no lethal. One looks for non-Bw^a males in the second generation.

rearrangements which stem from them, and at least X-rays induce point mutations as well. Most studies have been made with X-rays, the most accessible and widely used ionizing radiation for mutagenesis studies since the pioneering work of H. J. Muller, using *Drosophila*, and L. J. Stadler, using maize as experimental material.

Little can be said in detail about the mechanisms by which X-rays produce their genetic effects. It is probable that a number of kinds of damage to DNA are involved, some direct and some indirect in their causation. In only one respect is X-irradiation simple in comparison with other mutagenic treatments, and that is in showing a close proportionality between dose of radiation and yield of mutations. *Fig.* 12.3 shows the linear dose–yield relationship which has been found in the case of *Drosophila* sex-linked lethals, a class of mutations that is particularly convenient to assay (*see Box* 12.2). The roentgen unit, in terms of which X-ray dosage is expressed, is defined in terms of number of ionizations produced per unit volume, and the type of result shown in the figure implies that each mutation is the result of a single ionization. If, to take a simple alternative model, each mutation required two coincident ionizations, the yield of mutations would be expected to vary as the *square* and not the first power of the dose. In fact, chromosomal rearrangements, such as segmental interchanges and inversions, which do require two events (chromosome breaks) for their formation, *are* produced in proportion to the square of the dose, as *Fig.* 12.4 shows.

Another difference in dose–yield relationship as between point mutations and chromosomal rearrangements is the strong effect in the latter case, but not in the former, on the intensity with which the dose is given. It has been known for many years that, to a very good approximation, one obtains the same yield of point mutations from a given dose of X-rays whether it is given at a high intensity over a

Fig. 12.4

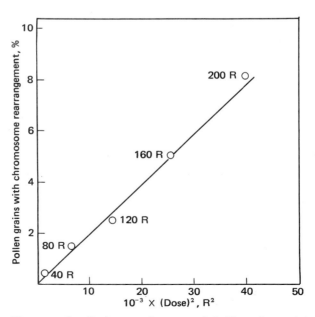

The proportionality between the square of the X-ray dose and the yield of chromosomal rearrangements in pollen grains of *Tradescantia*. Data of K. Sax (1940) cited by Auerbach.[14]

short time or, say, at a ten times lower intensity over a time period ten times as long. This is a matter of great importance since it affects predictions of the genetic effects of low levels of radiation spread over long periods, as might be experienced by workers in atomic industry or (the context in which the issue first became controversial) by people subjected to radioactive fall-out from tests of atomic weapons.

Experiments using very low dose rates are very laborious and expensive to perform on higher organisms because of the very large numbers required for the measurement of very low mutation frequencies. The most thorough study, carried out by W. L. and L. B. Russell on mice, led to the conclusion that a given dose of ionizing radiation induced three to four times as many mutations when given at a rate of 90 rad/min than when given at 90 rad/week.[15] A possible explanation of this difference is that, at very low dose rates, there is more opportunity for damage to DNA to be repaired.

Whatever the exact form of the relationship between mutation yield and dose rate, the prudent assumption to make is that there is no lower limit, or 'threshold', below which ionizing radiation has no mutagenic effect. If this assumption is correct, a proportion (probably a small proportion) of 'spontaneous' mutations and chromosome breaks is due to natural ionizing radiation, including cosmic rays and emissions from radioactive components of natural rock.

Chromosomal rearrangements are certainly induced more efficiently by a short intense dose of radiation than by a long low-level exposure delivering the same number of ionizations. This effect is readily explained by the tendency of broken chromosome ends to heal. For a rearrangement to occur it is necessary for two chromosome breaks to be present at the same time and, at low dose rates, the first break is likely to have healed to restore the original arrangement before the second break appears.

Indirect Effects: Involvement of Oxygen

A proportion of the damage caused by X-rays to genes and chromosomes is no doubt due to the direct effects of ionizations in the target DNA molecules themselves, with consequent breakage of chemical linkages. But there is evidence that the greater part of X-ray mutagenesis is more complex, requiring the presence of molecular oxygen for its completion. Both bean root tips and *Drosophila* pupae have been X-irradiated when in an atmosphere of either pure oxygen or pure nitrogen, and the yield both of chromosome aberrations in the former material and of sex-linked recessive lethal mutations in the latter was about three times greater under aerobic as compared with anaerobic conditions. It is supposed that free radicals formed from oxygen during the X-ray treatment are the immediate agents of much of the genetic damage or, alternatively, that such radicals inhibit some repair process.

Ultraviolet Light

Pyrimidine Dimers and Photoreactivation

Mutagenesis by ultraviolet light (UV) is a subject of great complexity, not because the direct effect of UV on DNA is complicated but rather because of the variety of ways in which living cells can respond to the damage caused. The primary effect of UV on DNA is indeed relatively simple and well understood, consisting to a very large extent of the formation of cross-links between pairs of pyrimidine residues occupying adjacent positions in the DNA chain. The linked pairs of pyrimidines, called **pyrimidine dimers**, have the following structure (the picture shows two linked thymines):

Thymine dimers are the most stable and plentiful type in UV-irradiated DNA, but cytosine dimers and mixed thymine–cytosine dimers also occur with frequencies depending on the DNA base composition.

In most organisms that have been investigated, ranging from *E. coli* to man, there is a very efficient mechanism for getting rid of pyrimidine dimers in DNA. This is dependent on a specific enzyme and on visible light, and is called **photoreactivation**. The enzyme simply catalyses the reversal of the dimerization and restores normal DNA structure, using visible light as a source of energy to drive the reaction. Photoreactivation enzyme is almost ubiquitous in cells or tissues which are exposed to light. Since, under natural conditions, ultraviolet always occurs together with visible light, the cure is virtually coincident with the injury. At least in bacteria and yeasts, where photoreactivation has been most thoroughly studied, it can prevent both lethal cell damage and mutation. From this it is inferred that virtually all UV-induced mutation results from pyrimidine dimers.

Excision Repair in Bacteria

Where photoreactivation cannot operate efficiently, as when a period of UV exposure is followed immediately by a period of total darkness, a more drastic means of removing pyrimidine dimers may be called into play. This is called **excision repair**, and involves the cutting out of a portion of DNA single strand containing the dimer and replacing it by new synthesis, using the probably undamaged complementary strand as template. The initial step in this 'cut-and-patch' type of repair is now somewhat clearer as a result of recent studies on the bacterium *Micrococcus luteus*.[9] In this species, an enzyme has been demonstrated which will cleave the glycosidic linkage of a dimerized pyrimidine residue. An endonuclease is then able to attack a phosphodiester linkage adjacent to the depyrimidinated deoxyribose, and digestion by an exonuclease can then remove a segment of single-stranded DNA including the dimer. The resulting single-stranded gap is filled by repair synthesis catalysed by DNA polymerase. It is conjectured that a similar mechanism exists in *Escherichia coli* and probably also in eukaryotic cells. The scheme is outlined in *Fig. 12.5*.

recA-dependent Mutagenic Repair in E. coli

Perhaps the most definite clue to the understanding of UV-induced mutagenesis in bacteria is the finding that *recA* mutants of *E. coli*, besides

Fig. 12.5

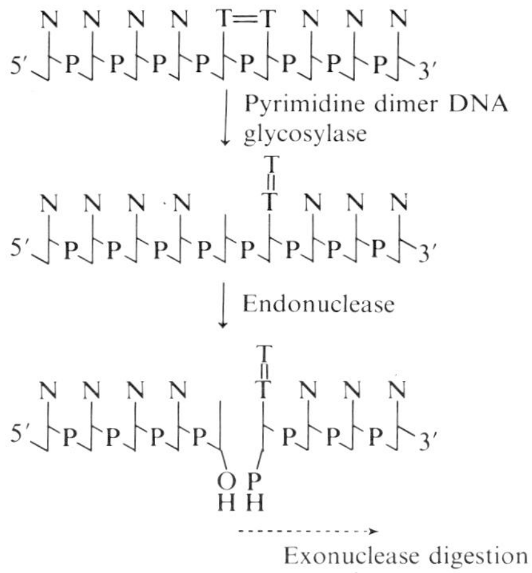

N N N N T=T N N N
5'—P—P—P—P—P—P—P—P—3'

    Pyrimidine dimer DNA
    glycosylase

        T
        ‖
N N N ·N   T N N N
5'—P—P—P—P—P—P—P—P—3'

    Endonuclease

        T
        ‖
N N N N   T N N N
5'—P—P—P—P   P—P—P—3'
        O P
        H H

- - - - - - - - - - - →
    Exonuclease digestion

            O
            ‖
N = any DNA base; —P— = —O—P—O— ; deoxyribose residues are indicated by
            O
vertical lines.   H

Mechanism for the removal of pyrimidine (shown here as thymine) dimers
from DNA.
After Heseltine et al.[9]

being recombination deficient, are unusually sensitive to the lethal effects of UV but immune to its mutagenic action. This at once suggests that a part of the repair of UV damage is both prone to error (mutation) and dependent on $recA^+$ function.

We have already, in Chapter 10, considered some of the evidence on the part played by the $recA^+$ protein product in recombination. It will be recalled that this protein binds to single-stranded DNA and can, under appropriate conditions, catalyse the annealing of complementary single strands or bring about the assimilation of a single strand into a homologous duplex, with unwinding of the latter. So far as *E. coli* is concerned, however, it is likely that the $recA^+$ protein is important less as a catalyst of recombination as such than as a part of an emergency ('SOS') system for repair of DNA damage by UV.

The 'SOS' response of *E. coli* comes into play following damage to DNA by UV and certain other agents. The signal for the response is probably the presence of DNA breakdown products resulting from excision of pyrimidine dimers. While it is thought to involve the derepression of a number of functions relevant to recovery from UV damage, its most obvious manifestation at the molecular level is the accumulation of large amounts of *recA* protein. The reason for this accumulation is now largely understood.[10] In undamaged cells, the transcription of *recA* is repressed by the protein product of another gene, called *lexA*. The *recA* protein, when 'activated' by combination with DNA breakdown products (probably deoxyribonucleotides), acts as a protease, cleaving the *lexA* repressor and thus unleashing the synthesis of its own messenger RNA. (In parentheses it may be remarked that activated *recA* protein also destroys lambda phage repressor, which is why ultraviolet light induces lambda lysogens—*see* p. 297.)

Once produced in large amounts, *recA* protein facilitates, probably by a mechanism similar to that pictured in *Fig*. 10.11, the filling ('patching') of single-stranded gaps in duplex DNA. These gaps are thought to arise following ultraviolet irradiation in at least two different ways. Firstly, there are the gaps which result from excision of pyrimidine dimers (cf. *Fig*. 12.5). These are for the most part filled by repair synthesis, probably catalysed by DNA polymerase I, before the DNA is replicated. More serious for the survival of the cells are the longer gaps created when replication occurs before all the pyrimidine dimers are excised. When DNA polymerase encounters a pyrimidine dimer on the template strand, its further progress is halted and synthesis has to be resumed at some more or less distant reinitiation site on the far side of the dimer. It seems to be for the repair of such long gaps (of the order of kilobases in length) that *recA* protein is essential.[11]

The mechanism of long-patch repair is still uncertain. It would be surprising if it did not involve the ability of *recA* protein to bind to single-stranded DNA (cf. p. 266). By so doing the *recA* protein may protect the single strands in the gapped DNA from nuclease attack until the gaps can be filled. Whatever the long-patch repair process is, it seems likely to be prone to a variety of kinds of errors.

It is clear from studies of mutant gene products, of the kind dealt with in the next chapter, that UV induces mutations of many different kinds. Base-pair substitutions (both transitions and transversions), deletions, frameshifts and chromosomal rearrangements all occur among UV-induced mutants. The simplest interpretation of the available data is that all of these originate through error-prone gap-filling.

Repair in Eukaryotes: Yeast and Man

The foregoing account of effects of UV irradiation was concerned entirely with bacteria and we saw how important clues can be gained from the UV-sensitive *recA* mutants and their immunity to UV-induced mutation. A great deal of work has been done on UV-sensitive mutants in other organisms also, especially in fungi. No summary of the results will be attempted here, except to say that they indicate great complexity in the mechanisms of repair of UV damage and in the relationships between these and mutation. In *Saccharomyces cerevisiae* a host of genes has been identified, all capable of mutating individually to give UV-sensitive phenotypes. Some of these mutations also result in increased sensitivity to ionizing radiation and some to sensitivity to alkylating chemicals as well (*see below*, p. 333). In several cases UV sensitivity is accompanied by reduced UV mutability but in a few others *enhanced* UV mutability has been reported. As yet there is no hard knowledge about what any of these genes codes for. The whole area provides a good illustration of the fact that the geneticist's standard approach to the understanding of complex processes, through the induction of mutations and the analysis of their effects, sometimes reveals more complexity than the investigator had bargained for.

A comparable complexity is emerging from studies of cells from human subjects with hereditary defects in DNA repair[16]. A number of such defects are known, but here we will mention only one class, known as **xeroderma pigmentosum** (XP). This syndrome is characterized by sensitivity to sunlight resulting in excessive skin pigmentation and skin lesions which frequently become cancerous. It is associated with failure to excise pyrimidine dimers from DNA. The genetic basis of the condition is homozygosity with respect to any one of a number of rare recessive alleles, assigned to seven different genes. The distinction between the seven different genetic classes was made on the basis of complementation tests by fusions of cultured fibroblasts (cf. p. 235). The implication is that dimer excision and subsequent DNA repair requires at least seven different gene products in human cells.

Cultured XP cells have been shown to have a much higher than normal UV-induced mutation rate and they are also hypersensitive to the effects of UV in inducing chromosome breaks and mitotic sister-chromatid exchanges (cf. p. 30) All of these effects could arise from the replication of DNA with dimers still present. It is known that such replication results in gaps in the newly synthesized strands, presumably because the DNA polymerase cannot synthesize continuously across the dimers. The enhanced mutation frequency could then be the consequence of error-prone gap-filling, and sister-strand exchange could be promoted by the presence of single-stranded regions and free single-strand ends, giving opportunities for the formation of hybrid DNA between chromatids (cf. p. 138).

In another human congenital condition, called Bloom's syndrome, the spontaneous frequency both of mutation and of sister-strand exchange is very greatly enhanced. The most obvious feature of this syndrome is again sensitivity to sunlight and it is thought that some defect of DNA repair is present, though the nature of the defect has not been identified.

Chemical Mutagenesis

The first chemical to be shown to induce mutations (by C. Auerbach in 1946) was the bifunctional alkylating agent bis(2-chloroethyl) sulphide, otherwise known as mustard gas. Since that discovery the list of chemical mutagens has grown enormously and a wide range of compounds has been shown to be mutagenic to different degrees. In this brief account we shall only consider some of the more potent, or informative, types of mutagen.

In order to be a mutagen a chemical has to fulfil one or other of three criteria. It must react chemically with DNA *or* it must be convertible metabolically into a compound which will so react *or* it must mimic in some way a normal DNA base or base-pair so as to interfere with replication. We shall see examples of all these types of action.

In research on chemical mutagens there has always been the hope that some specificity would be discovered—that a given mutagen would be found to act only on a specific base-pair or DNA sequence and/or induce one specific kind of mutation. On the whole, such hopes have been disappointed in that most of the best investigated mutagens have complex effects. However, some show a degree of specificity and are of some use for at least tentative diagnosis of the nature of mutational sites in DNA.

Alkylating Agents

Chemicals that will alkylate DNA bases include some of the most potent mutagens known. **Monofunctional** alkylating agents have just one alkyl group to donate, usually methyl or ethyl, while **bifunctional** agents have two reactive groups per molecule and hence can simultaneously link to two DNA bases and form cross-links between different DNA chains or different parts of the same chain. The formulae of some of the more effective and widely used alkylating mutagens are given in *Fig.* 12.6. The main targets for alkylation in the DNA are the nitrogen atoms of the bases: the N-7 position of guanine, the N-1 and N-3

Fig. 12.6

Monofunctional alkylating agents

		Alkylating group
Diethyl sulphate	$(C_2H_5)_2SO_4$	$-C_2H_5$
Ethylmethane sulphonate (EMS)	$C_2H_5-O-SO_2-CH_3$	$-C_2H_5$
Methylmethane sulphonate (MMS)	$CH_3-O-SO_2-CH_3$	$-CH_3$
N-Methyl-N'-nitro-N-nitrosoguanidine (MNNG)	$HN{=}C-NH-NO_2$ $O{=}N-N-CH_3$	$-CH_3$

O^6-*Ethylguanine (probably the main mutagenic product of ethylation)*

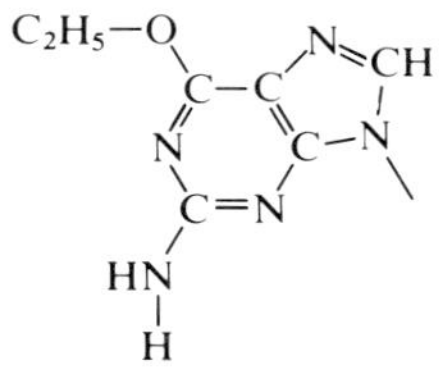

C₂H₅—O ... (structural formula)

HN
|
H

Bifunctional alkylating agent

Mustard gas

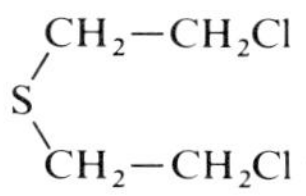

CH₂—CH₂Cl
/
S
\
CH₂—CH₂Cl

$X-CH_2-CH_2-S-CH_2-CH_2-X$
(cross-linking between bases)

Benzopyrene, a polyaromatic mutagen

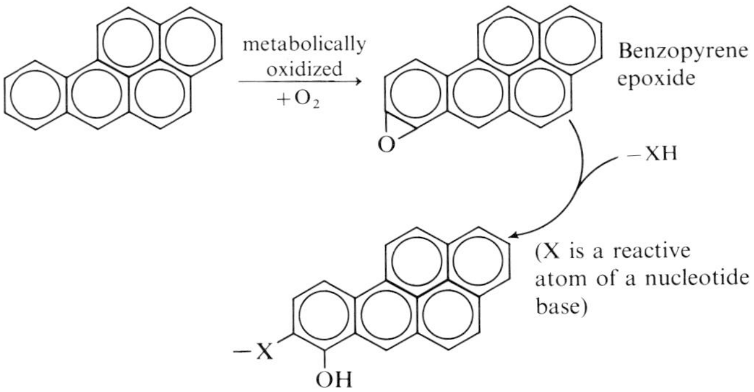

metabolically
oxidized
——————→
$+O_2$

Benzopyrene epoxide

—XH

(X is a reactive atom of a nucleotide base)

—X

OH

Some mutagenic chemicals and their modes of action.

positions of adenine and the N-1 position of the pyrimidines. However, the O-6 atom of guanine is also alkylated to a significant extent, and this is now thought to account for a large part of the mutagenic effect. The O-6 alkyl derivatives of guanine have the double-bond distribution of the *enol* form of the normal base, and hence have hydrogen-bonding properties similar to those of adenine, promoting G-C→A-T transitions on replication (cf. *Fig.* 12.1). If this were the only mechanism of action of monofunctional alkylating agents one would expect them to be specific for G-C→A-T transitions, and in *E. coli* there is good evidence that ethylmethane sulphonate (EMS) and *N*-methyl-*N'*-nitro-*N*-nitrosoguanidine (*see Fig.* 12.6) do induce such mutations preferentially. For example. EMS will readily mutate the amber chain terminating codon (TAG in DNA) to ochre (TAA) but not vice versa.

Alkylating agents can, however, have much more diverse effects. Base alkylation often appears to set in train repair processes which may have steps in common with the excision repair of UV-induced dimers and with much the same range of consequences. Alkylation by bifunctional reagents such as mustard gas is particularly disruptive since it results in cross-linking of bases. Mustard gas, in fact, mimics the effect of ionizing radiation in producing chromosome breaks, the effects of which tend to become worse with successive rounds of mitosis when anaphase bridges and consequent further breaks and losses of chromosome segments occur.

Polyaromatic Compounds and Metabolic Activation

A compound which has attracted much attention as a ubiquitous product of combustion of organic material and a potent mutagenic (and also carcinogenic) agent, is the polyaromatic compound benzopyrene. The correlation between mutagenesis and carcinogenesis is, incidentally, a very widespread one, and is notably exemplified by the alkylating agents; we shall discuss its possible significance in Chapter 17.

Unlike the alkylating agents, benzopyrene and other compounds of related structure are rather inert chemically and will not themselves react with DNA bases. However, they can be transformed by oxidative enzymes of the cellular respiratory system to epoxide derivatives, one example of which is shown in *Fig.* 12.6. The epoxide ring is very reactive and will attack vulnerable atoms in DNA—probably the same ones as are subject to alkylation.

The possibility, exemplified by the case of benzopyrene, of metabolic activation of a potential mutagen is one which is taken very seriously in the growing industry of testing chemicals for possible environmental hazard. The most convenient mutagen testing system, due largely to B. N. Ames, uses a range of different auxotrophic mutants of *Salmonella typhimurium* representing different types of mutation, and looks for induced back mutations to prototrophy. The bacterium cannot, however, be relied upon to carry out all the metabolic conversions of the compound under test that might occur in a human or animal cell, and so compounds are often tested for mutagenicity both directly and after incubation with a mammalian cell preparation.

Hydroxylamine and Nitrous Acid

These two compounds are examples of rather simple substances which react with DNA bases in known ways and which might be expected to have rather simple mutational effects. In the earliest experiments on chemical muta-genesis of bacteriophage T4, carried out by S. Benzer and E. Freese, hydroxyl-amine did indeed turn out to be the specific mutagen *par excellence*, apparently inducing only transitions and only in one direction. This is consistent with the fact that the compound attacks only cytosine among the DNA bases. It can be plausibly suggested that the cytosine derivative which is formed has the hydrogen-bonding properties of thymine, thus bringing about G-C→A-T transitions. Hydroxylamine is not an effective mutagen when used on cells, as opposed to free virus particles, probably because it is toxic in too many different ways, but its *O*-methyl derivative has been used with some success in fungi and may have a similar specificity to hydroxylamine itself.

Nitrous acid reacts with amino groups to form hydroxyl groups according to the equation:

$$R\!-\!NH_2 + HNO_2 \longrightarrow R\!-\!OH + N_2 + H_2O$$

Thus guanine can be deaminated to xanthine, adenine to hypoxanthine and cytosine to uracil. Of these transformations, the first may be non-mutagenic (or lethal), but the other two should lead to transition mutations, since hypoxanthine has the hydrogen-bonding properties of guanine and uracil those of thymine. On the basis of this reasoning, one would expect nitrous acid to be specific for transition mutations in both directions. This prediction has been borne out in experiments on mutagenesis in bacteriophage and also in tobacco mosaic virus. In the latter virus the isolated RNA is infective and can be treated with nitrous acid *in vitro* and then infected back into the plant host; amino acid replacements in the virus protein, brought about by nitrous acid-induced mutation have been identi-fied by chemical methods.

In fungi, on the other hand, nitrous acid has been shown to induce transversions as well as transitions. Simple deamination is not the only effect that nitrous acid can have on DNA bases; it appears also to cause a certain amount of cross-linking of bases, which could lead to various types of mutation through error-prone excision repair.

Base Analogues

We turn to mutagens which act through interference with replication of DNA. Base analogues have their effect through being incorporated into DNA in place of the bases whose structures they mimic. The two analogues which have been used extensively as mutagens for bacteria and bacteriophage are 5-bromouracil (and its DNA nucleoside 5-bromodeoxyuridine) which is a thymine analogue, and 2-aminopurine, which is an adenine analogue. Their mutagenicity is attributed to ambiguous base-pairing properties. 5-Bromouracil (5-BU), because of the electronegativity of the bromine atom which replaces the thymine methyl group, has a much greater tendency than has thymine to exist in the *enol* tautomeric form (cf. *Fig.* 12.1). 2-Aminopurine (2-AP) can probably use its 2-amino group (in the same position as in guanine) to pair with cytosine as well as with thymine.

If ambiguous base-pairing is the explanation for base-analogue mutagenesis we would expect the mutations induced to be base-pair transitions. The direction of the transition, whether A-T→G-C or the reverse, would depend on whether the analogue was 'mistakenly' paired with the 'wrong' base at the time of its incorporation or at some replication after incorporation. Detailed studies on the mutants induced in bacteriophage T4, mainly by E. Freese, confirmed that base analogues did induce transitions in both directions (a mutant induced by one analogue could usually be reverted to wild type by the same analogue), but that the frequencies in the two directions could be very unequal. The most marked inequality was shown by 5-BU, which had a strong bias towards inducing G-C→A-T transitions (the same direction as hydroxylamine, which was found to revert only a minority of 5-BU-induced mutants). It seems likely that the mutagenic effect of 5-BU is made predominantly through its occasional incorporation opposite guanine. Once incorporated it seems to function in replication as if it were thymine with very few errors. 5-BU can be incorporated into bacterial DNA to a high level and can, indeed, almost entirely replace thymine in bacterial strains in which the synthesis of the normal base is blocked by mutation. 2-AP, in contrast, is incorporated only to a small extent; presumably it gives a much larger number of errors after incorporation than does 5-BU.

Genes that have been thoroughly mapped and also extensively mutagenized by base analogues (the *rII* gene of T4 and the *lacI* gene of E. coli[4] are two which fit this description) show the phenomenon of mutational 'hot-spots'. These are sites which are exceptionally sensitive to mutation by one or other analogue and mutate over and over again in response to it. The 5-BU hot-spots are different from the 2-AP ones. It has been presumed that the likelihood of incorporation of a base analogue must depend on the adjacent nucleotide sequence (the 'context') as well as on the nature of the base-pair which is the actual site of mutation, since obviously only a small minority of base-pairs of each chemical type (A-T and G-C) are hot-spots.

In the case of 2-AP hot-spots in the *lacI* gene, however, a more precise explanation has now been given. It turns out that the susceptible base-pairs are those in which the cytosine is liable to methylation. We saw above (p. 321) that such base-pairs are at risk of spontaneous mutation, because of occasional deamination of methylcytosine to thymine. They are also especially mutable by 2-AP, supposedly because methylcytosine forms a more stable base-pair with 2-aminopurine than does cytosine itself.

Acridine Compounds

Proflavine, a member of the acridine family of compounds, was the first agent shown to induce frameshift mutations, the demonstration being made with T4 bacteriophage. It did not induce mutations of other kinds with appreciable frequency. This special mutagenic effect is plausibly explained by the known ability of the acridine ring system, which roughly mimics a hydrogen-bonded DNA base-pair in shape and hydrophobicity, to insert itself (**intercalate**) into the stack of base-pairs running up the core of the DNA double helix.

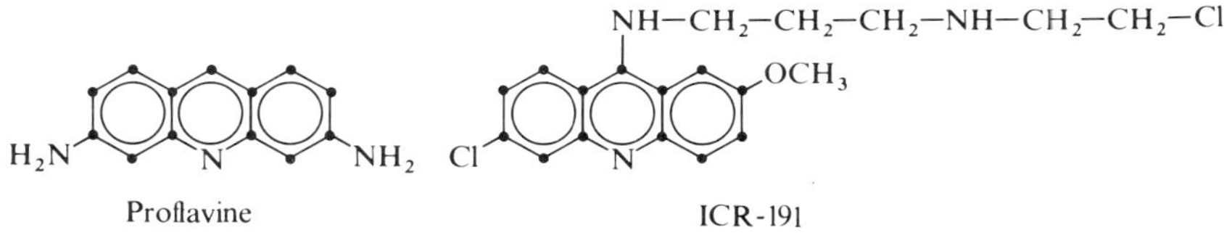

Fig. 12.7   Mutagens that induce frameshifts.

Proflavine is thought to induce frameshifts by stabilizing what would otherwise be very unstable strand associations, such as that between the two strands of the stem of a looped-out hairpin configuration of the type postulated in Streisinger's model for the origin of frameshifts (*Fig.* 12.2). To the extent that a duplex is held together by acridine molecules in its core, its stability does not wholly depend on a long sequence of perfectly complementary base-pairs.

Certain acridine derivatives are much more mutagenic in bacteria than proflavine is. A series of compounds originally designed for anticancer therapy was made by coupling acridines of different kinds to bifunctional alkylating agents. In these compounds, generally referred to by numbers prefixed by ICR (Institute of Cancer Research), one of the alkylating functional groups is left free for reaction with other molecules such as DNA. These acridine 'half-mustards', as they are sometimes called, probably owe their potency as mutagens to their ability to couple through their alkylating ends to a DNA base, thus bringing the acridine moiety into a position from which it has a high probability of insertion into the base-pair stack.

12.4 The timing of the appearance of induced mutations

Since the effects of mutagens are, at least by all the mechanisms that we partly understand, on just *one* strand of a DNA double helix, it might be expected that the establishment of a pure mutant clone of cells would be delayed for one replication cycle after the primary mutagenic event. If the mutagen is a base analogue, acting by 'correct' incorporations with a chance of 'incorrect' base-pairing at subsequent replications, the establishment of a mutation may be delayed for many cell divisions after the mutagenic treatment. *Fig.* 12.8 explains the two predicted modes of base-analogue mutagenesis; these have, in fact, both been demonstrated to occur in *Escherichia coli.*

On the other hand, there is strong evidence from a variety of organisms that all the descendants of a mutagenized cell may carry the same mutation. Thus when the progeny of X-irradiated *Drosophila* males has been examined for visible mutations, a proportion have been found to be fractional mutants, showing the mutant phenotype over only a part (most often approximately one-half) of their body. Rather more frequent have been 'whole-body' mutants, the occurrence of which suggests that *both* products of division of a mutagenized chromosome can carry the same mutation. This apparently paradoxical finding, recognized by H. J. Muller and his colleagues (1961) and called by them 'mutation of the already existing gene', has been repeated with different mutagens and different organisms, especially in fungi.

One trivial explanation would be that one of the first division products is commonly lost because of lethal mutation. On this interpretation the proportion of whole-body, or whole-colony, mutants to sectorials should increase with increasing dose of mutagenic treatment. This explanation may apply to some cases but there are others in which it has been ruled out. One is forced to consider the

Fig. 12.8

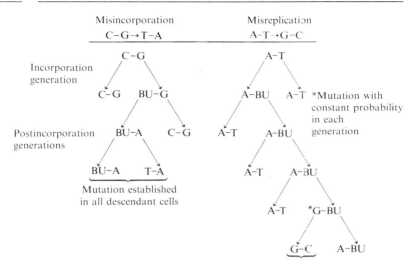

| Misincorporation | Misreplication |
| C-G→T-A | A-T→G-C |

Two different modes of mutation induction by 5-bromouracil.

possibility that a mutational change, originally induced in one DNA strand, can somehow be transmitted to the already existing partner strand before the next round of replication. The problem of possible mechanism here is formally the same as we faced in Chapter 5 in connection with meiotic gene conversion. In that context we considered an explanation in terms of mismatch correction by excision–repair. A similar mechanism, perhaps using the same enzymes or perhaps different ones, could well be at work in the establishment of whole-chromatid mutations.

A contrary phenomenon, the delay of appearance of a stable mutant phenotype for many generations after mutagenic treatment, has also been recorded from time to time. Both in *Drosophila* and in the fission yeast *Schizosaccharomyces pombe*, the pattern of appearance of mutants following mutagenesis sometimes makes it necessary to postulate that a special propensity to mutate, rather than an accomplished mutation, can be transmitted through an indefinite number of cell generations. This phenomenon has been termed **replicating instability**. In a number of examples, the unstable condition can change either to stable mutant or to stable wild type with approximately equal probabilities at each cell or sexual generation. A mechanism analogous to the second (misreplication) mode of base-analogue mutagenesis (*Fig.* 12.8) will not explain the observations because a chemically modified base, while it may persist through replication, cannot itself replicate by any established mechanism. One can certainly envisage ways in which secondary modifications of DNA bases might possibly be perpetuated through rounds of replication. It has been proposed, for example, that base methylation in a particular sequence might be self-perpetuating if the methylating enzyme had a propensity to methylate a newly synthesized strand where the template strand was already methylated; this is a mechanism which has been considered as one possible basis for cellular differentiation (*see* p. 511). There is, however, as yet no firm evidence for this or any other speculative model for replicating instability.

12.5   *In vitro* site-directed mutagenesis

The new techniques for *in vitro* cloning and manipulation of DNA molecules open up possibilities for the controlled induction of mutations of specific kinds at specific sites.

Access to specific sites in cloned DNA can be obtained by the use of restriction endonucleases of the type that makes staggered cuts in defined palindromic sequences. Under appropriate conditions, some of these enzymes can be used to nick only one strand of the target segment. For example, in the presence of ethidium bromide, EcoRI cuts between the G and A of just one strand of the

$$\begin{bmatrix} 5'\text{-GAATTC-}3' \\ 3'\text{-CTTAAG-}5' \end{bmatrix}$$

sequence. Having made such a nick, one can make a small single-strand gap adjacent to it by use of an exonuclease, or alternatively 'translate' the nick with DNA polymerase I of *E. coli*, which has the effect of replacing a string of nucleotides starting at the nick and proceeding in the $5' \rightarrow 3'$ direction. If a single-strand gap is produced, the exposed bases on the remaining strand are vulnerable to certain kinds of chemical mutagenesis, while nick translation can be used to introduce a mutagenic base analogue in the vicinity of the restriction site.

An example of the use of the nick-translation method is shown in *Box* 12.3, the target for mutagenesis being the single EcoRI site in cloned cDNA copied from β-globin mRNA (*see* Chapter 14, *Fig.* 14.2). The extent of nick translation was limited by the omission of dGTP from the reaction mixture; the replacement of nucleotides progressed until the first C was reached on the template strand and then it stopped. The unnatural analogue hydroxy-dCTP was used instead of dTTP. N^4-Hydroxycytosine is sufficiently similar to thymine in its hydrogen-bonding properties to be incorporated opposite adenine when no thymine is available, but at the following round of replication it is apt to pair with guanine instead of with adenine, thus bringing about an A-T to G-C transition.

Some of the potential mutations introduced into the cDNA by the manipulations summarized in *Box* 12.3 result in the effective loss of the EcoRI site. Selection was made for such mutations by reintroducing the manipulated plasmids into *E. coli* cells by calcium-mediated transfection (cf. p. 243), allowing them to replicate, re-extracting the plasmid DNA and treating it with EcoRI. The plasmid molecules still possessing the EcoRI cleavage site were rendered linear and non-infectious, and only those which were EcoRI resistant were recovered after a further round of transfection. The figure shows the different mutant base sequences that were found among clones derived from the EcoRI-resistant population of molecules. They code for a series of amino acid substitutions specifically at positions 121 and 122, near to the C-terminal end of the β-globin chain.

More and more DNA sequences are now being cloned, and the wide range of available restriction endonucleases makes it possible, in principle, to mutagenize them specifically at many different sites. Controlled site-specific mutation has become an important tool for the analysis of the functions of specific DNA sequences. It also has conceivable practical applications, since it is technically possible not only to propagate an artificially mutated higher organism gene in

Box 12.3   **Site-directed mutagenesis in a globin-coding sequence. Data of Müller et al.[12]**
see text.

Part of the cDNA, copied from β-globin mRNA and cloned in a plasmid, containing the sole EcoRI restriction site:

5′ G-C-A-A-A-G↓A-A-T-T-C-A-C-T-C-C-T-C-A-G-G 3′

3′ C-G-T-T-T-C-T-T-A-A↑G-T-G-A-G-G-A-G-T-C-C 5′

| Single-strand nicking by EcoRI in presence of ethidium
↓ bromide

a

5′ G-C-A-A-A-G A-A-T-T-C-A-C-T-C-C-T-C-A-G-G 3′

3′ C-G-T-T-T-C-T-T-A-A-G-T-G-A-G-G-A-G-T-C-C 5′

and

b

5′ G-C-A-A-A-G-A-A-T-T-C-A-C-T-C-C-T-C-A-G-G 3′

3′ C-G-T-T-T-C-T-T-A-A G-T-G-A-G-G-A-G-T-C-C 5′

| Nick translation by DNA polymerase in presence of
↓ dATP, dCTP and N⁴-hydroxy-dCTP

a

HO OH   OH   OH
| | | |
5′ G-C-A-A-A-G-A-A-C-C-C-A-C-C-C-C-C-C-A G-G 3′
- →

3′ C-G-T-T-T-C-T-T-A-A-G-T-G-A-G-G-A-G-T-C-C 5′

and

b

5′ G-C-A-A-A-G-A-A-T-T-C-A-C-T-C-C-T-C-A-G-G 3′

← - - - - - - - - - - - - - -
3′ C-G C-C-C-C-C-C-A-A-G-T-G-A-G-G-A-G-T-C-C 5′
   / | \ | \ |
HO HO OH OH OH

| Transfection into *E. coli* and replication *in vivo*
↓ —gives all permutations of the alternatives shown

a

5′ G-C-A-A-A-G-A-A-$^{C\,C}_{T\,T}$-C-A-C-$^{C}_{T}$-C-C-$^{C}_{T}$-C-A-G-G 3′

3′ C-G-T-T-T-C-T-T-$_{A\,A}^{G\,G}$-G-T-G-$_{A}^{G}$-G-G-$_{A}^{G}$-G-T-C-C 5′

and

b

5′ G-C-$^{G\,G\,G}_{A\,A\,A}$-G-$^{G\,G}_{A\,A}$-T-T-C-A-C-T-C-C-T-C-A-G-G 3′

3′ C-G-$_{T\,T\,T}^{C\,C\,C}$-C-$_{T\,T}^{C\,C}$-A-A-G-T-G-A-G-G-A-G-T-C-C 5′

Clones were selected as resistant to EcoRI. The variant DNA sequence found, and the amino acids coded for, were as follows.

Glu Leu	Glu Pro	Glu Ser
-A-G-A-A-C-T-C-A-	-A-G-A-A-C-C-C-A-	-A-G-A-A-T-C-C-
-T-C-T-T-G-A-G-T-	-T-C-T-T-G-G-G-T-	-T-C-T-T-A-G-G-

Gly Phe	Glu Phe
A-G-G-G-T-T-C-A	-A-G-A-G-T-T-C-A-
-T-C-C-C-A-A-G-T	-T-C-T-C-A-A-G-T-

Glu Phe
-A-G-A-A-T-T-C-A-

Normal sequence:

-T-C-T-T-A-A-G-T-

bacteria but also to reintroduce it into cells of the species from which it originally came.

We noted in Chapter 9 (p. 244) the recent development of a method for introducing cloned DNA into the mouse germ line. It is the difficulty of ensuring and controlling the expression of engineered and reintroduced genes that is likely to be the most serious technical obstacle to the application of genetic engineering to animal breeding or, indeed, to human medicine (*see* Chapter 19, p. 577).

12.6   Summary and perspectives

Spontaneous mutation is due to errors in DNA replication and, to some extent, to the inherent slight chemical instability of some of the DNA bases. All organisms possess means for correcting the DNA mismatches which arise from replication error or base instability before they can be replicated and established as stable mutations. Mutants defective in error–correction mechanisms of various kinds have enhanced mutation frequencies.

The spontaneous mutation frequency of an organism is to a large extent controlled by its own genotype. Apart from particularly mutation-prone sequences, such as tandem or inverted repeats in conjunction with clusters of A-T base-pairs, whole genomes may be more or less susceptible to mutation depending on the efficiencies of their error–correction mechanisms. It is reasonable to suggest that a certain low level of mutational 'noise' is favoured by natural selection. In the long run, situations will presumably arise in which the fittest member of a population is a new mutant. Spontaneous mutation frequencies may represent the best long-term compromise between conservatism and innovation.

Agents that induce mutations increase the probability of error during DNA replication, most usually by damaging the template strands which guide new synthesis. One of the most potent mutagens, ultraviolet light (UV), is a ubiquitous environmental hazard, and all organisms that live in the light have various mechanisms for protecting themselves from lethal UV damage. At least one of these mechanisms (excision–repair) also protects against certain kinds of chemical damage—for example by alkylating agents. Many kinds of mutation, therefore, can increase susceptibility to UV and/or alkylating agents. Two classes of mutants of *E. coli*, *recA* and *lexA*, are immune to UV mutagenesis, while being abnormally sensitive to UV killing. At least in bacteria, UV mutagenesis occurs as the result of an error-prone repair process.

Most chemical mutagens have some degree of specificity and a few appear to be quite highly specific. For example, some alkylating agents are highly selective for substitutions in G-C base-pairs while acridine compounds favour small insertions and deletions. Mutagen specificity can sometimes permit the tentative identification of the nature of mutation.

A newly recognized source of 'spontaneous' mutation, and a highly important one, is the tribe of movable sequences which seem to inhabit many or most genomes and can insert into, and inactivate, genes in an apparently random fashion.

As a consequence of the recent development of methods for molecular cloning and manipulation *in vitro* of genome fragments, it is now possible to bring about specific mutations at predetermined sites and to reintroduce the altered sequences into the cellular genome. These techniques considerably extend the scope of mutagenesis as a tool for investigating gene structure and function.

Chapter 12 — Selected Further Reading

Auerbach C. (1976) *Mutation Research*, 504 pp. London, Chapman and Hall.

Drake J. W. (1970) *The Molecular Basis of Mutation*. San Francisco, Holden-Day Inc.

Shortle D., DiMaio D. and Nathans D. (1981) Directed mutagensis. *Annu. Rev. Genet.* **15**, 265–294.

Witkin E. (1976) Ultraviolet mutagenesis and inducible DNA repair in *Escherichia coli. Bacteriol. Rev.* **40**, 869–907.

Chapter 12 — Problems

See Problems at end of Chapter 13.

13 Gene Function Investigated by Mutation

13.1   Hunting for mutants

Until the recent advent of methods for the physical identification and isolation of genes as pieces of DNA (*see* Chapter 14), the only way of defining the functions of genes, indeed the only way of knowing that genes existed, lay through the study of genetic variation. Given a clear-cut inherited *difference* in the phenotype of the organism, it can usually be mapped to one specific locus in the genome and, by detailed analysis of the phenotype, one can attempt to define the normal function which has been altered. Since already existing or new spontaneous genetic variants with sufficiently sharp effects are rather infrequent (though useful ones do turn up from time to time), and since the spontaneous frequency of new mutation is very low (perhaps of the order of 10^{-6} or less per gene per nuclear division), much of the geneticists' art lies in the devising of methods for inducing and selecting mutations of the desired types.

We dealt with the mechanisms of mutagenesis in the last chapter; in this one we shall refer as we go along to the origins of particular kinds of mutants, but a few general remarks about procedures for mutant hunts are in order at the outset.

Visual Methods—Replica Plating

Many mutants, for example those with altered pigmentation or morphology, can be distinguished by eye, and the use of the electron microscope can extend the class of visible mutants to submicroscopic objects such as bacteriophages. But if one is focusing attention on a particular gene or group of genes, it is generally preferable to examine an aspect of the phenotype which follows fairly directly from the primary gene function, and this means analysis at the level of metabolism or macromolecular synthesis.

A basic problem in devising procedures for selection of mutants is that, if one starts with the normal (wild-type) organism, most mutations with clearly scorable effects result in loss of function, so that the mutant is either completely unable to survive and grow, or can do so only under special conditions. How can one select for a more or less crippled organism?

In bacteria and yeast-like fungi which, on a growth medium gelled with agar, grow as colonies composed of free cells, a very convenient device known as **replica plating** can be used for visual identification of defective colonies among thousands of non-defective ones. In this method, a replicator, often a piece of sterile velveteen material clamped on to a flat round dish-sized wooden block, is used to transfer a pattern of cell colonies from one petri dish of agar medium ('plate') to another dish which may contain a different kind of medium. In a hunt for mutants, an array of colonies grown from mutagenically treated cells may be transferred from a 'master' plate to the surface of a clean replica plate which, because of the medium which it contains or because of the conditions of subsequent incubation, will not support growth of the kinds of mutant being looked for. Mutant colonies of the desired types on the master can be recognized by their failure to appear on the replica.

This procedure, which allows hundreds of colonies to be screened in a single operation, is particularly suitable for the recognition of **auxotrophic** mutants—that is mutants which require some nutrient additional to the minimal requirement of the **prototrophic** wild type. The net for auxotrophs can be cast narrowly or broadly depending on how many different nutrients are included in the medium in the master plate. The replication method can also be applied to the identification of temperature-sensitive, drug-sensitive, radiation-sensitive and many other kinds of handicapped mutants, the restrictive conditions in each case being provided in the replica plates.

Automatic Procedures for Auxotroph Selection

If particular kinds of auxotrophic mutants are required, it may be preferable to use some method of automatic selection so that one does not have to scan all the survivors of mutagenic treatment visually for what may be a rare type. The strategy is to devise conditions under which the function of the wild type works against its own survival. A comparatively crude but effective method, much used in *Neurospora* and applicable to other filamentous fungi that will form conidia, is **filtration enrichment**. In this method, mutagenized conidia are suspended and incubated in liquid minimal medium and the prototrophic mycelial growth is removed at intervals by filtration of the entire culture through sterile cheesecloth. After several cycles of growth and filtration almost all wild-type growth has been removed, and one can then rescue surviving auxotrophic conidia by plating them on medium supplemented with appropriate nutrients.

More subtle are methods which use conditions in which prototrophs destroy themselves by unbalanced growth while auxotrophs are saved through not being able to grow at all. The classical example is B. D. Davis' method of **penicillin enrichment**, which he applied to *E. coli*. Penicillin kills growing bacterial cells because, while it does not prevent general cell growth, it does inhibit cell wall synthesis. Cells which start to grow in its presence undergo lysis and die when they run out of cell wall. When mutagenized cells of *E. coli* are suspended in liquid minimal medium plus penicillin, the auxotrophs tend to survive because they are unable to start growing. After a time sufficient for the self-destruction of virtually all the prototrophs, the surviving auxotrophs can be removed from the penicillin medium by centrifugation and plated with nutritional supplements that will allow the desired types to grow and form colonies.

Another way of using the principle of unbalanced growth has been used for obtaining *Neurospora* auxotrophs. This method, called **inositol starvation**, starts with an inositol-requiring (*inl*) mutant and exploits the fact that such a mutant dies very quickly on minimal medium. The lack of inositol does not prevent protein or nucleic acid synthesis and so the bulk of the cell can continue to grow under conditions of inositol starvation. Inositol is, however, a major component of cell membranes and it is the depletion of these which leads to lysis and death. A mutation which superimposes a second major nutritional requirement will stop the unbalanced growth and so promote survival. Thus double auxotrophs can be efficiently selected by incubating *inl* conidia on minimal plates and, after a sufficient period of incubation, overlayering with agar medium supplemented both with inositol and with supplements that will permit growth of the desired types of double mutant. The secondary mutations can usually be easily segregated from *inl* by crossing to the wild type.

The Use of Drugs and Toxic Metabolites

The possibilities of making normal functions lethal are legion. Toxic analogues of normal metabolites are often relatively harmless in the absence of a function needed for their uptake or metabolism. The use of aza- or thioguanine and 5-bromodeoxyuridine to select HGPRT$^-$ and TK$^-$ mammalian cell lines was mentioned in Chapter 9. Toxic amino acid analogues, such as 4-methyltryptophan, p-fluorophenylalanine and canavanine which mimic tryptophan, phenylalanine and arginine, respectively, can be used to select mutants of bacteria and fungi that are defective in their ability to concentrate the corresponding normal amino acids from the growth medium. In *Drosophila*, the incorporation into the food of certain secondary alcohols, such as pentenol, automatically selects for larvae deficient in alcohol dehydrogenase (ADH), since this enzyme catalyses their oxidation to the corresponding aldehydes which are highly toxic.[1]

Many kinds of mutants with identifiable biochemical effects confer resistance to various drugs that inhibit protein synthesis. For example, bacterial mutants that are resistant to streptomycin, and fungal mutants resistant to cycloheximide, can readily be obtained by plating mutagenized cells on medium containing what would normally be totally inhibitory concentrations of the drug. Such mutants usually have modifications in one or another of the protein components of the ribosome.

Apparent Gains in Function

Although most mutants with distinctive phenotypes have losses of function, some, at least at first sight, seem to have gained an activity not present in the wild type. Particularly important are mutants that produce constitutively, i.e. under virtually all conditions, an enzyme which is produced by the wild type only under special circumstances. A famous and important example is the β-galactosidase of *E. coli*, the normal function of which is to hydrolyse lactose. In the wild type, the enzyme is present in perceptible amounts only when its formation is induced by lactose or a suitable analogue in the growth medium, but certain mutants produce the enzyme at high levels in constitutive fashion. Such mutants can be selected visually by spraying colonies grown on glucose medium with 5-bromo-4-chloro-3-indolyl-β-galactoside, a compound which is hydrolysed by β-galactosidase to give a non-diffusible blue dye but is not an effective inducer of the formation of the enzyme. Colonies that form the enzyme constitutively stain deep-blue with this reagent. On closer examination, constitutive mutants turn out to represent not gains of a new function but rather a loss of control. The constitutive β-galactosidase (*lacI*) mutants, for example, have suffered loss of, or critical alteration in, a repressor protein which, in the absence of inducer, prevents the transcription of the gene (*lacZ*) which codes for the enzyme (cf. p. 371).

Other examples of how loss mutations can sometimes appear to result in gains of function are provided by some kinds of **suppressor** mutants. These are recognized through their suppression of the metabolic defects resulting from mutations in another gene. A mutational block, or partial block, in one metabolic pathway can sometimes lead to the accumulation of a normal intermediate which is then shunted into another pathway, not normally operative, which can bypass the function of a second gene. *Fig.* 13.1 shows, in schematic form, how such effects operate in one area of *Neurospora* metabolism. This kind of suppression, **metabolic**

suppression as we may call it, should be distinguished from the fundamentally different **translational** suppression which we deal with later in this chapter and which results from changes in the decoding specificities of tRNA molecules.

Two-way Systems of Mutant Selection

For investigations of gene structure and function, it is a particular advantage if one can select both for loss and for gain of the gene function. The

Table 13.1	Some examples of two-way mutant selection in various organisms		
Organism	*Mutant*	*Selective system and principle of selection*	*Reverse selective system*
Escherichia coli	*galK*	Select *galE* mutant for *galE galK* double mutant on galactose–glycerol medium—double mutant does not accumulate toxic UDP-galactose[2]*	Isolate *galK* single mutant by recombination—select for ability to grow on galactose
Saccharomyces cerevisiae	*cyc1*	Resistance to chlorolactate, which is toxic to respiring cells[3]	Select for growth on non-fermentable substrates such as glycerol
Neurospora crassa	*pyr-3b*	Select in *arg-2* mutant on pyrimidine-suplemented medium for arginine independence—carbamoyl phosphate diverted from pyrimidine to arginine pathway[4]	Select for ability to grow without pyridmidine supplement
Drosophila melanogaster	*adh*	Select for resistance to pentenol—oxidized by alcohol dehydrogenase to toxic aldehyde[1]	Select for resistance to ethanol—toxic in absence of alcohol dehydrogenase
Mammalian cells†	*tk⁻*	Select for resistance to 5-bromodeoxyuridine—in presence of thymidine kinase this analogue is toxic because of incorporation into DNA	Select for growth on HAT medium—thymidine kinase needed for thymidine utilization
Mammalian cells†	*hgprt⁻*	Select for resistance to 8-aza-guanine—in presence of hypoxanthine–guanine phosphoribosyl transferase (HGPRT) this analogue is toxic because of incorporation into DNA	Select for growth on HAT medium—HGPRT needed for hypoxanthine utilization

* The galactose utilization pathway is:

$$\text{galactose} \xrightarrow{galK} \text{galactose-1-phosphate} \xrightarrow{galT} \text{UDP-galactose} \xrightarrow{galE} \text{UDP-glucose.}$$

† *See* Chapter 9, p. 235.

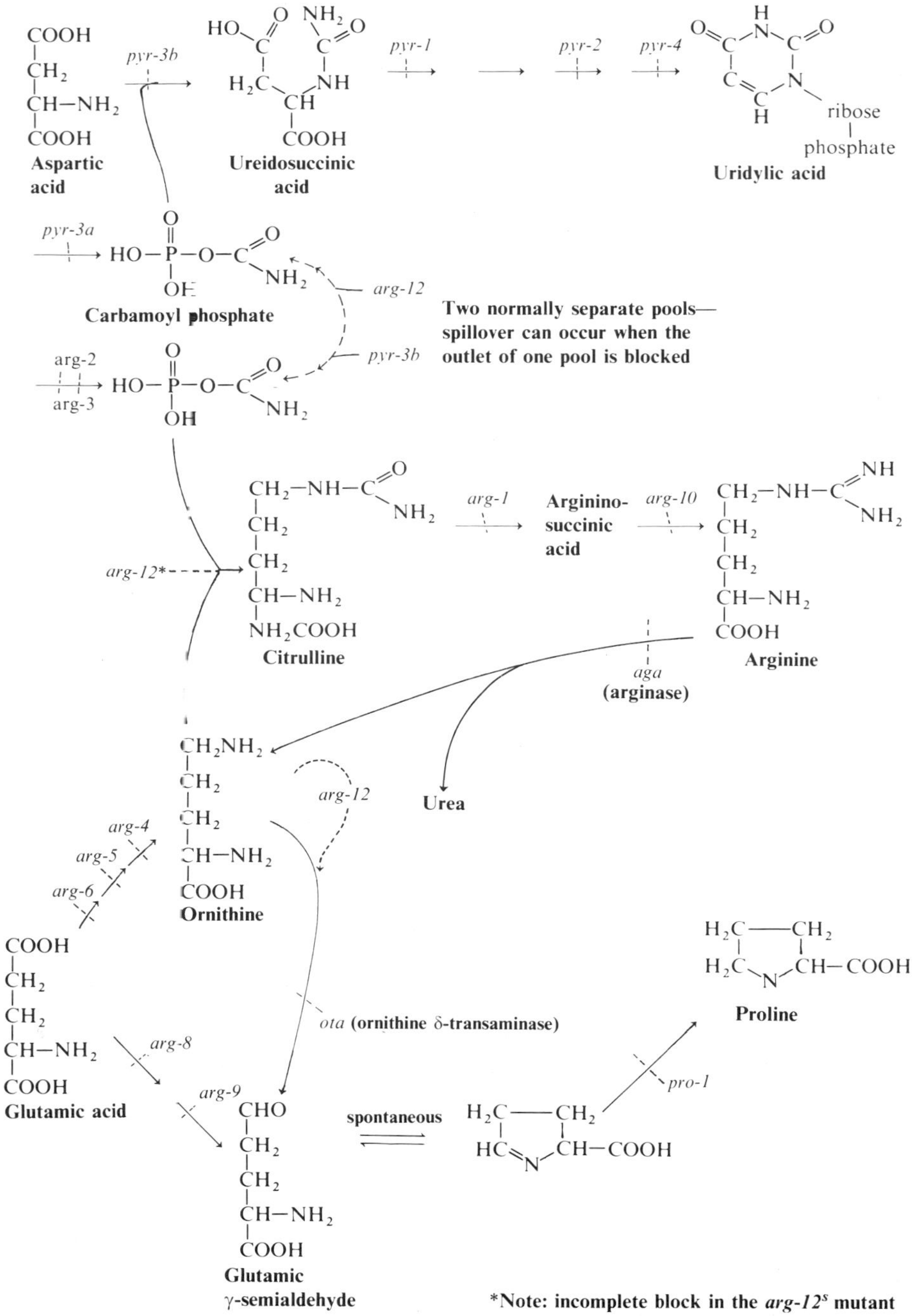

COOH
CH$_2$
CH—NH$_2$
COOH
Aspartic acid

pyr-3b

HO
C=O NH$_2$ C=O
H$_2$C NH
CH
COOH
Ureidosuccinic acid

pyr-1 *pyr-2* *pyr-4*

Uridylic acid
ribose
phosphate

pyr-3a

$$HO-\overset{O}{\underset{OH}{P}}-O-C\overset{O}{\underset{NH_2}{}}$$

Carbamoyl phosphate

arg-12

arg-2
arg-3

$$HO-\overset{O}{\underset{OH}{P}}-O-C\overset{O}{\underset{NH_2}{}}$$

pyr-3b

Two normally separate pools—spillover can occur when the outlet of one pool is blocked

arg-12*

CH$_2$—NH—C=O, NH$_2$
CH$_2$
CH$_2$
CH—NH$_2$
NH$_2$COOH
Citrulline

arg-1 **Arginino-succinic acid** *arg-10*

CH$_2$—NH—C=NH, NH$_2$
CH$_2$
CH$_2$
CH—NH$_2$
COOH
Arginine

aga (arginase)

CH$_2$NH$_2$
CH$_2$
CH$_2$
CH—NH$_2$
COOH
Ornithine

arg-4
arg-5
arg-6

arg-12 **Urea**

COOH
CH$_2$
CH$_2$
CH—NH$_2$
COOH
Glutamic acid

arg-8

arg-9

CHO
CH$_2$
CH$_2$
CH—NH$_2$
COOH
Glutamic γ-semialdehyde

ota (ornithine δ-transaminase)

spontaneous

H$_2$C——CH$_2$
HC N CH—COOH

H$_2$C——CH$_2$
H$_2$C N CH—COOH
Proline

pro-1

*Note: incomplete block in the *arg-12^s* mutant

Fig. 13.1   The pathways of synthesis of pyrimidines, arginine and proline in *Neurospora crassa*, showing the positions of the metabolic blocks in various auxotrophic mutants. Each arrow indicates a separate enzyme-catalysed step, but only those intermediates referred to in the text are specified. Of the *arg* mutants, *arg-4*, *5* and *6* grow if given either arginine or citrulline or ornithine, *arg-2* and

3 grow with arginine or citrulline, while *arg-1* and *arg-10* respond to arginine only (*arg-1* does not grow on argininosuccinic acid probably because this compound is not readily taken up by the mycelium). The mutants *arg-8* and *arg-9* are blocked in the early part of the proline pathway and respond to arginine, citrulline or ornithine because glutamic γ-semialdehyde, whose main pathway of synthesis is blocked in these mutants, can be formed from ornithine through transamination when the latter compound is present in excess. Ornithine is produced from supplied citrulline or arginine via arginase. **Suppressor** effects arise by interactions between pathways. Carbamoyl phosphate is made in two separate places in the cell, the nucleus and mitochondrion, and these two pools are used for pyrimidine and arginine synthesis, respectively. When one of these pathways is blocked immediately after carbamoyl phosphate, the accumulated intermediate can 'spill over' into the other pool, thus compensating for a block in carbamoyl phosphate synthesis in the other pathway. Thus *arg-12ˢ* suppresses the pyrimidine requirement of *pyr-3a* mutants so that the *arg-12ˢ pyr-3a* double mutant grows on minimal medium (*arg-12ˢ* is a 'leaky' allele causing an incomplete block which does not result in an absolute requirement for arginine). Conversely, *pyr-3b* suppresses the arginine requirement of *arg-2* and *arg-3*), so that a *pyr-3b arg-2* (or -3) double mutant requires pyrimidine only. The *arg-12ˢ* mutation also suppresses the proline requirement of *arg-8* or *arg-9* in double mutants, the explanation here being that the partial *arg-12ˢ* block results in the accumulation of ornithine, and hence of glutamic γ-semialdehyde, as well as of carbamoyl phosphate.
Note that the product of the *pyr-3* gene is a bifunctional enzyme with both carbamoyl phosphate synthetase and aspartate carbamoyltransferase activities; mutations in the gene can eliminate one activity without affecting the other. References may be found in Fincham et al.[28]

difficult part is to find a system for selecting in the loss direction; regain of function is usually easy to select for—the selective growth of prototrophic revertants from auxotrophs on minimal medium is an obvious example. *Table* 13.1 lists a number of important examples from diverse organisms.

A Note on Dominance

It is worth while at this point to reiterate a useful (if fallible) proposition regarding the dominance or recessivity of mutant alleles. In general, losses of function are recessive and gains (or selected-for modifications) of function are dominant. This simple rule can lead one astray in certain instances, mainly because of positive or negative interactions which sometimes occur between allelic products at the polypeptide level (*see* p. 365), but in general it is a good guide.

13.2 Effects of mutations on proteins

Identifying the Biochemical Lesion

The discovery of auxotrophic mutants led to a great deal of work aimed at the identification of the primary metabolic deficiency in each mutant. The clear conclusion which eventually emerged was that mutation in many genes caused simple changes in the polypeptide chains of proteins.

The earlier studies on the biochemical effects of mutations in *Neurospora* auxotrophs pointed to a one-to-one relationship between genes and enzymes. The arginine-requiring mutants of *Neurospora* provide one example (from among many others that could be cited) of the reasoning which led to this concept.

The *Neurospora arg* mutants fell clearly into different groups (eventually twelve in number) on the basis of complementation tests with heterokaryons (*see* p. 125). The mutants within each complementation group (gene) mapped, to a first approximation, to a single chromosomal locus (actually to a number of very closely linked sites) and these loci, which were given the gene symbols *arg-1* to *arg-12*, occurred in scattered positions over most of the chromosomes. *Fig.* 13.1 shows which step in arginine biosynthesis is attributed to each of the different genes. Two, *arg-8* and *arg-9*, are really blocked in proline biosynthesis, and respond to arginine as well as to proline because arginine, via ornithine, can serve as a secondary proline source. Several lines of evidence supported these assignments of genes to metabolic steps.

First, each mutant could use as a source of arginine, and hence as support for growth, any intermediate in the pathway coming after the step in which it was blocked, but could use no intermediate coming before that step. This is a principle of wide applicability to the analysis of auxotrophs. In the case of mutants concerned with arginine and proline biosynthesis one has to bear in mind the complication of the metabolic loop whereby ornithine can feed into the proline pathway. The nutritional responses of the various *arg* mutants are explained in the figure.

A second kind of evidence is provided by the unusual accumulations of metabolic intermediates found in some of the mutants. These can each be interpreted as due to a blockage in the metabolic reaction which would normally convert the accumulated compound to the next intermediate along the line. Thus *arg-1* and *arg-10*, which both respond to arginine but not to citrulline implying that both are blocked in the citrulline→arginine conversion, can be easily distinguished by the fact that *arg-1* strains accumulate citrulline in their mycelium while *arg-10* strains accumulate argininosuccinate; the latter compound is known from biochemical evidence to be an intermediate in the conversion. As expected, the double mutant *arg-1 arg-10* accumulates citrulline but no argininosuccinate; the earlier-acting mutation is epistatic to the later-acting one. We may note in passing that the accumulation of intermediary metabolites is the key to the understanding of many human metabolic disorders.

The final confirmation of the metabolic step affected in a given class of mutants comes from direct analysis of enzyme activities. It is possible to demonstrate the activity of an enzyme for each step in the metabolic pathway in wild-type cell extracts and just one of them, the one predicted from the nutritional and accumulation evidence, turns out to be absent in each mutant class. Thus *arg-1* mutants lack argininosuccinate synthetase and *arg-10* mutants lack argininosuccinase (which hydrolyses argininosuccinate to arginine and fumarate); each class of mutant possesses the enzyme which is lacking in the other.

13.3   Amino acid replacements in mutant polypeptides

The *E. coli* Tryptophan Synthetase Alpha Protein

Having established that a particular enzyme activity is missing in a mutant, the next step is to find out what has happened to the enzyme protein. A classic example is the investigation by C. Yanofsky and his colleagues[5] of the *trpA* mutants of *Escherichia coli*. The various tryptophan-requiring (*trp*) mutants of *E. coli* map in a number of genes which map in a tight cluster and are transcribed as a single unit (an **operon**, *see below*, p. 371). The *trpA* and *trpB* genes, which are quite distinct on the basis of their mutual complementation and lack of overlap on the fine-structure recombination map, are adjacent at one end of the operon. Mutation in either can result in loss of activity of tryptophan synthetase (TSase), which catalyses the reaction:

Indoleglycerol phosphate + Serine $\longrightarrow$
(IGP)

Tryptophan + 3-Phosphoglyceraldehyde.
(PGA)

This reaction takes place in two steps:

i. IGP $\longrightarrow$ Indole + PGA.
ii. Indole + Serine $\longrightarrow$ Tryptophan.

Indole, however, is formed only transiently as an enzyme-bound intermediate.

Inasmuch as two genes were apparently needed for formation of a single enzyme, this was an early exception to the 'one gene–one enzyme' rule. The situation was clarified by the finding that tryptophan synthetase was a tetrameric protein containing two kinds of polypeptide chain—two alphas and two betas. The correct formulation in this case was 'one gene–one polypeptide', since *trpA* mutants were defective only with respect to the α chain and produced normal β, while *trpB* mutants were defective with respect to β and produced normal α. Normal enzyme could, in fact, be formed simply by mixing extracts of *trpA* and *trpB* cells—an unusual example of readily demonstrable complementation in the test tube (*in vitro*).

A functional distinction between the two classes of mutants was first indicated by the fact that *trpB* mutants were able to use indole instead of tryptophan while *trpA* mutants were not. This was explained by the finding that *trpB* mutations, which eliminated the overall enzyme activity (IGP→trytophan), often retained a low level of activity for reaction (ii) above. Conversely, *trpA* mutants tended to retain some activity with respect to reaction (i). The α component seems to be primarily responsible for reaction (ii) and the β component for reaction (i), but only the combination of the two will catalyse the overall reaction.

Detailed studies have been made on both the *trpA* and *trpB* genes and their respective polypeptide products, but the *trpA*–polypeptide α relationship has been the most thoroughly worked out. The key discovery that permitted all the subsequent work was that many *trpA* mutants made a full amount of a polypeptide closely resembling the normal α but lacking its enzymic activity. When complexed to the normal β chain, these α-like mutant products formed an enzyme fully competent in reaction (i) but devoid of activity for reaction (ii) or the overall IGP→tryptophan conversion. The abnormal enzymes were purified by the same procedure as had been worked out for wild-type tryptophan synthetase, and the α-like chain was isolated from each. The amino acid sequence was then compared with that of normal α. The simple and important finding was that nearly all of the mutant chains differed from normal in just a single amino acid residue—a different substitution in each different mutant. At the same time the sites of the mutations were placed in a fine-structure genetic map by a combination of two different methods: (i) transductional mapping using bacteriophage Pl (*see* p. 295) and (ii) deletion mapping using a set of partially overlapping fragments of the *trpA/B* region present in a series of different φ80 specialized *trp*-transducing phages (closely analogus to the λd*gal* and λd*bio* phages described in Chapter 11, pp. 301–302). The sequence of mutational sites on the genetic map was the same as the corresponding sequence of amino acid replacements in the polypeptide chain, implying that the gene provided a linear code for the amino acid sequence. This discovery, of the **colinearity** of gene and polypeptide, did not come as a surprise, but it was nevertheless a major landmark in the development of molecular genetics.

In addition to the *trpA* mutants that produced mutationally altered polypeptides, there were many others that produced no detectable α-like product at all. This is the usual finding with series of mutants defective in specific proteins or polypeptides. The most commonly used procedure for distinguishing between the 'altered protein' and 'no protein' categories is through the use of a specific antiserum obtained by injecting rabbits with the purified wild-type protein. If a mutant protein is present in a cell extract it will almost always precipitate with the specific anti-wild antibodies. This reaction can be conveniently observed by a

Fig. 13.2

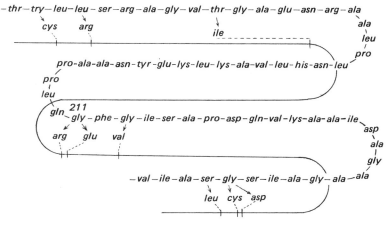

Comparison of the positions of mutant sites in the *E. coli trpA* gene, determined by recombination frequencies in intermutant crosses performed by transduction with bacteriophage Pl, with the corresponding amino acid substitutions in the polypeptide chain of tryptophan synthetase. The mutant sites on the genetic map (solid line) are spaced in proportion to recombination frequency. Only a portion of the complete polypeptide chain is shown. Redrawn from Fincham;[29] data from Yanofsky et al.[5]

variety of methods, the simplest of which is illustrated with reference to another gene–enzyme system in *Fig.* 13.3.

Mutant protein detected by reaction with antibodies formed against wild type is known as **cross-reacting** material (CRM), and the distinction between CRM$^+$ and CRM$^-$ mutants is commonly referred to. The most usual reason for a CRM$^-$ phenotype is premature termination of polypeptide synthesis because of the generation by mutation of chain-terminating codons—*see below*, p. 358.

The *cyc1* Mutants of *Saccharomyces cerevisiae*

The best known gene–protein system in a eukaryote, albeit a lower eukaryote, is the *cyc1*–cytochrome *c* relationship in baker's yeast. F. Sherman and J. W. Stewart and their colleagues accumulated a large collection of mutants unable to make iso-1 cytochrome *c*, an essential part of the electron transport chain linking respirable substrates to molecular oxygen. In fact, yeast has two cytochromes *c*, the second one, called iso-2, being a very minor component which we need not worry about in the present context. Like other cytochrome pigments, cytochrome *c* contains a molecule of the red iron–tetrapyrrole compound haem, which can be recognized in cells by its distinctive absorption spectrum. The mitochondrial-defective **petite** mutants, which we discussed in Chapter 7, lack cytochrome *b* and cytochrome a,a_3 (cytochrome *c* oxidase), and are left with cytochrome *c* as their sole major haem-containing component. Starting with a haploid petite strain, Sherman and his group screened a large

Fig. 13.3

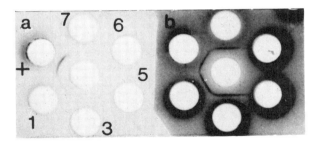

Distinguishing between CRM-positive and CRM-negative mutants in *Neurospora crassa*. The centre wells of Ouchterlony plates (containing agarose gel) were loaded with antiserum from rabbits immunized with purified NADP-specific glutamate dehydrogenase. The surrounding wells contained extracts of wild-type (+) *N. crassa* and five *am* mutants, three CRM$^+$ and two CRM$^-$. The antigen and antibody diffuse towards each other and form a precipitin band at the point where their ratio of concentrations reaches a critical value. Plate b was stained with Coomassie blue to show protein, while plate a was stained with a specific mixture (glutamate, NADP, the electron carrier phenazine methosulphate and nitro-blue tetrazolium) which precipitates an insoluble dye (a derivative of the tetrazolium) on the precipitin band if it contains active glutamate dehydrogenase (GDH). The continuity of the wild-type and CRM$^+$ mutant precipitin bands shows that they are identical in their interaction with the precipitating antibody molecules, but only the wild-type band shows enzyme activity.

number of colonies grown from mutagenically treated cells for variants that were almost or quite devoid of haem. Such variants could be identified either by individual examination with a spectroscope or by a reagent (benzidine + H_2O_2) which stains haem-containing cells blue. Later, a less laborious selective procedure became available with the discovery that chlorolactate kills respiring cells (cf. *Table* 13.1). Cytochrome c-deficient mutants were a common class among chlorolactate-resistant mutants induced in a non-petite strain.

Genetic analysis of the cytochrome c-deficient mutants selected by these various methods showed that they could be due to mutation in more than one gene, but there is good evidence that only one gene, *cyc1*, controls the amino acid sequence of the protein; the others probably affect the regulation of *cyc1* activity in ways which are not yet understood. A fine-structure genetic map was made of the mutational sites within *cyc1* by means of inter-mutant crosses. *CYC1* recombinants (using the yeast convention of capital letters for wild-type alleles) could be readily selected through their ability to grow on glycerol or lactate as carbon sources—it will be recalled that these compounds can only be respired and not fermented and are, therefore, absolutely dependent on cytochrome c for their utilization. The most definitive recombination data were obtained from crosses involving deletion mutants, identified as such by their inability to revert and their lack of recombination with sets of mutually recombinable point mutants. The principles of deletion mapping were explained in Chapter 5 (pp. 142–143).

Because of its unusually short polypeptide chain, only 105 residues, cytochrome c is a particularly favourable object for the identification of mutational substitutions. It was, at least at first sight, unfortunate that the great majority of the primary *cyc1* mutants failed to produce any detectable iso-1 cytochrome c. This can be understood on the basis that the commonly selected mutants will be those unable to attach the haem group to the protein, and that all that is needed for this attachment is the presence of the —SH groups of two cysteine residues placed close together not far from the N-terminus of the polypeptide chain. Especially when the screening is based on absence of haem pigment, rather than on cytochrome c function, the mutants recovered are likely to be those which suffer very early chain termination. This interpretation was borne out by extensive studies, outlined below (pp. 358–360), on secondary mutants obtained by mutagenizing primary mutants and selecting for restoration of respiratory function. As we shall see (*Figs* 13.5, 6), a large proportion of the primary mutants turned out to be due to failure to synthesize the chain beyond the N-terminal region, either because of mutationally generated chain-terminating codons or because of failure of initiation. Again the principle of colinearity holds in that the sequence of the termination codons in the mRNA mirrors the sequence of mutational sites in the genetic map.

Haemoglobin in Man

Experimental induction of biochemically defined mutants is difficult in higher organisms and is in any case ruled out in the case of man except in cell culture. In human genetics, however, effective hunts for mutants already existing in populations have been made in the course of medical surveys. These have turned up large numbers of variants in the major adult haemoglobin (HbA). Some of the variants came to light because they were associated with clinical

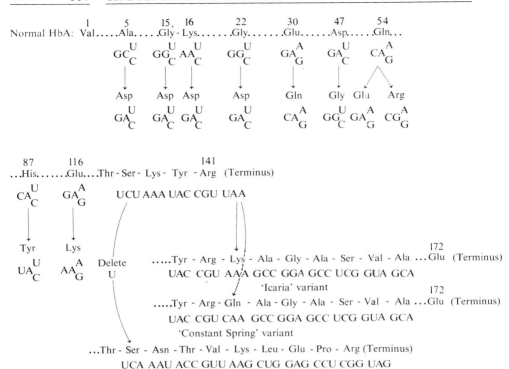

Fig. 13.4   Some of the mutational variations found in the human haemoglobin α chain. The amino acid residues are numbered from the N-terminal end of the chain and the possible mRNA codons are also shown. Note that with one exception all the amino acid substitutions could have arisen by single base changes. The elongated chains in the Icaria and Constant Spring variants can be interpreted as single base changes in the normal UAA termination codon. The Wayne variant must be due to a minus-one deletion causing a shift in the 'reading-frame' used in translation of the messenger.
For sources of data *see* Weatherall.[7]

symptoms, usually anaemia of various grades of severity, while others function normally but differ from the standard HbA in electrophoretic mobility (reflecting electrical charge).

Haemoglobin molecules in general consist of four polypeptide chains, two alpha chains and, in the case of HbA, two beta chains. In some of the population variants the peculiarity lies in the α chain, while in others it is the β chain which is abnormal. All are inherited as simple allelic differences attributed to two unlinked genes; the α chain is controlled by a gene on chromosome 16, and the β chain by a gene on chromosome 11.

Very nearly all the human haemoglobin variations, whether in the α chain or in the β chain, differ from the common type in single amino acid replacements. Furthermore, as was shown in *Table* 12.1, these replacements, with very few exceptions, are such as could be produced by single base changes in the messenger RNA codons. *Fig.* 13.4 shows some variants of the α chain, including one which must be due to a frameshift mutation (*see below*, p. 362).

Alcohol Dehydrogenase in *Drosophila melanogaster*

This enzyme is worth mentioning because it is one of the very few in higher eukaryotes (xanthine dehydrogenase in *Drosophila* is another) in which mutational changes have been induced and defined chemically in the polypeptide chain, and the colinear sequence of mutational sites has been mapped genetically in the corresponding genes.[6] The number of mutants analysed is as yet fairly small, but the case is an important one because of the amount of interest that has been attracted by the natural polymorphism shown by *Drosophila* alcohol dehydrogenase. We shall return to the population genetics aspect of this system in Chapter 18.

13.4   Polypeptide chain termination and chain initiation

Mutational Effects on Termination

Base-pair substitution mutations not only cause change of one amino acid codon to another (so-called **mis-sense** mutation), but are also liable to convert certain amino acid codons to chain termination codons (**nonsense** mutations). We know from studies on *in vitro* polypeptide synthesis, in which synthetic mRNA of defined sequence has been used, that three out of the sixty-four possible sequences of three bases code for no amino acid, and bring about chain termination when brought into register in the amino acid site of the ribosome. Many, though not all, of the amino acid codons are capable of conversion to one of these three (UAG, UAA and UGA) by a single base change; the two glutamate codons, GAA and GAG, which are particularly abundant in most protein-coding sequences, would be expected to be common sites of nonsense mutation.

Chain termination, if more than slightly premature, is likely to prevent the formation of anything recognizable as related to the normal protein product. Thus chain termination mutants are expected to be, and generally are, CRM$^-$ and many CRM$^-$ mutants (though not all, *see* Frameshift mutations *below*) can be shown to be chain terminators (*see* p. 359). Since, in selecting for sharply distinct mutant phenotypes, we are necessarily directing our attention to the most drastic effects on protein function, we would expect to find chain-termination mutants even more frequently than the proportion of 'nonsense' codons (3/64) would suggest, and this expectation is generally fulfilled.

Reversion of Nonsense Mutants

Mutants which are CRM$^-$, or shown by other methods to be totally deficient in the protein gene product, can sometimes be identified as having chain-terminator codons by a thorough analysis of the types of revertants that can be obtained from them. This type of analysis was first carried out in *E. coli*, but as good an example as any is provided by the *cyc1* mutants of yeast, which fail to form iso-1 cytochrome *c*. Some of the data are presented in *Fig. 13.5*.

Fig. 13.5

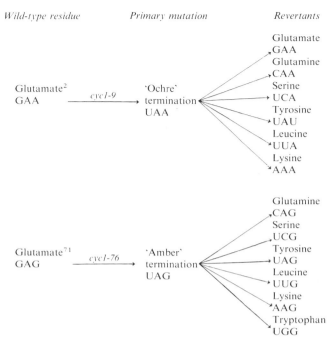

Wild-type residue   *Primary mutation*   *Revertants*

Identification of two mutations in the *cyc1* gene of *Saccharomyces cerevisiae* (coding for the polypeptide chain of iso-1 cytochrome *c*) as chain terminators. The *cyc1*-9 and *cyc1*-76 mutants, devoid of iso-1 cytochrome *c*, were induced to revert to respiratory competence, and the cytochrome *c* polypeptide produced by each of a number of revertants from each mutant was analysed. The *cyc1*-9 revertants showed a range of alternative amino acid residues at residue 2 from the N-terminal end, and the *cyc1*-76 revertants had the same range of alternatives, with the addition of tryptophan, at residue 71. Both positions in the wild-type sequence are occupied by glutamate. The figure shows how the revertants could all have been generated by single base changes from the chain-terminating codons UAG (in *cyc1*-76) or UAA (in *cyc1*-9). Data from Stewart et al.[8]

From each of several mutants, a family of revertants was isolated following mutagenic treatment (usually ultraviolet irradiation) and selection on glycerol medium. On purification and analysis of the cytochrome *c* that each revertant produced, it was found that most were not truly wild type but had a single amino acid substitution in one position. This position was always the same for revertants from a given primary mutant, but the amino acid substituted could be different in different revertants. These amino acids showed a family relationship in that all could have been produced by a single base change in the same codon. In several cases this putative progenitor codon was UAG, sometimes called the **amber** codon. In others it seemed more likely to be UAA (**ochre**), since tryptophan (codon UGG) was never found as the substitute amino acid in these cases, even though it was a common substituent in revertants from the presumed amber mutants. Only the occurrence or not of tryptophan permits the distinction between UAG and UAA; codons for the other substituent amino acids could equally well have been derived from either. The third termination codon, UGA, would yield a very different set of amino acids in revertants and should be readily identifiable by the same type of analysis, but it seems to be less common as a cause of mutational chain termination than the other two.

Amber and ochre mutants can be more simply identified by their susceptibility to suppression by certain specific mutations affecting tRNA structure. We return to this important phenomenon at the end of this chapter.

Initiation Mutants

Another mutational effect which one would expect, but of which there are rather few examples, is failure to initiate polypeptide synthesis because of a base change in the initiation codon AUG. Sherman and Stewart[9] were able to show that several mutants placed at the extreme left-hand end of the *cyc1* map were, in fact, initiation mutants, and they confirmed that the initiation codon was indeed AUG. They obtained from each of those mutants revertants in which polypeptide synthesis was restored by the generation of a new initiation codon, either to the right of the normal one (giving a slightly shorter chain) or to the left (giving a slightly longer one than normal). The first category of 'short-chain' revertants was evidently due to mutation in a lysine codon (AAG) which normally specifies residue 4 of the polypeptide. The 'long-chain' revertants varied depending on the primary mutant from which each had been isolated, but all appeared to have resulted from the generation of a new AUG codon outside, and immediately to the left of, the normal coding sequence of the mRNA. In these revertants, the translating ribosomes started reading the message one codon earlier than usual and read through the mutant codons occupying the normal initiation position. The nature of the mutation in each case thus stood revealed. A variety of different amino acids was found to be coded for by the mutant initiation codons; the only wild-type codon which could have given rise to codons for all of them by single base changes was AUG. The reasoning is explained in *Fig.* 13.6. It is interesting to note that the initiating methionine, normally removed immediately after the synthesis of the chain, remains in place in some of the mutants.

13.5   Frameshift mutations

Amino acid sequence analysis in revertants can reveal the nature of yet another important class of mutations and, incidentally, provide evidence of the exact codons used in the synthesis of at least part of a polypeptide chain. An excellent example is provided, once again, by the *cyc1* mutants of *Saccharomyces*. *Fig.* 13.7 shows the amino acid sequences in the N-terminal part of the cytochrome chain in a series of revertants isolated from one mutant. It will be seen that these revertants differ from the wild-type sequence in several adjacent amino acid residues rather than just one; the altered sequence extends, in different revertants, from different residues at the left-hand (N-terminal) end to a constant position on the right. These altered sequences, though at first sight quite haphazard, can all be related to the wild-type sequence by postulating a common insertion of a single nucleotide in the mRNA at the right-hand end, compensated for by a deletion, in a position specific for each revertant, to the left.

Fig. 13.6

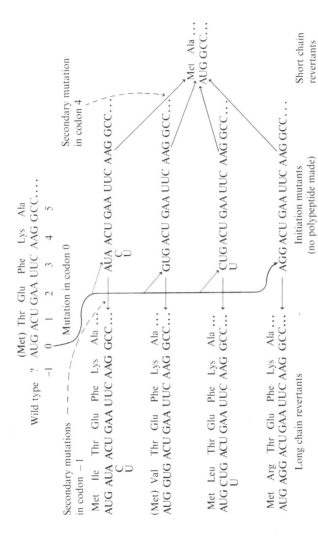

The analysis of revertants of *cyc1* mutants of *Saccharomyces cerevisiae* mapping at the extreme initiation end of the gene. The sequences of the cytochrome *c* polypeptide chain in the revertants show the original mutants to have been affected in the chain-initiation codon. The inferred codons in the primary mutant and revertant mRNA are written below the corresponding amino acids. Methionine residues in parentheses are believed to be cleaved from the chain after synthesis. Data of Stewart et al.[9]

Fig. 13.7

```
              1     2     3     4     5     6     7     8     9    10    11    12    13    14

(Met) - Thr - Glu - Phe - Lys - Ala - Gly - Ser - Ala - Lys - Lys - Gly - Ala - Thr - Leu -... Wild type

AUG ACX GAA UUC AAG GCC GGU UCU GCU AAG AAA GGU GCU ACA CUX

                                                                         A
                                                                                      (insertion of A)
(Met) - Thr - Glu - Phe - Lys - Ala - Gly - Ser - Ala - Lys - Lys - Arg - Cys - Tyr - Thr -...   (scrambled 183
                                                                                                   sequence)
AUG ACX GAA UUC AAG GCC GGU UCU GCU AAG AAA GGG UGC UAC ACU

                                                                                      Type 9 revertant
(Met) - Thr - Asp - Ser - Arg - Pro - Val - Leu - Arg - Lys - Gly - Ala - Thr - Leu -...          (deletion of A in
                                                                                                   codon 2)
AUG ACX GAU UCU AGG CCG GUU CUG CUA AGA AAA GGU GCU ACA CUX

                                                                                      183B (duplication of
(Met) - Thr - Glu - Phe - Lys - Ala - Gly - Ser - Ala - Lys - Lys - Met - Leu - Arg - Lys - Gly - Ala - Thr - Leu -...   sequence marked
                                                                                                   ⌐——in 183
AUG ACU GAA UUC AAG GCC GGU UCU GCU AAG AAA AUG CUA AGA AAA GGU GCU ACA CUX                         sequence above)

                                                                                      183A (deletion of
(Met) - Thr - Glu - Phe - Lys - Ala - Gly - Ser - Thr - Leu -...                                   sequence marked
                                                                                                   |- - - - - -|in 183
AUG ACU GAA UUC AAG GCC GGU UCU ACA CUX                                                            sequence above)
```

Frameshift analysis in *Saccharomyces cerevisiae cyc1* gene using amino acid sequences in revertants. The original frameshift mutant 183 is presumed to have had an extra A inserted in the messenger RNA for iso-1 cytochrome *c* in the position indicated; the scrambled amino acid sequence which presumably resulted was not actually isolated but is shown for illustrative purposes. The altered sequences in the various revertants were functional cytochromes which were purified and sequenced. Three examples are shown: the longest (type 9) frameshifted sequence between two compensating single base changes and two less common types involving more extensive deletion or duplication. Note that the analysis defines the wild-type messenger sequence.
Data from Sherman and Stewart.[10]

The consequence of such a double change would be a shift in the phase of reading of the mRNA (the 'reading-frame') one nucleotide to the right in the region backeted by the deletion and insertion in each case. The insertion common to all the revertants must have been present in the original mutant and must have caused out-of-phase translation through the essential haem-binding region and beyond, until termination occurred by an encounter with a chain termination codon generated by the frameshift. The various compensating frameshifts in the different revertants resulted in out-of-phase translation for a relatively short distance only, with the correct reading frame restored for the absolutely essential part of the sequence.

The two most widely separated mutually compensating frameshifts enable one to deduce unambiguously the nucleotides of the mRNA sequence lying between them. Messenger sequences cannot normally be fully deduced from amino acid sequences because of the multiplicity of different codons which correspond to each amino acid (with the exceptions of tryptophan and methionine). A frameshift of one base moves, in effect, the highly ambiguous third-position bases into the unambiguous second codon position, and this reduces the many different alternative messenger base sequences to one unique solution. This rather laborious method of deducing mRNA sequence is of more than merely puzzle-solving interest. In certain instances, it can lead to the identification of the gene in fragments of genomic DNA and then to its cloning and purification. The complete nucleotide sequence can then be obtained by direct chemical analysis. This whole programme has, in fact, been carried through in the case of the yeast *cyc1* gene, with the double-frameshift analysis as the essential starting point[11] (*see Fig. 13.7*).

A number of similar analyses, with comparisons of normal and frameshifted amino acid sequences, have been carried out in *E. coli* and there is at least one human haemoglobin variant which has been characterized as due to a frameshift (*Fig. 13.6*). But, historically, the most important piece of work on frameshift mutations did not involve amino acid sequence determination at all. Francis Crick and his colleagues in Cambridge arrived at the frameshift concept, and deduced the triplet nature of the genetic code, from observations of the mutually suppressive behaviour of certain *rII* mutants in bacteriophage T4 (p. 292). Two classes were identified, labelled plus and minus. Double mutants of plus–minus genotype were usually *rII*+ in phenotype [i.e. able to grow on K12(λ)], while double mutants of plus–plus or minus–minus genotype were *rII* in phenotype. The critical observation was that triple mutants carrying *three* mutations of the same sign, either triple-plus or triple-minus, usually showed the *rII*+ phenotype. In summary, a plus mutation suppressed a minus and vice versa, and a third mutation could suppress two others of the same sign. Crick's deduction was that three successive frameshifts in the same direction restored the reading frame to its proper register, and that the code must therefore be read in units of three. This evidence preceded the biochemical demonstrations of the functioning of triplet codons in cell-free systems.

It should incidentally be noted that the reason why not quite all plus–minus double-mutant combinations, and not quite all triple-plus or triple-minus combinations of mutant sites restored the wild-type function is that the mis-sense amino acid sequences generated between the first frameshift and the final compensating one are not always compatible with normal functioning of the protein product.

13.6 Effects of mutations on higher-order protein structure

Conformational Effects

The amino acid sequence of polypeptide chains is **primary** protein structure. The arrangement of the chains in helices or sheets of parallel chains is called **secondary** structure, and the higher order folding of these helices and sheets, together with 'random coil' polypeptide, to form globules or other compact three-dimensional forms, is **tertiary** structure. The packing together of different globules formed either from separate polypeptide chains, as in oligomeric proteins, or from different separately folded regions of the same chain, as in multiple-domain proteins (cf. p. 497), is **quaternary** structure. Mutations affect proteins primarily through changing their amino acid sequences, but the ultimate effects on protein function often require explanation in terms of higher-order structure.

As we have seen, mutations usually only change single amino acid residues in a polypeptide chain, and yet we know that in many cases this apparently small change knocks out the entire function of an enzyme. There are two types of explanation for such large effects. In some cases the altered amino acid is one with a specific and irreplaceable function—for example in forming a specific bond with a substrate molecule. It may be just as common, however, for the crippling mutation to affect the **conformation**, i.e. the higher-order structure of the enzyme, throwing other essential amino acid residues into positions in which they are no longer properly aligned for essential interactions with substrate molecules or with each other.

Allosterism

Particularly important in this connection is the phenomenon of allosterism in enzymes and other proteins. Allosteric proteins characteristically exist in two alternative conformational states in equilibrium with each other. One of these, the T-state, binds certain small molecules (ligands) very tightly and the other less tightly, or not at all. It follows from the laws of chemical equilibrium that addition of ligand will cause an overall conversion of the protein to the tight-binding state.

If, as is nearly always the case, the protein is an oligomer, containing two or several polypeptide components (**monomers**) each capable of binding the ligand(s), the curve relating proportion of protein in the T-state to ligand concentration is usually S-shaped—that is to say, at low ligand concentrations an increase in concentration has a disproportionately large effect on the protein conformation. This is because each monomer of an oligomeric protein is constrained to conform to the state of the others. It is, therefore, easier for all monomers of an oligomer to change state together than for them to do so separately. Thus the binding of ligand to just one monomer may hardly affect oligomer conformation, but the binding of two, or several, simultaneously may effectively tip the balance towards the T-state.

The effect of this **cooperativity** is to enable the protein to switch back and forth between its different states in response to small changes in the concentration of ligand (**allosteric effector**). Such ligand-controlled switching is important in the control of metabolism in various ways. For example, it is often found that the enzyme catalysing the first step in a biosynthetic pathway is switched to its inactive conformation in response to binding of the end-product of the pathway. In this way, the whole pathway is effectively shut down when the concentration of the end-product reaches a certain level. Proteins which act as repressors or activators of transcription of specific genes are also subject to allosteric control, as we see below (pp. 372–373).

Any allosteric protein is vulnerable to amino acid replacements, of which there are likely to be many, that stabilize one conformation to the virtual exclusion of the other. For example, the introduction of a new water-insoluble (hydrophobic or 'oily') residue will stabilize the form in which this residue is able to associate with other hydrophobic residues in the interior of the protein molecule and destabilize the form in which it is required to come into contact with surrounding water molecules. Certainly, some, perhaps a high proportion, of mutational inactivations of allosteric enzymes are due to amino acid replacements which stabilize the inactive conformation to such an extent that activation can no longer take place under normal intracellular conditions.

Allelic Complementation

Mutational enzyme inactivation due to stabilization of an inactive conformation is sometimes capable of being reversed when the conformationally 'frozen' mutant monomer is associated in the same oligomer with a monomer of another mutant variety. The second complementing mutant monomer may be unconditionally inactive in itself, but it may nevertheless be able to make a normal allosteric transition and to impart this normality in some degree to any mixed oligomer in which it finds itself. Mixed oligomers will be formed whenever different alleles coding for different mutant polypeptides of an oligomeric protein are present in a heterozygous diploid or a heterokaryon.

A model for allelic complementation by conformational correction, based on an example from *Neurospora*, is shown in *Fig. 13.8*. However, this is not the only way in which allelic complementation can occur. Cases are known, including some involving mutants of the *E. coli trpA* gene, in which a functional enzyme monomer can be pieced together from the 'good' sections of two mutant (sometimes incomplete) polypeptide chains having different and non-overlapping deficient regions. This mechanism of complementation, which may be more common than was at one time thought, involves the same specific and apparently very strong non-covalent intraprotein bonds (a mutually supporting array of hydrogen bonds, electrostatic attractions and hydrophobic associations) as stabilize the three-dimensional structure of the normal monomer.

Allelic complementation can be a confusing factor in the assignment of mutations to genes by complementation tests (cf. p. 125). In principle, allelic complementation should be distinguishable from complementation between different genes (as when an enzyme contains two different kinds of polypeptides) because the former will give abnormal protein (neither partner producing a normal polypeptide chain) while the latter will give at least a proportion of fully normal protein, each partner producing a normal version of the polypeptide defective in the other. In practice, this distinction is not always easy to make and the more generally useful hallmark of allelic complementation is its sporadic occurrence, especially its failure (with rare exceptions) to occur in combinations involving mutants of the CRM⁻ type.

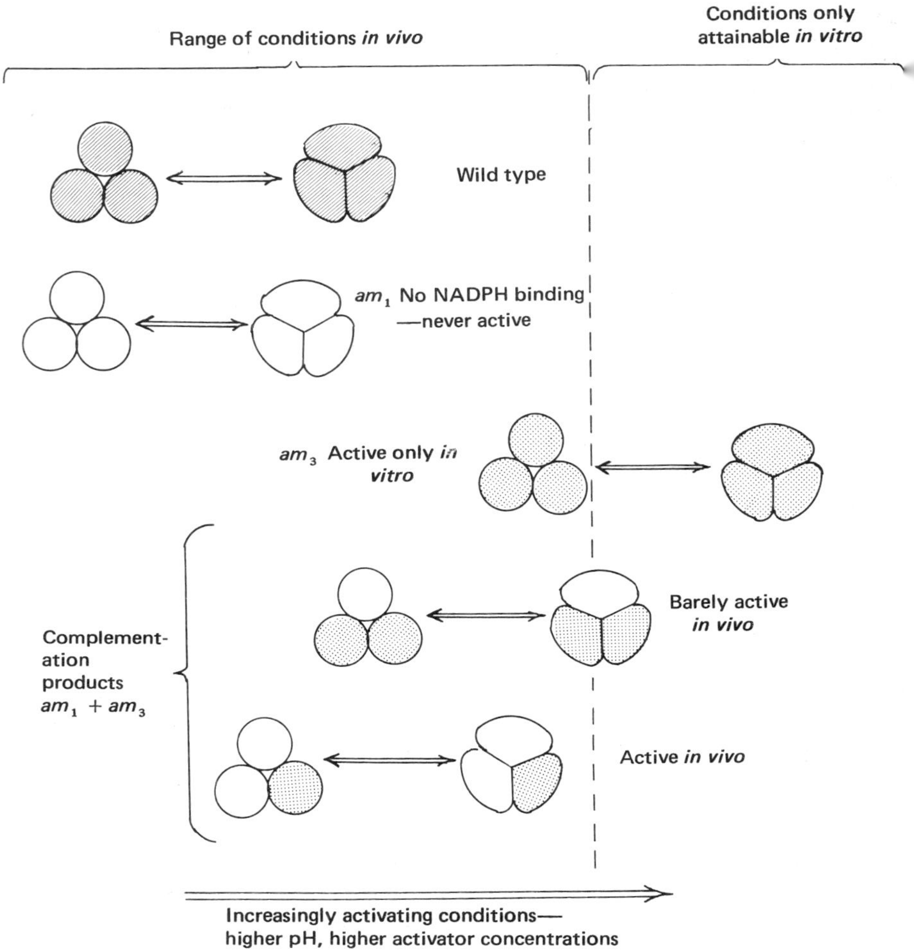

Range of conditions *in vivo*

Conditions only attainable *in vitro*

Wild type

*am*₁ No NADPH binding
—never active

*am*₃ Active only *in vitro*

Complement-
ation
products
*am*₁ + *am*₃

Barely active
in vivo

Active *in vivo*

Increasingly activating conditions—
higher pH, higher activator concentrations

Fig. 13.8   A diagrammatic representation of mutational inactivation of an enzyme by alteration of its allosteric equilibrium and of complementation of such a mutant by another producing unconditionally inactive but conformationally near-normal enzyme. The example referred to involves two mutants in the *am* gene of *Neurospora crassa*, coding for NADP-specific glutamate dehydrogenase; the model could apply to any enzyme with allosteric properties. Notes: *am*₁ and *am*₃ enzyme varieties have single amino acid replacements (Ser to Phe and Glu to Gly, respectively) in a polypeptide chain 452 residues long. The enzyme is a hexamer of normally identical chains, probably arranged as a stack of two trimers, and is in pH-dependent allosteric equilibrium. For the sake of simplicity the enzyme is shown here as a trimer, with only two kinds of mixed trimers—there will, in fact, be five different mixed hexamer compositions. The shapes of the monomers, circular in the inactive conformation and wedge-shaped in the active conformation, are imaginary, but serve to illustrate the principle of cooperativity, i.e. all monomers tend to change conformation in concert.

For further details and references *see* Fincham et al.[28]

13.7 Genes for transfer RNA— suppression of chain termination

Nonsense Suppression

Genes that we now know to control tRNA structure were first discovered in certain E. coli strains through their ability to suppress the phenotypic effects of amber mutations in T4 bacteriophage. Some E. coli amber suppressors were found also to be able to suppress T4 ochre mutants (cf. p. 287); amber-specific suppressors and suppressors acting on both ambers and ochres were found to map at the same loci. Amber and ochre mutations, susceptible to the same set of suppressors, were found in E. coli as well as in T4 genes. A proportion of the mutations in all genes coding for proteins were, it seemed, in one or other of the suppressor-sensitive categories.

Several kinds of biochemical evidence demonstrated the nature of this allele-specific, but gene-unspecific, type of suppression. First, suppressor-sensitive mutations in the T4 gene coding for head protein were found to produce, when infected into non-suppressing E. coli cells, a series of prematurely terminated head protein polypeptides. The length of the polypeptide produced by each mutant correlated with the position of the mutant site on the gene map—the farther from the N-terminal-coding end of the map, the longer the chain. Secondly, in a host carrying an appropriate suppressor gene, the polypeptide was restored to normal length. In the position at which premature termination would have occurred in the non-suppressed mutant, there was a particular amino acid residue, the nature of which was determined by the suppressor rather than by the mutant suppressed. Finally, it was possible to show that this suppressor-specific amino acid insertion was due to an altered anticodon in one of the tRNA species corresponding to the inserted amino acid. In the first example analysed, tyrosine-specific tRNA had its anticodon sequence changed from 5'-GUA-3' to 5'-CUA-3'. This change was one that would be expected to make the anticodon recognize the amber codon UAG rather than the normal tyrosine codons UAU and UAC.

We have already in the preceding section (p. 359) reviewed the evidence that the amber and ochre codons are UAG and UAA, respectively. The different specificities of the suppressor tRNAs in recognizing these codons is explained in terms of the general model for codon–anticodon pairing first put forward by Crick.[16] In this **wobble** hypothesis, third position pairing, i.e. that between the 3' base of the codon and the 5' base of the anticodon, is less precise than that in the first and middle positions. *Table* 13.2 sets out the alterations in the anticodon sequences that have been determined in a number of suppressor mutants of yeast, E. coli and bacteriophage T4 (it should be pointed out that this bacteriophage codes for a number of species of tRNA in its own genome).

Suppressors of chain-terminating mutations, then, are due to mutations in genes determining the base sequences of specific tRNA molecules. Different changes in the anticodon sequence of, for example, a tyrosine-specific tRNA can yield both amber-specific and (in yeast) ochre-specific or (in E. coli) amber/ochre suppressors. It is possible to induce mutation of one type of suppressor to another in one step. For example, S. Brenner showed that an amber suppressor allele of one of the E. coli tRNA^{Tyr} genes could be readily mutated to an amber/ochre suppressor allele

Table 13.2   Some anticodon changes found in tRNA molecules in various suppressor mutants

Organism	Normal amino acid specificity of tRNA	Normal anti-codon–codon pairing	Anticodon–codon pairing in suppressor mutant	Terminators suppressed	Reference
Phage T4	Glutamine	tRNA 3' G-U-U* 5' mRNA 5' C-A-A(A/G) 3'	→ 3' A-U-U* 5' 5' U-A-A(A/G) 3'	Amber (UAG) or ochre (UAA)	12
Escherichia coli	Tyrosine	tRNA 3' A-U-G 5' mRNA 5' U-A-U(U/C) 3'	→ 3' A-U-C 5' 5' U-A-G 3' 3' A-U-U 5' 5' U-A-A(A/G) 3'	Amber only Amber or ochre	13
Saccharomyces cerevisiae	Serine	tRNA 3' A-G-C 5' mRNA 5' U-C-G 3'	→ 3' A-U-C 5' 5' U-A-G 3'	Amber only	14
Saccharomyces cerevisiae	Tyrosine	tRNA 3' A-Ψ-G 5' mRNA 5' U-A-U(U/C) 3'	→ 3' A-Ψ-C 5' 5' U-A-G 3' 3' A-Ψ-U 5' 5' U-A-A 3'	Amber only Ochre only	15

U* is a methylated uridylate; Ψ is pseudouridylate, a structural isomer which has the uracil moiety linked to the ribose through a carbon atom (C-1) rather than the usual nitrogen (N-2).

According to the 'wobble' hypothesis of Crick,[16] which was based on stereochemical considerations, the pairing between the 5' base of the anticodon and the 3' base of the codon is less stringent than the pairing in the other two positions, so that codon G can pair with either U or C and codon U with either A or G in the anticodon. The reason why the 5'-U in the anticodon of yeast ochre-specific tyrosine-inserting tRNA does not 'wobble' between G (amber) and A (ochre) is not understood, but may have something to do with the pseudouridine in the adjacent anticodon position.

Note that the anticodon sequences are written (left to right) in the 3'-5' order and the codon sequences in the conventional 5'-3' order

Fig. 13.9

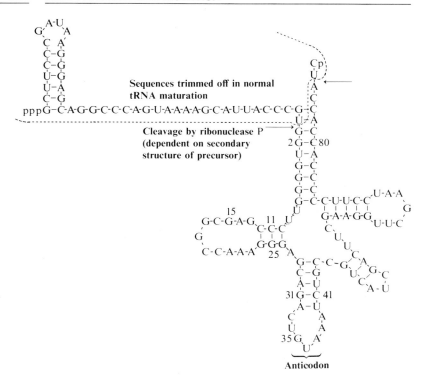

The structure of the precursor of one of the *E. coli* tyrosine-specific tRNA genes, mutations which have been defined within it, and their consequences. Abridged from Altman.[30] The secondary structure, with hairpin loops, is now known to be an oversimplification, but it is the best representation possible in two dimensions.

Key

Mutation	Effect
G2→A2	Neither tRNA nor precursor detectable—susceptible to nuclease degradation?
C80→U80	Reverses effect of G2→A2 by restoring base-pairing.
G15→A15	Little mature tRNA; precursor accumulated.
G25→A25	Little mature tRNA; precursor accumulated.
C11→U11	Reverses effect of G25→A25 by restoring base-pairing.
G31→A31	Little mature tRNA; precursor accumulated.
C41→U41	Reverses effect of G31→A31 by restoring base-pairing.
G35→C35	Anticodon now recognizes amber chain-terminating codon (UAG).
G35→U35	Anticodon now recognizes either amber (UAG) or ochre (UAA) chain-terminating codon.
Temperature-sensitive ribonuclease P	Accumulation of precursor at restrictive temperature.

by the mutagen ethylmethane sulphonate (EMS), consistent with the known tendency of this mutagen to induce G-C to A-T transitions (*see* p. 335).

Suppressors mapping at different loci may either represent genes coding for tRNA molecules of the same specificity (for instance there are eight tRNA^Tyr genes in yeast, each capable of mutation to give tyrosine-inserting suppressors) or they may affect tRNAs for different amino acids. In *E. coli* and yeast, amber suppressors may insert serine[14] or glutamine[12] as well as tyrosine.[13, 15] A considerable proportion of the genes for tRNA can, at least potentially, be identified in this way. However, where there is only one gene for a tRNA capable of inserting a particular amino acid, as is the case for tryptophan in *E. coli*, a mutation changing the coding specificity of that tRNA will usually be lethal. Mutant tRNA[Trp], suppressing the third chain-terminating codon UGA, is known in *E. coli* but only in strains in which the gene is duplicated.

Secondary Mutations in Suppressor tRNA—Effects on Precursor Processing

Where it is possible to select for loss of gene activity one can, starting with a suppressed mutant, obtain mutations in the suppressor tRNA gene which abolish or greatly reduce tRNA function. This approach was used by J. D. Smith, who was thereby able to isolate a number of different forms of *E. coli* tRNA[Tyr] with changes in parts of the molecule other than the anticodon. In several of these, the secondary structure of the molecule had been destabilized by base changes which disrupted base-pairing within one or other of the base-paired loops which are constant features of tRNA molecules and presumably essential for their function (*Fig.* 13.9).

Many mutants with lost tRNA function produce finished tRNA molecules only in low yield and accumulate a longer precursor molecule which, in normal strains, is converted to tRNA by a series of cutting and trimming reactions (*Fig.* 13.9). Mutants defective in RNA processing could have defects in the processing enzymes or in sequences within the RNA which these enzymes recognize. In the case of the *E. coli* tRNA-processing mutants both kinds have been demonstrated. (We defer mention of mutations affecting processing of mRNA precursors until the next chapter—p. 414.)

To summarize, mutations suppressing chain-terminating mutants, and mutations nullifying such suppression, first led to the identification of genes coding for tRNA molecules. Such genes can now be cloned and analysed even without mutations to mark them. Their ability to hybridize with tRNA permits their direct detection (*see* p. 186).

13.8 | Mutations in sequences controlling transcription

So far we have been dealing with effects of mutations on protein structure. Equally important in their implications, though generally less easy to find, are mutations which alter the level of gene expression without affecting the structure of the protein gene product. In this section we will review the first, and classic, example of quantitative control of gene action—the *lac* operon of *E. coli*, briefly mention variations on the *lac* pattern found in some other bacterial operons, and then briefly consider one of the systems of gene control which have been investigated in eukaryotes.

The *E. coli lac* Operon

The utilization of lactose as a carbon source by *E. coli* depends on the formation of two proteins—a specific **permease** protein in the cell membrane which mediates the uptake of lactose from the growth medium and β-galactosidase, the enzyme which hydrolyses lactose to glucose and galactose. The synthesis of these two proteins and also of a third one, thiogalactoside transacetylase, an enzyme whose physiological role is still not understood, is dependent on the presence of lactose or some β-galactoside analogue in the growth medium and is repressed by the presence of glucose. Mutants unable to use lactose as carbon source map in two adjacent genes: *lacZ* which is the structure-determining gene for β-galactosidase and *lacY* which determines the permease. The transacetylase is determined by *lacA*, which is adjacent to *lacY* (the gene order is Z–Y–A), but mutations in this gene have no affect on growth since the enzyme is not essential under any known conditions.

The activities of all three genes are coordinated, so that on addition of inducer the three products are produced in constant ratios. Biochemical studies have shown that all three proteins are translated from a single mRNA molecule which includes all three coding sequences, with their respective ribosome binding sites and starting and terminating codons in non-overlapping tandem array.

J. Monod and F. Jacob, who were mainly responsible for the observations and theory which opened up the *lac* operon as an experimental system, found two kinds of mutants which produced all three enzymes whether or not inducer was present, i.e. **constitutively**. The first type mapped in a locus called *lacI*, which was close to but not absolutely contiguous with *lacZ* (order I–ZYA). These *lacI⁻* mutants were recessive in their constitutivity—that is to say, heterogenotes with *lacI⁻* on the chromosome and *lacI⁺* on an F′-plasmid (or vice versa) showed normal inducibility and produced none of the *lac* proteins in the absence of inducer. This was taken to mean that the *lacI⁻* mutants lacked a repressor function, which was attributed to the wild-type *lacI⁺* gene. This is an example of the application of the simple presence/absence hypothesis for explaining dominance/recessivity—a hypothesis which may occasionally be misleading but is usually a good guide, as it was in this case. The second class of constitutive mutants, called O^c, mapped at the end of *lacZ* on the *lacI* side (i.e. the sequence was I–OZYA), and were distinguished by being *cis*-**dominant**. The *cis* prefix means that the O^c mutation conferred dominant constitutive expression of a ZYA gene cluster linked to it on the same chromosome (i.e. *cis*), but had no such effect on the same genes when they were in *trans* on a separate molecule of DNA, such as an F′-plasmid. Thus a heterogenote of constitution $\dfrac{O^c\ Z^+}{O^+\ Z^-}$ formed β-galactosidase constitutively, while one of constitution $\dfrac{O^+\ Z^+}{O^c\ Z^-}$ (which can be constructed by recombination between O and Z) expressed the enzyme only in response to induction.

The hypothesis advanced to explain these observations was that the O region, or **operator**, was a DNA-binding site for the *lacI⁺*-encoded repressor molecule, and that the bound repressor prevented transcription of the adjacent row of genes, which was dubbed an **operon**, meaning a contiguous set of genes transcribed as a unit and subject to a common transcriptional regulation.

Except that the repressor molecule was originally thought to be an RNA molecule rather than (as is now known to be the case) a protein, the model of Monod and Jacob has been entirely confirmed through intensive studies of the

Fig. 13.10

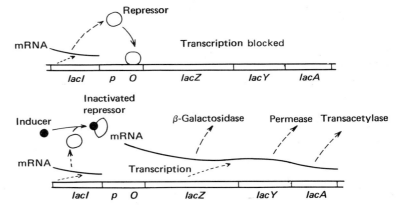

The *lac* operon of *Escherichia coli* and the effects of mutations on its expression.

Key:
Effects of mutations:

In *lacZ*:	Structural changes in β-galactosidase, many leading to loss of activity; chain-termination mutants depress the levels of expression of *lacY* and *lacA*.
In *lacY*:	Loss of activity (presumed due to structural changes) in β-galactoside permease; chain termination mutants depress levels of expression of *lacA*.
In *lacA*:	Structural changes in β-galactoside transacetylase, some leading to loss of activity; no effect on growth on lactose.
In *lacI*:	Structural changes in repressor protein; many (I^-) mutations, including all chain-terminators, give recessive derepression of the *lac* operon regardless of the presence of inducer; I^s mutants have altered repressor which represses even in presence of inducer, and this 'super-repressed' phenotype is dominant in either *cis* or *trans*; a few I^- mutations (I^{-d}) are also dominant, probably because of interactions within tetrameric I^+/I^{-d} hybrid tetramers resulting in the I^{-d} conformation being imposed on the I^+ monomers ('negative complementation').
In *O*:	Inability to bind repressor, resulting in permanent derepression of the adjacent operon; the effect is dominant, but it does not affect the repression of another *lac* operon in the same cell (i.e. *cis*-dominant but *trans*-recessive).
In *p*:	Decrease or increase in the level of expression of the entire operon without affecting control by repressor; in some cases an altered response to the CRP–cyclic AMP system (*see text*). For further details see Beckwith and Zipser.[31]

molecules involved. The *lacI*⁺ product has been purified as a protein, and its complete structure has been determined. The DNA of the *lac* operon has been isolated in F′-plasmids and in λ or φ80 transducing bacteriophages, and a large part of its nucleotide sequence is now known. The operator sequence has been isolated and the repressor protein has been shown to bind to it specifically and with extremely high affinity. The repressor protein has allosteric properties and its normal allosteric effector is *allo*lactose, an isomer to which lactose can be converted within the cell in a single metabolic step. *Allo*lactose stabilizes a conformational state of the repressor which has no affinity for operator DNA and thus brings about induction (**derepression** is a better description).

Several mutant forms of the repressor protein are known in different *lacI* mutants. The forms found in constitutive (*lacI*⁻) mutants have lost their affinity for operator DNA, or are permanently fixed in their non-binding conformation. Other mutants mapping in *lacI* have a dominant non-inducible (*lacI^d*) phenotype, and these produce forms of the repressor protein which are highly stabilized in their operator-binding conformation and cannot be shifted from it by inducer. The protein is a tetramer, and the mixed tetramer formed in *lacI^+/lacI^d* heterogenotes has conformational properties approaching those of the *lacI^d* homotetramer. This is an example of dominance through allelic interaction at the level of protein quaternary structure, with one very strongly stabilized mutant monomer restraining the normal conformational transition of wild-type monomers with which it is associated. It may be described as **negative complementation**.

The discovery of a further class of *cis*-acting mutations resulted in an important extension of the operon model. These mapped close to the operator mutations but a little further away from *lacZ*. They resulted in quantitative changes (either up or down depending on the mutant) in the expression of the operon without affecting its control by repressor. It was proposed that they occurred in a **promoter** sequence which binds RNA polymerase in initiation of transcription. This too has been confirmed, by *in vitro* studies of polymerase binding and RNA synthesis. The position of the operator region, a little way 'downstream' from the promoter, provides a satisfactory explanation of the effect of bound repressor protein in blocking transcription. The whole map of the *lac* operon can be summarized as *I–pOZYA*, with *p* the promoter. The close linkage of the *lacI* gene, coding for repressor, is not necessary for its function, but probably has an evolutionary significance. It means that a bacterium not already possessing the *lac* operon (not all coliform bacteria do) can acquire the whole operon together with its controlling gene in a single compact block of DNA.

Polarity in Operons

A characteristic feature of operons is that chain terminating mutants falling within any one gene of the operon tend to have **polar** effects on the expression of any gene in the same operon situated further away from ('downstream' of) the operator. Thus an amber or ochre mutation in *lacZ* will not only eliminate the function of that gene but will also, to different extents depending on the position of the mutant site, reduce the expression of the *lacY* and *lacA* genes. Chain terminating mutations in *lacY* reduce the expression of *lacA* but have no effect on *lacZ*. Thus the effect spreads in one direction—away from the operator. The most extreme polar effects are shown by mutants close to the operator end of their respective genes, and thus most distant from the start of the following gene.

Polarity has been studied in the *trp* operon in even more detail than in the *lac* operon, and has been shown to be due to degradation of the operon mRNA from a point corresponding to the termination codon proceeding in the operator-distal ('downstream') direction. Translation and transcription are closely coupled in prokaryotes and normally an mRNA molecule will become associated with ribosomes even as it peels off the DNA template. When a chain-termination codon is reached, the ribosomes travelling along the growing messenger chain will drop off and there will be a stretch of 'naked' mRNA from the premature terminator to the initiating codon of the following gene. The most likely explanation of the selective mRNA degradation is that any mRNA not protected by a covering of closely spaced ribosomes is vulnerable to attack by a nuclease enzyme, the vulnerability being the greater the longer the sequence uncovered. The idea is

Fig. 13.11

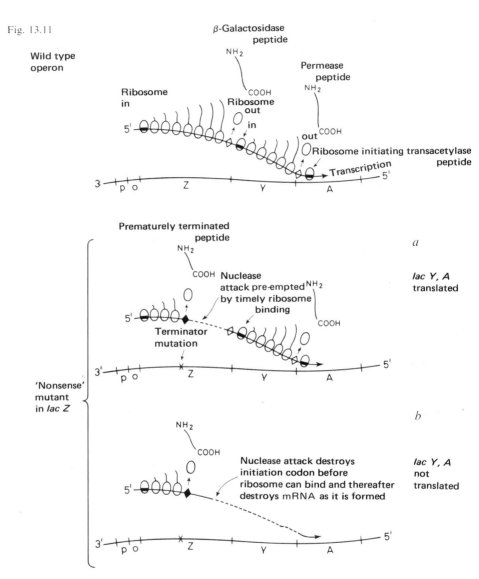

A model for the polar effect of chain-terminating mutations within operons. mRNA, following premature chain termination, is unprotected by ribosomes and is thought to be vulnerable to nuclease-catalysed degradation, pre-empting ribosome binding for translation of the following gene. The greater the distance between the mutational chain terminator and the initiation codon for the following gene, the greater the vulnerability of the mRNA and the greater the chance of outcome (*b*) rather than (*a*).

Symbols: (■) ribosome-binding and translation initiation sites; (◇) normal termination codons; (◆) mutant termination codon in *lacZ* reducing translation of both *lacY* and *lacA*.

illustrated in *Fig.* 13.11. Note that both transcription and translation proceed in the 5′ to 3′ direction with respect to the mRNA and so, presumably, does the mRNA breakdown.

Fig. 13.12

D-Xylulose-5-phosphate ←— L-Ribulose-5-phosphate ←— L-Ribulose ←— L-Arabinose

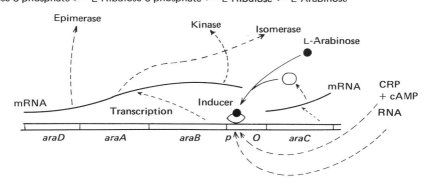

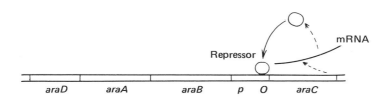

The *ara* operon of *E. coli* and the effects of mutation on its expression.

Key:

Effects of mutations:

In *araB*:	Structural changes in kinase, some leading to loss of activity; chain terminating mutations depress levels of expression of *araA* and *araD*.
In *araA*:	Structural changes in isomerase, some leading to loss of activity; chain-terminating mutations depress levels of expression of *araD*.
In *araD*:	Structural changes in epimerase, some leading to loss of activity.
In *araC*:	Recessive failure to induce any of the three enzymes (*araC⁻*); constitutive formation (i.e. without inducer) of all three enzymes (*araCᶜ*)—recessive to *araC⁺*, dominant to *araC⁻*. Some mutants increase or decrease the requirement for the assistance of CRP+cAMP in the initiation of transcription. Deletion of *araC*, leaving the *O* and *p* sites, causes an *araC⁻* phenotype which can be complemented by *araC⁺* in *trans*
Mutations in *O* and *p*:	Deletions covering *araC* and *O* give *araC⁻* phenotype, complemented by *araC⁺* in *trans* to give hyperinducibility for all three enzymes; deletions covering *araC*, *O* and *p* are *araC⁻*, non-complementable by *araC⁺* in *trans*: 'constitutive' mutations in *p* (*Iᶜ*) are hyperinducible in presence of *araC⁺* and *Iᶜ* mutants produce all three enzymes without induction when *araC* is deleted.[23,24]

Fig. 13.13

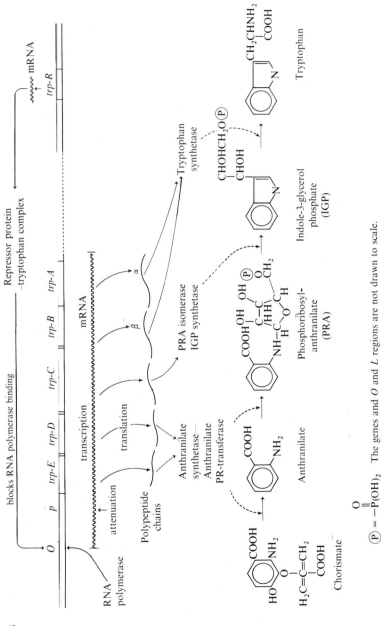

The genes and O and L regions are not drawn to scale.

$(P) = -P(OH)_2$

Fig. 13.13 *cont.*   The *trp* operon of *Escherichia coli*.
Key:
Effects of mutations:
trp-E, D,

C, B, A:	Changes in structure of corresponding polypeptide, in some cases leading to enzyme inactivation; chain-terminating mutations reduce expression of all genes more distant from the operator.
trp-R:	Changes in structure of repressor protein; some mutations, including all chain-terminating mutations, result in a recessive constitutive phenotype, i.e. the whole operon non-repressible by tryptophan.
L:	(Leader sequence) mutations in certain defined base-pairs decrease operon transcription by increasing attenuation (*see* Box 14.3).
O:	Mutations in certain defined base-pairs of the operator sequence lead to non-repressible (i.e. constitutive) formation of all the enzymes controlled by the operon; this constitutive phenotype is *cis*-dominant and *trans*-recessive.
Notes:	Both the anthranilate synthetase–phosphoribosyl (PR) trans-ferase complex and tryptophan synthetase contain two different polypeptide chains coded for by different, though adjacent, genes. The leader peptide and its role in attenuation of transcription, mediated by low levels of Trp-tRNA, are discussed in in Chapter 14 (p. 411). *See* Gunsalus and Yanofsky[25], Bennett and Yanofsky[26] and Zurawski et al.[27]

Carbon-catabolite Repression

Glucose represses the transcription of the *lac* operon, even in the presence of lactose, and the chemical signal through which it does so is adenosine cyclic $3':5'$-monophosphate (cAMP). This compound accumulates in the cell under conditions of carbon deprivation and falls to low levels when glucose is supplied. The action of cAMP is mediated by a cAMP-binding protein (CRP) which, when bound to cAMP, interacts with the promoter and with RNA polymerase to activate transcription of the *lac* operon. When it is not bound to cAMP, CRP has no such effect and transcription does not take place. Thus the CRP–cAMP (**catabolite repression**) system of control can override that exercised by that *lacI* repressor protein. CRP is an example of an important class of *positively* acting regulators of transcription. The effects of different kinds of mutations on transcription of the *lac* operon are summarized in *Fig.* 13.10. An account of their analysis in terms of DNA sequences is deferred until the following chapter.

Other Operons

The key features of the *lac* operon, i.e. operator, promoter and regulator protein, are also seen in a number of other *E. coli* operons. Two important variations on the *lac* model should be noted. Firstly, in some operons. of which that concerned with arabinose utilization (the *ara* operon) is the classic example, the allosteric regulator protein plays a *positive* transcription-activating as well as a repressor role (*Fig.* 13.12). One conformation of the *ara* regulator protein, stabilized by its allosteric effector arabinose, promotes transcription, while its alternative conformation, which prevails in the absence of arabinose, is a repressor, just like the *lacI* protein in its unliganded form. Thus different mutations within the *araC* gene, which codes for the regulator protein, can give constitutive or enzyme-negative phenotypes depending on how the protein is altered. If the protein is not produced at all the result is a recessive enzyme-negative phenotype. The *ara* operon (like the *gal* operon for galactose utilization, which generally conforms to the *lac* pattern) is subject to carbon-catabolite repression, mediated by cAMP and CRP.

The second major variation on the *lac* model is seen in operons controlling biosynthetic pathways, such as the *trp* and *his* operons which, respectively, code for all the enzymes in the tryptophan and histidine pathways. In these operons the allosteric effector which controls the state of the regulator protein is the end-product of the pathway, or perhaps in some cases an immediate derivative of it (such as the aminoacyl-tRNA). The effect is to stabilize the repressor form of the regulator protein rather than, as in the carbon-catabolic pathways, to stabilize the non-repressing conformation. Thus the end-products of these biosynthetic pathways act as negative regulators of the activities of the operons producing them. Some of the key features of the *trp* operon are summarized in *Fig.* 13.13.

Not all biosynthetic pathways in bacteria are controlled by operons; in *E. coli*, for example, only some of the many genes involved in arginine biosynthesis are linked in operons and the remainder occur in scattered locations. Operons tend to occur in pathways that are strictly channelled, without secondary inputs or outputs, so that no one enzyme encoded in the operon has any function apart from that of the others.

Control of Gene Transcription in Eukaryotes

Fungi

Operons are not easy to find in eukaryotes. There is at least one possible example in fungi,[17] but the general eukaryotic rule is for the genes coding for the enzymes of a metabolic pathway to occur in scattered locations and to be expressed through separate mRNA molecules. Nevertheless, some of the essential components of the operon model do appear to be relevant to eukaryotes—in particular the role of *cis*-acting transcriptional control regions in the DNA and the interactions of these regions with the products, probably proteins, of regulatory genes. Many examples could be given, but none is yet supported by anything approaching the degree of convincing detailed information that is available for the best known bacterial operons.

As good an example as any is the system of control of utilization of various nitrogen sources in *Aspergillus nidulans*. In this fungus, and also in *Neurospora crassa*, the formation of the enzymes necessary for the utilization of nitrate, acetamide, proline and purines, to name but a few possible nitrogen sources, is repressed by ammonia (or ammonium ions). The hypothesis is that a general regulatory protein acts as a positive activator of transcription of the genes coding for all these enzymes and that, in the presence of some derivative of ammonia, probably glutamine, this essential activator protein is converted to an inactive form. The gene that is thought to code for this protein in *Aspergillus* is called *are* (ammonia-regulation) and many different mutant alleles have been isolated.[18] Some (called *are^d*) show dominant repression of all the controlled genes, so that most nitrogen sources other than ammonia cannot be used. Others make production of one or several of the normally repressible enzymes non-repressible by ammonia. *Cis*-acting mutations are known that affect the response to *are* of adjacent genes. For example, a *cis*-dominant (*trans*-recessive) mutation mapping very close to the structure-determining gene for acetamidase (necessary for the utilization of acetamide) renders that gene non-repressible without affecting the regulation of any of the other ammonia-repressible genes.[19]

The *are* protein product has not been isolated from *Aspergillus*, but it seems that the analogous product in *Neurospora* may very well have been identified.[20] A protein has been detected which binds tightly to *Neurospora* DNA in the absence of glutamine but much less tightly in its presence. These are just the properties expected for an activator of transcription controlled by glutamine as a positive allosteric effector. The case is strengthened by the fact that a *Neurospora* mutant which seems to be analogous to the *Aspergillus are^d* class shows an alteration in this protein.

Higher Eukaryotes

In higher plants and animals there are a number of examples of mutations, either within or closely linked to and *cis*-acting on protein-structure genes, which affect the level and sometimes the timing of gene activity (*see* Problem 7 at the end of this Chapter). Paigen[21] has reviewed the rather extensive literature on regulatory mutants in mouse.

In maize, there are some interesting allelic series which show allele-specific variation in expression between different tissues and/or different times of development. For example, one of a series of alleles determining electrophoretically distinct forms of an esterase is expressed strongly in young endosperm but seems to be 'switched off' rather abruptly as the seed matures. Other alleles of the same gene maintain their activity.[33] In some flowering plants, for example *Antirrhinum majus*, one finds what may be termed **pattern** alleles, controlling the distribution of pigmentation over the flower or other parts of the plant.[22] It may be that these allelic variations are due to mutations in operator/promoter regions making gene expression more or less sensitive to variation between tissues with respect to regulatory proteins or their allosteric effectors.

Tissue-specific gene expression can, however, occur through a variety of mechanisms. At the simplest level it may result from different properties of allelic protein products. The Himalayan pattern in the rabbit, for example, has long been known to be due to a mutation in the gene coding for tyrosinase which makes this pigment-forming enzyme heat sensitive and hence inactive everywhere but in the colder extremities—the feet, nose and tips of ears. Much more recently, variations in the cutting and splicing of RNA precursors has been shown to be important (*see* Chapter 14, p. 415).

13.9 Summary and perspectives

The induction and mapping of mutations with specific effects on proteins have laid the basis for the understanding of gene function. Most genes, in the sense of functional units or cistrons, code for polypeptide chains. The gene–protein code, though established independently by biochemical methods, has been confirmed in practice by observations of amino acid replacements in mutant proteins, especially by frameshift analysis. Studies of genetic suppression of chain-terminating mutations have led to the identification of genes transcribed into tRNA molecules. Genes coding for regulatory proteins, and the genomic segments with which these interact to control transcription, have, especially in *E. coli*, been defined through the study of mutants with altered regulation of the synthesis of specific enzymes.

Powerful as it has been, the general method of defining gene function through the analysis of mutations has its limitations. It has had great success in bacteria and fungi, where efficient means exist for isolation and accurate mapping of large numbers of mutants of desired types. With rather more labour, the approach of 'saturating' a small genetic region with mutations can be applied to *Drosophila* also (*see* Chapter 15, p. 431). But in higher plants and animals it is hardly feasible. Furthermore, there are, at least in higher organisms, a number of essential steps in gene expression which we do not necessarily know how to recognize by the effects of mutation—certainly not by effects on protein structure. These include the specific initiation and termination of transcription at specific points, and the processing of the RNA produced. As we see in Chapter 14, processing in eukaryotes is complex and includes the cutting and splicing of primary transcripts as well as the 'capping' and 'tailing' of the messenger RNA molecules so formed.

Over the past decade much has been learned about these and other aspects of gene activity through direct studies of DNA structure made possible by 'gene cloning'. This new aspect of genetics is the subject of the next chapter.

Chapter 13 Selected Further Reading

Beckwith J. R. and Zipser, D. (ed.) (1970) *The Lactose Operon.* New York, Cold Spring Harbor Laboratory.

Fincham J. R. S., Day P. R. and Radford A. (1979) *Fungal Genetics,* 4th ed. Oxford, Blackwell Scientific Publications.

Problems for Chapters 12 and 13

1   In a classic experiment, S. Luria and M. Delbruck (*Genetics* 1943, **28**, 49) investigated the mutation of *E. coli* to resistance to bacteriophage T1. Two similar inocula of wild-type *E. coli* cells were grown up in parallel in similar volumes of liquid growth medium, the difference being that the first was divided equally between ten different tubes, while the second was contained in a single larger vessel. At the end of the growth period the contents of each tube were spread with an excess of T1 particles on a single plate, while the contents of the larger vessel were divided equally between ten plates, again with an excess of T1 particles. The number of T1-resistant mutant colonies growing up on each plate was counted. All the plates seeded from the single vessel produced the same number of resistant colonies within reasonable sampling error. The plates seeded from the separate tubes, however, showed great differences in numbers of colonies, far beyond what would be expected from sampling from a single population. What does this tell us about the origin of the mutants?

2   Of 140 spontaneous *lacI* mutants of *E. coli*, 94 appeared to map at the same site. These 94 mutants fell into two classes on the basis of reversion frequency; 78 reverted spontaneously with high frequency while 18 reverted rarely if at all. The region of the gene at which these mutations mapped corresponded to the sequence Arg-Leu-Ala-Gly-Trp-His in the *lacI* protein. Can you see any reason why the DNA sequence coding for this polypeptide sequence should be a mutational 'hot-spot'? What do you think the two classes of mutation in this region might be? (Farabaugh et al. 1978, *J. Mol. Biol.* **126**, 847.)

3   A number of bacteriophage T4 mutants, originally induced by ultraviolet light, were tested for revertibility with a number of chemical mutagens as well as for spontaneous reversion. The results are summarized in the table.

Mutant no.	1	2	3	4	5	6	7	8
Ethylethane sulphonate	−	−	−	−	+ +	−	+	−
Nitrous acid	+ +	−	−	+ +	+ +	−	−	−
Hydroxylamine	−	−	−	−	+ +	−	−	−
5-Bromouracil	+	−	−	+	+ +	−	−	−
Proflavine	−	−	+ +	−	−	+ +	−	−
Spontaneous	±	−	+	±	±	±	±	±

What can you infer as to the molecular nature of each mutant?

4   A chain-terminating mutation in the *am* gene of *Neurospora crassa* is capable of reversion to restore either the normal amino acid sequence of the gene product or an unusual but functional sequence which, in different cases, may have leucine or tyrosine in place of the normal glutamic acid residue at a certain position. It is efficiently reverted to give the 'tyrosine' version by nitroquinoline oxide, a mutagen thought to act specifically on G-C base-pairs. What is the likely nature of the chain-terminating mutation? (Seale et al. 1977, *Genetics* **86**, 261; P. A. Burns, unpublished.)

5   Some kinds of mutation are found more frequently than others. From what you know of the nature of different mutations, what would you expect, in general terms, to be the relative frequencies of the following: (i) mutation from wild type to loss of function of a particular gene; (ii) reverse mutation precisely restoring the wild-type gene product; (iii) suppressor mutation acting by blockage of a competing metabolic pathway; (iv) suppressor mutation reversing the phenotype of a chain-terminator through altering a tRNA species; (v) mutation restoring effective, but not necessarily completely normal, activity to a marginally active mutant protein?

6   It is possible to separate the two strands ('heavy' and 'light') of lambda bacteriophage DNA by virtue of their different proportions of purines and pyrimidines. Double-stranded DNA can be reconstituted by recombining the single strands. Suppose that a clear plaque C_1 mutant has a substitution of aspartic acid for asparagine in a certain residue of the C_1 protein product (repressor, permitting lysogeny). Suppose also that the separated strands of the DNA of this mutant were treated separately with hydroxylamine, recombined with their complementary (untreated) strands, repackaged to make infective particles *in vitro*, and plated in bacterial lawns to test for hydroxylamine-induced reverse mutation from C_1 to C_1^+ (turbid plaques). Would you expect to see reverse mutations following treatment of both strands or only one? Explain. How might the experiment tell you which strand was transcribed into the C_1 messenger? Would you expect to see entirely turbid plaques as the result of mutation, or plaques partly turbid and partly clear? What would you conclude from either observation?

7   The N-terminal sequence of the wild-type glutamate dehydrogenase of *Neurospora crassa*, coded for by the *am* gene, is:

N-acetyl-Ser-Asn-Leu-Pro-Ser-Glu-...... .

A revertant isolated from *am*[6], a CRM$^-$ mutant, was found to produce glutamate dehydrogenase with the unusual N-terminal sequence:

Met-Leu-Thr-Phe-Pro-Pro-Glu-...... .

The sequences represented by dots were identical in wild type and revertant. What is the simplest explanation of these observations? (Siddig et al., 1980, *J. Mol. Biol.* **137**, 125–135.)

8   In *Salmonella typhimurium*, genes coding for different enzymes necessary for histidine biosynthesis are arranged adjacent to one another in the sequence *GDCBHAF*. Normally the transcription of these genes is repressed when histidine is present in the medium. Certain mutations, called O^c, mapping in a segment immediately to the left of *hisG*, render all the genes non-repressible. Through the use of an F'-plasmid carrying the *his* region a series of strains was constructed with each strain having a different combination of *his* mutations on chromosome and plasmid. The repressibility of the *his* genes in these strains was determined with the following results:

$$F'\ \frac{O^+\ D^+\ B^+}{O^c\ D^-\ B^-}\qquad D \text{ and } B \text{ enzymes both repressible}$$

$$F'\ \frac{O^+\ D^-\ B^+}{O^c\ D^+\ B^-}\qquad D \text{ enzyme non-repressible, } B \text{ enzyme repressible}$$

$$F'\ \frac{O^+\ D^+\ B^-}{O^c\ D^+\ B^+}\qquad \text{Neither enzyme repressible}$$

What do these results tell us about the nature of the *his* gene cluster and the role of the *O* segment? (Fink and Roth, 1968, *J. Mol. Biol.* **33**, 547.)

9　In the mouse the kidney enzyme β-glucuronidase is synthesized in males in response to the male hormone testosterone, the primary effect of the hormone being on mRNA synthesis. Structural (charge) differences in the enzyme are due to allelic variation at a locus called *Gus-s*. Certain strains of mice show an unusually small stimulation of β-glucuronidase synthesis in response to testosterone, and this variant is attributed to a second locus called *Gus-r*. *Gus-s* and *Gus-r* are very closely linked and were for a long time not shown to be separable. Recently a recombinant was discovered in a population of mice exhibiting heterozygosity at both loci. Explain the importance of such recombinants in establishing the functional relationship between *Gus-s* and *Gus-r*. (For review, *see* Paigen, 1979, *Annu. Rev. Genet.* **13**, 433–449.)

10　The following are complementation maps of two *Neurospora crassa* loci, *arg-1* (Catcheside and Overton, 1959, *Cold Spring Harbor Symp. Quant. Biol.* **23**, 137) and *ad-3* (De Serres, 1956, *Genetics* **41**, 668). The number of mutants in each complementation class is indicated above the bar representing that class. What do the general forms of the maps suggest about the respective modes of organization of the two loci?

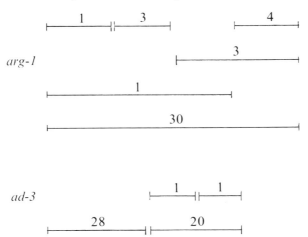

(Note that, in the convention for drawing complementation maps, mutants which do *not* complement one another are shown as overlapping bars and mutants which *do* complement one another as non-overlapping bars. The *arg-1* numbers are real; the *ad-3* numbers are invented but consistent with De Serres' much more extensive data.)

Genes as DNA Sequences

14.1 Introduction

That the genetic material is DNA has been known for over a quarter of a century. As we saw in the last chapter, combined genetical and biochemical analysis of mutants has confirmed the genetic code for amino acid sequence in proteins and led to many deductions concerning the spatial relationships of genes and of mutable sites within them. But until very recently, it appeared virtually impossible actually to isolate and determine the base sequence of the DNA of any particular gene. The reason for this pessimism was the enormous complexity of the DNA sequence in the genomes of even simple organisms and the lack of chemical means for distinguishing one part of the sequence from another. The problem seemed worse than that of the proverbial needle in the haystack—it was more like trying to find one particular strand of hay among millions of others of almost identical appearance.

The key to the DNA sequence identification problem was found in the specific complementary binding of the two DNA strands of a duplex, or of one DNA strand and the RNA molecule transcribed from it. Complementary nucleic acid sequences recognize each other with great precision. As we saw in Chapter 1 (p. 31), conditions have been worked out for annealing of complementary DNA single strands, and under somewhat different conditions (70 per cent formamide at 42 °C) DNA–RNA hybridization is favoured. Thus a length of single-stranded DNA or RNA, radioactively labelled, can be used as a **probe** for another DNA (or less often RNA) strand with complementary base sequence. Under relatively stringent conditions of hybridization, i.e. high temperature (say 70 °C) and low salt concentration, only long sequences (of the order of hundreds of bases), with nearly perfect complementary matching, will anneal securely. By relaxing the conditions, i.e. lower temperature and higher salt concentration, shorter and/or less well-matching probes will hybridize.

The use of specific nucleic acid probes can, in favourable instances, lead to the identification and isolation of the DNA of genes of known function. The DNA base sequences of these genes answer some questions as well as posing new problems.

14.2 Obtaining probes for genes

The Isolation of Specific mRNA

The most obvious molecular probe for a gene is the mRNA transcribed from it, and so it is necessary at this point to consider the available methods for fractionating RNA.

The first step in the preparation of nucleic acids from crude cell extracts is the removal of protein. This can be effected by extraction with phenol or some other organic solvent or solvent mixture in which proteins, but not nucleic acids, will dissolve. The RNA can then be precipiated with ethanol.

This leaves the problem of separating the different kinds of RNA from each other. One kind of fractionation is on the basis of size, for example by centrifuging the sample through a density gradient. A mixing device feeds into a centrifuge tube a mixture of 5 per cent and 35 per cent sucrose in steadily changing proportion, so that a gradient is established with the higher density at the bottom and the lower density at the top of the tube. The RNA sample is added as a top layer, and the RNA molecules are then sedimented through the gradient by centrifuging at 35 000 rev./min. The different molecules are sedimented at rates dependent upon their molecular weights—the larger the faster. After a period of sedimentation each molecular species has its maximum concentration at a particular level in the tube, with a more or less symmetrical distribution above and below that level. The tube contents are withdrawn dropwise, from the bottom of the gradient, and the nucleic acid concentration in each fraction can be conveniently monitored by measurements of light absorbance at 260 nm. Plotting absorbance against fraction number

Fig. 14.1

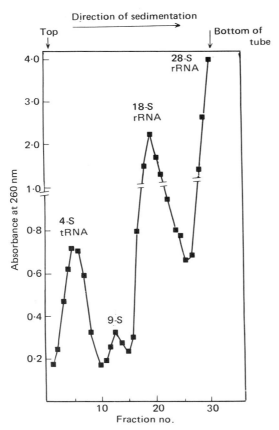

Fractionation of RNA, from polyribosomes of rabbit reticulocytes, by centrifugation at 27 000 rev./min for 25 h through a sucrose gradient—5 per cent sucrose at the top of the tube to 20 per cent at the bottom. The 9-S peak consists predominantly of α and β-globin mRNA molecules which can be further purified by adsorption to and elution from oligo(dT)-cellulose to which the mRNA binds through its poly(A) 3′ 'tails'. The α and β-messengers can be separated by gel electrophoresis.
Redrawn from Evans and Lingrel.[1]

shows different size classes as a series of non-overlapping or partially overlapping peaks.

Figure 14.1 shows the result of such fractionation applied to the RNA from reticulocytes (red blood cell precursors). The most conspicuous peaks are due to the two major RNA components of the ribosomes—28 S and 18 S, respectively, in terms of sedimentation coefficient expressed in S or Svedberg units. Much more slowly sedimenting, at 4 to 5 S, is a peak of much smaller RNA molecules consisting of tRNA and ribosomal 5-S RNA. In an intermediate position one sees a smaller peak at 9 S, and this can be shown to consist mainly of mRNA coding for the two peptides, α and β-globin, of adult haemoglobin. The predominance of these two messengers, to the virtual exclusion of the thousands of others encoded in the DNA, is an expression of the extreme differentiation and specialization of the reticulocyte. In other types of specialized cell one would find other kinds of predominant messengers, mostly larger than 9 S since most polypeptides are longer than α and β-globin.

A very useful further purification step for mRNA (indeed it can be used as an earlier step) makes use of the presence of poly(A) 'tails' at the 3' ends of nearly all eukaryote messengers. Poly(A) sequences will bind, by complementary base-pairing, to sequences of repeated deoxythymidylate [oligo(dT)]. Synthetic oligo(dT) can be chemically bound to cellulose powder which is then used as a matrix for packing a fractionation column. RNA molecules with poly(A) 'tails', but not those without them, will be bound to an oligo(dT) cellulose column at relatively high salt concentration and then, after the other RNA has been washed away, can be eluted from the column by a saline solution of lower concentration.

In the case of reticulocyte RNA, size fractionation followed by oligo(dT) binding and elution yields a fairly clean mixture of α and β-globin messengers. These two molecules of very similar length can then be separated from each other by gel electrophoresis; two closely spaced but still distinct bands are obtained which can be separately eluted from the gel.

Verification of the mRNA by *in vitro* Translation

That one really has isolated the desired species of mRNA can be verified by using it for translation into polypeptide in a cell-free system, i.e. *in vitro*. At least two such systems have been developed to a high pitch of efficiency. One uses an extract, containing ribosomes, soluble protein factors and tRNA molecules but no effective mRNA, from wheat germ, while the other uses rabbit reticulocytes as the source of an equivalent set of components. The addition of a complete mixture of amino acids as substrates, ATP and GTP (with Mg^{2+} ions) as energy sources, and mRNA as source of coded information, leads to the synthesis of kinds of polypeptides depending on the kinds of mRNA. The polypeptide product(s) can be made radioactive through the inclusion of [^{35}S]methionine or a ^{14}C-labelled amino acid in the reaction mixture, and they can be characterized in two ways.

First, their sizes can be determined by electrophoresis in polyacrylamide gel containing the detergent sodium dodecylsulphate (SDS). The effect of SDS is to prevent aggregation of polypeptides and to coat each one with a uniform density of negative charge, ensuring that they migrate towards the positive pole at rates determined only by their sizes (the gel exerting a sieving effect) and not on their detailed amino acid compositions. After electrophoresis, the gels are exposed to radiosensitive film to locate the positions of the synthesized polypeptides and, by comparison with a range of standards of known sizes, the molecular weights of the unknowns can be estimated. Where authentic protein samples are available for use as markers, the size of translation product will often indicate the identity of the messenger which has been isolated.

A more conclusive test, which can be combined with the first, utilizes specific antibodies raised in rabbits against the authentic purified protein. Where such an antiserum is available it can be used to precipitate the specific radioactive polypeptide from the *in vitro* translation mixture, and the precipitate can then be further analysed by SDS–gel electrophoresis. The antibody polypeptides are not radioactive and do not interfere with the analysis. Electrophoresis of immune precipitates can, indeed, be used to identify quite minor products of complex translation mixtures containing unfractionated mRNA, and this can be very useful for identifying cloned DNA sequences (*see* hybrid arrest, p. 397).

Fig. 14.2

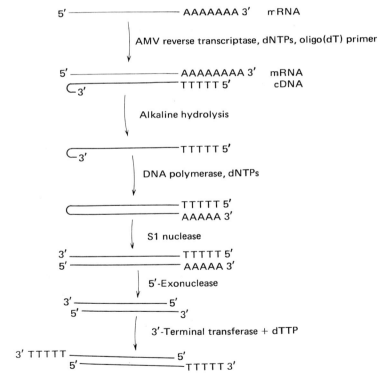

Procedure for synthesizing complementary DNA (cDNA) on a mRNA template, and preparing it for cloning. An unexpected feature of the reverse transcriptase reaction is the tendency for the completed cDNA strand to peel off the mRNA template at its terminus and to begin to grow 'backwards', using itself as a template, to form a short terminal 'hairpin'. This provides a built-in primer for synthesis of the complementary DNA strand. The final step is oligo(dT) 'tailing' for insertion into an oligo(dA)-tailed cloning vector (cf. *Fig.* 10.14).
After Maniatis et al.[2]

The Synthesis and Cloning of Complementary DNA (cDNA)

An important group of animal viruses, called **retroviruses** (cf. p. 504), have RNA as their genetic material in the infectious virus particle. However, the viral genetic information can be harboured in the host cell in the form of DNA integrated into the chromosomes. This DNA is transcribed from the viral RNA after infection by an enzyme called **reverse transcriptase**, which can be purified from virus particles. A good source of this enzyme is **avian myeloblastosis virus** (AMV). It can be used to synthesize a DNA complementary copy (cDNA) of any RNA molecule, and this copy can then be preserved and clonally propagated by insertion into a plasmid or phage-based cloning vehicle. One procedure, which was first devised by T. Maniatis and his colleagues for cloning cDNA coding for rabbit β-globin. is summarized in *Fig. 14.2*.

Chemically Synthesized Probes

Recently developed chemical techniques[3] make it possible to synthesize defined DNA molecules of up to tens or even hundreds of deoxy-ribonucleotides in length (*see below*, p. 416). Thus if the DNA sequence of a gene is known, even if only in part, it is possible to synthesize an effective probe for it. Normally, of course, one does not know the sequence; even if the amino acid sequence of the polypeptide gene product is known, the mRNA sequence cannot be fully deduced because of the different possibilities in the third-base positions of codons. Occasionally, however, a double frameshift analysis, of the kind reviewed in detail in the last chapter (*Fig. 13.7*), defines the mRNA sequence unambiguously for a sufficient distance.

The analysis set out in *Fig. 13.7* permitted B. D. Hall and his associates to synthesize a small DNA strand 13 bases long, matching a part of the yeast iso-1 cytochrome *c* messenger. This sequence was, given the comparatively small size of the yeast genome, sufficiently specific for use as a probe in the cloning of the *cyc1* gene by methods essentially similar to those described below for the rabbit β-globin gene.[5]

14.3 Use of a probe for gene detection and mapping without cloning: the example of the β-globin gene

An exemplary study, worth considering in some detail, was carried out by A. J. Jeffries and R. A. Flavell[6] on the structure of the rabbit β-globin gene. These workers isolated DNA from rabbit liver nuclei and digested it with each of six restriction endonucleases: BamHI, BglII, EcoRI, HaeIII, KpnI and PstI (cf. *Table 7.1*, p. 181). The enzymes were used both singly and in combination. Each of the digests was subjected to electrophoresis and the DNA fragments, separated according to size, were blotted from the electrophoresis gel on to nitrocellulose filters (the Southern procedure outlined on p. 186) and denatured to the single-stranded form by alkali. Then, fixed to the nitrocellulose by baking, they were allowed to react with ^{32}P-labelled β-globin cDNA. Those hybridizing to the radioactive probe were detected and positioned by exposure of the filter to X-ray film and their sizes were estimated from their respective electrophoretic mobilities. Putting all the results together, Jeffries and Flavell were able to construct a restriction fragment map (cf. *Box* 7.1) of genomic DNA 5-kb long, with sequences hybridizing with the cDNA probe accounting for about a quarter of this length.

The real surprise came when this genomic map was compared with the corresponding map of the cloned cDNA. A sequence of five restriction sites in the cDNA appeared in the same *order* as their counterparts in the genome but with their *spacing* drastically different in one place. The BamHI and EcoRI sites, only some 67 bases in the cDNA were, in the genomic clone, separated by about 700 bases. *Fig.* 14.3 shows the comparison.

As was later shown, the greater part of the long genomic interval between BamHI and EcoRI sites is not represented in the mRNA at all, but is an example

Fig 14.3

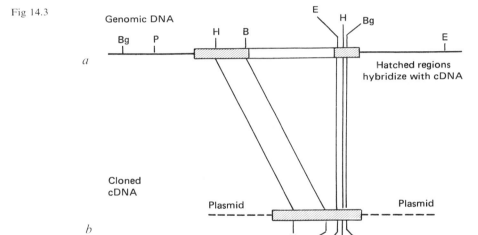

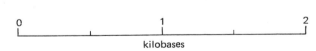

Restriction endonuclease maps of (*a*) cloned cDNA prepared from rabbit β-globin mRNA and (*b*) a DNA segment from the rabbit genome including sequences corresponding to the cDNA. The cDNA was made as shown in *Fig.* 14.2. The genomic map was deduced from the results of hybridizing Southern transfers of restriction fragments of total rabbit DNA with labelled β-globin cDNA (compare the mapping procedure explained in *Box* 7.1, p. 182).
Cleavage sites: B—BamHI, Bg—BglII, E—EcoRI, H—HaeIII, K—KpnI, P—PstI.
From Jeffries and Flavell.[6]
Note that there is a second, much shorter, intervening sequence, too small to be revealed by this procedure (cf. *Fig.* 15.9).

(one of the first to be established) of an **intervening sequence** or **intron**. Such a sequence is present in the primary RNA transcript but is in some way removed before this transcript is capped and tailed and exported to the cytoplasm as mature mRNA. We shall encounter even more striking examples of intervening sequences interrupting the protein-coding sequences of genes (pp. 436, 442).

14.4   Probes used for gene cloning

Procedures for Cloning

The type of analysis outlined in the preceding section is elegant and informative, but, in order to find out all that one wants to know about gene DNA, one needs to be able actually to isolate it in pure form and in adequate quantity for sequence analysis. This means, in practice, that the DNA must be cloned using some appropriate plasmid or phage vector system.

The first step in cloning a particular gene is usually the indiscriminate ('shotgun') cloning of the entire genome. We have already seen, in Chapters 9 and 10, how fragments of DNA from any source can be inserted into a suitable vector and replicated in *E. coli*. A collection of cloned fragments large enough to contain all genomic sequences with high probability is called a genomic 'library'.

In general, if one is aiming to clone a particular gene, it makes sense to work with the maximum size of fragment, since then there will be fewer clones to be screened and also a better chance of recovering the desired sequence all in one piece. There is also an advantage in using fragments with randomly located ends, especially when a lambda vector is being used. Such a vector imposes upper and lower limits on the cloned fragment sizes (because of the requirement for packaging in a phage head of fixed geometry) and a particular restriction enzyme, cutting at fixed points, may put the desired sequence into a fragment outside these limits.

It is not necessary here to review all the procedures which have been used successfully. One method of very general usefulness, devised by T. Maniatis and his colleagues,[4] is summarized in *Fig.* 14.4. This is a relatively sophisticated method and illustrates well several of the more ingenious tricks of the genetic engineer's trade.

Screening Genomic 'Libraries' for Specific Sequences

Filter Hybridization

The magnitude of the task of finding a particular item in a library increases in proportion to the size of the genome. Thus, with cloned fragments with a mean size of 20 kb, one needs to screen approximately 4000 clones from a *Saccharomyces* library in order to have a 99 per cent chance of finding one specific sequence. In the much larger genomes of *Drosophila* and rabbit (genome sizes of 1.7×10^5 and 3×10^6 kb, compared with 1.8×10^4 kb in yeast), the number of clones to be screened increases to 50 000 and 800 000, respectively. These numbers

Fig. 14.4

A general method for preparing large fragments
of DNA for cloning. After T. Maniatis et al.[4]
1. Induce random breakage by mechanical shearing, or near-rándom breakage by limited digestion with relatively unspecific nucleases.* Frac-
tionate by size in electrophoretic gel, and take the
fraction containing circa 20-kb fragments.
2. Methylate the EcoRI sites with EcoRI
methylase (the RI-encoded enzyme which protects
E. coli cells harbouring RI from degradation of
their own DNA by EcoRI endonuclease). This
step prevents cleavage of the 20-kb fragments
at step (4).
3. Add synthetic EcoRI 'linkers'

$$\begin{cases} \text{5'-GAATTC-3'} \\ \text{3'-CTTAAG-5'} \end{cases}$$

to ends of fragments using T4 ligase (see Fig.
8.14c). Linkers shown as broken lines.
4. Treat with EcoRI endonuclease to generate
mutually cohesive single-stranded ends

$$\begin{cases} \text{G-3'} \\ \text{CTTAA-5'} \end{cases} \text{and} \begin{cases} \text{5'-AATTC} \\ \text{3'-G} \end{cases}$$

5. Anneal and ligate into an EcoRI-cleaved site
in a cloning vehicle—Maniatis et al. used a
lambda phage derivative of the 'Charon' series
(cf. Fig. 11.10). The 'library' ready for infection into
E. coli for amplification and screening.
* The relatively unspecific endonucleases HaeIII
and AluI recognize sequences of only four base-
pairs and thus cleave DNA at many points. They
generate 'blunt-ended' fragments by making
opposite rather than staggered cuts. Maniatis et
al. used a short incubation with a mixtue of these
two enzymes, so that only a random sample of
the potentially cleavable sites was cleaved.

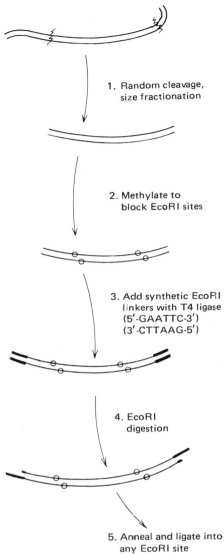

1. Random cleavage,
   size fractionation

2. Methylate to
   block EcoRI sites

3. Add synthetic EcoRI
   linkers with T4 ligase
   (5'-GAATTC-3')
   (3'-CTTAAG-5')

4. EcoRI
   digestion

5. Anneal and ligate into
   any EcoRI site

would pose an insuperable problem but for methods which have been devised for
screening many clones simultaneously.

Colony hybridization[7] is the usual procedure for screening hybrid plasmids in
E. coli colonies. Up to 1000 colonies on a plate can be sampled by imprinting on to
a nitrocellulose filter. The cells on the filter are lysed with detergent and the
released DNA is made single stranded with alkali and fixed to the nitrocellulose by
baking. The positions on the original plate of colonies containing DNA matching a
radioactive probe sequence can then be determined by hybridizing the DNA on the
filter to the probe, followed by autoradiography.

Plaque hybridization is the analogous procedure used when the library to be screened has been constructed in a lambda vector.[8] Using a lambda rather than a plasmid vector has some advantage in terms of efficiency of utilization of DNA. The lambda 'arms', bracketing ligated inserts of foreign DNA, can be artificially packaged in infective particles by incubating with a complete mixture of dissociated head and tail proteins. The DNA can then be introduced into *E. coli* by the highly efficient lambda injection mechanism.

Plaques formed by *in vitro*-packaged hybrid lambda genomes can be transferred to nitrocellulose filters and screened with labelled probe in much the same way as in colony hybridization. Up to 10^4 plaques per plate (at this density they are virtually confluent) can be screened by this method. Phage picked from the indicated positions may be initially mixed due to overlapping plaques, but a pure clone of the hybridizing phage can be obtained by replating at a lower density and a further one or two rounds of screening.

The power of the plaque hybridization method was well demonstrated by Maniatis et al.[4] who used it to isolate genomic clones of rabbit DNA that included the β-globin gene. They screened about 7.5×10^5 cloned 20-kb sequences distributed between 75 plates and identified four spots of strong hybridization with the β-globin cDNA probe. Four different clones including part or all of the β-globin gene were recovered from these spots and provided the basis for subsequent detailed investigation of the gene structure. The long intervening sequence identified by Jeffries and Flavell was confirmed and a second shorter one (cf. *Fig.* 15.9) was also demonstrated by analysis of the complete DNA base sequence in comparison with the known amino acid sequence of the β-globin polypeptide.

Similar methods have been used to clone a variety of other genes from various organisms. The common feature is the availability in each case of an abundant mRNA to serve as a source of cDNA for use as probe. Notable examples are the genes (for ovalbumin, conalbumin etc.) that are especially active in response to oestrogen in bird oviduct. We shall return to some of these in the next chapter (pp. 436, 437).

14.5   Selection of clones by functional complementation

Most genes are not very accessible through mRNA or cDNA probes since they are never transcribed in sufficient abundance for easy mRNA purification. An alternative approach, so far used successfully only for genes from fungi, is to select for expression in the bacterial cells used for cloning.

This approach, too, is subject to a serious limitation—it cannot work if the gene being sought contains intervening sequences (introns), since bacteria have no means of removing such sequences to form functional messengers. Fungal genes, however, seem so far to have few introns. The expression of yeast (*Saccharomyces*) genes in *E. coli* was first shown by B. Ratzkin and J. Carbon.[9] They cloned yeast DNA, reduced by mechanical shearing to fragments averaging 15 kb, using the plasmid ColE1 as vector. The plasmids, with yeast DNA insertions, were introduced into *E. coli* auxotrophic mutant cells of various types by the trans-

'ormation procedure (see p. 275). Deletion mutants were chosen for treatment since these were unable to mutate back to prototrophy. Hence any colonies growing on minimal medium from the treated cells could be attributed to gene functions supplied by the yeast DNA inserts. The simplest hypothesis was that the insert in each bacterial colony growing on minimal medium contained a copy of the yeast gene equivalent in function to the gene deleted in the *E. coli* auxotroph. Several yeast genes were selected in cloned DNA segments through their ability to complement *E. coli* auxotrophs.

14.6 Verification of DNA sequences after cloning

In vivo Translation

In cases where a gene is inserted into an *E. coli* cloning vector and is capable of expression in the host bacterium, one can look for the protein product of the gene and confirm that it is the right one. In order to see the plasmid products clearly it is necessary in some way to eliminate the synthesis of the bacterium's own proteins. This can be done in two ways.

The first strategem is to destroy the cell's capacity for transcription and translation by ultraviolet irradiation prior to infection with the DNA sequence to be tested, usually from a lambda vector. The second depends upon the fact that certain *E. coli* mutants tend to bud off small cells without DNA replication. These **minicells** have a full complement of cytoplasmic components but lack 'chromosomal' DNA. If, following infection with the hybrid plasmid to be tested, the recipient cells are allowed to form minicells, a part of the plasmid population will be segregated into the minicells which can then be separated from the normal cell population by their slow sedimentation in a sucrose density gradient. The minicells will be capable of synthesizing only those proteins encoded in messengers transcribed from the plasmid. An example of the use of minicells is shown in *Fig.* 14.5.

Using either of these two systems, it is possible to characterize the protein products of the plasmid by allowing protein synthesis to proceed in the presence of radioactive amino acid and then fractionating cell extracts on polyacrylamide–SDS gels as described above (p. 388). One may see polypeptides encoded in the plasmid vector as well as any due to the cloned DNA, but the former will be already well known and easily discounted (*see Fig.* 14.5).

In vitro Translation: 'Hybrid Arrest' and Hybrid Selection

When a cloned gene is incapable of *in vivo* expression in the bacterial cell, either because it lacks appropriate prokaryotic transcription/translation signals (*see below,* p. 406) or because of the presence of intervening sequences, confirmation of its product can often be obtained through the use of one of the well-established *in vitro* translation systems mentioned above in connection with mRNA fractionation (p. 388).

Fig. 14.5

Verification of a cloned plant chloroplast gene by
in vivo translation in *E. coli* minicells.
A 9·7-kb fragment of wheat chloroplast DNA
was inserted into the BamHI site of the cloning
plasmid pBR322 (*see Fig.* 10.15). *E. coli* minicells
containing either the hybrid plasmid or the
pBR322 as a control were allowed to synthesize
protein in the presence of [^{35}S]methionine and
extracts were electrophoresed in a SDS–polyacrylamide gel. The gel after electrophoresis was dried
and exposed to radiosensitive film to reveal bands
corresponding to the synthesized proteins. The
pBR322 translation products are seen in the left-
hand track and those of the hybrid plasmid in
the right. The doublet indicated by the arrow
corresponds to two slightly different forms of the
large subunit of ribulosebisphosphate
carboxylase—one probably a secondarily modified form of the other. Its identity was confirmed
by precipitation by a specific antibody. Several
other mitochondrial gene products also seem to
be encoded in the 9·7-kb DNA insert.
Photograph by courtesy of Dr Michael Saul.

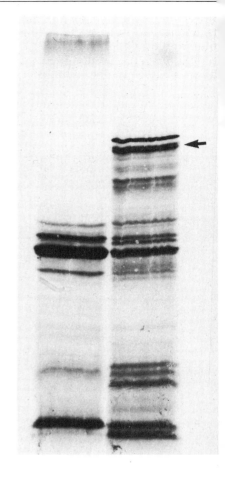

In one procedure, **hybrid selection**, the cloned DNA is used to isolate the mRNA
that matches it in base sequence, and this mRNA is then translated *in vitro* and the
polypeptide product characterized for size and by reaction with specific antiserum.
The DNA is converted to single-stranded form and chemically bound to specially
treated (diazobenzyloxymethylated) filter paper—DBM paper. The DBM paper
with the single-stranded DNA fixed to it can then be immersed in a solution of
total mRNA (isolated as poly(A)-tailed RNA) to allow hybridization to take place
between the DNA and any mRNA molecules of complementary base sequence.
After washing the paper to remove unbound RNA, the bound mRNA molecules
can be eluted and translated *in vitro*.

An alternative method, **hybrid arrest**, uses the cloned single-stranded DNA as an
inhibitor of the translation of the mRNA complementary to it. The translation
products of the total mRNA population (selectively precipitated with specific
antiserum when this is necessary to simplify the electrophoretic protein pattern)
are observed under two conditions of translation: (*a*) with the mRNA hybridized
with the DNA under test and (*b*) with the same hybridization carried out but with
the DNA–RNA hybrids melted apart by heating before the commencement of

translation. Under condition (*a*) just one polypeptide, that coded for by messenger taken out of commission by the hybridizing DNA, will be missing from the translation products. Condition (*b*) acts as a control, and is expected to allow translation of all the mRNA present. The results of one application of this hybrid arrest method are shown in *Fig.* 14.6.

Fig. 14.6

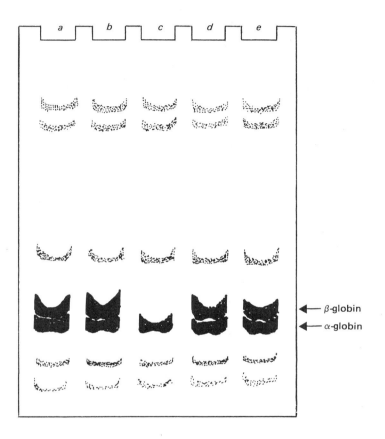

Identification of the β-globin coding sequence in a cDNA clone by the hybrid-arrest-of-translation method. cDNA synthesized on 9-S RNA from rabbit reticulocytes (*Fig.* 14.1) was inserted into the ColE1-derived plasmid pMB9 to give pβG1 (cf. pBR322, p. 274). The figure shows the radioactive polypeptide products of *in vitro* translation of reticulocyte mRNA, separated in a polyacrylamide–SDS gel and visualized by exposure to X-ray film. Before adding to the wheat-germ translation system the mRNA was allowed to hybridize with an excess of single-stranded DNA of pMB9 (*a, b*) or pβG1 (*c, d*). In (*b*) and (*d*) the DNA–RNA complexes were associated by heating immediately after hybridization. In (*e*) the mRNA was added without DNA. Note that α and β-globin are the predominant translation products and that translation of the β-globin messenger is arrested in (*c*), but not in (*a*), (*b*) or (*d*). Thus it is concluded that the cDNA sequence present in pβG1 (but not in the otherwise identical pMB9) hybridizes specifically with β-globin mRNA. Based on Paterson et al.[10]

Heteroduplex Mapping with mRNA or cDNA

A direct picture of the pattern of homology between cloned genomic DNA and the RNA transcribed from it can be obtained by observing DNA–RNA hybrid molecules with the electron microscope. This method, called heteroduplex mapping, shows intervening sequences, present in the DNA but not in the finally processed mRNA, as single-stranded loops flanked by double-stranded DNA–RNA heteroduplex. In one procedure, called **R-loop mapping**, the RNA is allowed to hybridize with double-stranded DNA in the presence of 70 per cent formamide. Under these conditions, DNA–RNA is favoured over DNA–DNA annealing, and the RNA is thus able to displace a loop or loops of DNA single strand from the duplex. Alternatively, the DNA can be first denatured to single-stranded form before hybridizing with the RNA. Pictures obtained by each of these methods are shown in *Fig.* 14.7.

DNA Base-sequence Determination

Now that rapid methods for DNA sequencing have been worked out, the most decisive way of confirming the nature of a cloned DNA segment is to show that its base sequence does indeed encode the amino acid sequence of the gene polypeptide product, when this is known. Sequencing can be started from any restriction endonuclease site within a restriction site map. *Box* 14.1 illustrates the method of A. Maxam and W. Gilbert, depending on specific chemical cleavage, and *Box* 14.2 shows F. Sanger's method, now coming increasingly into favour, which depends on specific interruption, by dideoxynucleotides, of enzymically catalysed synthesis. Either method, used in conjunction with the isolation and mapping of suitably overlapping restriction fragments, can generate tracts of sequence kilobases in length. Where the amino acid sequence of the gene product is known, it is found that an exactly corresponding sequence of codons (sometimes with intervening sequences) is present in the DNA of the gene.

Box 14.1*a*	Sequence determination of one end of IS1

The starting material was a single strand of a restriction fragment from R100 (*see Fig.* 8.3) overlapping one end of one of the IS1 sequences. The derived sequence:

5′ ATTAATTATAAATCG GGTGATGCTGCCAACTTACTGATTTAGTGTATGATGGTGTTTTT 3′

was from a fragment from a mutant plasmid with a deletion shown, by heteroduplex mapping, to end immediately adjacent to the IS1 sequence. The corresponding sequence from the normal plasmid was shown to be (in part):

5′ TTAAACAGGGAGGTGATGCTGCCAACTTACT 3′.

Thus the part of the sequence underlined is taken to belong to IS1.
For an explanation of the sequencing procedure *see Box* 14.1*b*.
Drawn from the sequencing autoradiograph of Ohtsubo and Ohtsubo.[32]

Box 14.1a contd

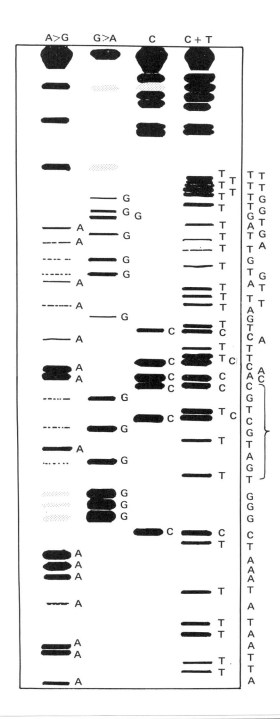

Fig. 14.7

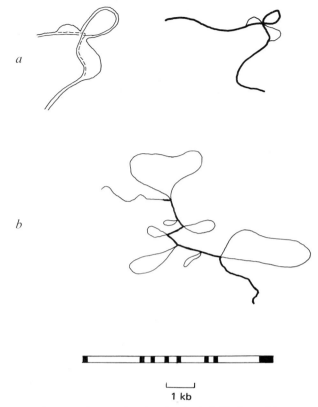

1 kb

Electron micrographs of molecular hybrids formed between mRNA and cloned
genomic DNA.
a. Double-stranded clone (in λgt) of rabbit β-globulin gene, hybridized with β-
globin mRNA. Under the conditions of hybridization used (70 per cent
formamide) RNA–DNA hybridization is more stable than DNA-DNA
hybridization, and the RNA displaces one of the DNA strands to
form what is called an 'R-loop'. Note the major intervening sequence which
interrupts the coding sequence of the gene and is not represented in the
mRNA. There is a small second intervening sequence which is hard to demonstrate by electron microscopy (*see Fig.* 15.9). In the interpretative diagram the
RNA is represented by a broken line. The staining of the nucleic acid strands
with cytochrome *c* was explained in Chapter 1 (p. 13).
From Tilghman et al.[11]
b. Single-stranded (heat-denatured) cloned chicken ovalbumin gene hybridized
with ovalbumin mRNA. Note *seven* intervening sequences. The diagram below
the tracing of the electron micrograph represents the consensus result from
measurments on ten hybrid molecules.
From Dugaiczyk et al.[12]

Box 14.1*b*   **Technique for the determination of nucleotide sequence in DNA (Maxam
and Gilbert[33])**

The DNA single strand to be sequenced is labelled at the 5′ end with [^{32}P]phosphate, and subjected to
four different chemical treatments. Each of these modifies one DNA base (or, in some cases, two bases)
in such a way that subsequent treatment can remove it from its sugar. The exposed sugar is then a weak
link in the DNA backbone and can easily be cleaved from its 3′ and 5′-phosphates. The first reaction,
the attack on specific bases, is allowed to proceed only for a limited time, so that only a small

roportion of the bases at risk are modified. The second step, cleavage of the chain at the positions of he modified bases, is taken to completion. The result is, in each reaction mixture, a family of adioactively labelled fragments each extending from the original 5' terminus to a position originally ccupied by the base (or one of the two bases) attacked by the specific reagent. The relative sizes of these ragments, as determined by electrophoresis in polyacrylamide gel followed by exposure to X-ray film, ndicate the distribution of the specific base(s) along the chain; a base cleaved near the 5' end will give a mall fast-running fragment, and one cleaved far from the 5' end will give a large slow-running one.

The four specific treatments are as follows:
1. Purine methylation with dimethyl sulphate, followed by cleavage of the chain at the adenylate residues (and to a lesser extent at the guanylate residues) by treatment with hot alkali.
2. Purine methylation as in (1) followed by cleavage at the guanylate residues (and to a much lesser extent at the adenylate residues) by heating in piperidine (an organic base).
3. Pyrimidine ring cleavage with hydrazine followed by chain cleavage by heating in piperidine. This cleaves equally well at thymidylate and cytidylate residues.
4. Cytosine ring cleavage with hydrazine in the presence of 2 M NaCl, followed by chain cleavage as in (3). This cleaves only at cytidylate.

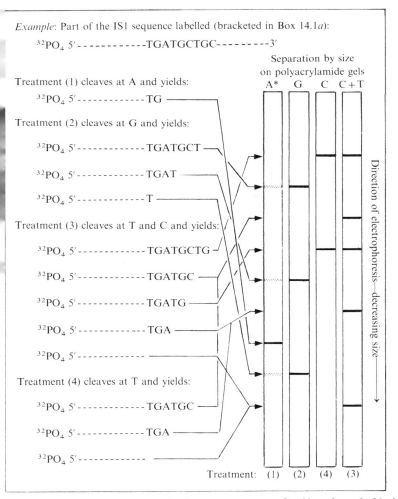

Example: Part of the IS1 sequence labelled (bracketed in Box 14.1*a*):

* Treatment (1) also cleaves to some extent at G residues—hence the faint bands corresponding to G in the treatment (1) lane.

Box 14.2   Chain terminator DNA sequencing (Sanger et al.[34])

The DNA fragment to be sequenced is cloned into a suitable vector—usually a derivative of the single-stranded DNA *E. coli* phage M13. A sequence complementary to the fragment is synthesized starting from a 'primer' complementary to the phage vector in a position adjacent to the cloning site. The synthesis is terminated with a certain probability at each occurrence of a specific base by incorporation into the polymerase reaction mixture of the corresponding dideoxynucleoside triphosphate. This analogue can be incorporated (with low efficiency) into the growing chain in place of the normal deoxynucleotide, but then prevents further elongation because of its lack of a 3'-OH.

Four polymerase reactions are run in parallel, each one with a different dideoxynucleoside triphosphate. The four sets of terminated chains, which are radioactive through the inclusion of [^{32}P]dATP in each mixture, are ranked in order of size by gel electrophoresis followed by autoradiography (cf. *Box* 14.1).

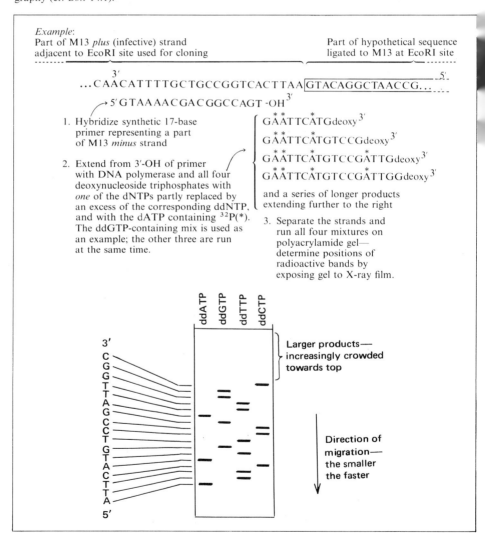

Example:
Part of M13 *plus* (infective) strand adjacent to EcoRI site used for cloning

Part of hypothetical sequence ligated to M13 at EcoRI site

3'   5'.
...CAACATTTTGCTGCCGGTCACTTAA GTACAGGCTAACCG...

5' GTAAAACGACGGCCAGT -OH$^{3'}$

1. Hybridize synthetic 17-base primer representing a part of M13 *minus* strand

2. Extend from 3'-OH of primer with DNA polymerase and all four deoxynucleoside triphosphates with *one* of the dNTPs partly replaced by an excess of the corresponding ddNTP, and with the dATP containing ^{32}P(*). The ddGTP-containing mix is used as an example; the other three are run at the same time.

GÅÅTCÅTGdeoxy$^{3'}$
GÅÅTCÅTGTCÅGdeoxy$^{3'}$
GÅÅTCÅTGTCÅGÅTTGdeoxy$^{3'}$
GÅÅTCÅTGTCÅGÅTTGGdeoxy$^{3'}$

and a series of longer products extending further to the right

3. Separate the strands and run all four mixtures on polyacrylamide gel—determine positions of radioactive bands by exposing gel to X-ray film.

ddATP ddGTP ddTTP ddCTP

3'
C
G
G
T
T
A
G
C
C
T
G
T
A
C
T
T
A
5'

Larger products—increasingly crowded towards top

Direction of migration—the smaller the faster

14.7 | New information from DNA sequencing

Confirmation of the Protein Code; Codon Usage

Comparisons of the base sequences of genes with the independently determined amino acid sequences of their polypeptide translation products provide direct evidence for the correctness of the genetic code, which was deduced in the first instance from experiments using simple synthetic nucleic acid sequences. They also show how frequently each of the alternative codons which exist for most amino acids is actually used in different genes and different organisms. The examples shown in *Table* 14.1 illustrate two general points: (*a*) all codons are in

Table 14.1	**Examples of preferential codon usage in different genes and different organisms** Strongly preferred codons are in heavy type						
	E. coli lacI *repressor*[13]	*Saccharomyces** 1	2	3	4	*Sea urchin All histone genes*[18]	*Mouse Ig heavy chain*[19]
Leucine							
UUA	5	0	2	1	1	0	0
UUG	5	**21**	**18**	5	3	5	2
CUU	2	0	2	1	0	8	2
CUC	3	0	0	0	0	14	8
CUA	1	0	2	1	0	4	2
CUG	**25**	0	0	0	0	11	9
Serine							
UCU	7	**13**	**14**	2	2	8	5
UCC	6	**12**	**12**	0	1	5	10
UCA	3	0	3	1	1	4	7
UCG	5	0	0	1	0	2	0
AGU	5	0	2	2	2	4	6
AGC	6	0	0	0	0	9	14
Isoleucine							
AUU	10	**9**	**14**	2	3	4	2
AUC	7	**11**	**16**	2	1	**25**	**12**
AUA	1	0	0	0	1	1	1
Glycine							
GGU	6	**25**	**29**	8	8	13	5
GGC	11	0	0	2	1	16	3
GGA	1	0	0	0	2	13	5
GGG	4	0	0	2	1	8	4

*Yeast genes coding for:
1. glyceraldehyde dehydrogenase;[14]
2. actin;[15]
3. iso-1 cytochrome *c*;[16]
4. iso-2 cytochrome *c*.[17]

fact used and (*b*) codon usage can be very strongly biased, the nature of the bias differing from one organism to another and sometimes even from one gene to another within the same organism. For *E. coli* there is evidence that the most commonly used codons are those served by the most abundant kinds of tRNA. In yeast there is a hint that codon usage is more strongly biased in abundantly translated genes (e.g. for actin and glyceraldehyde phosphate dehydrogenase) than in a more sparingly translated one (e.g. for iso-2 cytochrome *c*).

In one case, DNA sequence information has allowed the extension and correction of the previously accepted code for amino acids. Sequencing of mitochondrial DNA, both from yeast and humans, made it clear that in the mitochondrial genome, unlike the nuclear genome, UGA is *not* a chain terminator but (like UGG) codes for tryptophan.

Open Reading Frames

Even where there is no prior knowledge of the function of a cloned and sequenced segment of DNA, it is still possible to see whether it is likely to code for a protein. The essential requirement is an initiation sequence (5'-ATG-3') followed by a long succession of amino acid codons unbroken by a termination codon. Since (in nuclear genes) three out of the sixty-four possible sequences of three bases constitute chain termination signals, the chance that a random sequence of $3n$ bases, read off in threes from an initiation codon, will *not* include a terminator will be $(61/64)^n$—assuming 50 per cent (G + C), which is usually not so far from the truth as to spoil the calculation. This chance is about 8 per cent for $n = 50$ and 0·6 per cent for $n = 100$. So a long sequence of a few hundred amino acid codons suggests strongly that the DNA has been selected for a protein-coding function. Moreover, it is possible in such cases to predict the amino acid sequence of the polypeptide coded for (or at least the N-terminal sequence if the coding sequence is interrupted by introns), and to identify the polypeptide should it be found. We note below (*Box* 14.3) a case where the application of this kind of reasoning led to the prediction of a 'leader' polypeptide in *E. coli*.

Sequences Controlling Transcription in Prokaryotes

Signal Sequences in the *lac* Operon

One of the most thoroughly analysed parts of the *E. coli* genome is that including the *lac* operon and the closely linked *lacI* regulator gene. The entire DNA sequence from the transcription–termination ('downstream') end of *lacI* to beyond the transcription start of the *lacZ, Y, A* operon has been determined (*Fig.* 14.8). This sequence includes the genetically characterized promoter and operator segments (compare *Fig.* 13.10).

Several lines of evidence have been brought to bear in the elucidation of the functions of the different parts of this sequence. It was shown that each of the three proteins known to be involved in the transcription and the regulation of transcription of the operon, i.e. the *lacI* repressor product, the cAMP-binding protein (CRP) and RNA polymerase, bind with high affinity to a particular part of the sequence. The DNA segment bound by each of these three proteins could be identified because, when so bound, it was protected from digestion by DNAase. The evidence defining the repressor binding sequence is shown in *Fig.* 14.9. The

Fig. 14.8

DNA sequence of the *lac* transcription control region of *E. coli*—mutations and functions.
Data from Rosenberg and Court[20] and Goeddel et al.[21]

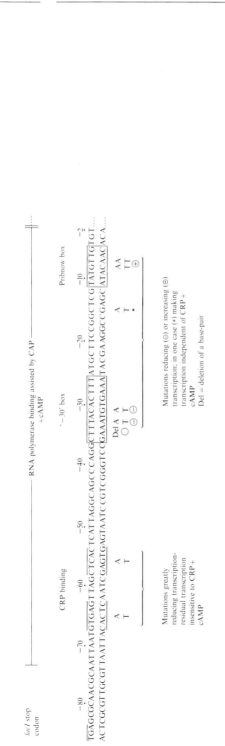

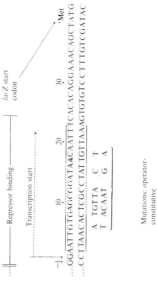

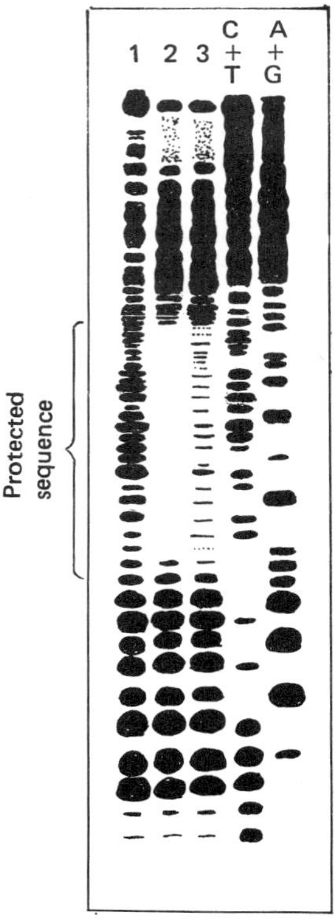

Fig. 14.9   The 'footprinting' method used to define the section of the *E. coli lac* promoter–operator region that binds to *lac* repressor protein. A DNA fragment including the operator–promoter region was purified from a plasmid clone, labelled at one 5′ end with ^{32}P and partially digested with DNAase I under three different conditions: (1) in absence of repressor protein; (2) in presence of repressor protein; (3) in presence of repressor protein plus an inducer (isopropyl-β-thiogalactoside) of *lac* transcription. The partial digests were run down an agarose electrophoretic gel in parallel with similarly end-labelled samples which had been partially degraded by the Maxam and Gilbert reactions 1 and 3 (*see Box* 14.1) so as to cleave with a certain probability at every purine (A + G) or at every pyrimidine (C + T). The repressor (track 2) protects a 27-base sequence from DNAase attack, so that no digestion products terminate within this sequence. The inducer partly counteracts the protective effect of repressor (it would counteract it completely if added at a higher concentration). The method is of general applicability in the investigation of protein-binding DNA sequences.
This figure is drawn from the photograph of Galas and Schmitz.[35]

technique, known as 'footprinting', is of general application in the investigation of DNA–protein interactions. The method for the identification of the starting point for transcription depends on the use of the enzyme S1 nuclease, which degrades single-stranded DNA or RNA but not DNA–RNA hybrid duplexes. A DNA restriction fragment, known to overlap a transcribed region and to include the transcription start, is converted to single-stranded form, annealed with an excess of messenger RNA and digested with S1 nuclease. The size of the surviving hybrid duplex, which can be determined on an electrophoretic gel, defines the distance from the transcription starting point to the downstream restriction site.

Another indication of the importance of particular sequences within the *lac* promoter–operator region is the presence of these sequences, or sequences very much like them, in the corresponding positions upstream of the transcription starting points of a number of other *E. coli* and bacteriophage genes. *Table* 14.2 lists some of these. The sequences are not absolutely the same from one example to another, but there are two regions where there is a reasonable consensus, one about ten and the other about thirty bases 'upstream' of the transcription start. The 'minus ten' sequence, boxed in the figure, is called the 'Pribnow box' after D. Pribnow[13] who first noticed it. The *lac* and *gal* operons also share a feature around sixty bases upstream of the transcription start marked by an interrupted three base-pair inverted repeat flanked by non-inverted T/A base-pairs:

TGTG .... CACT

ACAC .... GTGA

This appears to be the binding site for the cAMP-binding protein (CRP) which is essential for efficient transcription of the *lac*, *gal* and *ara* operons. Curiously, the *ara* promoter region does not show this feature, and so the sequence requirement for CRP binding is still not clear.

Confirmation of the significance of the 'consensus' sequences in the RNA polymerase binding (promoter) regions of the *lac* and other operons comes from the demonstration that a number of mutations that affect promoter function fall within or close to these sequences. Several of these are indicated in *Fig.* 14.8. The base-pairs concerned are shown by this mutation analysis to be of crucial importance for normal DNA–protein interaction in initiation and control of transcription, and their clustering within the regions already identified by the criteria of protein binding and 'consensus' puts the interpretation of the structure and function of the whole promoter region on a sound footing.

Sequences Controlling Transcription in Eukaryotes

Information on the base sequences upstream of transcription starting points is less abundant in eukaryotes. The picture is likely to be more complex than in prokaryotes, inasmuch as eukaryotes have three kinds of RNA polymerase, I, II, III, transcribing, respectively, rRNA, protein-coding genes, and tRNA and 5-S RNA. There are already strong indications that the DNA sequences recognized by these three classes of polymerase are different. Most

Table 14.2	Sequences common to several *E. coli* and bacteriophage DNA regions upstream of transcription starts					
Gene or operon	*CRP site*		*'−30' sequence*		*Pribnow box*	
lac	−68 TGTG	−58 CACT	−35 TTTACACTTT	−26	−11 TATGTT	−6
gal	−69 TGTG	−55 CACT	−39 TGTCACACTTT	−29	−12 TATGCT	−7
ara BAD	?		−39 CCTGACGCTTT	−29	−14 TACTGT	−9
trp	—		−36 GTTGACAATTA	−27	−13 TTAACT	−8
*λ*P$_R$	—		−36 TGTTGACTATT	−27	−12 GATAAT	−7
T7 early promoter	—		−36 GGTTGACAACA	−27	TACGAT	−8
'Consensus'*			TGTTGAC A - TT		TATAAT	

* Large capitals, strong consensus; small capitals, weak consensus; dash, no consensus. Based on a larger sample than shown here—*see* Rosenberg and Court.[20]

Fig. 14.10

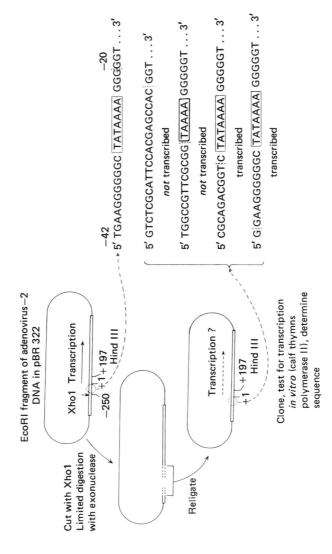

EcoRI fragment of adenovirus−2
DNA in pBR 322

Cut with Xho1
Limited digestion
with exonuclease

Xho1 Transcription

−250 +1 +197
Hind III

Religate

Transcription ?

+1 +197
Hind III

Clone, test for transcription
in vitro (calf thymns
polymerase II), determine
sequence

−42
5′ TGAAGGGGGGGC TATAAAA GGGGGT . . . 3′
                                           −20

5′ GTCTCGCATTCCACGAGCCAC GGT . . . 3′
not transcribed

5′ TGGCCGTTCGCGG TAAAA GGGGGT . . . 3′
not transcribed

5′ CGCAGACGGT c TATAAAA GGGGGT . . . 3′
transcribed

5′ G GAAGGGGGGGC TATAAAA GGGGGT . . . 3′
transcribed

Evidence from deletions produced by *in vitro* manipulation that the 'TATA box' is essential for transcription from an adenovirus-2 promoter.
Data of Corden et al.[23]

information is available about promoter regions for the action of polymerase II in animals.

A rather general feature of animal protein-coding genes is the presence, 20–30 bases upstream (5') of the translation initiation and mRNA 'capping' site, of a sequence of the form

$$5'\ TATA\begin{smallmatrix}T\\A\end{smallmatrix}A\begin{smallmatrix}T\\A\end{smallmatrix}\ 3',$$

the fifth and seventh positions having the common alternative bases indicated.[22] This 'consensus' sequence (the 'TATA box') has an obvious resemblance to the Pribnow box of prokaryotes, and may have a similar function as part of the polymerase-binding region.

What is lacking in the investigation of animal as opposed to bacterial genes is any ready means of inducing and selecting *in vivo* mutations affecting particular promoter functions. This handicap is now being overcome by the *in vitro* manipulation of cloned DNA sequences to introduce mutations (mostly deletions so far) in predetermined positions.

The procedure used by P. Chambon and his colleagues[23] is outlined in *Fig.* 14.10. By removal with exonuclease of sequences of different length, starting from a site some way upstream of one of the transcription initiation points in the DNA of an animal virus (adenovirus-2), they were able to show that the TATA sequence was essential for *in vitro* transcription from this starting point by RNA polymerase II.

Other short sequences (CCAAT or variants thereof), located from 70 to 90 bases upstream of transcription starting points, may also be of importance, since they are each common to several, though not all, animal and animal virus genes. The less-than-universal occurrence of these sequences suggests that they may be concerned with special rather than general expression, but little can yet be said about them. Further evidence, again derived from 'engineered' deletions, has shown that transcription can sometimes depend on sequences hundreds of bases upstream of the transcription start. It may be that these long-range effects are primarily on chromatin structure and only secondarily on binding of RNA polymerase or proteins immediately involved in the control of transcription.

Termination signals

Transcription termination sites in prokaryotes appear to be of two types, one dependent on a transcription–termination protein called rho (ρ) and the other not. The sites of action of ρ are characterized by the sequence 5'-CAATCAA-3', or a near relative of it, on the transcribed DNA strand.[31] ρ-Independent termination sites usually have fairly extensive inverted repeat sequences, capable in principle of forming stem-loop structures. Termination tends to occur about 20 bases downstream of the centre of the hypothetical loop, immediately following a run of A residues in the transcribed DNA strand. The structure shown in *Box.* 14.3*a* is typical.

In eukaryotes, terminator regions presumably require signal sequences not only for termination but also for the addition of poly(A) to the 3' termini of the messenger precursors. The sequence 5'-AAUAAA-3' has been found immediately to the 5' side of the poly(A) addition site in four animal messengers—for rabbit α and β-globins, for a mouse immunoglobin and for chicken ovalbumin.[30] The same sequence has subsequently been found in the corresponding position in adenovirus messenger and it will be interesting to see whether it is a general feature of the 3' termini of mRNA in all eukaryotes.

DNA Sequence and Conformation

Interesting possibilities concerning the possible role of specific DNA sequences in the control of transcription have been suggested by the recent discovery that sequences containing runs of alternating G and C residues are able to form *left*-handed double helix rather than the normal right-handed double helix (*Fig.* 1.3).[35] Left-handed duplex, called **Z-DNA**, is more stable at high salt concentrations or when the cytosine residues are methylated. One can imagine that the amount of Z-DNA in (G + C)-rich regions of chromatin might be controlled by binding or removal of proteins[37] with specific affinity for Z-DNA, or by methylation or demethylation. The stabilization of left-handed turns in a DNA domain not free to change its overall degree of winding (because it was in some way pinned at its ends, cf. *Fig.* 1.12c) would have to be compensated for by a loss of left-handed (negative) supercoiling (stabilization of stem-loop structure would have the same effect). Conversely, destabilization of Z-DNA will promote negative supercoiling, and the consequent torsional stress will tend to unwind the more weakly bonded [(A + T)-rich] regions of the primary duplex and make promoter sequences more accessible to RNA polymerase. In short, a transcription off–on switch could operate through stabilization–destabilization of Z-DNA. An attractive feature of this as yet speculative model is that it suggests a possible explanation for the negative relationship between methylation and transcriptional activity (*see* p. 512).

Left-handed helix was first discovered in synthetic DNA in solution, but it also occurs *in vivo*—or at least in isolated *Drosophila* polytene chromosomes.[36] Antibodies directed against Z-DNA and made fluorescent by chemical labelling have been shown to bind to the interband regions of the chromosomes but not to the bands, where the genes appear to be located (*see* p. 431). How, if at all, this fits with the model referred to in the previous paragraph remains to be seen.

Box 14.3   The leader sequence of the mRNA transcribed from the *E coli trp* operon

```
      5'          10         20         30         40         50
      pppAAGUUCACGUAAAAAGGGU AUCGACAAUG AAAGCAAUUUUCGUACUGAAAGG-

            60         70         80         90        100
      UUGGUGGCGCACUUCCUGAAACGGGCA GUGUAUUCACCAUGCGUAAAGCAAUC-

         110        120        130        140        150       160
      AGAUACCCAGCCCGCCUAAUGAGCGGGCUUUUUUUUUGAACAAAAUUAGAGAAUAA A

            170        180
      CAAUGCAAACACAAAAACCGACU........3'

      Met  Gln  Thr  Gln  Lys  Pro  Thr      ....(trpE polypeptide sequence)
```

Specific signals encoded in the primary structure are underlined: (i) probable ribosome-binding (Shine–Delgarno) sequences at 15–19 and 152–155; (ii) polypeptide chain initiation codons at 27–29 and 163–165; (iii) two consecutive tryptophan (UGG) codons in phase with the first initiation codon at 54–59; (iv) chain termination codon UGA in the same reading frame as the two UGG codons; (v) start of coding sequence for the known amino acid sequence of the *trpE* polypeptide at 163 onwards.

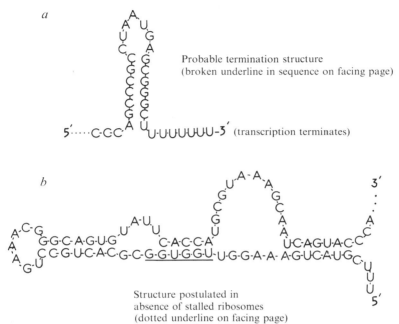

a

Probable termination structure
(broken underline in sequence on facing page)

(transcription terminates)

b

Structure postulated in
absence of stalled ribosomes
(dotted underline on facing page)
Protects terminator formation
and thus promotes termination

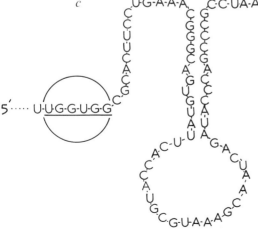

c

The 'pre-emptor' structure (overlined on facing page) postulated in presence of stalled ribosomes at the tryptophan codons. Terminator, as in (*a*), cannot be formed and transcription proceeds.

Sequence from Lee et al.[25]
Model after Keller and Calvo[26]

Attenuation Regulation in Bacterial Operons

As well as some DNA sequences that allow or prevent transcription initiation, there are others which exercise their control *after* transcription has been initiated. These latter were first identified in *E. coli*, and a good example is provided by the mRNA of the *trp* operon (cf. *Fig. 13.13*, p. 377). *Box* 14.3 shows the first 180 bases of the *trp* messenger from the 5′ end. This sequence was first determined in mRNA directly, and was afterwards confirmed by sequencing of the DNA from which it is transcribed.

A feature of the sequence, surprising at first sight, is the presence of a 162-nucleotide 'leader' preceding (upstream of) the initiation codon for anthranilate synthetase, the product of *trpE*, the first gene of the operon. The coding sequence for this enzyme is identified unambiguously by its perfect agreement with the independently determined amino acid sequence. It is preceded, 8–11 bases upstream of the initiating codon, by 5′-AGAGA-3′, which is one of a family of sequences first recognized by J. Shine and L. Delgarno[24] as present in roughly corresponding positions before the translation starting points of several *E. coli* genes. These sequences are capable of base-pairing, with at most one or two mismatches in each case, with 5′ ...ACCUCCUUA-OH 3′, the 3′-terminal sequence of *E. coli* small ribosomal subunit RNA (16-S RNA). This postulated base-pairing is thought to be important in mRNA binding to ribosomes.

Looking more closely at the leader sequence, one can see, starting at residue 27, a possible initiation codon (AUG) followed by an open reading frame extending for a further 14 codons. The AUG is preceded, at exactly the usual spacing, by another Shine–Delgarno sequence, this time 5′-AGGGU-3′. These features suggest quite strongly that the leader sequence codes for a 15-residue polypeptide with the most unusual feature, considering the relative scarcity of tryptophan in proteins, of two successive tryptophan residues (UGGUGG in terms of codons).

The significance of these tryptophan codons is reinforced by the results of analyses of leader sequences of a number of other *E. coli* operons concerned with amino acid synthesis. In each case one finds a number of consecutive codons for the amino acid that is the end-product of the pathway catalysed by the enzymes encoded in the operon. To take just three examples, the phenylalanine operon has a leader sequence coding for a polypeptide with three consecutive phenylalanine residues, while the leucine and histidine operons have leaders coding for peptides with four consecutive leucines and eight consecutive histidines, respectively.

The functional significance of the leader peptides is indicated by detailed studies, mainly by Yanofsky's group at Stanford University, on the transcription of the *trp* operon. Under conditions in which tryptophan is in adequate supply, transcription is terminated about 147 bases after initiation, so that a transcript of the *trpE* gene and the other genes of the operon is not made. The feature of the leader mRNA that appears to be responsible for the early termination is a sequence between bases 114 and 134 which is capable of forming a hairpin loop, with the bases of the double-stranded stem perfectly paired, and including six consecutive GC pairs which are expected to form an especially stable duplex segment. This hairpin structure is thought to block further transcription if it is allowed to form. Whether it forms or not depends, according to a plausible hypothesis, on whether or not a second hairpin structure is formed a little way upstream; this preceding hairpin, if formed, pre-empts the secondary structure and makes the formation of the terminating hairpin impossible, as shown in *Box* 14.3c. The pre-emptor structure, so the hypothesis states, can itself be pre-empted by another hairpin, overlapping with it but not with the terminator structure, which includes the two adjacent tryptophan codons. Under conditions of tryptophan starvation ribosomes supposedly become stalled at these codons for want of the tryptophanyl-tRNA which they need in order to proceed. Accordingly, these codons will be prevented from forming secondary structure, the pre-emptor hairpin will be free to form, the terminator structure will *not* be formed, and transcription will proceed into the enzyme-coding sequences. The whole scheme seems too neat not to be true, especially since the leader sequences of the other operons mentioned above, which show analogous runs of identical amino acid codons, also show potential hairpin-forming sequences in the appropriate places.

The leader sequences, in summary, seem to have the function of attenuating transcription in a controlled way—allowing it to proceed when the amino acid product of the operon is in short supply and ribosomes tend to become stalled at the repeated codons for that amino acid, and terminating it when there is enough of the amino acid in the cell to allow the ribosomes to move on.

It should be noted that the type of transcriptional regulation just discussed is not expected to exist in eukaryotes since it depends on the coupling of transcription and translation, processes which in eukaryotes occur on different sides of the nuclear envelope. However, one lesson from prokaryote leader sequences, that is the importance of secondary structure in mRNA, may well be important for eukaryotes too.

Signals for Excision of Intervening Sequences

The intervening sequences, or introns, which are so often found interrupting polypeptide-coding DNA sequences in animals[22] are, in fact, transcribed. They are represented in that major fraction of intranuclear RNA (heterogeneous nuclear RNA—hnRNA) which is rapidly synthesized and degraded within the nucleus and never exported, as such, to the cytoplasm. A proportion of the hnRNA, probably a different fraction in different cells and tissues, is processed to form mRNA. The processing involves 5′ 'capping', 3′ poly(A) 'tailing' and, most intriguingly, the removal of the segments transcribed from introns. The selection of RNA transcripts for processing is a second way at which gene action could be regulated (the first being the regulation of transcription itself), but as yet we have little idea as to how such selection might operate.

Examination of the DNA base sequences of split genes does provide some clues to possible mechanisms of intron removal. *Table* 14.3 lists a representative sample of the many sequences which have been determined across the junctions between introns and coding regions. The sequences in the interior of each intron are not listed, although they are now known in many cases; they display no regularities and are highly divergent even between obviously closely related genes such as those for mouse and rabbit β-globin. One hesitates to say that any piece of macro-molecular structure is without specificity, but it does seem that the nucleotide sequences in the interiors of introns hardly matter.

The boundary sequences, in contrast, show at least a fair approximation to a 'consensus' among split genes from a wide variety of animals. In terms of the RNA gene transcript, the sequences GU at the 5′ end of the intron and AG at the 3′ end are almost invariant, and the sequence CAGG overlapping each of the two intron ends is also a rather constant feature. This direct repeat means that, although the removed intron is of constant length, the precise cutting-and-splicing points can be

Table 14.3	Boundary sequences between introns and coding regions of various genes, and their relationship to part of the sequence of an abundant small intranuclear RNA molecule (abstracted from Lerner et al.[27])

Gene	5' Coding sequence / Intron / Coding sequence 3'
Mouse λ immunoglobulin chain	UUG'GUGAGA .... UCUCUCCACAG'U
κ immunoglobulin chain	AAG'GUUAAA .... UCCACUCCUAG'G
Mouse β-globin	CAG'GUUGGU .... CAUUUUCUCAG'G
Human β-globin (intron 2)	AAG'GUAGGC .... GUUUGCUCUAG'A
(intron 1)	CAG'GUUGGU .... CCCACCCUUAG'G
Human δ-globin	AAG'GUGAGC .... UUCAAUUACAG'G
Chicken ovalbumin (intron A)	CAG'GUACAG............ UUUCUAUUCAG'U
(intron B)	CCA'GUAAGU .... UUGCUUUACAG'G
(intron C)	AUG'GUAAGG............ CAUUCUUAAAG'G
(intron D)	GAG'GUAUAU............ UGGUUCUCCAG'C
(intron E)	CAG'GUAUGG............ UUUCCUUGCAG'C
(intron F)	AAG'GUACCU............ UUUUAUUUCAG'G
Silkworm fibroin	CAG'GUGAGU............ UUUUGUUUCAG'U
'Consensus' sequence	A U AG'GUAAGU............UUUU UUXCAG'G C C

5' end of U1 RNA: m³GppppmAmUACUUACCUGGCAGGGAGAUA...3'
Possible alignment of U1 sequence with intron ends:

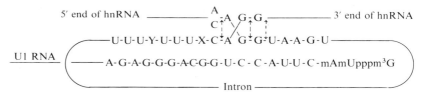

Note: m indicates a methylated base; Y indicates a pyrimidine (either U or C)—note that G can form a hydrogen-bonded pair with either C or U in RNA; X indicates a position at which there is no clear consensus. Near the hypothetical 'crossover' point various alternative positions of cutting/splicing are possible (↕).

imagined as variable over this four-nucleotide sequence without any variation in the structure of the spliced product.

If the base sequences at the ends of introns are essential for correct RNA processing, loss of gene activity should follow from certain intron mutations as well as from mutations within coding regions. There is already evidence that this is so. In one case,[31] an inherited deficiency in human β-globin (β-thalassaemia) was traced to mutation in the first intron, 21 nucleotides from its 3' end, changing 5'-CTATTGG-3' to 5'-CTATTAG-3'. The mutant sequence has the AG invariably found at the 3' ends of introns, and has a fair degree of resemblance to the true 3' end of this particular intron (5'-CCCTTAG-3', see Table 14.3). The result is that the mutant intron is, to a major extent, incorrectly spliced at this 'false' terminus, leaving the true 3' end in the finished messenger, where it effectively disrupts the reading of the coding sequence.

The highly conserved features of the junctions between introns and coding sequences suggest that there may be a universal splicing mechanism. Table 14.3 illustrates a recent conjecture as to the nature of this mechanism. Michael Lerner, Joan Steitz and their colleagues[27] have identified, among the small nuclear RNA molecules of a wide range of mammals and birds, an abundant molecule, which they call U1, which shows a striking base-sequence complementarity with the 'consensus' sequence at intron splicing junctions. As the diagram in the table shows, this U1 sequence can be envisaged as holding the two ends of the intron together in the correct orientation for cutting and splicing. The base-pairing would hardly be strong enough by itself, but the association between U1 and the intron sequences is thought to be stabilized by specific proteins, much as codon–anticodon associations are stabilized by ribosomes. The U1 sequence is indeed found bound to a specific set of proteins in a ribonucleoprotein particle sedimenting at 10 S, and the analogy with a ribosome subunit seems to be quite strong.

These observations encourage the hope that the mechanism responsible for the splicing together of fragmented coding sequences in hnRNA will soon be understood. The system of control that determines which transcripts will be processed to form functional messenger and which will not is now a main focus of interest.

How general are intervening sequences in eukaryotic genes? We consider the situation in birds and mammals in the next chapter. At present the best guess appears to be that most genes of vertebrates have them. They are known to be present in at least some *Drosophila melanogaster* genes, but in view of the complexity of *Drosophila* organization in relation to its rather modestly sized genome one would doubt that it had DNA to spare for many such extravagances. In *Saccharomyces*, which is one of the simplest of eukaryotes, it seems likely, on the basis of several genes which have already been cloned and fully sequenced, that introns are the exception rather than the rule.

Different Modes of Processing a Single Gene Transcript

Our final example of new insights gained from sequence determination is the discovery that the same gene transcript can be processed in different ways to serve different purposes. In the next chapter we see how different modes of RNA splicing can make the most of the limited coding capacity of some very small mammalian virus genomes. In animal cells similar versatility in RNA processing is associated, not with DNA limitation, but with the complexity of the functions which gene products have to fulfil.

One of the best studied cases concerns the various mRNAs coding for α-amylase in various mouse tissues.[28] Determination of the mRNA nucleotide sequences by analysis, in this case, of the RNA itself (rather than the DNA as in other examples used in this chapter) showed that the different messengers were identical except for the relatively short non-coding (leader) regions at their 5' ends. The major α-amylase mRNA found abundantly in liver had a 150-base leader sequence which was replaced in the mRNA species found in salivary glands by a totally different leader of only 48 nucleotides (including the 'cap' residue in each case).

It seems likely that the two kinds of α-amylase mRNA are generated by different ways of cutting and splicing the primary transcript of a single gene. The fact that both the coding and 3' non-coding regions of the two messengers are identical makes it seem rather unlikely that two different genes are involved. If the single gene hypothesis is correct, it may be that the significance of the two different leader sequences is that they contain differently controlled transcription initiation sequences, making it possible for the expression of the gene to be regulated in ways appropriate to the two tissues.

Another example of alternative modes of RNA processing, substantiated by both amino acid sequencing of the gene product and base sequencing of the genic DNA, explains the synthesis of two different forms of immunoglobin M in the mouse. The complex subject of immunoglobulin gene rearrangement is dealt with in Chapter 17. Here we need note only that the secreted (soluble) and membrane-bound forms of the M class of antibodies are due to two different forms of the μ polypeptide which is the heavy chain characteristic of this class. The membrane-bound type of μ chain differs from the soluble type only in having an extra C-terminal 'tail' of hydrophobic amino acids. The coding sequence for this tail is either included in the mRNA or not depending on the mode of cutting/splicing of the gene transcript.

Note that the variable splicing in this last case affects the coding sequence at the 3′ end of the mRNA rather than, as in the α-amylase example, the leader sequence at the 5′ end. The two kinds of protein product are used for different functions in the same cell rather than for the same function in different cell types in different tissues. In both cases, the versatility in splicing allows one gene to be used in two different ways.

14.8   The total synthesis of genes

With advances in the organic chemistry of deoxyribonucleotides and their derivatives, it has become possible to synthesize whole protein-encoding genes from their deoxyribonucleotide constituents. Apart from being a feat, such a total synthesis can have practical utility if the natural gene is hard to come by because of lack of a nucleic acid probe. Provided the amino acid sequence of the protein product is known, a functional gene (not necessarily the same in base sequence as the natural one) can in principle be synthesized.

The prime example at the time of writing is the total synthesis of an artificial gene, 514 base-pairs long, coding for the known 166-amino acid sequence of human interferon α, a protein conferring resistance to virus infections of various kinds. The synthetic sequence was ligated into the BamHI restriction site of a plasmid of the same general type as pBR322 (cf. p. 274). The plasmid had a built-in *E. coli lac* promoter sequence immediately adjoining its BamHI site, in correct orientation for transcription of the inserted sequence. When the plasmid, with the artificial interferon gene, was introduced into *E. coli*, interferon was synthesized in small but detectable amounts.

For details of the organic chemistry involved the interested reader is referred to the original paper.[29]

14.9   Summary and perspectives

This chapter has surveyed the new methodology that allows us to clone and determine the base sequence of virtually any gene for which a probe can be obtained—that is, in practice, any gene that is abundantly transcribed into RNA or for which a part of the sequence can be deduced from the protein product.

As a result of these new techniques, a number of features of genes which had been deduced from more traditional genetic analysis, i.e. the triplet character of the protein code, the presence of special sequences signalling initiation and termination of transcription and translation, have been confirmed and defined at the level of chemical structure. In addition some features which had not been previously suspected have come to light. In prokaryotes, the regulatory role of 'leader' sequences in mRNA has been deduced. In eukaryotes the major surprise has been the interruption of many coding sequences for single polypeptide chains by extensive non-coding intervening sequences (introns). No indication at all of the presence of introns had been obtained from conventional genetic analysis, yet they account, at least in animals, for a larger amount of DNA than do the coding parts of the genes.

In spite of the spectacular progress made through DNA sequencing, many questions relating to the control of gene transcription remain unanswered. It increasingly seems likely that an important, perhaps the *most* important, factor in such control is the local structure of the chromatin—the state of the nucleosomes and the degree of DNA supercoiling (*see* Chapter 17, p. 507). Much remains to be learnt about how chromatin structure relates to DNA sequence, and this remains one of the most important questions for future research.

In evaluating the roles of various recognizable sequences found in DNA it is a great help to have defined mutations in the regions under study. We have seen how useful a combination of genetically defined mutation and chemically defined sequence has been in, for example, the definition of key sequences in bacterial operons. We have seen two examples (pp. 341 and 408) of how chemical manipulation of cloned natural genes can yield defined mutations at predetermined sites. Indeed, methods of total chemical synthesis of DNA have now advanced to the point where a team of experienced chemists can synthesize the gene of their choice from deoxynucleotides without having any natural DNA as starting material.

The totally synthetic gene is still a novelty but it is likely to become much less so in the coming years. The precedent set by the synthesis of a gene for human interferon is likely to be followed by many other examples.

Chapter 14 | Selected Further Reading

The special issue of *Science* devoted to *Recombinant DNA* (Sept. 19, 1980; Vol. 209, No. 4663) contains important articles covering and extending the content of this chapter.

Maniatis T., Fritsch E. F., Lauer J. et al. (1980) The molecular genetics of human hemoglobins. *Annu. Rev. Genet.* **14**, 145–178.

Rosenberg M. (1979) Regulatory sequences involved on the promotion and termination of transcription. *Annu. Rev. Genet.* **13**, 319–354.

Chapter 14 | Problems

See Problems at the end of Chapter 15.

Genome Architecture and Evolution

15.1　Introduction

The main objective of genetics has always been as complete an understanding as possible of the genetic material, which we now know to be DNA. In recent years dramatic strides have been made towards this objective. The complete DNA nucleotide sequences of several small viruses and of some mitochrondria are now known. In higher eukaryotes, where the sheer size of the genome puts complete sequencing out of reach, more and more selected loci are being analysed at the DNA sequence level. Already it is possible to see, in general terms, how genomes of a wide range of sizes are constructed.

Our rapidly increasing knowledge of DNA sequence provokes, and to some extent answers, two kinds of question. Firstly, how far can we explain function in terms of DNA and to what extent can we account for the DNA in terms of function? Secondly, what do the sequences tell us about the processes of genetic divergence and the phylogenetic relationships of organisms? In this chapter we shall pursue the first of these aspects and touch upon the second with reference to a series of organisms at different levels of organization.

We start with *Escherichia coli*, as the best-known prokaryote. Here the genome is already very large and complex, but it seems possible to grasp the principles on which it is organized—principles which seem to be economical and rational. Next, the genomes of bacteriophages represent a step in the direction of specialization and what, in human technology, is called miniaturization. The smaller phages, seemingly faced with the problem of packing the information for increasingly elaborate machinery into a rigidly circumscribed space, use their DNA extremely economically to the extent of sometimes making the same sequence serve for two genes at the same time. Finally, and in complete contrast, the higher eukaryotes have their genomes arranged on relaxed and even extravagant lines, with a great deal of apparently inconsequential sequence and a certain amount of what appears to be wreckage or rubbish.

15.2　The *Escherichia coli* genome

The *E. coli* genetic map at present includes about a thousand genes, a substantial proportion of the maximum of a few thousand which would represent full utilization of the coding capacity of the 4×10^3 kilobase-pairs in the genome. While there are no doubt many more genes awaiting discovery, the map is probably already sufficiently complete to show us the main features of prokaryote genome organization. A simplified version of the map with a selection of the better characterized gene loci was shown in *Fig. 10.9*.

One obvious tendency, which is not seen to anything like the same extent in eukaryote genomes, is for functionally related genes to occur in clusters. The greater part of this clustering reflects organization into operons—teams of genes which are transcribed together (pp. 371–373). There is, in addition, a certain amount of clustering of different operons of related function. A well-studied example is the *lacZ, Y, A* operon, which is adjacent to the separately transcribed regulator gene *lacI*. The reason for the close linkage here is probably evolutionary rather than functional. One can imagine that the whole *lac* region has spread by transduction or plasmid transfer among species of enterobacteria as a portable self-contained kit for regulated lactose utilization. Some species, such as *E. coli*, have acquired it while others, such as *Salmonella typhimurium*, have not. It is less easy to think of a reason for the clustering of the several operons coding for protein synthesis functions (*Fig.* 15.1).

Within operons, genes are in general closely spaced, with gaps between them of only some tens of bases in most cases. Sequences which bind RNA polymerase and act as signals for transcription initiation occur only at the ends of operons and not within them, and so the intergene spaces within operons need only contain sequences for ribosome binding (remember that translation, as opposed to transcription, can start at the 5' end of any gene within an operon). In at least one analysed case (*Fig.* 15.2), the intergene space is not even sufficient for a ribosome binding site and the 'downstream' end of one gene has to serve two functions—to code for the C-terminal part of one polypeptide and to provide the ribosome-binding sequence for the initiation of synthesis of the following one. This must

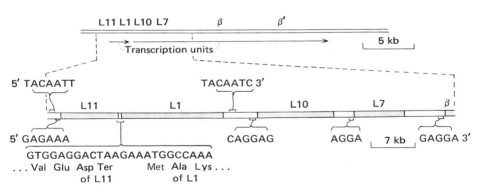

Fig. 15.1   Two adjacent *E. coli* operons concerned with ribosomal proteins and RNA polymerase.
The genes are labelled according to their products rather than with the conventional genetic symbols. L11, L1, L10 and L17 are four of the proteins of the large ribosomal subunit; β and β' are two of the subunits of RNA polymerase.
In the blown-up map in the lower part of the figure some key signal sequences are indicated: TACAATT is a 'Pribnow box' sequence, which is an essential part of the RNA polymerase-binding (promoter) site (cf. *Fig.* 12.8), while GAGG and AGGA are sequences complementary to the 3' end of 16-S ribosomal RNA and are thought to be involved in ribosome binding to messenger RNA. Note that the interval between the L11 and L1 genes is so short that the putative ribosome binding sequence for the L1 gene has to be accommodated within the coding sequence of the L11 gene All sequences are written left to right, 5' to 3'.
The gene cluster shown is located at 88' on the *E. coli* time map; there is another cluster of transcription/translation operons at 72' encoding several other ribosomal proteins, the α subunit of RNA polymerase and two protein factors involved in polypeptide synthesis.
From Post et al.[1] and Post and Nomura.[2]

Fig. 15.2

```
                    fMe t Tyr Leu...   xis sequence .......... A r g I l e A r g A s n G l y L y s L y s
5'..GGAGACTTTGCATGTAC..(198 nucleotides, 66 codons)..AGGATCAGAAATGGGAAGAAG
                                                                        fMe t G l y A r g A r g
```

```
A l a L y s S e r Te r
GCGAAGTCATGAGCGCCGG...(1029 nucleotides, 343 codons)..GAAATCAAATAATGA...3'
g A r g S e r H i s G l u A r g A r g..   int sequence ..............G l u I l e L y s Te r Te r
```

The overlapping *int* and *xis* genes of bacteriophage lambda. The underlined sequences are the putative ribosome-binding (Shine–Delgarno) sequences, complementary to part of the 3'-terminal sequence of 16-S ribosomal RNA (cf. p. 411). fMet is *N*-formylmethionine (cf. p. 42). Only one strand, the non-transcribed one, is shown; its nucleotide sequence corresponds directly to that of the mRNA (with T substituted for U).

obviously place some constraints on the amino acid sequence over and above those imposed by the need to preserve the function of the protein gene product. In general, however, the translation of each gene in *E. coli* is affected by the sequence of that gene alone.

Less is known about the spaces between different operons or between single genes that are individually transcribed. In at least some cases we know that the spaces must be rather short—*see*, for example, the spaces between the protein synthesis operons shown in *Fig. 15.1*. The *lacI–lacZ* space is also short (as shown in *Fig. 14.8*, p. 405) though this may be a special case as we noted above. There are a few regions of the *E. coli* genome which, to judge from spaces in the otherwise closely packed linkage map, may consist of some tens of kilobases of functionless material, although it would certainly be premature to conclude that this was so.

In a number of well-studied operons the direction of transcription, which is also the direction of the polar effects exercised by chain-terminating mutations (cf. p. 373), has been determined. This direction is in some cases clockwise on the conventional map and in others anticlockwise. In a few instances two operons of related function are transcribed divergently from a common control region; this has been shown to be the case in the *biotin* and *arginine* clusters. It is thus evident that both strands of the genomic DNA are utilized for transcription, but in different and generally non-overlapping parts of the genome.

The general impression that we get of the *E. coli* genome is that it is organized on rather compact lines, with sufficient space for all known or likely gene functions but without a great deal to spare. A few genes are duplicated. In *E. coli* K12, but not in several other strains of the same species, *argI*, which codes for ornithine carbamoyltransferase, is duplicated in near-identical form as *argF*. Genes for tRNA tend to occur in clusters, with several different kinds of tRNA in the same cluster and with some present in two or several copies. Genes for rRNA, less highly repeated than in eukaryotes, occur in seven different regions. On the whole, however, the *E. coli* genome shows very little sequence repetition. The same conclusion was reached on the basis of the kinetics of reannealing of *E. coli* DNA (Chapter 1, p. 18).

We dealt in Chapter 10 with the *E. coli* IS sequences, which have the ability to wander about the genome and are an important source of spontaneous mutation. They have no essential function from the point of view of the bacterium and are probably best regarded as DNA parasites. They are of a number of different kinds each of which occurs in a moderate number of repeated copies dispersed around the genome in variable positions. They make only a tiny contribution to the total quantity of DNA; their counterparts in eukaryotes (*see* p. 444) may bulk much larger in their respective genomes.

15.3 Bacteriophage genomes

Bacteriophage T4: a Close-packed Genome

This is the largest of the phages to have been studied genetically. Its genome, comprising between 160 and 170 kilobase-pairs, is too large to be a reasonable candidate for complete sequencing, but from the genetic point of view it is very close to being completely mapped. The use of temperature-sensitive and chain-termination ('amber' or 'ochre'-suppressible) mutants makes it possible to identify any gene which has an essential protein product, and there is reason to think that all of these have, in fact, been identified. In addition, there are a number of genes that are transcribed into various species of tRNA as well as some protein-coding genes whose products are not essential but may increase virus yield.

About 140 genes are listed in the authoritative review of W. Wood and H. Revel.[3] If genes averaged one kilobase in length there would be room for 160–170; those T4 protein gene products which have been characterized require, in fact, an average 1·2-kb gene to code for them, and so it is clear that the gene inventory must be almost complete. There is certainly no room for more than minimal spaces between genes. There is no indication of gene overlap, although rather little is known as yet about the precise nucleotide sequences between genes.

Clustering of functionally related genes is even more conspicuous than it is in the *E. coli* genome. Some organization of genes into multigene transcriptional units or operons does occur in T4, although it is not so thoroughly documented here as in *E. coli*. A possible additional reason for the clustering of a block of genes coding for head components, another for tail fibres, and so on, is that each such block has evolved as a set of precisely co-adapted functions which would be of little value if split up. We can imagine that such a set of functions might occasionally enter, through recombination in mixed infection, into new associations with other sets. Thus a given combination of head and tail structural components might become associated with a new type of tail fibre or a different set of genes controlling DNA synthesis. In this way a new phage type could arise with a particular morphology associated with a new host range. There is, in fact, strong evidence that a different group of *E. coli* bacteriophages, which includes λ and P22 and a range of related forms, does owe its diversity to just such a process of reshuffling of gene blocks.[4]

More Compact Bacteriophages: Multiple Use of DNA

One might suppose that bacteriophage T4 represented almost the maximum possible compression of the genetic information; the genes being closely packed end to end with only minimal spaces between them. What was not suspected until recently was that genes can overlap, with the same stretch of DNA being used to specify two different polypeptide chains.

Int and *xis* in Lambda

A relatively simple example is provided by a recent study of the molecular structures of the *int* and *xis* genes of lambda bacteriophage (cf. pp. 300 and 423). It turns out that the coding sequences of these two genes, which have different though related functions in lambda integration into and excision from the *E. coli* genome, overlap to the extent of 23 bases. This region of overlap is translated in one reading frame into the C terminus of the *xis* product and in another reading frame into the N terminus of the *int* product. One might wonder

Fig. 15.3

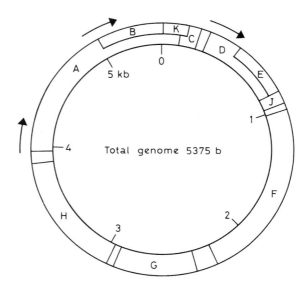

The genome of ϕX174 bacteriophage. Data of F. Sanger and various other authors. The numbering of nucleotides is according to Sanger et al.[5] In the designation of reading frame that of gene *A* is arbitrarily set at zero. The arrows mark the positions of known promoter regions and the direction of transcription from them.

Functions of genes: *A*, replication of double-stranded DNA and synthesis of single strands; *B, C, D*, single-strand synthesis and/or packaging; *F, G, H*, structural proteins of the phage coat (capsid); *J*, a small basic protein also part of the virus particle; *K*, unknown product; *E*, cell lysis.

Initiation and termination codons

Gene	Initiation codon	Termination codon	Reading frame
A	ATG 3973–3975	TGA 134– 136	0
B	ATG 5064–5066	TGA 49– 51	−1
C	ATG 133– 135	?	0
K	ATG 51– 53	TGA 249– 251	−1
D	ATG 390– 392	TAA 846– 848	−1
J	ATG 848– 850	TAA 962– 964	+1
F	ATG 1001–1003	TGA 2276–2278	+1
G	ATG 2387–2389	TGA 2912–2914	+1
H	ATG 2923–2925	TAA 3907–3909	0
E	ATG 568– 570	TGA 841– 843	0

how these two different functions are reconciled. The answer is not clear, but the suggestion has been made that the significance of the overlap lies in regulation of translation, in that, during bacteriophage induction, ribosomes translating the messenger in the *xis* reading frame may prevent other ribosomes from binding at the *int* initiation site; thus *xis* expression could restrict *int* expression, which may happen at the time of prophage excision. The precise amino acid sequence at the region of overlap may not be of critical consequence for one of the proteins.

Bacteriophage φX174

Much more extensive gene overlaps are found in a much smaller bacteriophage, which enjoys a distinguished place in the history of genetics on two counts. φX174 was the first bacteriophage to be shown to have *single*-stranded DNA in its infective particle, though it does, of course, replicate in double-stranded form. It was also the first virus for which the complete nucleotide sequence of the genome was determined.[5] By more or less conventional methods of mutation, recombination and complementation, at least ten genes (*A, B, C, D, E, F, G, H, J, K*) have been identified—more than one would expect in a genome of only 5·3 kilobases.

Close scrutiny of the nucleotide sequence has revealed the polypeptide initiation and termination codons for all the gene products. The results of this analysis are summarized in *Fig.* 15.3. Two genes are completely included within other genes—*B* within *A* and *E* within *D*, two different reading frames being used in the same DNA sequence in each case. One gene, *K*, overlaps *A* and *C*, three different reading frames being used for these three genes. There are only five clear spaces between coding sequences and two places where the termination codon for one gene overlaps the initiation codon for the next. Three promoter sequences have been recognized and it appears that, in this bacteriophage, all the transcriptional units are multigenic.

We meet here the same paradox that we encountered above in the case of the lambda *int* and *xis* genes, but in much more extreme form. If the same RNA is translated into two different polypeptides through the use of two reading frames, the amino acid sequence of the second polypeptide is, though not completely determined, as least very greatly constrained by the amino acid sequence of the first. Given the first sequence, only the third bases in the codons (and not all of those) are free to vary to suit the function of the second gene product. The best attempt at rationalizing the situation invokes a special feature of φX174 codon usage. An unusually high proportion of the codons have uracil in the 3′-terminal position, corresponding to thymine in the DNA. In the region of the *D* gene which overlaps *E*, 39 per cent of the codons end in thymine. When, by use of the *E* reading frame, these thymines are moved into the second codon positions, the code decrees that hydrophobic amino acids will be coded for—phenylalanine, leucine, isoleucine, methionine or valine. The function of gene *E* is to do with lysis of the host cell membrane, and it is conjectured that a hydrophobic N-terminal sequence, which could dissolve in the membrane, is just what is required for this purpose. In the case of *A–B* overlap the position is reversed, it being the included gene *B* which has the high usage of codons with thymine in the third position (33 per cent). Here it may be that the *A* product acquired a possibly advantageous hydrophobic C-terminal extremity by 'reading-through' out of phase into the *B* gene, the ancestral genes having been non-overlapping. It is implied by these conjectures that, in some parts of some polypeptide chains, the precise amino acid sequences are less important than the overall hydrophobic character, which may well be true when the function of the regions in question is to dissolve in lipid. Such explanations of how overlapping genes may have come about is plausible but rather special, and could only apply to genes of certain kinds of function and to genomes with certain kinds of non-random codon usage.

However it arose, the φX174 genome appears as a masterpiece of compact packing of information. There is a good reason why the genomes of viruses that have capsids of a fixed geometry should tend to become overcrowded with genes. Short of entirely changing the system whereby the protein subunits of the capsid fit together, which may be virtually impossible if several precisely fitting components have to be changed simultaneously, the space available for DNA cannot be increased. Consequently any new genetic functions, which may in themselves have a high value in the extremely competitive microbial world, must somehow be found without increasing genome size.

MS2: An RNA Bacteriophage

A second completely sequenced bacteriophage genome is that of the RNA bacteriophage MS2. Here the genetic material is linear single-stranded RNA, and it is only 3569 nucleotides in length. It is free of some of the complexities of the only slightly larger φX174 genome which we have just considered. The genes, which are only three in number, do not overlap. Since the protein coat of MS2 is a narrow hollow tube, which could be easily extended linearly to accommodate a larger genome, there would not appear to be the same restraint on size as there is in a spherical virus. The genomic RNA is used directly as messenger and there is no evidence that it requires any splicing or other processing for this purpose.

The nucleotide sequence is subject to a constraint connected with the regulation of the translation of the phage RNA. The growth cycle of the phage in the *E. coli* host cell consists, roughly speaking, of two phases: RNA replication and coat protein synthesis. The first phase depends on the synthesis of a special enzyme, RNA replicase, the structure of which is encoded in the MS2 replicase gene. The other two genes code, respectively, for the coat protein and for an 'A' protein which is not part of the infective particle but seems to be essential in some way for its assembly. As the amount of coat protein in the cell builds up, the translation of the replicase gene is shut off as a result of specific binding of the coat protein to a segment of the RNA sequence overlapping the replicase initiation codon. This binding depends on the secondary structure of the RNA which, in the region of coat protein binding, tends to be double stranded through 'fold-back' of complementary sequences. It seems that the coat protein recognizes, binds to, and hence stabilizes this double-stranded section.

Fig. 15.4 shows the relevant features of the nucleotide sequence. It will be noticed that there are two other regions of probable secondary structure overlapping the replicase and A protein genes; the significance of these is less clear, but the degree of self-complementarity is too good to be due to chance.

It seems, therefore, that the protein-coding sequences of MS2 RNA are constrained by the need for a degree of complementary base-pairing with each other. We have already considered one other example (the *E. coli trp* leader sequence—p. 410) of important potential secondary structure within coding sequences. There may well be others awaiting discovery in the relatively unexplored territory of control of gene activity in eukaryotes.

Fig. 15.4   The genome of the MS2/R17 group of single-stranded RNA bacteriophages, and an important region of secondary structure in the genomic RNA.

a. The total genome of MS2—3395 nucleotides.
From Fiers et al.[6]

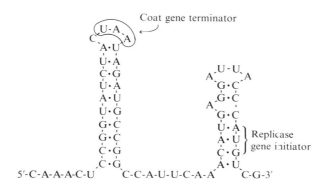

b. The sequence at the 3′ end of the coat protein gene of R17 and the 5′ end of the replicase gene to which the protein binds, and its secondary structure. From Bernardi and Spahr[7] and Gralla et al.[8]
The sequence shown is that which is protected from ribonuclease digestion by bound coat protein. Hyphens indicate nucleotide bonds; dots indicate hydrogen bonds connecting base-pairs. The corresponding sequence in MS2 is very nearly the same.

15.4   SV40: a compact animal virus

An interesting comparison with the bacteriophages is provided by the genome structure of a small animal virus, simian virus 40, usually called 'SV40.' This virus, whose natural hosts are monkeys, happens to have a circular genome nearly the same size as that of ϕX174. Like most other DNA viruses, but unlike ϕX174, the DNA in the infective particle is double stranded.

The functions of the different parts of the genome have been worked out by biochemical rather than genetical methods. The complete nucleotide sequence of the genome has been determined, the RNA molecules transcribed from various parts of the DNA have been characterized and their relationship to the DNA determined by heteroduplex analysis. The mature mRNA molecules are derived

Fig. 15.5   Transcriptional and translational map of the monkey virus SV40.
                   After Fiers et al.[9]

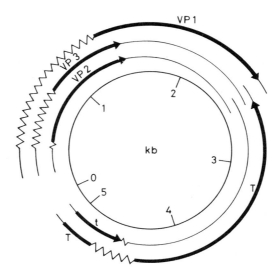

a. The circular genome, consisting of 5224 base-pairs, is calibrated on the inner
circle in kilobases. The surrounding arcs indicate the different species of transcribed RNA; all are initiated near the position marked zero on the kilobase
scale and extend for different distances clockwise on one DNA strand for the
late-expressed functions (virion proteins 1, 2 and 3) and counter-clockwise on
the complementary DNA strand for the early expressed functions (t and T
antigens). Heavy lines with arrowheads indicate regions of mRNA encoding
amino acid sequence, thin lines indicate mRNA not translated into polypeptide, while zig-zag lines indicate RNA regions spliced out during
maturation of the messenger RNA.

```
                                   VP2/3
  ThrSerAlaLeuLysAlaTyrGluAspGlyProAsnLysLysLysArgLysLeuSerArgGlySerSerGlnLys
5'...ACTTCTGCTCTAAAAGCTTATGAAGATGGCCCCAACAAAAAGAAAAGGAAGTTGTCCAGGGGCAGCTCCCAAAAA
              (Met)LysMetAlaProThrLysArgLysGluSerCysProGlyAlaAlaAlaProLysLy
                                   VP1
```

```
                                   VP2/3
  ThrLysGlyThrSerAlaSerAlaLysAlaAlaHisLysLysArgArgAsnArgSerSerArgSerTer
  ACCAAGGGAACCAGTGCAAGTGCCAAAGCTCGTCATAAAAGGAGGAGTAGAAGTTCTAGGAGTTAAAACTGGAGTAGA...3'
  sProArgGluProValGlnValProLysLeuValIleLysGlyGlyIleGlyValLeuGlyValLysThrGlyValAsp
                                   VP1
```

b. Nucleotide sequence of one DNA strand in the region of the VP2/3–VP1
overlap. Note that VP1 is read from the same DNA sequence in a different
reading frame.

from the primary transcripts by cutting and splicing reactions, and they too have been aligned with both the primary transcripts and the DNA. The picture which has been obtained is summarized in *Fig.* 15.5.

Five 'genes' have been defined by their mRNAs and protein products. Three specify proteins which form the coat of the infective virus particle (VP1, 2 and 3), while the other two specify proteins (T and t) which are recognized as antigens present in animal cells which have been infected or transformed (*see* p. 503) by the virus. These proteins are somehow involved in cell transformation, but their precise functions have yet to be determined.

Like ϕX174, SV40 shows gene overlap, using in one part of the genome (the VP1–VP2/3 overlap) the same DNA sequence in two different reading frames. It also displays an entirely different method of generating different gene products from the same stretch of DNA—the cutting and splicing of the same primary transcript in two or more different ways so as to form different mRNA molecules (cf. pp. 415–416). Thus the mRNAs for VP1, VP2 and VP3 are all derived from the same primary transcript. The nature of the protein formed is determined by the position of the AUG codon at which translation is initiated, and this is determined by the length of the intervening sequence which is removed during processing. Thus, in the VP3 messenger the VP2 initiation codon has been removed, while in the VP1 messenger both the VP2 and VP3 initiation codons are absent. VP3 is, in fact, merely a shortened version of VP2, but VP1 is read in a different frame and has an entirely different amino acid sequence with a different terminator codon.

The mRNAs for the two antigens T and t are also generated from the same primary transcript but in a rather different way. Here the T mRNA results when a substantial intervening sequence in the primary transcript is removed. In the t messenger most of this intervening sequence is retained and translation, initiated at the same codon as in the case of T, reads into this sequence. Unexpectedly, this makes the t product smaller than the T product since the intervening sequence contains a terminator codon which T translation escapes. Thus the t and T proteins have a common amino acid sequence at their N termini but soon diverge into two entirely different sequences, that of T being much the longer.

In its extensive use of splicing in the formation of its mRNAs, SV40 conforms to the practices of the cells which it inhabits. Such splicing is unknown in prokaryotes and bacteriophages.

It will be noticed that ϕX174 and SV40 also differ substantially in their respective repertoires of genes. SV40 appears to have no encoded information for its own DNA synthesis and to rely entirely on the enzymes of its host in this respect. Since it inhabits the cell nucleus, it presumably makes use of the cellular apparatus of chromosome replication; indeed it itself forms a small circular piece of chromatin with typical association with histones and nucleosome structure (cf. p. 20).

We return to the topic of the role of viruses in cancer in Chapter 17 and, in this context, will consider a quite different group of animal viruses, the retroviruses, which have RNA as their infective genetic material.

15.5   Lower eukaryotes

It has become customary to distinguish between **lower** and **higher** eukaryotes, the former including the fungi and simple algae, and the latter more complex plants and animals. To what extent their respective modes of genomic organization really differ in any fundamental way it is a little early to say. What is clear, however, is that the lower eukaryotes have much less DNA in their genomes than the higher eukaryotes which, as noted in Chapter 1, differ greatly among themselves in this regard. It seems likely that the lower eukaryotes have fewer genes and also have these genes arranged in a more compact way and interrupted by fewer intervening sequences (*see Fig.* 15.9). Within the next few years a great deal will undoubtedly be learned about the spacing of chromosomal genes in *Saccharomyces*, but anything written at this moment would rely heavily on

Fig. 15.6

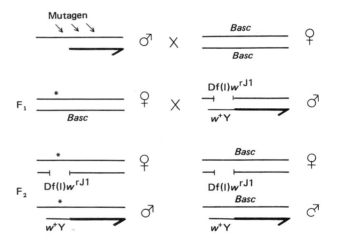

a. The scheme used by Judd et al.[10] for saturating a defined region of the *Drosophila melanogaster* X-chromosome with mutations. Males mutagenized with X-rays or a chemical mutagen (ethylmethane sulphonate or N-methyl-N'-nitro-N-nitrosoguanidine) are mated to females homozygous for the 'balancer' chromosome *Basc* (carrying a crossover-suppressing inversion and the dominant *Bar* marker). F_1 females are mated to males with a deficiency (Df, indicated in the figure as a gap) in the w–z region of the X compensated for by the presence of the same segment transposed to the Y-chromosome (w^+Y). Absence of females in the F_2 generation indicates an induced lethal mutation in the w–z region of the treated X; a visible mutant phenotype in all the F_2 females indicates the induction of a viable visible mutation in the same region. In either case the lethal mutation is preserved in the non-*Bar* F_2 males and can be kept in males in a stock with attached X-chromosomes and the w^+Y-chromosome, and in females in another stock together with a suitable 'balancer' X. Complementation tests could then be performed by crossing $X^{*1}w^+$Y males with $X^{*2}/X^{balancer}$ females ($*1$ and $*2$ signifying two different mutations) and seeing whether viable X^{*1}/X^{*2} females of wild phenotype appear in the following generation. If they do, the mutants complement each other and are assigned to different complementation groups.

speculation. There is at present more concrete information about genome structure in *Drosophila* (a small higher eukaryote) and in mammals, and these organisms will be used as the main examples in this section. There is also increasingly detailed knowledge of the genomes of eukaryote organelles, especially of mitochondria, and these have features distinct from those either of prokaryote or chromosomal eukaryote genomes and are dealt with in Section 15.8.

15.6   *Drosophila* species

Polytene-chromosome Bands—How Many Genes?

As a result of the availability of the giant polytene chromosomes of the salivary glands and some other glandular tissues, genes in *Drosophila* species can be assigned to visible chromosomal loci with greater precision than in any other type of organism. Each polytene chromosome arm has of the order of a thousand distinct bands (cf. p. 33) and, through complementation tests using chromosome deletions covering known sets of bands, it is possible to locate most genes within single bands. A question of great interest is whether each band contains only a single gene or several genes. Measurements of the amounts of

Fig. 15.6 (*contd*)

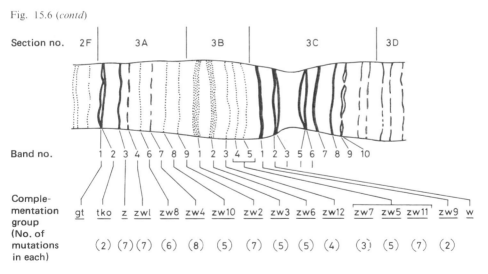

b. Assignment of induced mutations to polytene-chromosome bands in the *w–z* segment of the X-chromosome of *Drosophila melanogaster*.
Notes: The deletion [Df(1)w^{rJ1}] used in isolating the mutants extends from 3A2 to 3C2. The genes *gt* and *tko*, like *w* and *z*, were already known; mutation in them results in defects in larval development and neurological function, respectively. Genes (complementation groups) inferred from analysis of the new recessive lethal mutations between *z* and *w* are named *zw1, zw2* etc. Each gene (complementation group) corresponds to a single band except for *zw7, zw5* and *zw11* which correspond to a region seen to contain only two bands. Figure adapted from Judd et al.[10]

DNA present in individual bands, using photometric estimates of the amount of colour in each band following specific staining of the DNA, have given estimates of an average of about 20 kilobase-pairs of DNA per band. This, in bacteriophage T4 or *Escherichia coli*, would be enough for 10 to 20 genes. In fact, several detailed studies on particular short segments of *Drosophila* chromosome have led to the general conclusion that there is a close approximation to a one-to-one relationship between bands and genes, at least when genes are defined through clearly demonstrable individual functions.

The first evidence pointing to the conclusion that there was, in general, one gene per polytene chromosome band was obtained by B. H. Judd and his colleagues.[10] Their study was focused on a short segment of the X-chromosome defined by a deletion which removes the *w* and *z* loci (which interact in the formation of eye pigment) and all the loci between them. Their approach was to 'saturate' the segment with mutations so that there was a good probability that every gene had been hit by mutation at least once.

Fig. 15.6*a* summarizes the procedure used. All mutations within the *w–z* interval which, when present in females opposite the deletion, cause either a clear visible or a lethal effect, could be identified. The great majority of the mutations recovered were recessive lethals which could be maintained only by being complemented by a wild-type allele of the mutated gene. Two kinds of stocks were established of each mutation. In one the recessive lethal was transmitted through the female line, always 'balanced' against another X-chromosome with a different recessive lethal and a dominant visible marker. In the other it was transmitted only through males with a complementary X-fragment transposed to the Y-chromosome; females of these stocks had attached X-chromosomes to force father-to-son transmission of the paternal X in each generation (cf. *Fig.* 5.1). By crossing females carrying one recessive lethal with males carrying another and examining the females of the next generation it could be determined whether or not the mutant *w–z* segments showed complementation and hence whether the two mutations affected different genes or the same gene.

The striking conclusion of the analysis was that, save for one section with three genes and two bands, each visible chromosome band harboured one gene and no more. The fact that each of the complementation groups of recessive lethal mutations had at least two members made it improbable that more than one or two essential genes of similar mutation frequency could have been missed. Later work has not, in fact, turned up any further groups of recessive lethals in the *z–w* region although three groups of mutants have been recovered which affect behaviour but apparently (since they include no recessive lethal members) do not represent essential genes.

Thus, although the one gene–one band generalization may not hold exactly, it does seem to be a useful approximation. We need not be too disconcerted by finding slightly more genes than bands since the staining of individual bands varies from intense to barely detectable, and there could be a few which escape microscopic detection.

The Organization of DNA Sequences in Bands

If, in general, each polythene-chromosome band (chromomere) contains just one gene, how can we account for the 10–20 kb of DNA per band? Nobody has, as yet, reported the complete nucleotide sequence across a band, but the DNA of at least two bands, both in *D. melanogaster* and in some other species of the genus have been subjected to restriction site mapping. These are the third-chromosome bands 87A and 87C, which contain repeated copies of the structural gene for one of the major proteins produced by larvae in response to a heat shock.

When *Drosophila* larvae are transferred from their normal culture temperature of 25 °C to the near-lethal temperature of 37 °C they respond by making, in large amounts, a special set of proteins which can be characterized on SDS–polyacrylamide gels (cf. p. 388). Furthermore, within minutes of the heat shock, nine different salivary gland chromosome bands, including 87A and 87C, become 'puffed'—that is to say, they become enlarged and diffuse, and can be demonstrated, by radioactive labelling and autoradiography, to be very active sites of RNA synthesis (*Fig.* 15.7).

Fig. 15.7

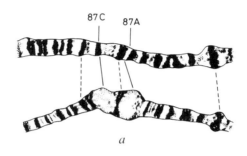

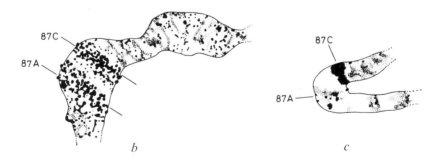

The 87A and 87C bands of the right arm of chromosome III of *Drosophila melanogaster*.
a. Unpuffed (above) and puffed (below) states of the bands, seen in salivary glands from untreated and heat-shocked larvae, respectively.
Drawn from the photograph of Ashburner and Bonner.[11]
b. Active RNA synthesis in isolated heat-shocked salivary glands shown by administration of radioactive (tritium-labelled) uridine. The regions of RNA synthesis are seen as concentrations of black silver grains (deposited close to the source of the short-range tritium radiation) in the radioactive film to which the salivary gland preparation was exposed.
Drawn from the photograph of Bonner and Pardue.[12]
c. Salivary chromosome bands hybridized to tritium-labelled RNA formed after heat shock. Note that the 87C band is the more heavily labelled; this reflects three copies of the gene for the 70 000 dalton heat-shock protein (70 K) in 87C as compared with two in 87A, and also the presence in 87C, but not in 87A, of the transcribed (but apparently not translated) αβ sequences (*see Fig.* 15.8).
Drawn from the photograph of Spradling et al.[13]

Messenger RNAs coding for several of the major heat-shock proteins have been isolated and cloned as complementary DNA (cf. p. 389), which has then been used as a probe to identify, by *in situ* hybridization of radioactively labelled cDNA to preparations of chromosomes (*Fig.* 15.7*b*), the loci from which the messenger was transcribed. cDNA from mRNA for one of the most abundant heat-shock proteins, one with a polypeptide chain of 70 000 daltons, hybridized strongly to both 87A and 87C. Another species of RNA molecule, which was also produced in response to heat shock but did not appear to be translated into protein, was found to hybridize strongly to 87C but not at all to 87A. Cloned fragments of *D. melanogaster* DNA have been identified as hybridizing with one or other of these cDNA probes and, from restriction endonuclease analysis of these and of other cloned genomic fragments found to overlap them, coherent restriction maps representing at least most of the 87A and 87C bands have been constructed. The positions within these maps of the genes coding for the 70 000 dalton protein (Hsp70) and of the 87C region transcribed into the non-translated RNA have been defined.

The very curious arrangement of sequences that was revealed is shown in *Fig.* 15.8. There turn out to be five copies of the gene for Hsp70, two in 87A and three in 87C. The pair in 87A are close together in inverted orientation, while 87C has a pair of closely spaced tandem repeats and a more distant inverted copy. From the

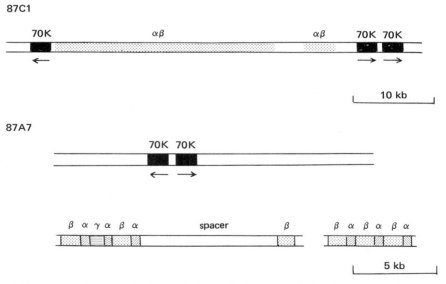

Fig. 15.8   The organization of the transcribed sequences in two *Drosophila melanogaster* 'heat-shock' loci corresponding to polytene-chromosome bands 87A7 and 87C1.
In the two upper diagrams (after Ish-Horowicz and Pinchin[14]) the sequences coding for the 70 000 dalton protein induced by heat shock are shown as black blocks, and their orientations are indicated by arrows. Segments containing (but not entirely consisting of) the repetitive αβ sequences are cross-hatched. In the lower diagram (after Lis et al.[15]) the fine structure of two pieces of the αβ-containing regions is shown on a larger scale. The nature of the spacer is unknown. Lis et al. estimate that there are at least 12αβ and at least 6αγ units in the 87C1 band, but there is none in the 87A7 band and none in either of the corresponding bands in the closely related *D. simulans*. The repetitive sequences are transcribed in response to heat shock but it is not known whether they are translated into protein.

similarity of the arrangement of the inverted repeats of the Hsp70 gene it looks as if the two bands originally arose by duplication of a single band though, since all *Drosophila* species so far examined show the same apparent duplication, it must have arisen very early in the evolutionary diversification of the genus. The close similarity maintained between the two gene copies within each inverted repeat raises interesting questions—*see below*, p. 450. The third gene copy in 87C seems to be the result of a later duplication event, since it is not present in several related species.

The 87C *D. melanogaster* region from which the non-translated RNA is transcribed proves to be composed of a series of repeated short sequences called αβ, which alternate with some sequences of another kind, αγ, interspersed between them. The αβ repeats are absent from 87A and also from the equivalent of 87C in certain closely related species such as *D. simulans*. It looks as though a DNA segment has become transposed into the 87C band from some other part of the genome, a possibility which is supported by the finding that αβ occurs in numerous copies at many different loci, especially in the centromeric regions of the chromosomes. The αβ sequence evidently belongs to the category of dispersed repetitive sequences which occupy a significant proportion of the genomes of higher eukaryotes of all kinds.

The structure-determining genes for the 70 000 dalton heat-shock protein and the αβ and αγ sequences account, between them, for only a part of the 87C and 87A DNA. The remainder ('spacer') is apparently unique-sequence DNA but it has no known function.

The finding that nearly all bands (chromomeres) of polytene chromosome in *Drosophila* harbour single loci, as defined by single groups of non-complementing recessive lethal mutations, argues against the idea that most of them carry multiple gene copies. It they did so we would not expect to be able to eliminate the function by single mutations, unless all the mutants recovered had suffered deletion of all the gene copies, which seems improbable. In carrying, respectively, three and two copies of the same protein-coding gene, the 87C and 87A bands may well be atypical. Yet, in another way, they may be fairly representative of *Drosophila* chromatin. They include single-copy coding DNA, single-copy non-coding DNA and transcribed DNA of intermediate repetitiveness and doubtful function in proportions which may not be very different from those in the genome as a whole, leaving aside the highly repetitive 'satellite' DNA (p. 449) which is mostly concentrated in large blocks in the centromeric regions of the chromosomes. The significance of the transcription of the αβ sequences is still obscure, and it is not yet known how general a property this is of *Drosophila* middle-repetitive DNA; at least some of the 'wandering' repetitive sequences discussed in the final section of this chapter are transcribed and may code for their own transposition functions.

The example of chromosome organization just discussed is important in pointing to some control of transcription at a higher level than that of the individual gene. In response to heat shock, the whole of the chromatin in 87C becomes puffed, and all its known transcribable sequences, the αβ segments as well as the gene coding for the 70 000 dalton protein, are transcribed into RNA simultaneously. The transcription is into separate RNA molecules presumably from several different promoters, but all are made available to RNA polymerase together, seemingly as a consequence of the chromatin unfolding made visible in the puffing. What signals the puffing, and what sequence or sequences within 87C respond to the signal, is unknown, but it is evident that the unit of puffing, i.e. the polytene band, constitutes a level of control of gene action higher than that of the unit of transcription.

15.7 Gene organization in birds and mammals

The Prevalence of Intervening Sequences

During the past few years a considerable number of genes from birds and mammals (mostly from chickens, mice, rats and humans) have been cloned, and their sequences compared by analysis of heteroduplexes formed with the messenger RNAs transcribed from them. At first it seemed that these analysed genes might not be typical of genes in general, since they all shared the feature which made them accessible to detection in the first place—that of being abundantly transcribed in specialized cells and tissues. The addition to their number of the gene coding for dihydrofolate reductase,[16] however, takes away some of the force of this reservation, since the gene for this enzyme is not known to be especially strongly expressed in any differentiated tissue though it is, as we see on p. 453, capable of extraordinary magnification under special conditions in cell culture *in vitro*.

Representative or not, the bird and mammalian genes studied up to now nearly all agree in containing a large amount of intervening sequence. *Fig.* 15.9 shows a

Fig. 15.9

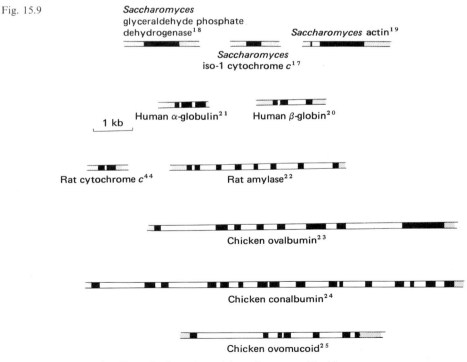

Saccharomyces glyceraldehyde phosphate dehydrogenase[18]   *Saccharomyces* actin[19]

Saccharomyces iso-1 cytochrome *c*[17]

1 kb   Human α-globulin[21]   Human β-globin[20]

Rat cytochrome *c*[44]   Rat amylase[22]

Chicken ovalbumin[23]

Chicken conalbumin[24]

Chicken ovomucoid[25]

A gallery of eukaryote protein-coding genes. Flanking sequences are cross-hatched, coding sequences are black and intervening sequences open boxes. Sources indicated by numbers on the figure.

gallery of examples, with a few yeast genes for comparison. The globin genes, the first to be determined at the DNA sequence level, are comparatively moderate in their content of intervening sequences (**introns**), having only two which together more or less double the gene length. However, the chicken gene coding for ovalbumin (the major egg white protein) was found to have six introns, considerably exceeding the coding sequences in aggregate length. Another protein, conalbumin, produced in the oviduct in response to the same steroid sex hormones as evoke ovalbumin synthesis, turned out to be coded by a gene split into no fewer than seventeen fragments by introns, stretching the gene by a factor of more than five.

Other genes (*see Fig.* 15.9) show an even higher degree of fragmentation, and it is becoming clear that it is by no means unusual for genes to be spread out over ten times or more the length that they would occupy if they were in one piece. So far as we can judge from known examples, lower eukaryotes like yeast, and even a small higher eukaryote like *Drosophila*, have much more compact genes with fewer and shorter introns. With their relatively small total quantities of DNA they could hardly afford introns on the lavish scale that one finds in larger eukaryotes.

The functional significance of introns is at present unclear. One possible long-term evolutionary advantage which, it has been suggested, might result from the fragmentation of genes by intervening sequence is the possibility of reshuffling parts of genes by transposition, which would be much more likely to produce functional results if coding sequences did not necessarily have to be continuous. There is some evidence for this proposition in that, at least in mammalian genes coding for immunoglobulins (cf. pp. 495–502), functionally distinct domains of polypeptide chains tend to correspond to continuous coding sequences in the gene while being separated from each other by long introns. The idea that proteins may have been assembled in evolution from modular components, corresponding to originally widely separated segments of coding DNA, is analogous to that, already discussed above, of the modular construction of genomes (p. 421). What is envisaged in each case is the bringing together of components from different sources to form new and fortuitously well-adapted combinations of qualities.

Whatever the functional significance of introns (and we will return to this question in connection with mitochondrial genome organization on p. 442) they go quite a long way towards accounting for the apparently excessive amounts of DNA in the genomes of higher eukaryotes. Instead of estimating the 'average gene' to be around 1 kb in length, which is the size required to encode an 'average' polypeptide chain of 333 amino acids, we must reckon on the basis of something like 5 kb per gene, if genes in general are as riddled with introns as the ones so far analysed. We still do not understand why rather closely related organisms, for example different amphibians or different flowering plants, differ from one another so much in *C*-value (cf. Chapter 1, p. 17); it would be surprising if the prevalence of introns were found to be very different among different species of generally similar organisms. Such differences can be partly accounted for by different degrees of repetition of repeated sequences whose functions, if any, are even more problematical. We return to the subject of repeated sequences below (p. 444).

Gene Clusters

A striking result of recent research on gene organization has been the demonstration, at the level of DNA sequence, of clusters of genes of related function. We have encountered gene clusters in two other contexts already in this book. In *Escherichia coli* they reflect, for the most part, the existence of multigene transcription units (operons). In many bacteriophages they may reflect the assembly of blocks of coordinated functions from different sources. In eukaryotes neither of these explanations seems to apply; multigenic transcription seems to be absent or, at least, unusual and the exchange of blocks of genes between different species hardly occurs. We can conjecture that those genes which are clustered in eukaryotes are so for reasons other than common transcription or trans-specific transmission, and these reasons may have to do with the existence of regions of chromatin which, in some as yet unknown way, act as regions of coordinated gene action, even though the genes in each region are separately transcribed. Here we will take two examples of very different kinds. The first is that of the histone gene cluster, which has been analysed in different species of sea urchin (Echinoderms, a group of invertebrates). The second is really a connected pair of examples, the alpha and beta 'families' of globin genes of mammals.

The Histone Gene Cluster of Sea Urchins

Fig. 15.10 shows the arrangement of a cluster of five genes coding for histones in the sea urchin species *Psammechinus*. These genes, unlike most of our other examples in this section (but like the recently analysed human interferon genes[17, 19]—cf. p. 416), have no intervening sequences, and are arranged in tandem (in the sense that all are transcribed in the same direction) with spaces between them roughly equal in length to the coding sequences themselves. This array of genes, contained within the space of less than seven kilobases, is repeated several hundred times within the same chromosome region, all the copies being closely spaced in tandem orientation. The base sequences of the genes themselves are very highly conserved, both between different copies within the same species and between different species of sea urchin. The spacers between the genes, on the other hand, have been found to vary markedly both within and, even more, between species. All the genes are transcribed into separate mRNA molecules and so the significance of their close linkage is not obvious. However, since the histones, which are the products of the individual genes, are needed in stoichiometrically equal amounts, it seems likely that there is some kind of linkage-dependent coordination of the activities of the genes. The nature of this is not at all clear, but it may well be that the whole chromatin region containing the histone–gene complex undergoes a concerted conformational change so as to permit transcription of whole compex at the appropriate times—presumably during the S-phase of mitosis. The selective activation for transcription of a chromosome region is a phenomenon of which we have already seen an example in the 'heat-shock' chromomeres of *Drosophila*, and there is no reason to think that it is confined to species in which it makes itself visible in the form of polytene-chromosome puffs.

Globin Genes

Using the readily isolated globin mRNAs from reticulocytes as probes, it has been possible to identify and clone globin genes from several mammalian species including man. All the species examined, which include man, mouse and rabbit, have given the same general picture. There are two groups of genes encoding globin polypeptides. In one closely linked group there are one or more genes encoding the beta polypeptide of adult haemoglobin and other genes encoding polypeptide chains that substitute for beta in embryonic and fetal haemoglobins. In the other groups there are one or more copies of the gene encoding the alpha chain of adult haemoglobin and, at least in humans, genes for the zeta polypeptide which partly replaces alpha in embryonic haemoglobin. The arrangements of the genes have been most thoroughly worked out in the case of the human haemoglobins, and *Fig.* 15.10 illustrates these arrangements. The sequence

Fig. 15.10

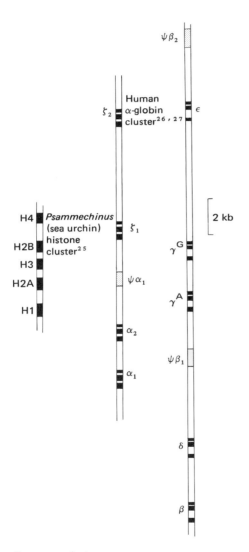

Some gene clusters.
The direction of transcription of the genes is from the top to the bottom of the diagram. Cross-hatched blocks are 'pseudogenes'—*see text.* γ^A and γ^G are duplicated gamma-globin genes differing only in a glycine/alanine substitution. α_1 and α_2 are exact duplicates.
Sources: β-cluster, Fritsch et al.[27] and Kaufman et al.;[28] α-cluster, Pressley et al.[29] and Proudfoot and Maniatis;[30] histone cluster, Schaffner et al.[31]

of expression of the different genes from the embryonic through the fetal to the adult stage of humans is summarized in *Fig.* 15.11.

Unlike clustered histone genes, globin genes are expressed at different times. In the case of the human alpha–zeta cluster the duplicated zeta genes, at one end of the cluster, are expressed early in embryonic development, while the duplicated alpha genes come into action a little later and then remain active throughout fetal and adult life, when the zeta genes are entirely 'switched off'. In the beta cluster, there are at least three stages of differential activation; epsilon is expressed only in the early embryo, the two gamma genes, which are not quite identical (there being a glycine in haemoglobin gamma-G replacing an alanine residue in gamma-A), become active a little later and remain so during the fetal stage, and finally the single delta and beta genes take over to the exclusion of the others around the time of birth. The linking order, as will be seen from the figure, seems to reflect the activity order; but this is probably just a coincidence since a similar relationship is not found in all other species in which globin gene clusters occur.

In Chapter 17, when we come to consider the differential expression of genes encoding immunoglobulins, we discuss the connection of the expression of these genes and their linkage order. Here the meaning of the connection is becoming clear—it makes possible the controlled deletion of DNA segments which enforces the changes in gene expression. But in the case of the globin gene clusters it has been shown that there is no change in the restriction site map of the DNA during the differentiation of reticulocytes—the globin-producing cells. It may be that the clustering of the globin genes merely reflects their origin by tandem duplication followed by functional divergence. Within each cluster all the genes show rather close homology, and it seems likely that they all arose from a common progenitor gene by an initial tandem duplication followed by the generation of further repeats

Fig. 15.11

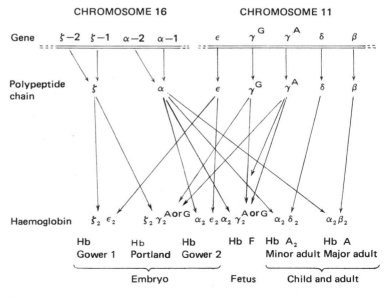

The sequence of formation of human haemoglobin peptides at different stages of development.
Adapted from Weatherall et al.[32]

by unequal crossing-over (cf. p. 323). Such an origin would be compatible with the rather regular spacing which they show (*Fig.* 15.10).

The multiplication process appears to have produced a number of genes in excess of that which can be maintained by natural selection. Several sequences have been found which, though homologous to globin genes by the criterion of molecular hybridization, appear to have fallen into decay. In man, each of the two globin gene clusters has at least one of these **pseudogenes**. The pseudo-alpha gene, the complete sequence of which has been determined, is obviously non-translatable, since it has a mutation in the initiation codon and also frameshift mutations which would throw translation out of phase even if it were initiated. The pseudo-beta gene is also untranslated, for similar reasons.

It has recently been shown[33] that the mouse has a cluster of three active alpha genes on one chromosome and two pseudo-alpha genes on two other chromosomes. We may note in passing that one of the two pseudogenes has lost its introns, as if it were a reverse-transcribed copy of mRNA (cf. Alu, p. 449). Leaving aside the question of how the transpositions took place, the observation that the transposed sequences have been relegated to the status of pseudogenes, while three genes in the main cluster have remained active (one in the embryo, two in the adult), does tend to suggest that the chromosomal location of the globin genes is of some functional significance. As is so often the case, one is led to suspect a role for higher-order chromosome structure without, as yet, being able to show how it works.

Perhaps the most reasonable view to take of the globin gene clusters and their significance is that they arose by tandem duplication occurring without immediate selective advantage, but that sometimes (not necessarily always) the situation has led to natural selection of functionally advantageous gene divergence and systems of activity control that depend on chromosomal position.

A number of other examples of gene clusters in mammals and birds could have been described, but we must let them go with only a brief mention. The genes coding for the protein components of the major histocompatibility system in mammals (which functions in the recognition by tissues of foreign proteins, and brings about rejection of grafted tissue from unrelated animals) are clustered. The chicken ovalbumin gene is closely linked to two other genes with which it shares a large proportion of its base sequence; these genes, called *X* and *Y*, do not have any important function so far as is known and they may well be 'decayed' versions of the ovalbumin gene proper. We have already referred to the duplicated and triplicated heat-shock protein genes of *Drosophila*; these, unlike the bird and mammal examples reviewed above, show not only tandem but also inverted orientation.

15.8   Mitochondrial genomes

The best known mitochrondrial genomes are those of the yeast *Saccharomyces cerevisiae* and man. The genetic and physical mapping of yeast mitochrondrial DNA was outlined in Chapter 7. The human mitochondrial genome is considerably smaller than that of yeast—about 15 as compared to 75 kb; its nucleotide sequence has now been completely determined, but it is less well characterized genetically. In each case the mitochondrial DNA codes for a complete set of tRNA molecules—albeit a minimal set in that, with very few exceptions, there is only one tRNA for each amino acid (cf. p. 167). In yeast, about nine polypeptides of the inner mitochondrial membrane are also encoded (*Table 7.2, p. 185*) and the situation in human mitochondria is probably similar. The human mitochondrial DNA molecule seems a reasonable size for the encoding of this amount of information (the coding sequences must occupy about one-half of the whole), but the yeast mitochondrial DNA sequence would seem to be somewhat excessive in relation to its known functions. Recent detailed investigations begin to show us how this apparently superfluous DNA can be accounted for. They have certainly confounded any expectation that mitochondrial DNA would turn out to be like a prokaryotic genome in its organization.

About one-half of yeast mitochondrial DNA consists of repetitious sequence extremely rich (90 per cent or more) in A-T base-pairs. Within the long stretches of (A + T)-rich sequence there are some shorter repeats rich in G + C, and it has been shown that these latter repeats are 'hot-spots' for the intramolecular rearrangements which result in the greatly simplified mitochondrial genomes found in many *petite* mutants (*see* Chapter 7, pp. 179–184). Far from appearing to have any useful function, this repetitious DNA, by providing sites for intramolecular homologous recombination, seems to be an actual hazard to the maintenance of the unique and functionally essential sequences.

Fig. 15.12

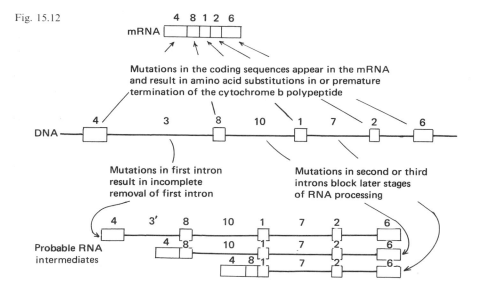

The structure of the yeast mitochondrial gene *cob* and the effect of mutations in its coding and intervening sequences. The numbers were originally given to different groups of mutants defined by mapping and complementation criteria; they were later assigned to coding sequences and introns as shown. Mutations in the coding sequences are either mis-sense or nonsense; the latter have the expected consequence of shortening the cytochrome *b* polypeptide chain to an extent depending on their respective positions. Mutations in the introns can lead to failure of RNA processing and the consequent accumulation of processing intermediates.
After Church et al.[34, 35]

Another finding that helps to explain the comparatively large size of the yeast mitochondrial genome is the presence of extensive introns in some of the protein-coding genes. In this respect the yeast mitochondrion has a genetic organization more like that of a higher eukaryote than that found in the yeast nuclear genome, so far as the latter is yet known. Most attention has been given to two genes: *oxi3*, coding for subunit 1 of cytochrome oxidase and *cob*, coding for the polypeptide of cytochrome *b*. The *oxi3* gene is expanded by introns to a length of about 10 kb, as compared with the approximate 1 kb which is required for coding the polypeptide product. (We may note in parentheses that in this respect *oxi3* contrasts with *oxi1*, which codes for subunit 2 and contains no introns.) The *cob* gene is also expanded about ten-fold and in this case the introns, four in number, have been mapped by heteroduplex formation between the DNA and cytochrome *b* messenger RNA. Analysis of the effects of mutations falling within the *cob* introns provided some of the most definite clues to the problem of intron function. Some of these effects are summarized in *Fig.* 15.12.

The following are the most interesting features of the *cob* intron mutants. Firstly, each group of respiration-deficient mutants falling within a *cob* intron constitutes a complementation group; mutants in different introns complement each other as well as any mutant in any part of the coding sequence (mutants in the coding sequence never complement with each other). From this it is inferred that each intron (or at least each of the first three introns numbered 3, 10 and 7) has some essential function in *cob* expression and, moreover, can provide this function in *trans*, as if it were due to a diffusible gene product.

The second important feature is that the intron mutants accumulate relatively large amounts of large RNA molecules which are virtually undetectable in the wild type; at the same time they fail to produce any mature cytochrome *b* mRNA. It seems clear that the accumulated RNAs are partly processed derivatives of the primary transcript from the *cob* region. Thus the intron mutations are blocking the normal processing of these intermediates—in other words, mutations within introns prevent removal of introns. In fact, the sizes of the accumulated molecules in relation to the gene map are consistent with the notion that each intron is responsible for a product which brings about its own removal or that of the previous intron, such removal proceeding one intron at a time, from left to right (as the map in *Fig.* 15.12 is drawn).

The third important observation is that the intron mutants also accumulate unusual polypeptides. These are of different molecular weights depending on the intron in which the mutation has occurred, and there is good evidence that at least some of them have the N-terminal sequence of cytochrome *b* polypeptide joined to another and longer sequence not found in cytochrome *b*. The interpretation of this finding is that these unusual polypeptides are the translation products from incompletely processed RNA in which the N-terminal coding RNA sequence is joined to intron sequence.

The model proposed to accommodate all these observations is as follows.[35] In the wild type, polypeptide translation products from incompletely processed RNA (with part of their sequence the same as that of cytochrome *b* and the remainder coded by intron sequence) are instrumental in the removal of introns, thus destroying their own templates. Obviously they will be present only in very small amounts. In the intron mutants, whether these are mis-sense or chain termination mutants, the intron-coded part of these polypeptides is defective so that they no longer act to remove introns; consequently the RNA intermediates accumulate and go on being translated into the defective processing polypeptides.

Assuming that further experiments support this model, the next and even more fascinating question will be whether the introns of higher eukaryotes have similar functions. In itself, the postulated 'function' may seem a rather paradoxical one; the intron merely removes its own nuisance. It is conceivable, however, as Slonimski has suggested, that the removal of introns is an integral part of some other essential process, for instance the transport of messenger through the nuclear membrane or, in the case of mitochondrial messengers, the positioning of the RNA in the mitochondrial membrane.

15.9 Repetitive and movable sequences—major components of eukaryote genomes

In Chapter 1 we saw how measurements of the rates of reannealing of denatured DNA gives information about the prevalence of repetitive DNA in the genomes of higher eukaryotes. We also distinguished between highly repetitive sequences, which can sometimes be recognized as 'satellite' sequences with distinct buoyant densities, and sequences that are only moderately repeated—a few hundred rather than up to hundreds of thousands of times as in some of the satellites. Both of these categories of repetitive sequence can differ drastically between otherwise very similar species, both in number and in location.

Dispersed Repetitive Sequences

In Chapter 10 we reviewed the properties of IS elements and transposons in bacteria. Soon after the discovery of these transposable sequences, evidence began to accumulate for the existence of surprisingly similar elements in eukaryotes. It now appears that potentially mobile sequences, dispersed throughout the genome in multiple copies and at apparently random locations, account for an appreciable and perhaps a major part of the intermediate-repetitive DNA in yeast and *Drosophila* and presumably in other eukaryotes as well.

The Ty1 Element of *Saccharomyces*

This element came to light when it was found that a specific EcoRI-generated DNA fragment carrying a tyrosine-tRNA gene differed in size by 5·6 kb from one *S. cerevisiae* strain to another. The difference was shown to be due to the insertion in one strain, close to the tRNA gene, of a 5·6-kb sequence (**Ty1**) which was present at many other sites within the genome. The structure of Ty1 was investigated by restriction site mapping and partial nucleotide sequence determination, and it turned out to be surprisingly similar to a bacterial transposon. Like a typical transposon, Ty1 is flanked by repeated sequences, but in direct orientation rather than in the inverted orientation found in most, though not all, known transposons. The Ty1 flanking sequence is 0·3 kb in length and is called **delta**; its resemblance to an IS sequence is enhanced by the fact that it occurs in many positions in the genome unaccompanied by the remainder of the Ty1 element. The genome of the yeast strain investigated was found to harbour at least 100 copies of delta but only about 35 of Ty1, implying about 30 or more 'free' deltas. Another typical feature of bacterial IS elements, the presence of near-perfect inverted repeats at their ends, finds only a faint echo in delta, which has just two base-pairs at one end repeated in inverted orientation at the other.

The resemblance of Ty1 to a bacterial transposing sequence is impressively enhanced by the finding that it occurs flanked by directly repeated five base-pair sequences derived by duplication of the sequence of the recipient chromosome. This strongly suggests that Ty1 is inserted into the chromosome by a mechanism similar to the one which operates on transposons in bacterial cells (cf. *Fig.* 10.13, p. 269).

In spite of its transposon-like appearance, Ty1 seems in general to be rather stably fixed at the chromosomal loci which it occupies. However, it can be a cause of mutation and chromosomal breakage. Some gene inactivations have been traced to the insertion of Ty1 into the coding sequences, and some cases of deletion and transposition have been explained as due to chromosome breakage occurring precisely at Ty1 termini. Here again, the parallel with IS elements and transposons is striking.

Movable Sequences in *Drosophila*—copia etc.

The *D. melanogaster* element known as **copia** was first identified in a cloned DNA fragment which was selected as hybridizing to an abundant ('copious') species of cellular RNA. The copia sequence was subsequently shown to occur dispersed in the *D. melanogaster* genome in about 20–40 copies, the number depending on the geographical origin of the flies. It is about 5 kb in length, and the base sequences have been determined for considerable distances in from its termini. It will by this time probably come as no surprise to the reader to learn that the terminal regions are repetitious; in this case the terminal repeats are 276 bases in length and are in direct, rather than inverted, orientation. Very much like bacterial IS elements, the 276 base repeats have short inverted repeats at their termini—17 base-pairs long, with a few mismatches. A rather astonishing feature of these inverted repeats, considering the almost total lack of any obvious similarity between *Drosophila* and yeast, is their close similarity to one end of the Ty1 delta sequence. The resemblances between copia and Ty1 are shown in *Fig.* 15.13. Like Ty1 and the bacterial transposable elements, copia insertions are found flanked by direct duplications (seven base-pairs long in this case) of a sequence which was present in single copy at the insertion site before copia was inserted.

There are two lines of evidence leading to the conclusion that copia is a movable sequence. Firstly, its transposition to a new site in the genome has been directly demonstrated.[38] In one experiment, in which *Drosophila* cells were cultured *in vitro*, it was found that, after a period of culture, the number of copies of copia per genome had increased about three-fold as compared with the embryonic cells from which the cell line had been derived (the abundance of the sequence was determined by reannealing kinetics, *see* Chapter 1, p. 18). A BamHI fragment containing the copia sequence was cloned from the cultured cells and compared with the homologous BamHI fragment cloned from embryo DNA. The two fragments were found to differ in size by 5 kb, and this was shown by restriction site mapping to be due to the insertion of a copia element in the cell culture fragment. Some of the evidence is summarized in *Fig.* 15.14.

The second strong indication of copia mobility comes from comparisons

Fig. 15.13

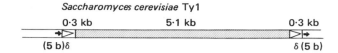

Saccharomyces cerevisiae Ty1

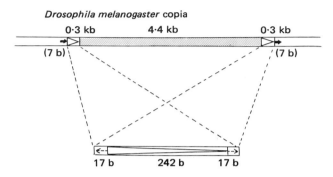

Drosophila melanogaster copia

Some movable repetitive DNA elements in eukaryotes.
After Eibel et al.[36] and Levis et al.[37]
Comparison of the termini of the copia 276 base-pair repeat and the yeast
delta sequence.

copia   5' TGTTGGAATA -----CGGAACAACA 3'
           3' ACAACCTTAT -----GCCTTGTTGT 5'

Tyl delta 5' TGTTGGAATA ---------------TCA 3'
           3' ACAACCTTAT ---------------AGT 5'

between *Drosophila* strains of different geographical origins. By the technique of *in situ* hybridization, in which salivary gland chromosome preparations were exposed to ^{3}H-labelled single-stranded copia DNA, it was shown that copia was present at different sites in different strains. In this respect, copia resembled two other kinds of dispersed repetitive sequence called **297** and **412** after the numbers of the cloned fragments in which they were first identified. *Fig.* 15.15 shows the different chromosome 2 bands in which sequence 412 is integrated in American and Japanese strains; the comparison is facilitated by the side-by-side alignment of the two second chromosomes in F_1 hybrid larvae.

There is some evidence that copia is associated with certain extraordinary cases of localized genetic instability. One such case involved an unstable state of the eye-colour mutation *white–apricot* (w^a), which was observed to undergo transposition to, for example, various different sites on the second chromosome. W. Gehring and R. Paro[40] were able to show the presence of a copia element close to w^a in its normal position on the X-chromosome and also at the second chromosome site to which, in the case under investigation, w^a had been transposed. It is plausibly conjectured that chromosome breaks generated by the copia element were involved in the observed transpositions and that the same element, and probably other kinds of movable sequences, are the cause of other 'hot-spots' for transpositions and deletions which have been reported in *Drosophila* from time to time.

More recent work indicates that there are many more kinds of dispersed repetitive sequences in *Drosophila*, and that most of them are comparatively short

Fig. 15.14

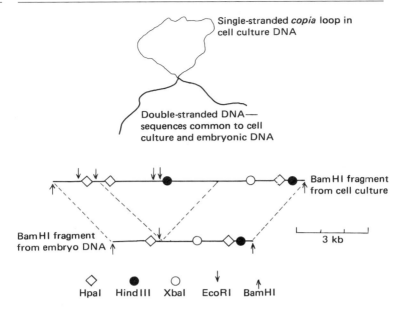

Acquisition of a copia element by a segment of *Drosophila melanogaster* chromosome during *in vitro* cell culture. A BamHI fragment from DNA of a cell culture was selected on the basis of its hybridization with a copia DNA probe. The corresponding BamHI fragment was identified and cloned from the DNA of embryos and shown not to contain the copia element both by hetero-duplex analysis and by restriction site mapping.
After Potter et al.[38]

Fig. 15.15

Sites of integration of the movable sequence 412 in the second chromosomes of two strains of *Drosophila melanogaster*.
From Strobel et al.[42]
Salivary gland chromosomes from larvae from the cross Oregon × Seto were hybridized with ³H-labelled single-stranded DNA including the 412 sequence (*see text*). The silver grains induced in the applied radiosensitive film (arrowed) are seen superimposed on the bands where the 412 sequences reside. The Oregon and Seto second chromosome are closely juxtaposed (through the normal somatic pairing of polytene nuclei), and differences as well as similarities in sites of 412 integration can be seen.

(of the order of 1 kb) and tend to occur in clusters;[41] the clustering might have been expected from the observation of 'long-period' interspersion of unique and repetitive sequences in the *Drosophila* genome. The αβ and αγ sequences referred to above (p. 434) are probably examples.

The *Drosophila* genome also contains many repetitive sequences consisting largely of pairs of long inverted repeats.[39] This accounts for the 'foldback' DNA fraction recognized by its almost instantaneous reannealing (cf. Chapter 1, p. 18). One such element has had its base sequence completely determined,[39a] and turns out to be a little over 4 kb in length including inverted flanking repeats of 500–600 base-pairs. The sequence between the flanking repeats contains an open reading frame which could code for a protein—perhaps a 'transposase', since the element must be supposed to be movable in order to account for its dispersal over the genome. The long inverted repeats are reminiscent of some of the bacterial transposons.

The presence of so many potentially movable sequences in *Drosophila*, even if individually they move very seldom, could well account for the greater part of the spontaneous mutation frequency (*see* Chapter 16, p. 478).

One of the manifestations of movable DNA sequences in *Drosophila melanogaster* is **hybrid dysgenesis**.[52] This term covers a syndrome of effects seen in progenies of crosses between certain pairs of strains of different geographical origin. The effects may include low fertility, high spontaneous mutation rate, a tendency to chromosome breakage, and the occurrence of some recombination in the male (recall that in *Drosophila*, recombination is normally confined to the female sex). All of these effects have been attributed to movable sequences present in the male parental strain but not in the female. It seems that the sequences concerned (which may be of different kinds in different cases) are relatively stable in a strain that is habituated to their presence, but become destabilized, i.e. subject to frequent movement with accompanying risk of chromosome breakage and mutagenic insertion at new sites, when introduced into the egg of a female lacking the sequences. The dysgenic effect is usually seen only when the cross is made in one direction; when a female already harbouring the movable sequences is crossed to a male lacking them there are no untoward consequences. One interpretation is that the potentially dysgenic sequences code for repressors of their own transposition, and that these repressors have to be already present in the egg if the disruptive movement is to be held in check. It is easy to think of analogies in prokaryotic systems—two examples are the self-regulated transposition of certain transposons (p. 270) and the self-repression of lambda prophage (p. 490).

The movable sequences (called 'P-elements') associated with one form of hybrid dysgenesis have now been cloned and their infectious and mutagenic properties clearly demonstrated.[55]

'Controlling Elements' in Maize

Some of the properties which have been demonstrated for bacterial transposons, suspected for some of the *Drosophila* dispersed sequences, were deduced by B. McClintock in the 1940s and 1950s from her genetic analyses of some highly unstable states of genes in *Zea mays* (i.e. corn or maize). The instabilities that she studied took the form of numerous spots or sectors of tissue in which gene function had been restored, the plant having started its development with an inactive mutant allele. In several instances (when the restoring mutation occurred in the germ cells) it was possible to associate the restoration of gene activity with the movement of a so-called controlling element from the locus of the originally non-functional gene to some other site in the genome. In this respect, the maize elements differ from bacterial transposons which, as we saw on p. 269, transfer by replication to new sites without vacating the old ones. The maize workers were able to distinguish three different families of movable elements on the basis that members of the same family could interact by mutual regulation or complementation, whereas members of different families behaved quite independently of one another. It was as if each type of element carried the information necessary for its own transposition and also certain controls of the frequency and timing of the transposition. The complex evidence upon which these conclusions were based has been reviewed elsewhere.[43] Other less thorough investigations on other cultivated plants suggest strongly that the McClintock phenomena are not confined to maize. Considerable attention is now being given to the investigation of the maize transposing elements at the DNA level.

Dispersed Repetitive Elements in Mammals

Dispersed repeats certainly exist in mammalian DNA and there is an extreme example in man. Reannealing of human single-stranded DNA fragments leads, under appropriate conditions, to the rapid formation of many double-stranded segments, a high proportion of which are about 300 bases in length. When these are isolated, the bulk of them are found to be cleavable at a site about 170 bases from one end, by the restriction endonuclease AluI. These 'Alu' sequences are substantially homologous to one another and extremely numerous.[44] It is estimated that they comprise as much as 3 per cent of the total DNA of the genome—that is to say, there are about 300 thousand copies per haploid chromosome complement. The base sequences of the members of the Alu family are not constant, but they share substantial homology and significant common features,[53] most notably in being flanked by short direct repeats. They are usually capable of being transcribed by RNA polymerase 3; the transcript includes RNA complementary to the whole Alu element and terminates a little way beyond it. It has been conjectured[53] that this transcription is an essential part of the Alu transposition mechanism, the transcript being copied back into DNA by reverse transciptase and the reverse DNA copies then being reinserted into the genome at more or less random locations. This scenario casts Alu in a role akin to that of a retrovirus genome (see Chapter 17, pp. 504–506). It is obvious that the movability of the Alu elements must be held at a very low level—otherwise we would hardly be able to tolerate them. It is not known whether they perform any essential function.

Satellite DNA

This, as we saw in Chapter 1, is the name given to relatively short and very highly repetitive sequences which, by chance, may happen to deviate significantly from the average base composition of the total DNA and thus be separable as 'satellite' components with distinct buoyant densities. The sequences in question may be extremely short. To take just two examples, the major mouse satellite consists predominantly of tandem repeats of the sequence

5'-GAAAAATGA-3'

3'-CTTTTTACT-5'

while in certain species of crab one finds simply

5'-ATATAT-3'
3'-TATATA-5'

or

5'-ATCCATCC-3'
3'-TAGGTAGG-5'.

In each case the simple sequence is repeated millions of times. In other cases, for example in species of the field mouse (genus *Apodemus*) and in humans, the highly reiterated sequences are a few hundred base-pairs in length—still short by comparison with most protein-coding sequences. In size, degree of repetitiveness and dispersal over all chromosomes, these longer satellite sequences are not clearly distinct from some abundant dispersed repetitive sequences, for example the Alu element in the human genome. But they differ from the dispersed Alu elements in being closely clustered in tandem arrays in heterochromatin—most conspicuously in the centromere regions of the chromosomes (cf. *Fig.* 1.18 for a demonstration of centromeric localization).

Two striking facts about satellite DNA sequences which demand explanation are their great variability between otherwise closely similar species and their relative uniformity within each species. The interspecific variation may take the form of an enormous multiplication on one species of a sequence present in only relatively few copies in another. The second remarkable finding is the presence of the same abundant satellite sequences on *all* the chromosomes of a given species. We touch again on these facts in Chapter 18 (p. 550).

Much thought and investigation have been devoted to the problem of the functions of satellite sequences. Several possible functions have been suggested but some of these have been discredited and not one has been clearly substantiated. At least in *Drosophila*, heterochromatin, consisting largely of satellite sequences, shows little or no crossing-over and presumably renders any genes embedded within it immune to recombination, but it is not apparent that this confers any advantage on the organism. An argument against satellite DNA having any significant function is its great variability between otherwise similar species, both in the kinds of sequence present and in their quantity.

Conformity among Repetitive Sequences

During the past few years evidence from several sources has suggested a new factor affecting genomic architecture, with interesting evolutionary implications.

We saw in *Fig.* 15.8 (p. 434) the arrangement in *Drosophila melanogaster* of the five repeated copies of the gene coding for the 70 000 dalton heat-shock protein (Hsp70). They occur as two inverted duplications, one in polytene-chromosome band 87A and one in band 87C, plus an additional single copy in 87C. The two inverted duplications, though not the additional singleton, also occur in 87A and 87C in the closely related species *D. simulans* and *D. mauritiana*. Leigh-Brown and Ish-Horowicz[49] made restriction site maps of the relevant DNA sequences from all three species, using a method essentially similar to that described in *Fig.* 14.3 (p. 391). A very striking regularity emerged. The two genes of an inverted pair have identical restriction site maps in five out of the six cases (87A and 87C in three

Table 15.1	Similarities and differences in repeated gene copies between different chromomeres and different *Drosophila* species			
	Restriction sites within the Hsp70 coding sequence			
*Comparison**	*In common*	*In 1st not in 2nd*	*In 2nd not in 1st*	*Total differences*
D. melanogaster				
87A(L)/87A(R)	10	0	0	0
87C(L)/87C(R)	9	0	0	0
D. simulans				
87A(L)/87A(R)	9	0	0	0
87C(L)/87C(R)	8	0	0	0
D. mauritiana				
87A(L)/87A(R)	9	0	0	0
87C(L)/87C(R)	7	0	3	3
D. mel. 87A/*D. sim.* 87A	8	2	1	3
D. mel. 87A/*D. mau.* 87A	8	2	1	3
D. sim. 87A/*D. mau.* 87A	9	0	0	0
D. mel. 87C/*D. sim.* 87C	8	2	1	3
D. mel. 87C/*D. mau.* 87C(L)	6	3	1	4
D. mel. 87C/*D. mau.* 87C(R)	8	1	2	3
D. sim. 87C/*D. mau.* 87C(L)	7	1	0	1
D. sim. 87C/*D. mau.* 87C(R)	8	0	2	2
D. melanogaster				
87A/87C	8	2	1	3
D. simulans				
87A/87C	8	1	0	1
D. mauritiana				
87A/87C(L)	7	2	0	2
87A/87C(R)	9	0	1	1

*(L) and (R) refer to the 'left' and 'right' inverted repeats each polytene-chromosome band, 87A or 87C (cf. *Fig.* 15.8). Five different restriction enzymes were used to identify numbers of sites listed.
Data of Leigh-Brown and Ish-Horowitz.[49]

species), and this in spite of the fact that the interspecific comparisons showed differences in all but one case. There were also differences of a roughly similar magnitude between the 87A and 87C genes of the same species (*Table* 15.1).

Since the inverted repeats must already have been in existence at the time the species diverged, one would have expected more differences, rather than fewer, to have accumulated between two members of an inverted duplication within each species than between the representatives of the same member in different species. The authors concluded that there must be some mechanism for maintaining similarity within the linked pairs and they proposed that this mechanism might be the base-pairing and correction of mismatches between inverted repeats, probably by looping-out of a stretch of single-stranded DNA as shown in *Fig.* 10.5 (p. 257).

Another example, this time involving tandem rather than inverted duplication, has been inferred from data on the base sequences of three human γ-globin genes[50] (cf. *Fig.* 15.10). Two of these, one copy of γ^A and one of γ^G, were cloned from the DNA of one individual and could be shown to represent sequences that had been linked on the same chromosome. The third was the γ^G gene from the *other* chromosome of the same person. The two γ^G genes from different chromosomes of the same individual has a surprising number of differences (indicating considerable natural polymorphism) and, with respect to many of the varying sites, there was much more similarity between γ^A and γ^G on the same chromosome than between the two γ^G genes on different chromosomes. Again some mechanism promoting intrachromosomal uniformity is suggested.

That conversion *can* occur between tandemly repeated genes has been demonstrated directly in *Saccharomyces*.[51] By transforming protoplasts of a histidine-requiring mutant strain (*his4C*⁻) with the DNA of a plasmid carrying an inserted *his4*⁺ gene it was possible to select for histidine non-requirers which had the *his4*⁺ allele as an extra gene copy integrated in tandem close to the *his4C*⁻ allele. Presumably the integration of the transforming DNA occurred by a single crossover between the yeast segment of the hybrid plasmid and the homologous segment of the yeast chromosome, just as in the analogous case of λd*gal* in *E. coli* (*see Fig.* 11.7, p. 302). By ordinary meiotic recombination, the investigators replaced the *his4*⁺ by a second *his4*⁻ mutant, *his4A*⁻, so that a histidine-requiring strain was obtained with two *different his4* mutant alleles in tandem— *his4A*⁻ *his4C*⁻. The interesting property of this duplication strain is that it yielded a spontaneous frequency of about 2×10^{-4} *his4*⁺ mitotic recombinants. These recombinants retained the duplication and were of constitution *his4A*⁻ *his4*⁺ or *his4*⁺ *his4C*⁻. Had they arisen by reciprocal crossing-over, their formation would have been accompanied by the excision of one of the two tandem repeats. In fact, it was shown that in most cases the duplication was retained. This strongly implied that the recombination event was of the conversion type, probably involving the formation of hybrid DNA between the repeated genes (paired by intrachromosomal loop formation, cf. p. 323) and correction of base-pair mismatches.

Concerted changes in clustered repetitive gene copies may occur when the sequences involved are functional in the coding sense and when they are apparently non-functional satellite sequences. Such **concerted evolution** of DNA sequences, whether it is due to random 'drift' or is driven by selection or some other force, such as directional gene conversion (cf. p. 92), is an interesting possible mechanism for relatively rapid sequence divergence of recently separated species. The aspect will be further discussed in Chapter 18 (p. 550).

The Concept of 'Selfish' DNA

In 1980 two papers[45,46] appeared advancing what was, to some, the rather outrageous notion that a part of the DNA which we find in organisms, particularly higher eukaryotes, is there not because it has been selected as conferring fitness on the organism carrying it but rather because it is able to maintain itself and proliferate on its own account, like a molecular parasite. This idea is, perhaps, startling at first sight, but on further consideration it seems quite reasonable. We find no difficulty in accepting that virus DNA, integrated as it often is into the host genome, is there 'for its own sake' and not because it confers a benefit on the host cell. Bacterial transposons, except that they spread only within the cell and are not released as extracellular infective particles, are rather like viruses and, at least in some cases, code for protein products that catalyse their own replication and accompanying transposition (cf. p. 270). It seems likely that the more recently discovered eukaryotic elements—the copia-like and inverted repeat elements of *Drosophila*, Ty1 in yeast and possibly Alu in man (and homologous sequences found in other animals)—are similarly autonomous in their ability to move. We need not suppose that such sequences, even if they are 'selfish' in origin, are necessarily always without function for the organism. Origin and current use are two different things and it is very possible that natural selection at the organism level, ever opportunistic, will sometimes 'recruit' a hitherto 'selfish' sequence and make it perform some useful function.

Satellite sequences are in a different category from transposon-like elements in that they have no special features conducive to their own proliferation, other than the fact that they often occur in tandemly clustered repeats. Nevertheless, their cross-chromosome distribution does suggest that they have the capacity to multiply differentially and to transpose, and it seems possible that their tandem repetition is, in itself, a basis for 'selfish' behaviour. As we saw on p. 323, unequal sister-strand exchanges involving tandemly repeated sequences leads to an increase in the number of copies on one chromatid but to decrease on the other. This process might result in the extension of blocks of satellite sequences if selection acted in favour of such an increase, but in the absence of selection at the organism level it is hard to see why unequal crossing-over should lead to the state of abundance we see in many satellites.

Another more speculative possibility is that intrachromosome recombination, involving members of tandemly repeated series, could lead to the excision from the chromosome of circular elements composed of a number of repeats, and that such an element, if it included a replication origin, might fill the nucleus with copies of itself which might then undergo reintegration at several different chromosomal sites. The first part of this scenario is, in fact, very like what is known to happen in the magnification of rRNA genes in the oocytes of many different animals (cf. p. 323). The final part of it, the reintegration step, is not so fanciful as it might have seemed before the transformation experiments with animal cells described on pp. 243–244 had been performed. It seems that DNA, once in the nucleus, can be integrated even into chromosomal loci which have no extensive homology to the incoming sequences. Once a foreign sequence has gained a foothold in a chromosome in which it did not previously occur, it could provide a point for integration of further sequences of the same kind, since much more efficient homologous recombination could then come into play. If these speculations have any substance, they imply that once a sequence acquires a certain abundance, in the form of tandem repeats, it may occasionally spread in almost epidemic fashion. DNA which can proliferate differentially merely because of its repetitiveness, rather than because of anything special about its sequence, has been termed 'ignorant' rather than 'selfish'.[47] The distinction is a useful one.

There must, obviously, be a limit to the amount of repetitive and functionless sequence that a genome, or the organism possessing it, can tolerate, but the wide variations which we see between species suggest that the limits of tolerance may be fairly wide. Indeed, having learnt how to live with a large amount of satellite sequence, an organism could even become dependent upon it.

We know of at least one instance in which a DNA sequence can very quickly become highly reiterated so as to confer a benefit on the cell. Animal cell cultures selected stringently for resistance to methotrexate, a drug specifically inhibiting the enzyme dihydrofolate reductase, develop a few enormously enlarged chromosomes padded out by a several hundred-fold reiteration of a DNA segment which includes the coding sequence for this enzyme. Even though this is a very special case, seldom if ever seen in nature, it nevertheless conveys some general lessons. It is impressive testimony to the existence of a fast-acting mechanism, which may also

Fig. 15.16

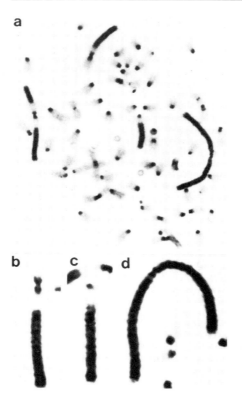

a. The karyotype of a line of mouse cells selected in culture for resistance to methotrexate. The metaphase chromosomes were stained by the C-banding procedure (cf. p. 36) to show centromeric heterochromatin. Note the four giant chromosomes, not present in the normal mouse karyotype, consisting very largely of repeated blocks of centromeric heterochromatin.
The enlargements of three of the chromosomes in (*b*), (*c*) and (*d*) show clearly the regular periodic structure, suggestive of repeated blocks of heterochromatin. The gene coding for dihydrofolate reductase, extending, with introns, over about 40 kb is included within a segment of heterochromatin of perhaps as much as 1000 kb, and it is the whole segment which has become reiterated to several hundred copies during selection for methotrexate resistance.
From Bostock and Clark[48] reproduced by permission.

have created the arguably functionless satellite sequences, whereby populations could restructure their chromosomes in response to intense selection pressure. It also suggests that DNA sequences which in some contexts may appear selfish or ignorant in their proliferation may in other circumstances render the cell a real service.

It seems appropriate in this context to mention **B-chromosomes**, which are accessory chromosomes without essential function; they can be present in variable numbers or absent altogether without any obvious effect on the phenotype. They are commoner in plants than in animals, and have been especially studied in grasses. Their origin is unknown and their DNA sequence content not well understood. It is tempting to regard them as 'selfish chromosomes', but there is evidence that they sometimes confer some additional fitness on plants carrying them and it is not clear how otherwise they would be able to maintain themselves in populations in the face of accidental loss. The large body of literature on B-chromosomes has been fairly recently reviewed.[54]

It is worth noting that when the idea of selfish DNA was first put forward it was thought that intervening sequences might be included in this category. This now seems very doubtful. Intervening sequences are not, in general, repetitious and there is nothing to show that they can spread through a genome. They seem, in the genes as they presently exist, to play some essential role in mRNA precursor processing (cf. p. 414), although, on the other side of the argument, it could be that the processing mechanism has become adjusted to and ultimately dependent upon the introns rather than the other way round.

The 'selfish' DNA concept is a valuable one insofar as it counteracts naive teleology (argument based on presumed 'purpose'). Its possible danger is that it may serve as a device for avoiding thought—once a particular class of DNA has been labelled 'selfish' we may excuse ourselves from worrying any more about possible functions. A product of scepticism, it should itself be viewed sceptically.

15.10   Summary and perspectives

The goal of the complete analysis of the genome has been attained in the case of certain small viruses. In the genomes of some of these, one sees an extremely compact organization of genetic information with some sequences being used twice over by translation in different reading frames or different modes of transcript processing.

In prokaryotes, the task of obtaining complete base sequences of all genes and intergene spaces is beyond our present resources, but in those parts of bacterial genomes that have been analysed one sees a rather consistent pattern of non-overlapping genes with minimal spacing between them. At least in the enteric bacteria exemplified by *E. coli*, functionally related genes are often physically contiguous, with units of transcription (operons) each including several cistrons. There is very little repetitive DNA and what there is is either functional (e.g. rRNA sequences) or belongs to the families of transposable (IS) sequences which seem to be molecular parasites.

Eukaryote nuclear genomes also have a certain amount of clustering of functionally related genes, but these are virtually always separately transcribed. The clustering may be merely a consequence of evolutionary history, but there are some hints that coordination of gene activity at the chromosome level may be involved. Eukaryotes (especially higher eukaryotes) frequently have their coding sequences interrupted by intervening sequences (introns), which, not uncommonly, may stretch the genes to ten or more times the coding length. In higher eukaryotes the spaces between genes are usually large. The spaces are filled with a variety of material of dubious function or none—repeated sequences of various kinds (including some movable sequences which may simply be parasitic), 'decayed' genes which are no longer functional, and unique sequences whose functions, if any, appear not to be sensitive to mutation.

Whereas closely related species are usually, so far as we know, nearly identical or at least very similar in the base sequences of their protein-coding genes, they can be very different with respect to their highly repetitive sequences. The occurrence of 'concerted evolution' of repetitive sequences may be a factor in the early divergence of genomes after species formation.

Mitochondrial genomes, if those of yeast and man can be taken as representative examples, show, on the whole, eukaryotic rather than prokaryotic features with a gene organization that is not particularly compact. In yeast mtDNA some of the genes have extensive introns and much repetitive sequence is present.

The total analysis of the genome is an achievable objective with prokaryotes and already an achieved objective in the cases of several small viruses. So far as eukaryotes are concerned, it is at present beyond reach. However, we can expect, over the next few years, to read detailed accounts of many local excursions into selected parts of eukaryote genomes. Given a point of entry, such as may be provided by any gene or other sequence for which one has a probe, it is possible to 'walk' along the chromosome by the use of overlapping restriction fragments, using each one for the detection and cloning of the next. Each step will normally only advance the walk by a few kilobases, but occasional 'leaps' can be accomplished through the use of fortunately placed inversions. A sequence identified at one end of a segment inverted in a mutant strain may be picked up perhaps hundreds or thousands of kilobases further along the chromosome in DNA from a structurally normal gene. Given knowledge of the genetic map, tactics such as these can render most parts of a eukaryotic genome susceptible to sequence analysis starting from a limited number of entry points.

Chapter 15 — Selected Further Reading

Comings D. E. (1978) Mechanisms of chromosome banding and implications for chromosome structure. *Annu. Rev. Genet.* **12**, 25–46.

Dover G. S. and Flavell R. B. (ed.) (1982) *Genome Evolution*, 382 pp. London, Academic Press.

John B. and Miklos G. L. G. (1980) Functional aspects of satellite DNA and heterochromatin. *Int. Rev. Cytol.* **58**, 1–114.

Long E. O. and Dawid I. B. (1980) Repeated genes in eukaryotes. *Annu. Rev. Biochem.* **49**, 727–764.

Spradling C. and Rubin G. M. (1981) Drosophila genome organization: conserved and dynamic aspects. *Annu. Rev. Genet.* **15**, 219–264.

Problems for Chapters 14 and 15

1 A 9·2-kb HindIII fragment of genomic DNA was cloned and selected on the basis of its hybridization with a cDNA probe copied from mRNA. The fragment was cut out of the vector and subdigested with EcoRI, BamHI and PstI singly and in pairwise combinations. The resulting fragments were transferred to nitrocellulose and tested for hybridization to the probe. The following were the sizes (in kb) of the fragments; * indicates hybridization to the probe.
EcoRI: 4·5*, 3·9*, 0·8.   PstI: 4·2*, 2·8*, 2·2.
BamHI: 4·5*, 3·2*, 1·5.   EcoRI + PstI: 3·9*, 2·3*, 2·2, 0·5, 0·3.
EcoRI + BamHI: 3·2*, 2·4*, 1·5, 1·3*, 0·8.
PstI + BamHI: 2·7*, 2·2, 1·8*, 1·5, 1·0*.
Make a restriction site map of the cloned fragment and locate the sequences within it that hybridize to the cDNA probe. What can one conclude about the structure of the gene?

2 The rabbit β-globin gene consists of mRNA-coding sequences of lengths (in order from 5′ to 3′) 146, 222 and 223 bases, with intervening sequences of 126 and 573 bases between the first and second, and second and third, respectively. Polyadenylated RNA isolated from rabbit bone marrow cells was hybridized to single-stranded DNA from a genomic clone containing the gene, and the single DNA strand which was not hybridized was digested away with nuclease S1. The surviving DNA fragments were separated from the RNA and their sizes were determined (with an accuracy of within 5 per cent). Five fragments were distinguished of estimated sizes 1250, 1085, 1000, 222 and 141. What does this suggest concerning the processing of the gene transcript? (Grosveld et al., 1981, *Cell*, **23**, 573.)

3 Suppose that genomic DNA is randomly sheared and fractionated by size to give a population of fragments of approximate length 10 kb which are then cloned in some suitable vector. How many clones will need to be screened in order to give a 95 per cent chance of finding, in a single clone, the whole of a 5-kb single-copy sequence from (i) *Saccharomyces*, (ii) *Drosophila* and (iii) mouse? Note that if one wants to reduce the chance of failure to less than 0·05 (a chance given by e^{-m}, the zero term of the Poisson distribution), one has to aim at a number of successes roughly three times the mean expected number ($e^{-3} \simeq 0·05$).

4 The following DNA sequence was found in a cloned fragment overlapping the upstream end of a ribosomal protein gene of *E. coli*:

5′ ATATTCTTGACACCTTTTCGGATCGCCCTAAA-
ATTCGGCGTCCTCATATTGTGTGAGGACGTTTT-
ATTACGTGTTTACGAAGCAAAAGCTAAAACCAG-
GAGCTATTTAATGGCAACAGTT... 3′

What significant features can you recognize in this sequence? Where would you expect transcription and translation to start? What would you predict as the first few amino acid residues at the N terminus of the gene product? (Post and Nomura, 1980, *J. Biol. Chem.* **255**, 4660.)

5   The following sequences were found at the 3′ ends of the mRNAs for (a) a mouse immunoglobulin light chain and (b) rabbit β-globin:

    a. 5′ CUAAUAUUUGAUGUCGCAGAAAAUAUUCAAUAA-
AGUGAGUCUUUGCACUUGCAAAAAA-polyA 3′
    b. 5′ AGCCCCUUGAGCAUGACGGCUGCUAAUAAAGG-
AAAUUUAUUUUCAUUGCAAAAA-polyA 3′

What features do these two sequences have in common? (Proudfoot and Brownlee, 1974, *Nature*, **252**, 359.)

6   The heavy chain of mouse immunoglobulin M exists in two forms: membrane-bound (m) and secreted (s). The two forms differ only at their C termini, where a 20-residue relatively water-soluble sequence in the s form is replaced by a 41-residue highly hydrophobic sequence in the m form. The two kinds of chain are translated from different mRNAs (produced in differently differentiated cells) and cDNA from each kind of mRNA has been cloned. A genomic DNA clone, including the whole of the coding sequence for the C-terminal half of an s chain and a considerable stretch of 3′-flanking sequence, is available. How could you investigate the possibility that the two kinds of messenger were derived by two ways of processing the same primary transcript? (Early et al., 1980, *Cell*, **20**, 313.)

7   Considerable attention has been given to the question of whether operons exist in eukaryotes. A possible example was investigated in *Neurospora* (Rines et al., 1969, *Genetics,* **61**, 789). The synthesis of the aromatic ring is effected by a series of seven enzyme activities, of which numbers 2–6 can be eliminated individually or together by mutations which cluster closely at an *aro* locus. A sample of 470 *aro* mutants included the following types: (a) 256 each lacking only one enzyme activity (35 enzyme 2, 5 enzyme 3, 96 enzyme 4, 39 enzyme 5 and 81 enzyme 6); (b) 132 lacking all enzyme activities and able to complement none of the class *a* mutants; (c) 41 lacking all activities but able to complement class *a* mutants defective in enzymes 2, 5 or 6; (d) 3 lacking all enzyme activities but able to complement class *a* mutants lacking enzymes 2 or 6; (e) 38 lacking all enzyme activities but able to complement class *a* mutants lacking enzyme 6. A number of the mutants in classes *b* to *e*, but none in class *a*, have been shown to be chain-termination mutants by the criterion of suppressibility. Does this look like an operon? If not, what could it be? What additional information do you need to answer the question?

8   Sucrose synthetase, an abundant enzyme of maize endosperm, is encoded by the gene (*sh*) which, when homozygous in its recessive mutant form, gives a *shrunken* phenotype. The mRNA for the enzyme has been isolated, and cDNA copied from it has been cloned. McClintock has obtained genetic evidence for a transposable element, or family of elements, which can be inserted at many loci, including *sh*, to cause highly unstable inhibition of gene expression. How could the cDNA clone help in the elucidation of the molecular nature of these wandering elements? (Burr and Burr, 1981, *Genetics*, **98**, 143.)

9   Hitherto the basic unit of heredity, the gene, has been defined in various ways on the basis of genetic methodology—recombination and complementation. Now that, in many examples, we have access to DNA sequences and the RNA sequences transcribed from them, as well as to the protein products if any, it should be possible to define the genetic unit in biochemical terms. How should we define the biochemical gene?

Genetic Controls of Cellular Determination and Differentiation

16.1   Introduction

The identification of genes and their products takes us only part of the way in our attempt to understand how the genotype determines the phenotype of a complex multicellular plant or animal. Just as important, and a great deal more difficult to elucidate, are the controls which ensure that specific genes become active and others inactive at particular times and in particular places. It is clear that cell and tissue differentiation is the result of selective and precisely programmed switches, either on or off, of gene activity or of a *potential* for activity. We saw in Chapters 12 and 13 how genetic controls of transcription operate in bacteria and reviewed some of the evidence that somewhat similar mechanisms may be important in eukaryotes also. But in multicellular organisms there is at least one level of control which is hardly understood at all. Not only are there controls of transcription, but the *setting* of the controls often has the property of heritability through successive cell divisions, so that a whole cell lineage can be determined to show some pattern of gene activity, even though the manifestation of the pattern may be delayed until some future time and stage of development. The mechanism, or mechanisms, of **determination** and of its maintenance through cell division is arguably the most important unsolved problem for geneticists. This question of molecular explanation will be taken up again in the next chapter. In this chapter we shall consider how genetic approaches help to define the phenomena to be explained.

Genetic methodology can advance our understanding of cellular determination and differentiation in two ways. Firstly, genetic mutations can be used to 'mark' clones of cells, so that their fates can be followed through development. This is often essential for a description of how the developmental programme unfolds and for determining the extent to which its successive steps are irreversible. Secondly, one can study the effects of mutation in genes which seem likely to be involved in developmental switches, in the hope that light will be shed on the modes of action of such genes.

This chapter will be very much concerned with *Drosophila melanogaster*, which is the species in which the genetic control of development has been studied in the greatest depth. Examples will also be drawn from the opposite ends of the range of multicellular complexity—the nematode worms and the mammals.

16.2   Rigidity and flexibility—mosaic versus regulatory development

A Nematode Worm

The most complete descriptions of the development of an animal have been achieved in nematode worms, which include some of the simplest of multicellular animals. The favourite species for investigation in recent years has been *Caenorhabditis elegans* which has advantages in small size, ease of culture and convenience for genetic analysis. The adult animal has only about 800 cells, and on hatching from the egg it has only about 550. Using continuous monitoring by videotape, and a light microscope with interference optics, it has been possible to trace the origin of every cell through a lineage extending back to the fertilized egg.

Fig. 16.1 shows the pattern of cell descent up to the stage where the embryo has 182 cells—that is two to three further rounds of cell division from maturity. The striking feature of this pattern is its reproducibility. The fate of any cell at any stage of development can be predicted with certainty. Each branch of the cellular 'family tree' seems to have its own timing control of cell division—some undergo more divisions and some less, and all at predetermined times—and each has certain special capacities for further differentiation.

That the potential of each branch of the cell lineage is determined from an early

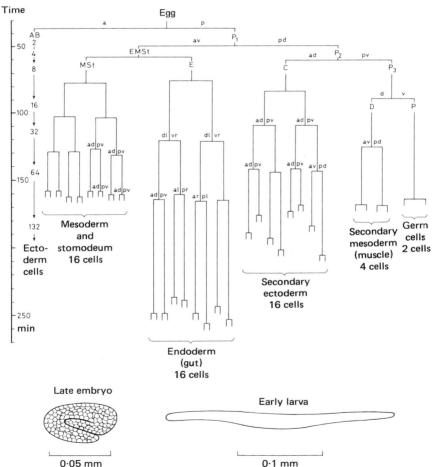

Cell lineage in the development up to the late embryo stage of the nematode worm *Caenorhabditis elegans*.
After Deppe et al.[1]

stage and not easily changed has been demonstrated by experiments in which single cells were isolated from the others and left to develop and divide on their own. W.B. Wood and his associates[2] developed a method for bursting the shell of the developing egg and culturing the cells thus released in a medium that would support several further rounds of cell division. Very often the bursting procedure resulted in the death of every cell except one, and it was then possible to see whether that cell continued to follow its own predetermined programme or whether it could adjust to take over the functions of its dead sister cells. The results were all consistent with maintenance of the cell programme. Thus the AB cell, which is the anterior cell resulting from the very first cleavage of the egg, divided to give a partial embryo which showed no sign of any differentiation of muscle or gut, whereas the posterior product of the first cleavage, the P_1 cell, was able on its own to develop into a partial embryo which did show some early development of these tissues (cf. *Fig.* 16.1). The potential for developing into gut (recognized by the presence of characteristic granules in prospective gut cells) appeared to segregate at the following division of P_1; an isolated P_2 cell did not show this capacity, which is normally a property of the other P_1 product, the EMSt cell. The separated cells of disrupted embryos did not, in fact, go on to develop into perfect replicas of parts of adult worms; not surprisingly, the products of cell division lost their normal orientation with respect to one another in the partial embryos. But at least a considerable part of each cell's determination to differentiate in a certain way and no other seemed to be maintained in spite of the drastic change in cellular environment.

The development of the *Caenorhabditis* egg is a classic example of what has been called the mosaic type—each cell has its own developmental fate and is not interchangeable with any other cell. The contrasted type of development is described as regulatory, with some possibilities for reprogramming in response to experimental manipulation.

Mouse Embryos

Regulatory capacity is seen in vertebrate early embryos dissected experimentally into individual cells. In an experiment illustrated in *Fig.* 16.2 single cells were isolated from the four-cell stage of the early embryo of a white mouse, and placed in a culture medium which allowed them to undergo further division. The pairs of cells formed by one more cleavage were again separated, and each cell of a pair was associated by manipulation with four other cells isolated from eight-cell embryos of black mice in such a way that the 'white' cell was on the inside and the 'black' cells were on the outside. The aggregates were then further cultured and each was found to develop into a normal-looking **blastocyst**, i.e. the hollow ball of cells which goes on to invaginate and form the layered **gastrula**, which is the next stage of embryogenesis. These synthetically derived blastocysts developed into normal late embryos and young mice when each was implanted into the uterus of a pregnant female.

The important conclusions drawn from these experiments were as follows:

1. The whole mouse which developed from each aggregate of five cells was descended from the inside ('white') member of the aggregate; the four outside cells ('black') divided to make the placenta.

2. Both members of each pair of sister cells at the eight-cell stage were individually capable of developing into an entire animal. There is clearly no irreversible segregation of potentialities up to the four-cell stage in normal

Fig. 16.2

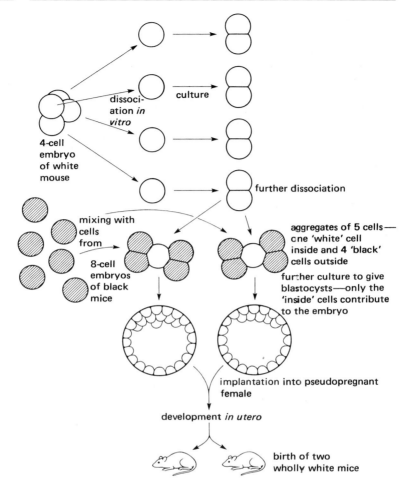

4-cell
embryo
of white
mouse

dissoci-
ation *in
vitro*

culture

further dissociation

mixing with
cells
from

8-cell
embryos
of black
mice

aggregates of 5 cells—
one 'white' cell
inside and 4 'black'
cells outside

further culture to give
blastocysts—only the
'inside' cells contribute
to the embryo

implantation into pseudopregnant
female

development *in utero*

birth of two
wholly white mice

An experiment testing the totipotency of cells in the early mouse embryo.
After Kelly.[3]

mouse development nor after one further cell division in culture. The development
mental fates of cells at this stage seem to be determined by their *positions* within
the cell aggregate rather than by their immediate ancestry.

In a more recent study,[14] mouse embryos were disaggregated at the eight-cell
stage into single cells which were then allowed to divide once more in culture. The
usual result of this further division was the formation of two morphologically
distinct cells, one resembling the normal sixteen-cell stage peripheral cells, which
go to form the trophoectoderm (potential placenta) and the other resembling the
sixteen-cell stage inner cells, which together develop into the mouse. Thus, it
appears that cells which, as we saw above, are totipotent at the four-cell stage give,
after one more division in the intact embryo, cells programmed to differentiate
when they in turn divide to give the sixteen-cell embryo.

Thus some degree of developmental programming does occur rather early in mammalian embryogenesis, but it occurs later than in nematodes; we do not know how irreversible it is.

16.3   The use of genetic markers for tracing cell lineages

In the example just described a mixture of embryonic cells from albino and black mice was used to show the ability of a particular cell to develop into a whole mouse. The more usual use of mixtures of genetically distinct cells (**chimaeras**) is to show which parts of whole organisms are derived from cells of the same lineage (i.e. clones), and how early in development such clones originate.

Chimaeras in Mammals

Two kinds of chimaera, one naturally formed and one artificial, have been studied in mammals. The naturally occurring kind is exemplified by all female mammals that are heterozygous with respect to an X-linked visible marker. As we see in the following chapter, random inactivation of one of the two X-chromosomes occurs in every somatic cell of the female mammalian embryo. This is the basis for the mosaic pattern of the (always female) tortoiseshell cat, and for a number of striped or 'tabby' phenotypes in female mice and guinea-pigs due to X-linked coat-colour mutations. Each patch of uniformly coloured fur in the coat of such an animal must stem from a single cell present in the embryo at the time the X-inactivation occurred (*see Fig.* 17.13, p. 510).

Rather similar mosaic patterns, but not confined to sex-linked markers and usually more fine grained, can be obtained in **allophenic** mice which develop from blastocysts constituted artificially from random mixtures of cells of different genotypes, determining distinct coat phenotypes.[4] Beyond showing that considerable areas of skin can be colonized by single clones of pigment-producing cells (melanocytes), studies of mammalian chimaeras have not been very helpful in mapping mammalian development; the melanocytes, which provide the observable surface variation, are no guide to the distributions of cells in internal tissues. As we see below, the situation is more favourable in *Drosophila*, where the visible surface phenotype is very rich in information.

Fate Mapping in *Drosophila*

Whereas *Caenorhabditis* is a very simple invertebrate animal, *Drosophila* is a very complex one, and it is a much more difficult problem to describe its development in terms of the fates of individual cells. Demonstrable determination of cell fate does not occur until more than ten rounds of mitosis have occurred, giving a number of nuclei comparable to the total cell number in the nematode worm.

The first nine nuclear divisions in the egg produce a cluster of nuclei not

Fig. 16.3

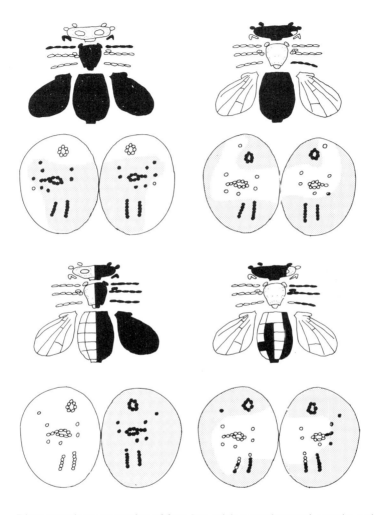

Diagrammatic representation of four *Drosophila* gynandromorph mosaics and their presumed origins. Blastocysts are shown as split into left and right halves with the primordia of imaginal discs shown as circles—black for wild type (female) and open for mutant (male). The random positioning of the boundaries between wild type (stippled) and mutant (open) areas in the blastocysts is due to the random orientation of the spindle of the first mitotic division in the fertilized egg.
Redrawn from Hotta and Benzer.[5]

separated by cell membranes and these migrate to the egg periphery where they undergo a few more divisions and form the blastocyst, which is essentially a hollow ball of undifferentiated cells. Although it is difficult to rule out, there is nothing to indicate that the fates of nuclei are in any way determined before the blastocyst stage, and it is generally believed that their respective developmental programmes become fixed by positions which they take up in the blastocyst. The positions corresponding to the determination of all the external features of the fly have been mapped by Y. Hotta and S. Benzer, using an ingenious idea due originally to A. H. Sturtevant.

The Hotta–Benzer method depends on the production of 50 : 50 mosaics through frequent loss of one X-chromosome from one of the products of the first mitotic division of the egg nucleus. A particular kind of ring X-chromosome, when present in a fertilized egg together with a normal X, has a high probability of loss at this division, the effect being to leave one daughter nucleus with only one X-chromosome. At subsequent mitotic divisions, this trouble apparently does not recur and so the larva, and subsequently the whole adult fly, will be a female/male mosaic (**gynandromorph**) consisting of 50 per cent XX (female) and 50 per cent XO (male) tissue. If the structurally normal X-chromosome carries recessive mutations, the effects of which can be seen virtually all over the body surface (*yellow* body colour and *singed* bristles fulfil this requirement), with the ring-X carrying the dominant wild-type alleles, it is possible to see precisely which parts of the surface derive from one daughter nucleus of the first division and which from the other. The two territories are usually of about equal extent and are each rather coherent with little intermingling; the boundary between them can fall anywhere and run in any direction. It thus appears that the orientation of the first mitotic division spindle is at random in relation to the egg surface and that, when the time comes for the nuclei to migrate to the surface, they do so in a fairly orderly way without too much mixing of different cell lineages. This explanation is illustrated in *Fig.* 16.3.

The reasoning leading to the construction of the fate map is as follows. If each structure or 'landmark' on the surface of the fly stems from a particular region of the blastocyst, then the probability that the male y/female y^+ boundary will run *between* two such landmarks will be proportional to the distance separating the corresponding regions of the blastoderm. This distance can be calculated in 'sturts' (named after A. H. Sturtevant), one sturt equalling a 1 per cent probability that, in a 50 : 50 gynandromorph, the two adult landmarks will appear in differently marked sectors (i.e. on different sides of the gynandromorph boundary). Through observations of the precise boundary positions in many hundreds of flies, the distances in sturts for all possible pairs of landmarks were estimated. All the distances together were consistent with a map which, over moderate distances, was two-dimensional, but overall was three-dimensional like the map of the globe. The surface features of the left and right-hand sides of the fly could be represented by two mirror-image two-dimensional maps, presumably corresponding to the left and right surfaces of the ellipsoidal blastocyst (*Fig.* 16.4). Distances could also be determined along lines connecting the two mirror-image maps on their dorsal and ventral edges, corresponding presumably to the dorsal and ventral surfaces of the blastocyst.

As first made, on the basis of surface features, the fate map included large vacant spaces; these were thought to correspond to parts of the internal anatomy of the fly. This supposition was confirmed when Hotta and Benzer managed to fate-map a behavioural (leg-shaking) abnormality; the blastoderm focus of this trait was found to be in a hitherto empty area which, to judge from its position, probably represents the ganglion controlling the movement of the leg in question.

Fig. 16.4 shows the *Drosophila* blastocyst fate-map based on surface characters. It will be seen that the disposition of the various determined regions of the blastoderm bears some relation to the arrangement of parts in the adult fly (the imago). However, the more direct relationship is between blastoderm and **imaginal discs**, the nature of which we shall now discuss.

Fig. 16.4

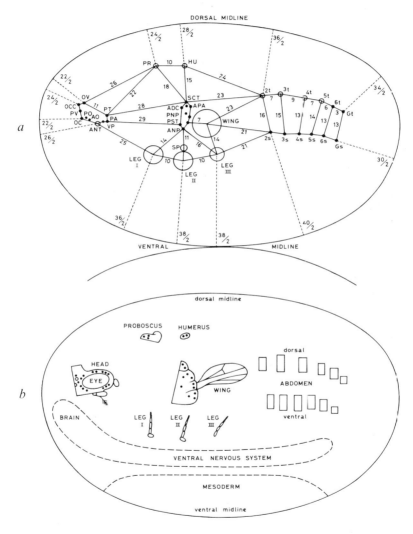

The use of gynandromorph mosaics for fate mapping in *Drosophila melanogaster*.
a. Fate-map with positions of blastoderm cells determined to differentiate into various features of the adult fly, spaced according to their separation in 'sturt' units (*see text*).
b. A diagrammatic representation of (*a*).
After Hotta and Benzer.[5]

16.4 Determination in *Drosophila* development

Imaginal Discs

Flies, like many other kinds of insects, have a larval phase which is totally unlike the adult. In the pupa, which intervenes between larva and adult, the organism is entirely reconstructed from groups of cells that are carried through the larval phase without showing any sign of differentiation. Several of these groups are clearly visible in the larva as so-called **imaginal discs** (*Fig.* 16.5). Each disc is destined to proliferate and differentiate in the pupa into a particular structure in the adult fly. Thus there is a labial disc, programmed to produce mouth parts, a genital disc for the structures involved in mating, and paired eye, wing, haltere (rudimentary second wing) and leg discs. Other parts of the fly are provided for in other, less conspicuous, groups of cells.

The fascinating thing about imaginal discs is that each kind carries a preset developmental programme which is maintained through cell divisions with a good deal of stability, even when the cells are induced to proliferate indefinitely in a quite unnatural environment. Thus, as E. Hadorn[7] first showed, it is possible to transplant a disc into the abdominal cavity of an adult fly and to culture it and its descendent cells without differentiation through successive subtransplants to further adult hosts. When reintroduced into a larva, these artificially multiplied cells will, on pupation, usually differentiate into the structure corresponding to the type of disc from which the cell line originated. Clearly, determination occurs in *Drosophila* well in advance of overt differentiation and is capable of being maintained through cell divisions even when the opportunity for putting the determined programme into effect is indefinitely deferred.

Hadorn also found that the stability of determination, while impressive, was not absolute. Sometimes a subcultured disc fragment would switch its developmental aim from one adult structure to another—for example from genitalia to antenna, or from antenna to leg. Many of these occasional **transdetermination** events can occur in either direction, but others are unidirectional—the change from haltere to wing determination, for example, is fairly frequent while the reverse change has never been observed. With prolonged culture, there is a tendency for all cultured disc fragments to stabilize finally in the state potentiating leg structure.

We thus get the impression of a high degree of self-perpetuation of each distinct development programme but with a tendency to 'slip' from time to time into another programme, the slippages occurring more readily in some directions than in others. We shall see later in this chapter how the controlled selection of one programme rather than another is dependent on certain identifiable genes.

The phenomena just outlined have been known for more than 20 years, but there is still virtually no knowledge of the mechanisms underlying either the stability of the determined state of undifferentiated cells or its occasional lapses from stability. However, some further understanding of the internal structure of imaginal discs has been achieved. It is now apparent that the discs are themselves mosaics of cells of different determination, and the following section deals with some of the relevant evidence.

Fig. 16.5

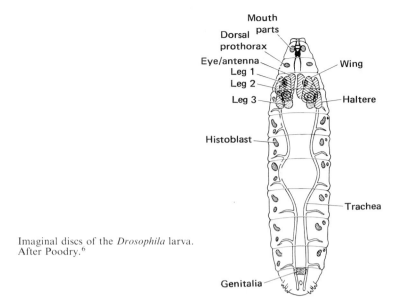

Imaginal discs of the *Drosophila* larva.
After Poodry.[6]

Compartments in *Drosophila* Development

We saw in preceding pages how genetic mosaics established by loss of a ring-X at the first cleavage division of the *Drosophila* egg can provide data from which fate-maps can be deduced. In order to follow the destinies of clones of cells initiated at later stages of development, it is desirable to have a means of inducing the expression of visible genetic markers in cells at *any* desired stage. Such a means is available in X-ray-induced mitotic crossing-over, which was dealt with in Chapter 9 in the context of parasexual genetic analysis (pp. 232–234).

Starting with a phenotypically wild-type larva heterozygous with respect to a recessive visible mutation, one can use X-rays to initiate clones of cells homozygous for the mutant allele at any time. With the marker mutations commonly used, such clones can be seen only when they occur on the surface of the fly, but recently P. A. Lawrence[8] has obtained a mutation causing loss of succinate dehydrogenase, an enzyme whose presence or absence can be determined by specific staining cell by cell throughout the fly, thus providing a basis for detecting the clones of cells in internal tissues.

If the fate of a cell is already determined at the time that a mitotic crossover is induced within it, the resulting marked clone of cells will be confined in the adult fly to one tissue or structure. If, on the other hand, it is not yet determined, then the marked clone can contribute to diverse parts of the fly which stem from different subsequent determination events. X-irradiation of eggs at the blastoderm stage gives clones that sometimes, for example, overlap wing and second leg both of which are structures of the middle segment of the thorax (mesothorax). This shows that, at this stage, cells are not determined to contribute to any one specific imaginal disc although they *are* committed to a specific segment. The determination of segments apparently occurs very early, and clones are not observed to over-run segmental boundaries.

We may conclude that determination occurs in steps—first fixing segments and later discs within segments. Indeed, clonal analysis has established a step intervening between these two. Clones induced in the blastocyst are often seen to be confined to **compartments** comprising the front (anterior) or, alternatively, the posterior half of a leg or wing, indicating that the two halves are separately determined and that this determination occurs relatively early. Sometimes a clone will contribute to the anterior (or alternatively posterior) halves of *both* the wing and the leg, from which it is deduced that the separate determination of wing and

Fig. 16.6

a. Imaginal disc in larva

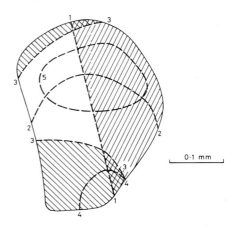

0·1 mm

b. Upper (dorsal) side of wing and notum

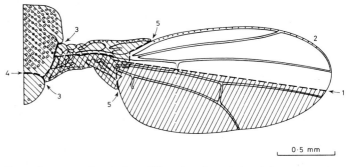

0·5 mm

Successive compartmentation of the wing disc—deduced from comparison of the compartment boundaries in the adult wing (determined by the method of induction of marked clones by Garcia-Bellido) with a fate-map of the disc (deduced from the developmental potential of each part when introduced into a larva). The compartments separated by boundaries 1 and 3 are marked by different cross-hatching. Boundary 3 runs out of sight over the edge of the disc, but is envisaged as a circle, concentric with boundary 5.
Based on Garcia-Bellido[9] and Bryant.[10]
For further discussion of the theory of developmental compartments, *see* Kauffman et al.[11] and Morata and Lawrence.[12]

mesothoracic leg disc cells follows an earlier anterior/posterior decision. This, combined with the careful recording of the boundaries of clones induced at later times, leads to a picture of successive stages of determination, each one representing a further narrowing of developmental options. *Fig.* 16.6 illustrates this hypothesis as applied to the wing discs which are determined to produce not only the wings but also that part of the dorsal mesothorax (the notum) to which the wings are attached.

Compartments once formed are surprisingly exclusive, and cells determined to belong to one compartment do not normally intrude into another. This property of exclusiveness was well shown by an experiment in which mitotic crossing-over generated clones of cells that were at a considerable competitive advantage by comparison with their neighbours. The flies in this experiment were heterozygous with respect to the partially dominant mutation *Minute*, which causes a great reduction in bristle length and a considerable reduction in cell division rate. Clones of cells homozygous for the wild-type allele, generated by induced somatic crossing-over, have improved growth and tend to contribute disproportionately to the tissues subsequently formed. However, the expansion of a non-*Minute* cell clone is abruptly halted when it comes up against an intercompartment boundary. For example, when such a clone arose in that part of the wing disc destined to form the anterior wing compartment, the cells of that clone came to occupy a large area of the anterior half of the wing, but never extended beyond the straight-line boundary into the posterior compartment.

Imaginal disc cells determined to produce different structures will efficiently sort themselves out into mutually exclusive groups when artificially mixed together and reimplanted into a larva. For example, genetically marked leg and antenna disc cells do not, after *in vitro* mixing and reimplantation, give rise to structures in which leg and antenna features are mixed haphazardly. Nor, in general, do the cells change their determination; rather they sort themselves out and give rise to coherent leg structure with one genetic marker and antenna structure with the other. Evidently an important result of compartmental determination is the establishment of some means, the nature of which is quite unknown, by which cells of like kind can recognize one another and associate together.

The factors involved in the establishment of compartments are not yet understood. In view of the clonal transmission of compartment specificity after it is established, one might have expected that compartments would be clonal in origin, i.e. that each compartment would result from a change in a single cell in the blastocyst which, through further division, would give rise to all the cells of the compartment. It seems that this is not the case. It is not possible to trace back a compartment to any single cell, and the favoured view is that determination of a given kind occurs simultaneously in a *group* of cells which then multiply to give a cell population which is not a clone but rather a **polyclone**. Presumably what the groups have in common, to induce them all to become programmed in the same way, is their occupancy of a particular area of the blastoderm. The concept is that there is something about the position, perhaps the concentration of some chemical signal compound or series of compounds, which conveys information. Nobody, however, supposes that such **positional information** includes any detailed instructions. The developmental instructions must already exist as a series of alternative options in the blastocyst cells; the positional information merely selects from among these options.

16.5 | Homoeotic mutants of *Drosophila*

Selection of Alternative Developmental Programmes

If, as has just been suggested, developmental determination is a matter of cells being switched, by signals dependent on position, into one or another of a series of alternative preprogrammed pathways, we would expect to find mutants in which the switching process breaks down so that the wrong option is taken at some stage of development. Many such mutants are now known in *Drosophila melanogaster* and some are shown in *Fig. 16.7*. Most of these show transformation of a set of structures in one segment of the adult insect into a different, but analogous, set of structures which properly belong in another segment. Such mutants are called **homoeotic**.

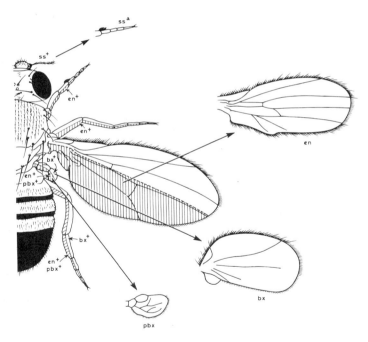

Fig. 16.7   The effects of some recessive homoeotic mutations in *Drosophila melanogaster*. The effects shown are those of *spineless-aristapedia* (*ss*^{a}) which converts the arista of the antenna to a structure resembling the tarsal segment of the first leg, *bithorax* (*bx*) which converts the anterior haltere compartment to anterior wing, *postbithorax* (*pbx*) which converts posterior haltere to posterior wing, and *engrailed* (*en*) which converts posterior wing to mirror-image anterior wing (and has an analogous effect on haltere and legs). The dotted line on the wild-type wing indicates the boundary between the anterior and posterior compartments; note that the anterior and posterior wing margins are clearly distinguished by the type of bristles, coarse or fine, which they bear.
Modified from the figure of Morata and Lawrence.[12]

A few mutants show failure of normal compartmentation *within* segments. For example, flies homozygous for the recessive mutation *engrailed* (*en*) have the anterior wing compartment transformed into a mirror-image of the posterior wing compartment. Although this mutant phenotype affects a boundary within a segment, its abnormality is not confined to that segment since a similar posterior-to-anterior transformation occurs in the halteres as well. This confirms the conclusion already reached on p. 472 that posterior/anterior compartmentation precedes the commitment of cells to discs. In another mutant, *wingless* (*wg*), there is a breakdown in a later compartmentation step (step 3, *Fig.* 16.6)—that which separates the outer (distal) part of the dorsal mesothorax, which produces the wing, from the inner (proximal) part (the notum). The effect is failure to produce wing. Again there is a parallel effect in the next segment back, the metathorax, where the halteres are absent.

The mutant *engrailed* was the subject of an important study which established the connection between the maintenance of compartment boundaries and the morphological development of compartments. In the preceding section (p. 472), it was mentioned that a clone of cells formed by mitotic crossing-over during development of one-half of the wing disc was unable to spread across the compartment boundary into the other half. When, however, the crossover results in homozygosity with respect to *engrailed*, the resulting clone of cells no longer recognizes the anterior–posterior boundary and can extend from the posterior into the anterior compartment. It is thus clear that the effect of homozygous *en* is not only to switch the developmental programme governing visible structure (*see Fig.* 16.6), but also to change those surface properties of the cell that are responsible for the recognition of boundaries.

What do the genes that can mutate to give homoeotic effects actually do? From the observation that several of them have parallel effects on more than one segment, it is clear that they are not concerned with specifying the fine anatomical detail which distinguishes one segment from another. Nor do they seem to be responsible for those general architectural features which different segments have in common. Other genes can be identified as playing this kind of role; for example, mutations that change the number of joints in the tarsus (i.e. the lower segment of the leg) do so whether the tarsus is a normal one or whether it is the product of homoeotic transformation of the arista of the antenna (in *ss^a* homozygotes, *see Fig.* 16.7). In other words, the gene that mutates to bring about the developmental switch is not responsible for the detail of the structures formed after switching. The various alternative developmental programmes for the different segments and intrasegment compartments are laid down independently of the homoeotic genes, whose role is confined to selection of one programme rather than another. For this reason such genes are sometimes called **selector genes**. It would be premature, however, to conclude that the genes so far recognized carry the whole responsibility for selection. They may be components of a much more complex system of control.

A Cluster of Selector Genes—*bithorax*

The homoeotic mutants that have provoked the most interesting speculations are those that map in the *bithorax* gene cluster in chromosome 3. Here different members of a series of genes seem to act successively in bringing about a series of developmental switches from one segmental programme to another, proceeding in an anterior-to-posterior direction. A deletion of the entire gene

Fig. 16.8

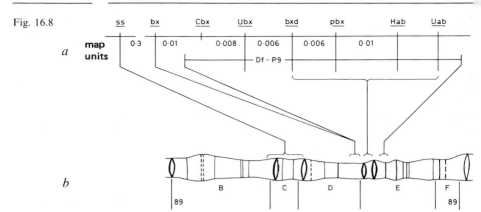

Maps of the *bithorax* region of the third chromosome of *Drosophila melanogaster* (from Lewis[13]).
a. Genetic map.
b. Cytological map (section 89 of chromosome 3).

cluster results, when homozygous, in arrest of development so that the ill-fated larva consists entirely of segments resembling those of the normal middle segment of the thorax (mesothorax). Thus one can infer that some product or products of genes in the *bithorax* region divert development from a basic mesothoracic pattern into the modified pathways leading to the other kinds of segment. It is even possible to speculate that the mesothoracic pattern was ancestral in evolution and that something like it is, or was, present in all segments in 'primitive' insects like centipedes which have legs (if not wings) on all segments.

Mutations at different loci within the region tend each to affect only one segment, or sometimes two adjacent segments. The recessive mutations *bithorax* (*bx*) and *postbithorax* (*pbx*) each, when homozygous, converts metathoracic to mesothoracic structures, *bx* being specific for the anterior and *pbx* for the posterior compartments of these segments. Another mutation, *bithoraxoid* (*bxd*), is a recessive which tends (with some variability of penetrance) to convert the whole metathorax to mesothorax; in this case the developmental switch also affects the adjacent first abdominal segment, which takes on some of the features of metathorax, sometimes including the development of halteres and an extra pair of legs.

Ultrabithorax (*Ubx*) is a dominant mutation which, even in single dose, tends to impose mesothoracic development on both metathorax and the first abdominal segment; this mutation is lethal when homozygous. Two other dominants, *Contrabithorax* (*Cbx*) and *Hyperabdominal* (*Hab*, previously called *Contrabithoraxoid, Cbxd*), act in the opposite direction, pushing development away from the supposedly primitive mesothoracic level to metathorax and, in the case of *Uab* (*Ultra-abdominal*), from metathorax to first abdominal segment. Lewis,[13] who has been responsible for most of the detailed work on these mutants, suggests that they are analogous to *cis*-dominant constitutive mutants in bacteria (cf. p. 371). The effect of *Cbx* appears to be on the expression of *Ubx*$^+$, and it shows only when the two alleles are coupled in *cis*. The *cis/trans* comparison gives a clear result. *Cbx* + / + *Ubx* shows the transformation of meso- to metathorax, while

$Cbx\,Ubx/ + +$ is virtually wild in phenotype. Some of these features of the *bithorax* mutants are summarized, with interpretation, in *Table* 16.1.

What is the significance of the close clustering on chromosome 3 of mutations affecting segmental development? First, we have to ask whether the mutations really are in separate genes in the functional sense defined by complementation (cf. p. 125). The fact that recessive mutants mapping at the three loci *bx*, *bxd* and *pbx* are fully complementary in *trans* double heterozygotes in all combinations strongly suggests that the three loci really are separate units of function. Nevertheless they are all within 0·01 map units and, as was shown by a recently constructed restriction

---

Table 16.1 **Effects of mutations in the *bithorax* region of the *Drosophila melanogaster* third chromosome**

Segment of fly	*Switches of programme attributed to selector genes*
Mesothorax (MS) (Wings, second pair of legs)	MS $\xrightarrow{\ Cbx\ }$ MT
Metathorax (MT) (Halteres, third pair of legs)	(anterior) bx^+ (posterior) pbx^+ MS $\xrightarrow{\ Ubx^+\ }$ MT $\xrightarrow{\ Hab\ }$ AB2
1st abdominal segment (AB1)	MS $\xrightarrow{\ Ubx^+\ }$ MT $\xrightarrow{\ bxd^+\ }$ AB1 $\xrightarrow{\ Hab\ }$ AB2 $\xrightarrow{\ Uab\ }$ AB3
2nd abdominal segment (AB2)	AB2 $\xrightarrow{\ Uab\ }$ AB3

Key:

$\Longrightarrow$ Normal switches

$\longrightarrow$ Abnormal switches in mutants

Genes: *bithorax* (*bx*), *postbithorax* (*pbx*), *Contrabithorax* (*Cbx*), *Ultrabithorax* (*Ubx*), *bithoraxoid* (*bxd*), *Hyperabdominal* (*Hab*), *Ultra-abdominal* (*Uab*).

Interpretation: Lewis[13] puts forward the following hypothesis. The basic programme is that leading to mesothoracic structure. Ubx^+ (in conjunction with *bx* and *pbx*) and *bxd* are required for the normal switches of programme in the metathoracic and first abdominal segments. Other hypothetical genes are responsible for further switching to the programmes leading to the other abdominal segments. These genes are supposed to produce positive switch signals in the appropriate segments, and most mutations causing loss of signal and consequent failure to effect the switch indicated (*bx*, *pbx*, *bxd*) are recessive. *Ubx*, which is a dominant (lethal when homozygous), is an exception; in this case a signal at 50 per cent of the wild-type level is inadequate for normal switching. The dominant mutations, *Cbx*, *Hab* and *Uab*, which carry the successive switches of programme beyond what is required for normal development in particular segments are *cis*-acting constitutive mutations, rendering switching genes active in segments in which they are normally repressed. *Cbx* is thought to derepress Ubx^+, and *Hab* and *Uab* to derepress two hypothetical genes involved in switching to abdominal development. In the table the symbols MS, MT, AB1, AB2, AB3 refer to the presence of some or all features of normal mesothorax, metathorax and first, second and third abdominal segments, respectively.

site map, within a DNA length of about 100 kb. (The same restriction map was used to show that most of the mutations known in the *bx* region are due to insertions of movable elements[15]—cf. p. 448). The *ss* locus, affecting the eye/antenna segment, is clearly separated by the complementation criterion and is, in any case, too far away (0·3 map unit) to be a plausible part of any *bithorax* gene; nevertheless its relative closeness seems unlikely to be due merely to chance.

It may or may not be merely coincidental that the linkage order of the loci of the *bithorax* complex tends to correspond to anterior-to-posterior sequence of the segments affected (*pbx* being the only locus out of line). Lewis has postulated that there is a gradient along the chromosome with respect to the ease of derepression of the gene; Ubx^+ is supposed to be normally derepressed in all segments to the rear of the mesothorax, while the gene controlled by *Uab* is normally only expressed in the third and more posterior abdominal segments. It is possible to imagine a progressive chromatin unfolding, and concomitant gene unmasking, travelling along the chromosome. This unfolding could be controlled by either time or position-dependent chemical signals. Such speculations are at present uncertain and rather vague, but there is a growing suspicion among geneticists that the clustering of functionally related genes, which has now been observed in several cases, must have some functional significance and that this has to do with higher-order chromatin structure. The control of such structure is still but poorly understood.

16.6   Summary and perspectives

This chapter has been concerned with *Drosophila* more than with any other organism because it (in common with other highly evolved insects) has two features which make it especially suitable for study. The first is its segmental structure. The different segments are partly homologous—variations, as it were, on a theme—and they can be interchanged to some extent without complete disruption of the whole organism. It is this feature which permits the identification of viable homoeotic mutants, which promise to tell us something about the control of programme selection. No comparable mutants are known in vertebrate animals. This may mean that, in vertebrates, there does not exist the kind of homology between different parts that we see between the different segments of an insect. The metaphor of selection between alternative modular programmes may be misleading so far as vertebrates are concerned. It may be, on the other hand, the greater degree of integration in vertebrate, as compared with insect, development (for instance in the much more centralized nervous system) that precludes viable homoeotic variants. In that case, the *Drosophila* results may still give some clues as to the way in which vertebrate development is controlled.

The second convenient feature of *Drosophila* is its two-phase development, with a larva harbouring determined but as yet quite undifferentiated nests of cells which will later give rise to the adult fly. This enables a very clear distinction to be made between determination and differentiation, and this is not possible in vertebrates. This may mean either that the same kind of determination does not occur in vertebrates, or that the progression through determination to differentiation occurs in vertebrates too rapidly for the two stages to be separated experimentally.

Drosophila is the best-studied complex eukaryote from the point of view of the genetic control of its development. Yet it has to be admitted that we are nowhere near to an understanding of how this control is exercised. The phenotypic effects of homoeotic mutants are very precisely described, and the structure of the genome at the DNA-sequence level is beginning to be analysed to an extent which a few years ago would hardly have seemed possible. But we are very far from tracing the connections between primary gene transcription and the developmental programme or even from seeing exactly how one is going to unravel the threads.

Some (though not many) biologists have argued that processes of known kinds, ultimately analysable in physical and chemical terms, can never account for the properties of living systems and that other forces of an unknown or even miraculous nature must be assumed to exist. Such a conclusion is unwarranted. It is not difficult to suppose that the developmental programme could be encoded in the genotype. One should not simply consider the information provided by some 5000 genes (in *Drosophila*, more in mammals) in isolation. One also has to take into account the enormous number of *interactions* between gene products. At the protein level, one gene product may activate or inhibit another, and such interactions can be linked sequentially in 'cascades', so that initially small signals can be greatly magnified. Direct products (RNA or protein) and indirect products (hormones or other metabolites) of gene activity can interact with the promoters of the same or other genes, or with regions of the chromatin controlling its local conformation, to promote or inhibit transcription. With interactions of known kinds, involving feedback and amplification, it is entirely conceivable that control networks of any degree of sophistication could be constructed. The problem is not one of lack of sufficient complexity in the genotype. It is rather that there is *too much* complexity in the interactions of the products of several thousand genes for easy experimental analysis.

The elucidation of the molecular mechanisms underlying cellular differentiation is a daunting enterprise, but a start on it can be, and is being, made. The following chapter attempts to review some of the more promising developments.

Chapter 16 Selected Further Reading

Leighton T. and Loomis W. F. (ed.) (1980) *The Molecular Genetics of Development*, 478 pp. New York, Academic Press. (Articles by Edgar R. S. on *Caenorhabditis*, Yund M. A. and Germeraard S. on *Drosophila* and Paigen K. on mouse.)

McLaren Anne (1976) Genetics of the early mouse embryo. *Annu. Rev. Genet.* **10**, 361–388.

Chapter 16 Problems

See Problems at end of Chapter 17.

Mechanisms of Cellular Differentiation

17.1 Is DNA relevant to differentiation?

We saw in the last chapter how, at least in animals, cells become programmed to fulfil different roles as the organism develops. The programming is not necessarily irreversible, but it often has a considerable degree of stability and may be transmitted faithfully through many cycles of mitosis. Moreover, the programme may be determined many cell generations before it is actually put into operation. Differentiation involves progressive cell specialization with commitment to specialization sometimes occurring well in advance of its actually happening. What can be the nature of such stable and apparently heritable changes in the cellular programme? One obvious possibility is that it involves controlled changes in DNA structure. Such a suggestion runs counter to a principle, some would say a dogma, long accepted by most geneticists, namely that of the **constancy of the genome**. Nevertheless, there are now rather numerous examples in which cell differentiation is dependent upon genomic *inconstancy*. In this chapter we review some of these examples and consider what other possible molecular explanations there may be for differentiation.

17.2 Cell differentiation in unicellular organisms

The Macronuclear System of Ciliates

In unicellular organisms, in which each cell is capable of propagating the species (a property known as **totipotency**), any cellular differentiation involving irreversible genome changes might appear to be ruled out. But the Ciliates, one of the most important groups of unicellular eukaryotes, confound this expectation by having a dual nuclear system. As was explained in Chapter 7 in the case of *Paramecium aurelia* (*Fig.* 7.1, p. 175), the immediate cell phenotype is determined by the macronucleus while the long-term genetic continuity of the species is the responsibility of the micronucleus. At every meiosis, the previously existing macronucleus is discarded and a new one is formed by division of the micronucleus.

In one group of Ciliates, typified by the genus *Oxytricha*, up to 95 per cent of the DNA sequence present in the micronucleus is lost during differentiation of the macronucleus, and the remainder becomes greatly amplified in the form of separate 'gene-sized' (several kilobases) pieces.[1] In *Paramecium*, such wholesale changes in genome organization are not known to occur, but nevertheless macronuclei of some species become irreversibly committed to determining one of two alternative mating types soon after they are formed (Sonneborn[39]). Mating type cannot be altered until, following meiosis, the macronucleus is discarded and replaced by a new one derived from the micronucleus.

a. Host-range switch in bacteriophage Mu

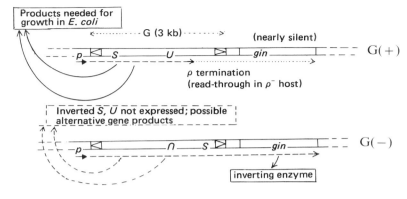

b. Flagellar antigen switch in *Salmonella*

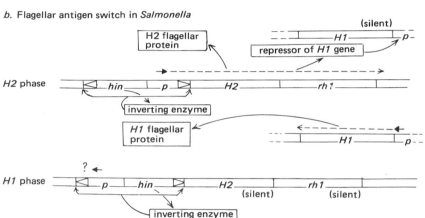

Fig. 17.1

Switches of gene expression due to invertible DNA segments.
Notes:
a. p is a promoter just outside the invertible G segment. In the G(+) orientation the genes S and U, necessary for growth in *E. coli*, are transcribed from the correct strand and transcription is terminated by *E. coli* termination factor rho (ρ) before it reaches the *gin* (G-inversion) gene, located outside G on the side away from the promoter. In *E. coli* mutants defective in rho (ρ^-), transcription may run into the *gin* gene, thus permitting G inversion and occurrence of the G(+)→G(−) switch; normally this occurs only very rarely in *E. coli*. In the G(−) orientation the S and U genes are transcribed on their 'wrong' strand and their protein products are not formed, although alternative polypeptides, perhaps necessary for infection by *Citrobacter*, may be translated from the 'wrong-strand' transcript. In the G(−) phage *gin* is transcribed (because the transcription–termination site within G does not function in reverse orientation), and the G(−)→G(+) switch can thus occur with fair frequency.
Information from van de Putte et al.[2]
b. H1 and H2 genes are not closely linked; *rh1* (repressor of H1) is closely linked to H2 and the two genes are transcribed together from a promoter *p* located within the adjacent invertible segment. The gene *hin* (H-inversion) is transcribed from a promoter within the inverted segment and is itself confined within that segment, and so its transcription occurs in both orientations.
Information from Simon et al.[3, 4]
In both diagrams the arrowheads flanking the invertible sequences indicate short inverted repeats.

Switches in Gene Action due to Inversions of DNA Segments

Certain bacteria and bacteriophages reconcile differentiation and totipotency by means of specific and *reversible* DNA segmental inversions.

The Bacteriophage *Mu* Host-range Switch

Bacteriophage *Mu*-1 is a medium-sized (genome 38 kb) temperate bacteriophage with the unusual property of being able to integrate as a prophage virtually anywhere in the bacterial genome. When the prophage is integrated within a gene-coding sequence the function of that gene is inevitably destroyed and the effect is that of an inactivating gene mutation. It was because of its mutagenic properties in *Escherichia coli* that *Mu*-1 first achieved fame, and it is to these properties that it owes its name.

Infective *Mu* particles can be obtained in two ways—directly from a lytic infection and indirectly by lysogenic induction. Unlike lambda prophage, wild-type *Mu* prophage is not induced by ultraviolet light and it is convenient to work with a mutant which, because it produces an unusually thermolabile repressor, can easily be induced to lysis from the prophage state by brief heat treatment.

Mu phages produced directly from a lytic infection of *E. coli* are, with exceedingly rare exceptions, able to infect only *E. coli* and a few closely related species. In striking contrast, the phages liberated after induction of an *E. coli* lysogen are of two kinds—one with the *E. coli* specificity and the other, about equally numerous, unable to infect *E. coli* but capable instead of infecting another set of bacterial species, of which *Citrobacter freundii* is a representative. The infected *Citrobacter* cells yield progeny phage which, for the most part, are again of the *Citrobacter*-infecting type; a minority, however, are found to have switched back to the *E. coli* specificity.

Physical studies of the DNA of the two forms of *Mu* (performed by heteroduplex analysis and restriction site mapping) have revealed a consistent difference with respect to a 3-kb segment close to one end of the linear genome. This segment, called the 'G-segment', is inverted in one form relative to the other. *Mu* prophage evidently undergoes fairly frequent inversions of the G-segment, so that induction releases a 50:50 equilibrium mixture of the two orientations. During vegetative growth of the phage, on the other hand, inversion is either substantially reduced in frequency (in *Citrobacter*) or almost entirely eliminated (in *E. coli*).

There is little doubt that the G-segment inversion is responsible for the switch of gene action which is reflected in the alteration in host specificity. The mechanism through which inversion occurs is not yet fully understood, but it is presumably connected with the presence at the termini of the G-segment of short inverted repeats—a feature which we have learned to regard as a hallmark of movable sequences.

The currently favoured hypothesis for explaining the effect of the switch is summarized in *Fig.* 17.1.[2]

Flagellar Antigen Determination in *Salmonella typhimurium*

The phenomenon of antigenic phase variation in *Salmonella typhimurium* has been known for more than 30 years. The important antigens on the surface of these bacteria are carried on the flagella, and infected animals form antibodies against the particular type of flagellum present on the infecting bacterium. When large populations of *S. typhimurium* are exposed to immune serum a proportion of cells, about 10^{-3}–10^{-5}, are found to escape inactivation and, when these surviving cells are propagated, they can be shown to have switched to the production of another flagellar antigen type. The alternative states are called H1 and H2, and they are interconvertible.

Insight was obtained into the mechanism of this reversible transition when DNA fragments including the gene specifying flagellar protein H2 were cloned in ColE1 and selected on the basis of their ability to confer motility on an initially flagella-less strain. Some of the cells transformed by plasmids carrying the cloned H2 fragment were of H1 serotype and others were H2. The plasmids determining the two different flagellar types were compared by heteroduplex mapping and were found to be fully homologous except for a 1·5-kb region which was inverted in the one relative to the other.

The gene specifying the H2 flagellar protein has been mapped almost immediately adjacent to the invertible segment. A second gene closely linked to H2 and transcribed together with it codes for a product which represses the transcription of the distinct gene specifying the alternative H1 protein. *Fig.* 17.1 summarizes the situation. Close to one end of the invertible segment there is a 'Pribnow box' sequence which can be taken as a strong indication of a transcription promoter. In one orientation, this is directed towards the adjacent H2 gene and the H1-repressor gene. When the segment is in the other orientation, the promoter is directed away from these genes with the result that H2 messenger is no longer formed and transcription of H1 is no longer repressed.

The ability to shift antigenic type is of obvious value to a bacterial parasite in allowing it a further line of attack against a host already immune to one of its alternative forms. Another point of general interest illustrated by this case is the coordination of the expression of one gene with the repression of another. Such mechanisms, but involving sets of genes rather than single-gene alternatives, very probably play an important part in cellular differentiation in higher organisms. A more complicated example of mutually exclusive systems of promotion and repression is provided by the lysis/lysogenization switch in the *E. coli*–lambda system (*see below*, p. 488).

Switches in Gene Action due to Segmental Transpositions

Reversible cell differentiation controlled by DNA segmental inversion has not so far been described for any eukaryote. There is, however, one very fully worked out example of control by segmental **transposition**.

The *Saccharomyces* Mating-type Switch

The mating types of *Saccharomyces cerevisae* (p. 51) have, in some strains, the appearance of being controlled by two stable alleles at a single locus. Other strains, however, appeared to be homothallic (sexually self-fertile in pure haploid culture); on closer analysis these strains were shown to contain cells of both mating types undergoing frequent interconversion. In 'homothallic' strains, each time a cell buds it has about a 50 per cent probability of switching mating type, although the daughter cell retains, for the moment, the original mating type of the mother cell.

Extensive genetic analysis of a variety of strains showing different mating reactions revealed four genetic loci involved in mating-type determination and mating-type switching. The mating type actually expressed at a given time is controlled by one locus with two alleles, *MATa* and *MATα*. On the same chromosome as *MAT*, but not closely linked to it, are two other loci, respectively to the left and to the right, called *HML* and *HMR*. These are necessary for mating-type switching and each can exist in two states: *HMLα* and *HMRα* promote switching of *MATa* to *MATα*, while *HMLa* and *HMRa* promote the opposite switch. Most yeast strains among those investigated carry *HMLα* and *HMRa*. Another gene, *HO*, is essential for switching in either direction, and a haploid carrying its recessive allele, *ho*, is stable in whatever mating type it happens to be. The postulated actions of these three loci, and of a fourth one which we need not discuss here, are depicted in *Fig.* 17.2.

Genetic analysis of a range of mating-type variants and observation of the switching behaviour led to the formulation of the following hypothesis: *HML* and *HMR* are repositories (or 'cassettes') of spare mating-type information, and switching of the mating type expressed at the *MAT* locus is due to transposition of either *a* or *α*-determining DNA sequence to *MAT* from either *HML* or *HMR*, the sequence previously present at *MAT* being displaced.

The transposition or 'cassette' hypothesis was soon supported by impressive genetic evidence. A number of different mutant alleles of *MATα* and *MATa* are known which are associated with impaired mating ability. In one experiment,[5] a mutation was induced in the *HMLα* gene and identified as such by an ingenious procedure which we need not detail. The defective sequence was shown to be afterwards transferred to the *MAT* locus, which was thereby changed from *MATa* to (defective) *MATα'*.

What could be the advantage to the organism of such an elaborate system? The answer, in so far as there is one, is probably to be found in the finely balanced choice, faced by all sexually reproducing organisms between inbreeding and outbreeding (*see* Chapter 18). The yeast *HO–MAT–HML–HMR* system allows the retention of both options.

When it was first proposed, the 'cassette' hypothesis seemed a very adventurous one since, although it was consistent with the genetic observations, there was no

Fig. 17.2

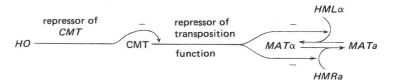

Genetic control of mating-type switching in *Saccharomyces cerevisiae* (after Hopper and Hall[6]). *CMT* stands for *change-of-mating type* and is supposed to code for a repressor and to itself be repressed by the *HO* product. Evidence: *HO* and *HMT* each have recessive mutant (presumed non-functional) alleles. Haploids of genotype *ho CMT* cannot switch—mating type is stable. Those of genotype *HO cmt* or *ho cmt* switch mating type normally (i.e. *cmt* is epistatic to, or suppresses, *ho*). *HMLα* (or the rarer allele *HMRα*) is necessary for *a→α* switching; *HMRa* (or the rarer allele *HMLa*) is necessary for *α→a*.

direct confirmation of the extraordinary things that were supposed to be occurring at the DNA level. Such confirmation is now available as the result of the successful isolation of DNA sequences corresponding to *MAT*, *HML* and *HMR*.

A sequence determining α-mating type was isolated in the following way. A 'library' of yeast DNA BamHI fragments was cloned in an *E. coli* plasmid which also carried a wild-type yeast *leu2*⁺ gene. DNA from this library was used to transform yeast cells of a haploid *ho* yeast strain carrying two linked mutations—it was leucine requiring because of a *leu2* mutation and had no functional mating type because of a mutation in *MAT*. Some yeast colonies, which were selected as having been transformed to ability to grow on leucine-free medium, were found also to have been transformed to a functional (α) mating type. The plasmid harboured by these colonies was used as a hybridization probe to detect

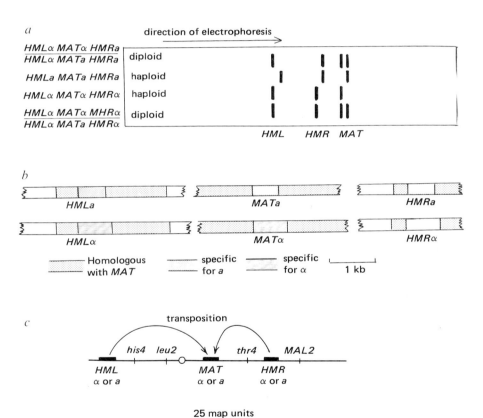

Fig. 17.3   Structure of the *MAT*, *HML* and *HMR* loci of *Saccharomyces cerevisiae*.
a. HindIII fragments, separated according to size by electrophoresis, hybridizing to a BamI–HindIII fragment selected as containing the information for α mating type.
Redrawn from Hicks et al.[7]
b. Structures deduced from heteroduplex and sequence analysis.
Information from Nasmyth and Klar et al.[8, 9]
c. Map of chromosome 3 and the transpositions postulated in the 'cassette' model for mating-type switching. The lengths of *MAT*, *HML* and *HMR* are exaggerated.

homologous sequences in HindIII fragments of DNA isolated from various yeast genotypes and the tetrads obtained from crosses between them. *Fig.* 17.3 shows some of the results.

Three fragments, differing widely from one another in size, were found to hybridize to the probe in all genotypes and each of these fragments was present in one or other of two slightly different sizes, depending on the genotype with respect to the *MAT, HML* and *HMR* loci. The largest fragment varied in size according to the allele at *HML*, with the *HMLα* version about 150 b larger than the *HMLa* one. The second largest fragment showed size variation correlated with the state of the *HMR* locus; the *HMRα* variant is again about 150 bases larger than the *HMRa* variant. The smallest HindIII fragment varies according to the expressed mating type at the *MAT* locus; mating type α is associated with a fragment about 150 b larger than that in mating type *a*. The conclusion was inescapable that the two *HM* loci were indeed repositories of mating-type information, and that the DNA sequences determining the α and *a* mating-type reactions differed in size by about 150 base-pairs.

Why did the cloned sequence, which apparently carries α information, hybridize not only to the α but also to the *a* alleles at the *MAT, HML* and *HMR* loci? This question has been answered through more detailed physical mapping of the DNA sequences identified by the probe. It turns out that the sequence specific for mating type is comparatively limited—about 600 base-pairs for *a* and about 750 for α. Flanking these specific sequences are regions apparently identical at all three loci and for both states of each; these constant sequences amount to about 1 kb on one side and 1·5 kb on the other. It is presumably by homologous recombination involving these common sequences that transposition of *HM* sequence to *MAT* is brought about. Little more can be said about the mechanism of the transposition at the time of writing.

Switching of Antigenic Type in Trypanosomes

It is too early to say how general the controlled transposition of DNA sequence may be as a cause of differential gene action in unicellular eukaryotes. It is, however, already clear that it is not unique. The second example is that of switching of antigenic type in *Trypanosoma brucei*, a fly-transmitted protozoan parasite of animals and man in Afria. The switch in this species, which replaces one protein antigen on the cell surface by another, is strongly reminiscent of the flagellar antigen switch in *Salmonella* reviewed above (p. 484) and has the similar effect of enabling the parasite to evade, at least temporarily, the immune defences of the host animal.

It has been possible to isolate cDNA corresponding to mRNA for each of a series of alternative antigens. Using these cDNAs as probes in Southern transfers of restriction fragments of DNA isolated from trypanosomes expressing different antigens, it has been shown that, in several cases, the DNA sequence coding for a certain antigen appears in a new context, i.e. in a new restriction fragment, in cells in which that antigen is being produced. Rather as in the yeast mating-type cases, there are apparently certain sites at which each specific sequence is conserved irrespective of whether it is being expressed; switching-on seems to require transposition to a special 'expressing' site. More information will undoubtedly soon be available on this important system.[10]

Stabilized Cell States not Dependent on DNA Rearrangement

That many kinds of cellular differentiation could depend on clonally inherited changes in DNA structure (in base sequence, in secondary base modification or in conformation) is still a relatively novel idea and, consequently, is currently attracting great interest. It should not, however, distract attention from the existence of many possibilities for stabilizing cellular states in a quasi-heritable manner by means not involving any kind of change in DNA or chromatin structure. Put briefly, the general idea here is that the products of certain genes can maintain themselves by promoting activities of the genes that produced them—positive feed-back, in other words. The same products might, directly or indirectly, suppress the expression of other 'rival' genes. To what extent this kind of mechanism is responsible for cellular differentiation in higher organisms is not known, but there are several examples in prokaryotes and unicellular eukaryotes. Here we shall mention just two of them.

Self-maintenance of Induced Permeases

A simple example concerns the acquisition by lac^+ E. coli cells of the ability to use very low concentrations of lactose as carbon source. It is difficult for cells not possessing β-galactoside permease (cf. p. 371) to take up enough lactose to induce the permease formation. However, once the first molecule of permease is formed it will bring lactose into the cell, leading to the induction of more permease (and β-galactosidase); thereafter the capacity to use lactose is self-maintaining. The synthesis of enough permease to set the cycle of induction in motion is largely a matter of chance and happens in a random fashion to only a proportion of the cells in the population within a limited span of time; consequently, after short times of exposure to low concentrations of lactose, there is a differentiation between clones of cells which can use the sugar and clones which cannot do so. The positive feed-back loop in this example consists of the permease promoting the uptake of its own inducer.

Lambda Lysogeny in E. coli

A much more complex example is provided by the establishment and maintenance of the lysogenic state in E. coli cells infected with lambda bacteriophage. The essentials of this system, which in the subtlety of its controls may well mimic some modes of eukaryote cell differentiation, are set out in Fig. 17.4.

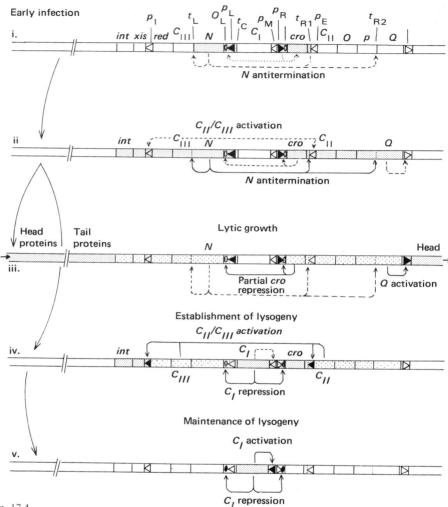

Fig. 17.4

Control of developmental pathways in bacteriophage lambda.[11, 12, 40]
i. Shortly after infection the lambda genome becomes circular (left and right ends of the map must be imagined to be joined). Transcription starts from p_L and p_R and terminates at t_L and t_{R1} or t_{R2}.
ii. The gene N protein product promotes transcription through t_L, t_{R1} and t_{R2}. The C_{II} and C_{III} gene products can now help to establish lysogeny while the Q gene product tends to drive the vegetative multiplication of phage and cell lysis.
iii. The Q product activates promoter p_R, which initiates transcription of all the genes coding for lambda head and tail proteins and cell lysis enzyme. Meanwhile the cro repressor reduces transcription from p_L and p_R.
iv. The alternative to (iii). The C_{II}/C_{III} protein complex activates transcription from p_E ('establishment' promoter) and p_I (integration promoter). The C_I protein, produced as a result of p_E activity, begins to repress transcription from p_L and p_R by binding at the two operators (0). It also begins to activate p_M. The int protein enables the phage DNA to be integrated by recombination between the lambda and bacterial att sites.
v. C_I repression of p_L and p_R now complete. Transcription is now solely of the C_I gene from p_M; the C_I protein maintains p_M activity and confers immunity to further lambda infection. The control of p_M and p_R depends in a subtle way on the absolute and relative concentrations of the cro and C_I proteins; the p_M/p_R operator region contains several binding sites with different affinities for one or other protein or both.
Transcription indicated by stippling.

17.3 | Are somatic nuclei in multicellular eukaryotes totipotent?

Major Chromosome Changes—*Ascaris*

Generally speaking, there is very little microscopical evidence for changes in chromosomes during the development of multicellular eukaryotes. So far as external appearance goes, the genome usually retains the content and structure which it had in the germ cells. There are, however, a few notable exceptions to this rule.

For example, in some nematode worms of the *Ascaris* family, a considerable fraction of the material of the chromosomes is lost early in development from those cells destined to give rise to the somatic tissues, and is retained only in the cells of the germ line. *Fig.* 17.5 shows the dramatic changes seen in the chromosomes of *Parascaris equorum* during the first cleavages of the fertilized egg. The bulky distal parts of the two original chromosomes, one inherited from each parent, are lost in three of the cells at the four-cell stage and the central region of each chromosome undergoes fragmentation. Only the fourth cell, which forms the germ line, retains the chromosome constitution of the zygote.

Recent work[13] has shown that the portion of the genome which is discarded in the soma consists largely or entirely of highly repeated sequences and thus may contain little or no specific information. Nevertheless, the loss of chromatin is presumably sufficient to prevent an *Ascaris* somatic cell from ever acting as a zygote and starting development again from the beginning. Such early loss of totipotency certainly does not occur in all multicellular eukaryotes.

Flowering Plants

Among higher eukaryotes, it is the flowering plants which show the property of totipotency most clearly. The capacity possessed by many flowering plants to regenerate whole fertile plants from pieces of stem, leaf or root is part of the traditional lore of horticulture. In recent years it has been shown that, in some species, it is possible to culture individual cells from, for example, leaves and, with appropriate hormonal stimulation, to get them to regenerate roots and shoots and ultimately whole plants (Chapter 6, p. 161). So far, most of the successful experiments have been done with members of the family Solanaceae (such genera as *Nicotiana* and *Petunia*). However, it seems most unlikely that there is any fundamental difference in developmental controls between solanaceous plants and others, and much more probable that the difficulties experienced with such groups as the cereal grasses are technical in nature.

Frogs

Nobody has ever claimed to be able to regenerate a higher animal from a single differentiated somatic cell. There is, however, evidence that the *nucleus* of a differentiated cell, when introduced into an egg whose own nucleus has been inactivated, is at least sometimes capable of supporting the whole course of

Fig. 17.5

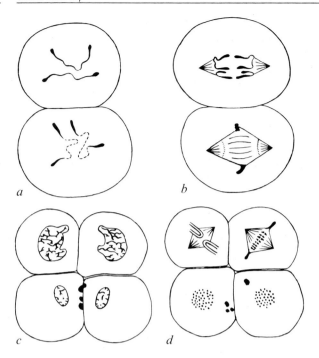

Chromosome differentiation of somatic from germ-line nuclei in the parasitic nematode worm *Parascaris equorum* (redrawn from White,[41] originally from Boveri).

a. Metaphases of second mitotic divisions of the fertilized egg. The germ cells each contribute only one chromosome, with centromere function distributed over the central region. The club-shaped chromosome ends are heterochromatic. In one of the two cells multiple breaks appear in the central chromosome region.

b. Anaphases of second mitotic divisions. In one cell the pair of large chromosomes remains intact, and two chromosomes disjoin to each pole. In the other the heterochromatic termini are left on the periphery of the spindle equator while a number of separate minute chromosome fragments, each with centromere (spindle attachment) function, disjoin to the poles.

c. Four-cell stage, with two cells with the full genome of the fertilized egg and two with diminished genomes—terminal fragments left close to the cell membrane.

d. Metaphases of the third round of mitotic division. The two lower cells (polar view) have the diminished fragmented chromosome complement. Of the upper cells, the right-hand one is undergoing the same fragmentation and diminution as seen in the lower cell in (*a*) and (*b*). The upper left cell gives rise to germ cells.

development. This evidence is complex and, to some extent, contradictory and it is worth describing in a little detail.

The pioneering work in this area was performed by T. J. King and R. Briggs[14] using the frog *Rana pipiens* as the experimental animal. They enucleated (i.e. removed the nuclei of) eggs by a microsurgical technique after first having activated the eggs for cell division by pricking with a needle (a procedure that mimics the effect of fertilization). They then replaced the removed nuclei by microinjection of nuclei taken from cells from embryos at different stages of development.

The outcome depended very much on the stage of the embryo that served as the nuclear donor. Nuclei from the **blastula,** at which stage the embryo is just a hollow ball of apparently undifferentiated cells, were able in the majority of cases to substitute adequately for the egg's own nucleus in supporting normal development at least up to the swimming tadpole stage (it is not clear whether mature frogs would always have been obtained had the tadpoles been allowed to develop). Significantly different results were obtained when the donor nuclei came from the gastrula stage, when the hollow ball of cells invaginates to form the different cell layers (ectoderm, mesoderm and endoderm) from which the various tissues of the developing animal are derived. Most transplanted late gastrula nuclei supported development only up to the blastula or gastrula stage or failed to undergo mitosis at all. Most of the transplants that got beyond gastrulation developed some abnormality in later development and only a few produced normal swimming tadpoles.

To test the idea that the nuclei of the gastrulae were beginning to undergo *heritable* changes that limited their developmental potential, King and Briggs developed a method of 'cloning' such nuclei by serial transplants. A gastrula nucleus would be inserted into an enucleated egg and the transplant allowed to develop only to the blastula stage, at which point the donor nucleus had multiplied by mitosis but had, it was hoped, not undergone any further irreversible changes. Nuclei from the blastula were then transplanted into a series of enucleated eggs to give a family of developing embryos with nuclei all descended from the single original gastrula nucleus. The remarkable finding was that the members of each such family all tended to develop, or fail to develop, in the same way and differently from other families. One experiment is summarized in *Fig.* 17.6. The implication was that some, at least, of the gastrula nuclei had undergone permanent losses of developmental potential differing in extent from one nucleus to another and transmitted more or less faithfully to daughter cells during development from egg to blastula.

A few years later the same line of enquiry was taken up by J. B. Gurdon and his colleagues, using the African frog *Xenopus laevis* instead of *Rana pipiens.* The advantage of *Xenopus* is that it has larger cells, so that the nuclear transplantation is easier. Gurdon achieved a somewhat higher frequency of development to swimming tadpoles following transplantation of early embryo nuclei, but his most striking success was in obtaining full development with nuclei taken from differentiated cells of tadpole intestine.[15] In one experiment (others were less successful), a total of 120 intestinal nuclei were transplanted and most of these were allowed to multiply up to the blastula stage and transplanted again. Altogether seven frogs were obtained, five of which were shown to be sexually fertile; these five stemmed from just two different original intestinal nuclei. The possibility that this full development was due to the original egg nucleus not having been destroyed (in these experiments X-rays rather than microsurgery were used to inactivate the egg nucleus) was ruled out by the presence of a recognizable marker in the donor nuclei—the deletion of a nucleolus organizer, resulting in a reduction in the number of nucleoli from the usual two to one. The five fertile frogs all showed single nucleoli, thus confirming their descent from the intestinal donor. They were also diploid, whereas the unfertilized egg nucleus would have been haploid.

The numbers in the critical experiments are small but they do, nevertheless, appear to justify the conclusion that at least *some* nuclei of differentiated cells remain totipotent and that irreversible changes in the genome cannot be the

Fig. 17.6

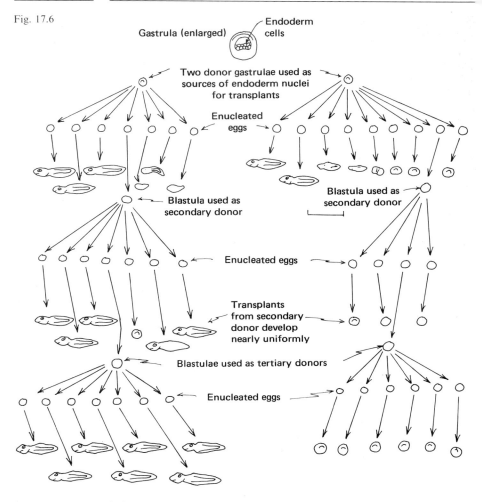

Serial transplantation of nuclei into enucleated eggs of the frog *Rana pipiens*.
Redrawn from King and Briggs.[14] *See text.*

universal cause of cellular differentiation. The difficulty in demonstrating that
anywhere near *all* nuclei of differentiated cells are totipotent has been attributed to
accidental damage to the nuclei during the transplantation procedure. According
to this interpretation, the clonally transmitted developmental limitations demonstrated by King and Briggs were the results of experimental accidents and not part
of any developmental programme inherent in the genome.

The matter cannot, however, be considered as fully resolved, and it remains
possible that at least some of the clonally inherited changes demonstrated by King
and Briggs had a natural origin. This possibility is given rather more weight
following the recent demonstrations that certain kinds of cell differentiation
undoubtedly *are* based on genomic rearrangement.

17.4 The mammalian immune system—differentiation based on irreversible DNA rearrangement

A direct approach to the question of genomic constancy is to compare the structure of a gene in germ cells or embryonic tissues with its structure in differentiated cells. It is also of interest to compare two different kinds of differentiated tissue, one in which the gene is expressed and one in which it is 'silent'. Now that we know about the prevalence in higher eukaryotes of introns within coding sequences, an obvious question is whether there are any changes in introns correlated with the switching on or off of gene activities during cell differentiation.

These questions can in principle be answered whenever mRNA or cDNA corresponding to the gene in question is available for use as a probe. By treating total DNA from the tissues to be compared with a suitable restriction endonuclease, separating the resulting DNA fragments on an electrophoretic gel, and hybridizing Southern transfers (cf. p. 186) from the gel with the radioactively labelled probe, any genomic rearrangement that alters the sizes of restriction fragments by more than a few tens of bases can be detected. Most of the relatively few experiments of this kind that have been performed have given results consistent with the hypothesis of genomic constancy. However, one very important class of mammalian genes, those encoding the polypeptide chains of antibodies (immunoglobulins), shows clear evidence of rearrangement during development.

Different Lymphocyte Clones Produce Different Antibodies

Immunoglobulins are produced by cells that originate in the bone marrow (B-cells) and later differentiate into lymph cells (lymphocytes). B-cells can become differentiated to produce a very large number of alternative antibodies, and one or more is available to bind to virtually any foreign protein or other macromolecule that may be introduced by infection or (in experimental immunology or clinical immunization) by injection. Clones of cells collectively producing all possible types of antibody are believed to arise spontaneously among the huge population of B-cells, but each individual clone produces only one. The response to a foreign antigen is thought to consist in the selection and differential multiplication of those clones which are already differentiated to produce appropriate antibodies. One can make a distinction between primary and secondary immune response—a second challenge with an antigen to which the animal has already been exposed evokes a stronger and more rapid production of antibodies than on the first occasion, because selected clones of cells persist from the first challenge.

Most of what we know in detail about antibodies, and the genes encoding them, stems from studies of cancers of the lymphoid system. Malignant transformation to give a **myeloma** cell line undergoing unrestrained proliferation can occur (though with very low probability) in any B-cell. Each myeloma produces one specific antibody drawn at random from the enormous repertoire of antibodies. Unlike a normal lymphocyte clone, it can be grown indefinitely in culture. In mice, particularly, different myeloma cell lines have been used as sources of both specific immunoglobulins and the mRNAs from which they are translated.

Monoclonal Antibodies

In practice, it is not possible to find a separately originating myeloma cell clone for each of the many antibodies that one might wish to have. The **hybridoma** technique, due to C. Milstein,[16] makes it possible to use one myeloma cell line for making any antibody. Mice are immunized by injection with a particular protein and their spleens are used as a source of lymphocytes, which will be relatively rich in clones producing antibodies directed against the protein injected. The isolated spleen cells are then induced (with polyethylene glycol) to fuse with cells of a myeloma line that has lost its own ability to make antibody and has the additional property of being unable to grow on HAT medium because of HGPRT deficiency (cf. p. 236). The hybrid cells can be efficiently selected on HAT medium since they combine the properties of the two parent cell types—the rapid and unlimited growth of the myeloma and the HGPRT$^+$ character of the spleen lymphocyte. Each hybridoma line also inherits from its lymphocyte parent the ability to produce antibody of one specific type. A number of rapid methods has been devised for picking out the clones producing antibody of the desired specificity. **Monoclonal antibodies** from single hybridoma clones are extremely powerful analytical tools.

Antibody Structure

Antibody molecules (immunoglobulins, abbreviated to Ig) are tetrameric, each containing two identical 'heavy' (H) chains of about 500 amino acid residues, and two identical 'light' (L) chains of about 250 residues in length. In man, mouse, rabbit and other mammals, there are two kinds of light chain called **kappa** (κ) and **lambda** (λ) and at least seven kinds of heavy chain—alpha (α), several kinds of gamma (γ_1, γ_{2a}, γ_{2b}, γ_3), delta (δ) and mu (μ). Ig molecules are named according to the types of H chain which they contain—IgG antibodies have gamma chains of various classes, IgM μ chain and IgA α chains, and each of these types can contain either kind of light chain.[17]

Each class of Ig chain, whether heavy or light, consists of two distinct parts (*Fig. 17.7*). The C-terminal part (about one-half of the whole for the L chains and about three-quarters for the H chains) is constant and characteristic of its class. The remaining (N-terminal) sequence is extremely variable (so-called V-region). Inspection of the large number of V-regions of H and L chains that have been sequenced reveals the existence of a number of different V-region classes; the many different V_H sequences can each be combined with any of the classes of H chain constant regions (C_α, C_μ, C_{γ_1}, etc.) and there are separate sets of variable sequences, called V_κ and V_λ, respectively, for combination with the constant moieties of the lambda and kappa chains.

Within each V_H, V_κ or V_λ sequence class there is a fair degree of homology and the variability is largely confined to a few **hypervariable** short regions. Studies on the three-dimensional structure of Ig molecules and the way they combine with antigens have shown that it is the variable portions of the heavy and light chains which combine to form the antigen-binding part of the antibody and that, within

Fig. 17.7

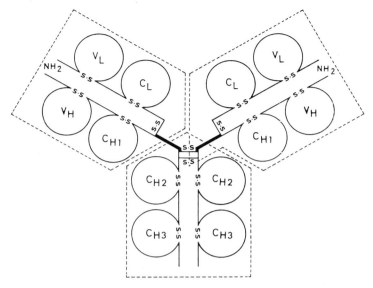

Structure of immunoglobulin G.
The diagram shows in very simplified form the separately folded domains (four for the heavy chains, two for the light), each looping out from a disulphide (cysteine–cysteine) bridge. S–S bridges join the heavy and light chains and the two heavy chains to each other. The so-called hinge regions of the heavy chains are shown as heavy lines. The three domains of the C-region are each specified by a continuous coding sequence in the DNA; the introns (see Fig. 17.2) separate the coding sequences for different domains. All four C domains (C_{H1}, C_{H2}, C_{H3} and C_L) show considerable amino acid sequence homology and are thought to have arisen in evolution through gene duplication and divergence.
Redrawn from Amzel and Poljak.[17]

the V-regions, it is particularly the hypervariable short sequences which line the 'pocket' of the molecule within which the antigen is specifically bound.

The Arrangements of Ig Genes in Germ-line and Differentiated Cells

The physical investigation of immunoglobulin gene structure was made possible by the availability of mRNAs for specific heavy and light chains which are produced in quantity by myeloma cell lines. These messengers, or cDNA copies from them, can be radioactively labelled and used as hybridization probes for corresponding sequences in restriction digests of DNA prepared from germ-line (sperm) or embryo cells on the one hand and myeloma cells on the other. S. Tonegawa was the first to show (using a split messenger to probe separately for the two halves of the kappa chain gene) that the C- and V-coding sequences were contiguous in the antibody-producing cell but separated in the germ-line. Such a result had been predicted on the basis of genetic evidence combined with amino acid sequencing; C-regions showing the same genetically determined amino acid sequence variant were found combined with different V-regions in the same individual. Subsequent work has shown that V–C splicing at the DNA level is responsible for generating the coding genes for both light and heavy chains.

By the use of nucleic acid probes it has become clear that, at least in mice, the genes in question are distributed over three different chromosomes. On one chromosome (identified as mouse chromosome 6) a gene is located which encodes the C part of the lambda chain and some distance away on the same chromosome is a cluster of V_λ-coding sequences—a hundred or more in all, made up of multiple copies of each of a number of distinct classes. On another chromosome there resides a gene encoding the C part of the kappa chain and, again, some distance away on the same chromosome, is a multiclass cluster of some hundred or more V_κ-encoding sequences. On the third chromosome involved in the system the situation is a little more complex. All of the heavy-chain constant regions (α, γ_1, μ etc.) are encoded in a cluster of single copy sequences linked (though distantly in molecular terms) with a repetitive multiclass cluster of V-encoding sequences, this time specific for heavy chains. It thus seems that the developmental cutting and rejoining of the DNA always occurs within a chromosome and not across chromosomes.

We see below that light-chain V-region DNA sequences each include a third spliced segment (one of the J or junction sequences) and that heavy-chain V-regions include not only C, J and V sequences but another short sequence (D) as well.

The Generation of Ig V-regions by DNA Splicing

For the sake of brevity we shall concentrate on the DNA rearrangements which effect the heavy chains of mouse immunoglobulins, with only passing reference to the parallel changes which occur in the light-chain DNA sequences.

In the putting together of genes coding for heavy chains there are two kinds of rearrangement. The first generates the sequence coding for a particular V_H region and the second associates this V_H sequence with a particular C_H sequence. The rearrangements involved in each of these stages have been defined by detailed comparisons, using restriction mapping and nucleotide sequencing, of cloned DNA sequences from antibody-producing (usually myeloma) cells and cells of other kinds (e.g. germ cells, embryonic cells or reticulocytes). For example, comparisons using an IgM mRNA probe and an IgM-producing myeloma have led to the picture, shown in *Fig. 17.8a*, of the genesis of the V_H-coding sequence for a μ chain.

In undifferentiated (embryo or germ-line) DNA, nearly all of the sequence coding for variable regions is found divided between a relatively long main V_H segment and a much shorter J_H (**junction**) segment which is sited between the V_H and C_H clusters, but much closer to the latter. There are, in fact, three different J_H segments in the mouse germ line; they show many differences from one another and it is thought that any one of them can be spliced to any one of the distant main V_H sequences. In addition, as P. Early, L. Hood and their colleagues showed, there is a short sequence in the mRNA between the regions encoded by V_H and J_H and represented in neither of the latter DNA segments. This adventitious sequence was attributed to yet another germ-line DNA segment (called D, for **diversity**) located, perhaps in multiple copies, between V_H and J_H. This hypothesis has now been confirmed by cloning and sequence determination of the DNA on the 5′ side of the J segments.[38]

The many different possible V–J combinations, together with the possibility of further variation due to multiple D segments, go quite a long way towards explaining antibody diversity. Yet further variation is evidently generated by what appear to be alternative ways of splicing these DNA sequences together (*see Fig.* 17.9). In the study by Early et al.[19], comparisons of the V_H–D–J_H encoded regions of antibody chains of independent origin (though of similar specificity) revealed that, although they seemed to have been assembled from similar V_H and J_H segments, there was a considerable amount of variation at or close to the

Fig. 17.8

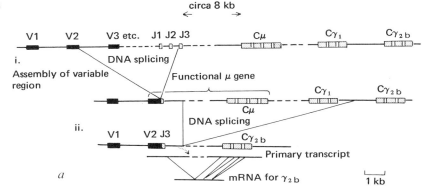

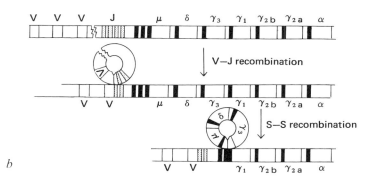

Origin of the coding sequence for the immunoglobulin heavy chain.
a. The two steps of DNA splicing: (i) assembly of the V-region and (ii) heavy-chain class switching.
After Davis et al.,[18] Early et al.[19] and Maki et al.[20]
Notes: The V-cluster and J-cluster are separated by a large and as yet undetermined distance along the chromosome; the C_μ–C_{γ_1} distance is shorter but again not known exactly. The J cluster is separated from C_μ by about 8 kb (distance shortened in the diagram). Other distances are to scale. Early et al.[19] present evidence for a contribution to the V-region from one of another cluster of D (for diversity) segments; these are very short and located between the V and J clusters in germ-line DNA.
b. Model for heavy-chain switching, after Kataoka et al.[21]
Notes: The solid black rectangles represent short sequences (S) involved in deletion by a recombination process. From one to six segments can be deleted in a single step or in sequential steps. The δ segment corresponds to a minor species of heavy chain not referred to in the text.

intersegment junctions. Variability in the DNA splicing process may thus be a part of the explanation for V-region diversity. It seems from other evidence, however, that some kind of high-frequency somatic mutation must also be involved.

A rather similar, though simpler, picture has been deduced for the derivation of the variable part of coding sequences for lambda and kappa chains. It is simpler because, although there are separate junction (J) segments as there are for heavy chains, all the information for the light-chain sequences is accounted for in the V and J segments, with no need to postulate an analogue of D.

The Heavy-chain Class Switch

The second phase of DNA rearrangement determines the kind of constant heavy-chain segment which is to be associated with a processed and selected V segment. There is, in fact, evidence that the same V segment can become successively associated with different C segments. The primary immune response generally involves the formation of IgM antibodies, but a secondary response typically consists in the massive production of IgG antibodies, and IgA antibodies may eventually be formed as well.

It appears that this H-chain class switching, as it is called, is due to successive deletions of DNA. The evidence comes from the use of mRNA or (more usually) cDNA probes for assaying the C-segment content of different myeloma cell lines. Cells differentiated to produce IgM have been found to retain all the segments that can be probed for—γ_1, γ_{2b}, γ_{2a}, γ_3 and α as well as μ itself. In cell lines producing IgG$_3$ antibody the segment encoding C$_\mu$ is apparently missing—at all events it has been displaced from its germ-line position closest to the J segments by the rearranged C$_{\gamma_3}$ segment. However, all other C-region genes are still present in normal quantity in IgG$_3$-producing clones. Myeloma cells forming IgG$_1$ on the other hand show loss of *both* C$_\mu$ *and* C$_{\gamma_3}$ segments. Expression of other kinds of heavy chain is associated with further deletions: expression of C$_{\gamma_{2b}}$ correlates with loss of μ, γ_3 and γ_1, that of C$_{\gamma_{2b}}$ with loss of μ, γ_3, γ_1 and γ_{2b} and that of C$_\alpha$ with loss of μ, γ_3, γ_1, γ_{2b} and γ_{2a}. This sequence of deletions agrees with the order in which the C-region coding sequences are expressed during the development of immunity. The model proposed to explain the observations is depicted in *Fig. 17.8b*. Deletions of different length bring different C genes close to the same V–J compound sequence to form a number of alternative sequences coding for molecules with a common antigen specificity.

One feature of antibody-producing cell differentiation that stands in need of explanation and is relevant to the heavy-chain switch is **allelic exclusion**. When, because of the presence of heterozygosity with respect to genetic markers, it is possible to distinguish the contributions of two homologous chromosomes to antibody formation, it almost always turns out that only one of the two is expressed. One might suppose either that the splicing needed to construct an active gene occurs only on one (randomly selected?) chromosome, or that splicing occurs on both but with a high frequency of error, so that only rarely are both chromosomes successfully rearranged in the same cell. In some of the studies that detected deletions correlated with choice of heavy chain, the dosage of the 'deleted' sequences had been reduced to a half, as if the first alternative mentioned above were true. In others, the deleted segments appeared to be missing altogether, which seems to be more compatible with the second hypothesis.

Possible Mechanisms for DNA Rearrangement

The DNA base sequences adjacent to the V and J segments for both L and H chains provide some clues to the mechanism of V–J joining. Adjacent to each V segment on the 'downstream' side is a six base-pair sequence (labelled X in *Fig.* 17.9) which is repeated, in inverted orientation adjacent to each J segment, on the 'upstream' side. A further pair of inverted repeats (labelled Y in *Fig.* 17.9) occurs 10 or 11 base-pairs 'downstream' of each V–X and this appears again as a reverse repeat 20–22 base pairs 'upstream' of each X′–J, so that the overall arrangement is V–X–(11)–Y–.....Y′–(22)–X′–J, where the primed symbols X′ and Y′ indicate inverted orientation. Hood et al.[22] have proposed that the cutting–rejoining enzyme has two subunits, recognizing and binding to X–(11)–Y and X–(22)–Y, respectively. The significance of the spacing may be that 10–11 base-pairs correspond to one turn of the DNA double helix and 20–22 to two turns. If the enzyme clamps to one side of the DNA duplex it may be able to recognize, in the major groove, only base-pairs presented on that side. Whether this suggestion is valid or not, the '11–22 rule' does seem likely to have some significance for mechanism. All of the considerable number of V and J flanking sequences so far determined conform to it.[22]

Heavy-chain class switching appears to depend upon recombination of some kind between short partly homologous sequences, called S, situated 'upstream' of all C-region DNA segments and 'downstream' of the J segments to which they

Fig. 17.9

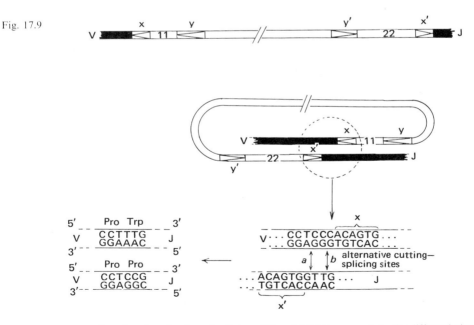

Postulated mechanism of splicing of V and J DNA segments in the differentiation of genes for immunoglobulin light chains. The sequences shown, and the different consequences of recombination between them, were determined for mouse kappa DNA sequences by Sakano et al.[23] Similar sequences, and variable positions of splicing, appear to be involved in the differentiation of the V–J regions of lambda and heavy-chain genes.

become joined. The homology between different S-sequences has been shown[44] to extend over 49 nucleotides, with only 13 variable positions (*see Fig.* 17.8 and Problem 7).

It must be remembered that even after the functional immunoglobulin H-chain and L-chain genes have been assembled by DNA rearrangement, they still contain extensive intervening sequences between their C and V-coding regions. Cutting and splicing of the primary RNA transcripts therefore still have to occur before coherent mRNA molecules can be formed. In fact, further diversification of antibody molecules occurs at the stage of RNA processing (*see* p. 416).

17.5 | Pathological DNA rearrangements—cancer

The Role of Genomic Change in the Origin of Cancer

Tumours are due to the accidental acquisition by cells of the capacity for unlimited and uncontrolled division. When, in addition, the cells are able to become detached from the tissues in which they first arise and migrate to other parts of the body, the tumour becomes malignant—in other words, a cancer. Tumour formation is a property of clones of cells and is thus an example of cellular differentiation—or, some would say, of dedifferentiation, since the capacity for unlimited multiplication is a property that properly differentiated cells have lost.

We have no room in this book for a general review of the cancer phenomenon, but one point of relevance to the theme of this chapter should be made. There is strong and growing evidence that many and perhaps all cancers are due to genomic changes, ranging from major chromosomal rearrangements to invisible changes which may be point mutations. The strongest argument for the mutation theory of cancer is the high correlation between mutagenicity and carcinogenicity over a very wide range of chemical and physical agents. In fact the connection is well-enough established for the demonstration that a compound induces mutations in a bacterial assay (usually one of the battery of tests devised by B. N. Ames—*see*. p. 335) to be taken as good reason for regarding it as a potential carcinogen, to be handled with caution and kept out of the environment.

Certain kinds of human cancer are strikingly correlated with specific kinds of chromosomal rearrangements, though whether as cause or effect is not certain.[26] To take just two examples, one form of human leukaemia (chronic myelogenous leukaemia) has been found to be associated with a deletion in the long arm of chromosome 22 (giving the so-called Philadelphia chromosome) and translocation of the deleted segment to chromosome 9.[24] A number of human breast cancer cell lines (seven out of seven studied) were shown to have a segmental interchange involving the long arm of chromosome 1 and another chromosome, different in different cases.[25]

Attempts have been made to analyse the genomic basis of tumorogenicity by cell hybridization and subsequent chromosome loss (cf. p. 236).[26] The general rule is that hybrids between normal and malignant cells are non-malignant to start with, but that some subclones revert to malignancy. Such reversion can sometimes be correlated with losses of specific chromosomes, which is what one would expect if the malignant property were due to recessive chromosomal mutation of some kind.

Transforming DNA Viruses

One kind of structural change which can lead to the transformation of an animal cell to a state of uncontrolled and indefinite proliferation is the integration within it of all or part of a virus genome.

In Chapter 14, we considered in some detail the genome structure of the monkey virus SV40. This virus proliferates to produce more virus in monkey cells. It is unable to do so in cultured human cells, but *is* able to transform them. In the abortive infection of the human cell it seems that a virus function which, in the monkey cell, would contribute to production of virus, instead stimulates cell

Fig. 17.10 *a*

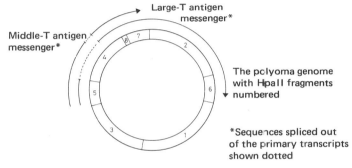

b EcoRI fragments of DNA from transformed rat cell lines

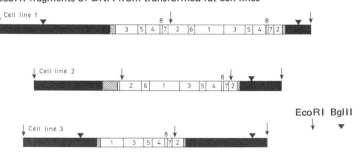

Fragments of polyoma virus integrated into the genomes of transformed rat cells.
a. The polyoma genome, with HpaII fragments numbered.
b. EcoRI fragments of DNA from transformed cell lines. Hatched regions could be either rat or polyoma sequence; black regions are rat. The fragments were mapped with respect to several kinds of restriction sites; only the EcoRI and BglII sites are shown.
NB The host cell sequences flanking the integrated polyoma DNA are different in each cell line. Cell lines 1 and 2 include fractionally more than one copy of the polyoma genome, with ends cyclically permuted; cell line 3 contains an incomplete copy. All cell lines can produce middle-T polyoma antigen but only line 1 can produce large-T.
Data from Lania et al.[27]

proliferation. Transformed human cells can be shown to contain SV40-specific DNA sequences integrated, at quite variable positions, into their chromosomes.

A closely related virus is **polyoma**, whose natural host is the mouse; rat cells will not reproduce the virus but, like human cells infected with SV40, can be transformed by it. *Fig.* 17.10 shows the results of an analysis of cloned fragments of DNA isolated from polyoma-transformed rat cells, selected on the basis that they contained integrated fragments of the polyoma genome. The physical mapping of these DNA clones, in comparison with the map of the complete virus genome, shows that each transformed cell clone contains a particular fraction and arrangement of the virus genome, including in all cases an uninterrupted copy of a virus gene encoding one of the virus-specific antigens.

An obvious hypothesis is that this virus gene product is in some way responsible for the maintenance of the transformed cell phenotype. There is some evidence that the polyoma antigen, like the analogous antigen determined by SV40 in cells transformed by that virus, is a protein kinase, which may affect the function of one or more host cell proteins by phosphorylating them. Protein phosphorylation is an important mechanism for the modification of enzyme function and may be a very potent means of changing cell type.

Transforming Retroviruses and Oncogenes

Retroviruses, which have been studied particularly in rodents and birds, have RNA as their infective genetic material, but this is reverse-transcribed (by a virus-encoded enzyme, **reverse transcriptase**) into single-stranded DNA after infection and then replicated as double-stranded DNA, probably in circular form. Their genomes have the important property of being able to become linearly integrated as double-stranded DNA into the chromosomes of their host cells and, at least in many cases, of being able to induce in their host cells a state of unrestrained proliferation (oncogenic transformation) leading, in different cases, to leukaemia or solid tumour formation.

Some of the key features of retrovirus genome structure are shown in *Fig.* 17.11. There are some striking resemblances to the prokaryotic and eukaryotic transposons, which were surveyed in earlier chapters (pp. 268 and 444). In their integrated (provirus) form, all analysed retrovirus genomes have long direct terminal repeats (LTRs) and these each have short inverted repeats at their ends (compare Tn9 and its flanking IS sequences, *see* p. 268). Furthermore, integration of the provirus generates short *direct* repeats in the host genome at the site of integration. It seems very likely that the mechanism of integration is transposon-like. The LTRs themselves seem, from their structures, to be compounded from the two ends of the infective genome via circularization.

Recent work indicates two ways in which retroviruses can transform their host cells. The first, exemplified by avian leukosis virus (ALV), depends upon the (presumably) chance integration of the virus genome adjacent to a host **oncogene** which, when transcribed excessively, can cause transformation. A promoter in the provirus located in the right-hand LTR causes abundant transcription of the host-cell oncogene; the corresponding promoter in the left-hand LTR is presumably responsible for the transcription of the virus genes.

There are a number of kinds of oncogenes in vertebrate cells; some, at least, code for kinases that phosphorylate tyrosine residues of proteins the normal function of which may be to stimulate controlled growth in response to appropriate developmental signals. Placing one of them under the control of a powerful virus promoter takes cell growth out of normal developmental control.

The second way in which retroviruses can transform cells—and a much more efficient way—is through incorporating a cellular oncogene into their own

Fig. 17.11

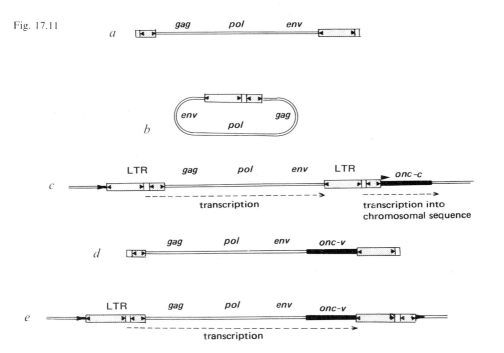

Structures of two related retrovirus genomes, in their infective and host-integrated states.
a. Structure of the infective RNA of avian leukosis virus (ALV) and of the linear DNA copied from it by reverse transcriptase after infection.
b. Circularized double-stranded DNA derivative found in infected cells.
c. Proviral ALV DNA integrated into the host genome adjacent to a host cell 'oncogene' which is abundantly transcribed from a viral promoter in the right-hand long-terminal repeat (LTR). The same promoter in the left-hand LTR initiates transcription of the virus's own genes, *gag, pol* and *env*, which code for a viral glycoprotein, reverse transcriptase and the virus envelope protein, respectively.
d. Infective RNA or copy DNA of avian sarcoma virus (ASV).
e. Proviral ASV DNA integrated into the host genome. Since ASV has acquired its own *onc* gene at some time in the past from one of its hosts, it can bring about cellular transformation regardless of where it integrates.
Key: Hatched, open and stippled boxes represent three components of the proviral long direct repeat (LTR) and are, respectively, 229, 21 and 80 bases long. The short arrowheads represent 15-base near-perfect inverted repeats at the ends of the 229 and 80 base elements. The longer arrowheads indicate 8-base direct repeats generated on insertion of the proviral DNA from an 8-base sequence present only singly in the uninfected host chromosome (compare transposon integration, p. 269).
The drawings are not to scale; the LTR regions are each only 330 bases and the rest of the genome is 8–9 kb.
Data from Swanstrom et al.[28] and Hayward et al.[29]

genomes. The mechanism of such gene acquisition and its selective value for the retrovirus are not very clear, but numerous cases have now been described in which it has undoubtedly happened. A diversity of *onc* genes, all homologous to genes present as single copies in uninfected cell genomes, have been found in different potent transforming retroviruses. Several of these induce solid tumours (sarcomas) with high efficiency. The structure of avian sarcoma virus (ASV), which is obviously closely related to ALV, is shown in *Fig.* 17.11. Some retroviruses carrying host-derived oncogenes are defective, the *onc* gene having displaced part of the normal viral genome. In order to produce infective particles such viruses need to be complemented by a competent 'helper' retrovirus.

The Relevance of Tumour Viruses to Cancer Generally

Tumour viruses are not the only cause of tumours—indeed, in humans, there is not much evidence that they are an important cause of cancer at all. Nevertheless, they may give an important insight into the general nature of cancer.

The mechanism by which such viruses as ALV transform cells involves, as we saw above, the integration of the viral genome, with its powerful promoter, directly adjacent to one *onc* gene which probably has a normal stringently regulated function in cell growth.

Recent work has implicated oncogenes in the determination of spontaneous tumours, as well as those provoked by retroviruses. A new experimental approach was opened with the discovery that DNA isolated from spontaneous human or chemically induced animal cancers was capable of transforming mouse fibroblasts (skin cells) to tumour cells. It is possible to recover from transformed cell lines the specific foreign sequences responsible for their transformation. One way of doing this is exemplified by the recent isolation of the oncogenic DNA sequence from a human bladder carcinoma.[42] This depended upon the fact that any fragment of human DNA of more than a few kilobases in length is likely to contain one or more copies of the specifically human version of the dispersed highly repetitive *Alu* element (cf. p. 449), identifiable by molecular hybridization with a labelled *Alu* probe. DNA isolated from the bladder carcinoma cells was used to transform mouse fibroblasts, and a 'library' of EcoRI fragments from a transformed cell line was prepared in a lambda cloning vector. Screening of the library with an *Alu* probe showed that one 25-kb fragment contained human sequences, and this fragment by itself was shown to be capable of transforming more fibroblasts. The transforming ability was, in fact, confined within a 6·6-kb subfragment which did not contain any *Alu* sequence but did hybridize strongly to the oncogene of a mouse retrovirus (Harvey murine sarcoma virus), which is closely similar to a gene present in the normal genome.[43] Further analysis[45] has shown that the protein product of the bladder carcinoma oncogene differs from the normal cellular gene in position 12 of the polypeptide chain, valine having been substituted for glycine. It is thought that it is this replacement, resulting in altered or enhanced catalytic activity of the gene product, that potentiated the cancerous condition.

All human and animal cells appear to contain a limited number of potential oncogenes,[46] the normal alleles of which are expressed at low levels to provide functions essential for cell growth and normal properties of the cell surface. Some of the gene products (though not, so far as is known, the Harvey virus-related oncogene) are protein kinases, and may be capable of profoundly changing the properties of other proteins by phosphorylating them. The current view is that many cancers, if not all, are due to hyperactivity of single oncogenes, either because of mutational changes in their structure (as in the case of the bladder carcinoma mentioned in the last paragraph) or because of enhanced transcription of the normal gene. In some animal cancers caused by retroviruses the enhanced transcription is due to the presence next to the oncogene of an active viral promoter. In those kinds of human cancer that are consistently associated with specific kinds of chromosome rearrangement it is likely to be the change in the chromosomal environment of an oncogene, perhaps from a domain of low transcriptional activity to one of high transcriptional activity, that is responsible for the cellular transformation. In the bladder cancer example described above there was no gross chromosomal rearrangement, but there was a small change in the oncogene protein-coding sequence. In other cases; small changes in oncogene promoter sequences may be involved. We may expect to see, in the near future, detailed comparisons of the sequences in and around normally controlled and hyperactive oncogenes.

There are no doubt many kinds of structural change in the DNA adjoining oncogenes that could promote their excessive activity. We might expect any mutagen or inducer of chromosomal rearrangements to be carcinogenic in some degree, and that is indeed what tends to be found.

Whether the oncogene hypothesis is valid for most forms of cancer is a question that should be answered rather soon.

17.6   Chromatin structure and gene expression

Even if it turns out that DNA rearrangements account for only a few special cases of stable cell differentiation, there still remains the possibility of self-perpetuating changes in DNA or chromatin at a different level.

DNAase Sensitivity of Transcribed Regions

In Chapter 1 (p. 39) we noted the general DNAase I sensitivity of transcriptionally active, in contrast to transcriptionally inactive, chromatin.[37] Some conformational difference between active and inactive nucleosomes is thought to be responsible. Superimposed on this general sensitivity is a special *hyper*sensitivity of single sites, or clusters of sites, some hundred or even thousands of bases upstream of transcribed or potentially transcribed sequences.[36] Hypersensitivity may be an indication that nucleosomes are locally absent, widely spaced or easily detachable. It might also (though this is speculation) be in some way connected with Z-DNA or DNA methylation, the possible relevance of which for control of transcription was referred to in Chapter 15.

The significance, and the inter-relationships, of these different features of chromatin structure are still largely unresolved, but there is little doubt that some of them, or some combinations of them, are of vital importance in determining whether genes are transcribed or not. The question of interest in the context of the present chapter is whether they can be self-perpetuating through cycles of cell division, and thus provide a basis for stable cellular differentiation. In the remainder of this chapter we shall review some of the evidence that some states of chromatin *do* seem to be self-perpetuating, even though their molecular explanations are still for the most part unknown.

Position—Effect Variegation in *Drosophila*

In most organisms, chromosome segmental rearrangements have no phenotypic effects provided that they are balanced and do not cut into DNA sequences that are transcribed into functional RNA. In *Drosophila*, however, there is a well-known kind of **position effect** which is frequently observed when, because of a rearrangement, a gene which is normally located in a euchromatic part of a chromosome is brought close to heterochromatin. The effect seen is of gene inactivation (or failure of normal activation)—not in all cells but in certain clones which seem to arise in a more or less random way at the early larval stage of development. The adult fly thus comes to show a variegated pattern. In one well-known example, the wild-type (w^+)allele of the *w* gene, normally near the tip of the X-chromosome, is brought through a long inversion to a position adjoining heterochromatin at the base (centromeric region) of the X. In the male, or when in

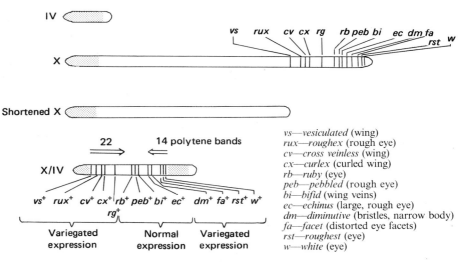

vs—vesiculated (wing)	
rux—roughex (rough eye)	
cv—cross veinless (wing)	
cx—curlex (curled wing)	
rb—ruby (eye)	
peb—pebbled (rough eye)	
bi—bifid (wing veins)	
ec—echinus (large, rough eye)	
dm—diminutive (bristles, narrow body)	
fa—facet (distorted eye facets)	
rst—roughest (eye)	
w—white (eye)	

Fig. 17.12　　An example of position–effect variegation in *Drosophila melanogaster*, after Demerec.[30]

The chromosomes diagrammed are the relevant part of the genome of females heterozygous with respect to an insertional translocation of a distal segment of the X-chromosome into the heterochromatin of chromosome IV.

Heterochromatic regions are shown as stippled. The structurally normal X carried a subset of the recessive markers shown—different subsets in different experiments since not all mutant phenotypes can be scored at once. The genes showing variegated expression in the X/IV compound chromosome are indicated.

The phenotypes associated with the marker mutations are as shown above.

heterozygous combination with the non-functional recessive w in the female, the inverted X-chromosome gives a red/white mottled or sectored eye phenotype, as if some cell clones had undergone inactivation of w^+.

This apparent clonal inactivation of genes unaccustomedly close to transcriptionally inactive heterochromatin has been attributed to a self-perpetuating conformational change spreading from heterochromatin into the adjoining euchromatin. The idea of a spreading effect receives support from studies of cases where a euchromatic chromosome segment carrying a number of marker genes is inserted into a block of heterochromatin. In such cases, the genes near each end of the inserted segment have been found to be inhibited in many sectors of tissue. But genes nearer the centre of the segment, and hence more distant from the flanking heterochromatin, are affected in fewer clones the greater their distance from the nearest breakpoint. A classic example due to M. Demerec[30] is described in *Fig. 17.12*.

X-chromosome Inactivation in Mammals

In both *Drosophila* and mammals, the X-chromosomes carry many genes whose functions are not particularly related to sexual differentiation but which are present at twice the dosage in the female as in the male. In both kinds of animal, devices exist which have the effect of equalizing the effects of X-linked genes between the two sexes—so-called **dosage compensation**. The mechanisms, however, are quite different. In the flies, both X-chromosomes are active in the female, but each X contrives, in some way not yet understood, to have only one-half of the effect of the unpaired X in the male sex. In the mammals, one of the pair of Xs in every cell of the female embryo is inactivated and becomes heterochromatic and visible in the non-dividing nucleus as a dense contracted **Barr body** (first described by M. Barr).

Thus female mammals that are heterozygous with respect to X-linked coat colour markers show a mosaic pattern, with some patches showing the effect of one X-chromosome and others that of the other. This is the **Lyon effect**, so called after M. F. Lyon, who was the first to realize its significance. In typical mammals, X-chromosome inactivation occurs in the early embryo at random, so that the two kinds of cell clones in mosaic females are about equal in frequency (in marsupials, in contrast, it is always the X of paternal origin which is inactivated). It occurs in all cells and not just in those which contribute to the surface pigmentation; for example it can be demonstrated in blood cells by using a sex-linked glucose 6-phosphate dehydrogenase variant as a marker (cf. p. 239).

B. Cattanach and others have shown that in female mice in which a portion of an autosome is translocated to, or inserted into, the X-chromosome, genes present in the autosomal segment can be inactivated through a spreading effect from the X-chromosome to which they are linked. As in the parallel case of position–effect variegation in *Drosophila*, linked genes are inactivated in a sequence corresponding to their distances from the junction with heterochromatin (*Fig. 17.13*).

The *Drosophila* and mouse examples just considered differ from one another in that, in the *Drosophila* case, the heterochromatin is **constitutive**, the fourth chromosome region in question always being heterochromatic, while in the mouse case it is **facultative**; a given X-chromosome may or may not be inactivated. The two examples resemble each other, however, in that, in both cases, the heterochromatic state is not only inherited vertically through cycles of mitosis but is

Fig 17.13

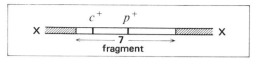

a

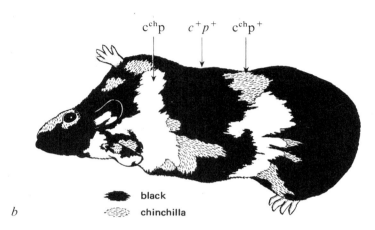

b

Coat colour variegation in the mouse due to insertion of a fragment of chromosome 7, carrying the coat-colour genes c^+ and p^+, into the X-chromosome.
a. Structure of the insertion.
b. Appearance of a mouse possessing two structurally normal chromosomes 7, both carrying the recessive alleles c^{ch} and p, plus the extra chromosome 7 fragment, carrying $c^+ p^+$, inserted into the X. Arrows indicate areas of phenotype typical of $c^+ p^+$ (black in this mouse strain), $c^{ch} p^+$ ('chinchilla') and $c^{ch} p$ (white). The pale phenotype characteristic of $c^+ p$ is not seen. The pattern is as if, in the inserted chromosome 7 fragment, inactivation affects c^+ alone or c^+ and p^+ together—never p^+ alone. This suggests that inactivation extends into the insert progressively from the left-hand side.
After Cattanach.[31]

transmitted *laterally* along the chromosome into regions not normally heterochromatic when the latter are brought into unaccustomed proximity to the heterochromatin. Heterochromatization acquired by lateral transmission is then maintained through mitotic divisions.

The mechanisms underlying the establishment and maintenance of heterochromatin are not at all clear, though it seems inescapable that there must be alternative self-stabilizing states of chromatin. The stabilization may come from cooperative interactions and constraints among the chromatin proteins; the pre-existing established conformation may be imposed on neighbouring molecules, whether these are newly synthesized at loci already heterochromatic in the parental chromosome, or at loci not previously heterochromatic but immediately adjacent to established heterochromatin. We may expect further information to give more definite form to these ideas within the next few years. They may be of great relevance to the problem of cell differentiation.

Self-perpetuating conformational changes may be possible for small chromosome domains and not just for the relatively large blocks which become visible as heterochromatin. The idea of coordinated clusters of genes nested in separate chromatin domains, with each domain subject to a common conformational control so as to permit or prohibit transcription, may turn out to be extremely important. We may already have examples in *Drosophila* polytene 'puffs' (*see* p. 433). More evidence, however, is required.

Secondary Changes in DNA Bases— Methylation

The replacement of one base-pair by another is not the only possible change in the covalent structure of DNA. Various secondary changes can occur to DNA bases and, of these, the most studied and possibly the most important is methylation. The base that is most frequently methylated is cytosine, the product being 5-methylcytosine. In mammalian DNA it is the cytosines that happen to be adjacent to guanine on the 5′ side,

$$\text{i.e.} \quad \begin{matrix} 5' \ \overset{*}{C}\text{–G} \ 3' \\ 3' \ \text{G–C} \ 5' \\ {}_{*} \end{matrix} \quad \text{sequences}$$

which are vulnerable to methylation, because of the specificity of the methylating enzyme. Not all such cytosine residues, however, are affected. Some are never methylated, while others appear to be methylated only in certain tissues. A plausible mechanism has been suggested for the self-perpetuation of states of methylation or demethylation.

The principle of the idea is illustrated in *Fig.* 17.14. All one needs for self-maintenance is a methylating enzyme which acts preferentially on *half*-methylated sites. Such an enzyme could ensure that each already fully methylated G–C sequence remains methylated through cycles of DNA replication, while no previously non-methylated site becomes so. There is evidence that cells contain methylases with the required specificity.

There is also evidence of a more direct kind for the replication of the methylated (or non-methylated) state. It depends on the use of two restriction endonucleases; HpaII which will cut

$$5' \ C\overset{*}{C}GG \ 3'$$

$$3' \ GGCC \ 5'$$
$${}_{*}$$

sequences but only when the starred cytosines are non-methylated, and MspI which cuts the sequence whether it is methylated or not. M. Wigler et al.[34] introduced a cloned DNA sequence including a chicken thymidine kinase gene into *tk*⁻ mouse cells by transformation (*see* p. 243) and investigated the state of methylation 25 cell generations later. They did this by isolating the total DNA, digesting it with HpaII or with MspI, separating the fragments by size by gel electrophoresis and determining the sizes of the *tk*⁺-containing sequences by Southern blotting (cf. p. 186) and hybridization with a *tk*⁺ radioactive probe. The two restriction enzymes were expected to give the same set of *tk*⁺-containing fragments if the integrated chicken sequence was non-methylated; if it were completely methylated the result should be the same with MspI, but HpaII should yield only a single larger fragment.

The result obtained depended on the methylation state of the DNA used for transformation. If it started unmethylated it remained largely unmethylated, while

Fig. 17.14

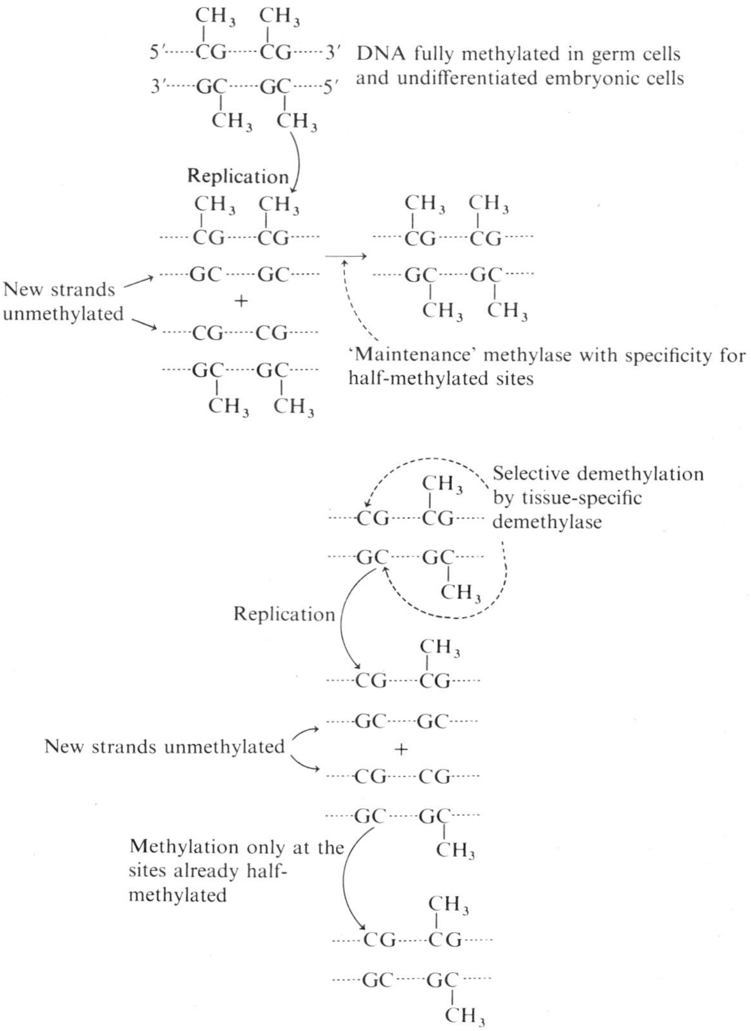

$$CH_3 \quad CH_3$$
$$| \qquad |$$
$$5'\text{-----}CG\text{-----}CG\text{-----}3'$$ DNA fully methylated in germ cells
$$3'\text{-----}GC\text{-----}GC\text{-----}5'$$ and undifferentiated embryonic cells
$$| \qquad |$$
$$CH_3 \quad CH_3$$

Replication

$$CH_3 \quad CH_3 \qquad\qquad CH_3 \quad CH_3$$
$$| \qquad | \qquad\qquad\qquad | \qquad |$$
$$\text{-----}CG\text{-----}CG\text{-----} \qquad \text{-----}CG\text{-----}CG\text{-----}$$
$$\text{-----}GC\text{-----}GC\text{-----} \qquad \text{-----}GC\text{-----}GC\text{-----}$$
$$+ \qquad\qquad\qquad | \qquad |$$
New strands unmethylated $$\text{-----}CG\text{-----}CG\text{-----} \qquad\qquad CH_3 \quad CH_3$$

$$\text{-----}GC\text{-----}GC\text{-----}$$
$$| \qquad |$$
$$CH_3 \quad CH_3$$

'Maintenance' methylase with specificity for half-methylated sites

Selective demethylation by tissue-specific demethylase

$$CH_3$$
$$|$$
$$\text{-----}CG\text{-----}CG\text{-----}$$
$$\text{-----}GC\text{-----}GC\text{-----}$$
$$|$$
$$CH_3$$

Replication

$$CH_3$$
$$|$$
$$\text{-----}CG\text{-----}CG\text{-----}$$
$$\text{-----}GC\text{-----}GC\text{-----}$$
$$+$$
New strands unmethylated $$\text{-----}CG\text{-----}CG\text{-----}$$
$$\text{-----}GC\text{-----}GC\text{-----}$$
$$|$$
$$CH_3$$

Methylation only at the sites already half-methylated

$$CH_3$$
$$|$$
$$\text{-----}CG\text{-----}CG\text{-----}$$
$$\text{-----}GC\text{-----}GC\text{-----}$$
$$|$$
$$CH_3$$

A model for the self-replication of DNA methylation. After Holliday and Pugh,[32] and Razin and Riggs.[33]

if it had been artificially methylated at CCGG sequences by treatment with a specific methylase before introduction into the cells, it tended to remain methylated at these sites through the 25 cell divisions.

An attractive hypothesis is that cellular differentiation is triggered by the action of site-specific demethylating enzymes which come into play individually at the appropriate times. There is as yet no evidence for these postulated enzymes, but they might be difficult to find since they would need to be present only transiently and in very small amounts—they would be needed only to initiate differentiation and not to maintain it.

That the state of methylation of DNA in chromatin may indeed have something to do with its transcriptional activity is strongly suggested by the results of several recent studies. For example, it has been shown that *certain* sites, not *all* sites, close to globin genes in the rabbit are unmethylated or at least undermethylated in cells committed to haemoglobin production, but are totally methylated in other cells.[35]

The methylation hypothesis is an attractive one for explaining the self-maintaining nature of cellular differentiation, but it is very unlikely to be the whole answer. Even if it is true within its own limits it does not explain the distinction, demonstrated so well by *Drosophila* imaginal discs, between overt cellular differentiation and a covert commitment to differentiate in a certain way at some future time (in any case *Drosophila* DNA shows very little methylation). It seems likely to be an important *component* of a future more comprehensive theory, the formulation of which will require more knowledge of the various states of chromatin than we currently possess.

17.7   Summary and perspectives

In this chapter we considered the possible molecular mechanisms responsible for the clonal transmission of differentiated cell states. The quasi-hereditary character of cellular differentiation obviously suggests the possibility of controlled changes in the structure or organization of the genetic material. But, in fact, it is possible to have self-stabilizing and transmissible cell states even without any such changes. In prokaryotes, we know of a number of control systems in which a gene product maintains the activity of its own gene and/or represses the activity of other genes; the lysis/lysogenization switch of bacteriophage lambda provides some particularly good examples.

Nevertheless, cellular differentiation does sometimes involve DNA or chromosomal changes, which can occur at at least three different levels. Rearrangements of DNA primary base sequence account for several examples of *reversible* differentiation in unicellular organisms and for one example of irreversible differentiation, that of immunoglobulin-producing cells, in vertebrates. The evidence that not all differentiated cells in vertebrates have undergone DNA rearrangements comes in part from direct molecular investigations and in part from demonstrations, using nuclear transplantation, that the nuclei of at least some cells from differentiated tissues are totipotent when transplanted into eggs. The question of how many, if any, kinds of vertebrate cellular differentiation, other than that of the immune system, are associated with DNA rearrangements is still open.

Even with a constant primary DNA base sequence there are possibilities for self-maintaining secondary changes in the covalent structure of the DNA. One, for which there is a clear mechanism and some evidence, involves selective methylation or demethylation of cytosine residues; the self-maintenance here depends on methylases acting specifically on half-methylated sites in duplex DNA.

Another possibility, or set of possibilities, involves higher order chromosome structure. Blocks of chromatin, obviously larger than single units of transcription, can become transcriptionally active or inactive as a whole; *Drosophila* polytene 'puffs' (*see* Chapter 15, p. 433) provide examples of activation, and *Drosophila* position–effect variegation examples of apparently clonal inactivation of chromosome segments hundreds of kilobases in length. Female mammals show clonal inactivation of entire X-chromosomes. Molecular explanations of self-maintaining and (at least sometimes) mitotically transmissible states of chromatin are not available at this time, but they may not be long in coming. There is a growing conviction that controlled local changes in chromatin structure play an important part in the unfolding of the developmental programmes of eukaryotes, and much research effort is currently being directed into this area.

Chapter 17 **Selected Further Reading**

Leighton T. (ed.) (1981) *The Molecular Genetics of Development—An Introduction to Recent Research on Experimental Systems.* New York, Academic Press.

Leighton T. and Loomis W. F. (ed.) (1980) *The Molecular Genetics of Development*, 478 pp. New York, Academic Press (particularly articles by Echols H. on lambda bacteriophage and Herskowitz et al. on yeast mating type).

Sonneborn T. M. (1977) Genetics of cellular differentiation: stable nuclear differentiation in eukaryotic unicells. *Annu. Rev. Genet.* **11**, 349–367.

Problems for Chapters 16 and 17

1   In the mouse, agouti/non-agouti mosaic coat patterns are observed in allophenic mice resulting from artificial aggregation of homozygous agouti (A/A) and non-agouti (a/a) embryos at the 16-celled blastocyst stage. The patterns seen in mice heterozygous with respect to sex-linked alleles Ta/ta (tabby) have been considered to be very similar in the mean sizes and distributions of differently coloured striped and patches. If one accepts this assessment (it has been questioned) what conclusions can one draw concerning (a) the nature of the mosaicism seen in Ta/ta mice and (b) the timing of X-chromosome inactivation? (Mintz, 1974, *Annu. Rev. Genet.* **8**, 411.)

2   In *Drosophila melanogaster*, y/y flies have yellow body surface and yellow bristles (chaetae) wherever these are formed. Flies homozygous for h have chaetae on the wing in place of the normal trichomes (small hairs). In a certain stock, doubly heterozygous y^+/y h^+/h, a chromosomal rearrangement had brought the two loci together on to the same chromosome arm, so that mitotic crossing-over could result in the formation of clones of cells which were y/y h/h. It was found that X-irradiation of larvae earlier than 8 hours prior to pupation led to the appearance of yellow patches of cuticle on the thorax and to occasional yellow chaetae on the wings, presumably due to induced mitotic crossing-over. When the X-ray treatment was given within 8 hours of pupation, yellow patches (smaller ones) still appeared on the thorax but no chaetae were ever found on the wing. In another experiment with larvae heterozygous y^+/y but homozygous h/h, X-irradiation within 8 hours of pupation did lead to occasional yellow chaetae among the non-yellow ones on the wings. What do these results imply regarding the time of action of h^+ in development? (Garcia-Bellido and Merriam, 1971, Proc. Natl Acad. Sci. USA **68**, 2222.)

3   Mitotically recombinant clones, when induced sufficiently early in *Drosophila* larval development, sometimes extend between two compartments without necessarily filling either. Why is this information important for understanding the origin of compartments? Clones induced at the blastoderm stage sometimes cross between left and right prothoracic legs, between wing and mesothoracic leg or between dorsal metathorax and metathoracic leg. However, clones induced at the same stage have never been observed to overlap different thoracic segments or the posterior and anterior halves of the same leg. What does this tell us about the sequential formation of compartment boundaries? (Kauffman et al., 1978, *Science*, **199**, 259.)

4   In *Drosophila*, a semidominant sex-linked mutation *Hyperkinetic* (*Hk*) causes shaking movements of all legs when flies are under ether anaesthesia; the effect is much stronger when the mutation is homozygous or hemizygous. By the use of the unstable ring-X chromosome, gynandromorphs were obtained in which approximately half of the tissue was heterozygous hk^+ y^+/Hk y and the other half hemizygous Hk y. Examination of many such gynandromorphs, with classification of each leg by colour and by behaviour, showed that a leg could fall on one

side of an XX/XO boundary by the one criterion and on the other side by the other. Thus the colour and the behaviour of a given leg could be mapped to different foci on the fate map. The following were some of the estimated fate map distances in 'sturts', C and B indicating colour and behaviour, respectively and I, II and III first, second and third legs on one side of the fly: IC–IB 14, IIC–IIB 11, IIIC–IIIB 10, IB–IIB 17, IIB–IIIB 14, IC–IIB 12, IIC–IB 19, IIC–IIIB 13, IIIC–IIB 15. The fate-map distance between the behavioural foci of corresponding left and right legs was determined as 40, 36 and 40 for first, second and third legs, respectively. With reference to *Fig.* 16.4*a*, determine as closely as you can the positions of the foci governing the behaviour of each leg. What do you think these foci might correspond to? (Hotta and Benzer, 1972, *Nature* **240**, 527.)

5 Certain 'varieties' (probably they should be regarded as distinct species) of *Paramecium aurelia* show a peculiar mode of inheritance of mating type. Conjugation of a cell of mating-type I with another of mating-type II gives two exconjugant cells undifferentiated as to mating type; these divide to give four cells (two from each exconjugant) which each gives rise to a clone of stable mating type—either I or II with equal probability. A single conjugating pair can give in different cases 4:0, 3:1, 2:2, 1:3 or 0:4 ratios of mating types after the first exconjugant division. The progeny mating types do not correlate at all with the mating types of the conjugants from which they derived their cytoplasm. In one experiment involving conjugating cells from a single pair of clones, the five above ratios were found in respective frequencies 1, 18, 21, 13 and 3. Referring to *Fig.* 7.1 for details of the nuclear events accompanying conjugation and the following cell fission, speculate on the possible basis of mating-type determination in this variety of *P. aurelia*. (Sonneborn, 1942, *Am. Nat.* **76**, 46.)

6 A deletion of a part of the mouse chromosome 7, including the gene c^+, which is necessary for pigmentation, is recessive lethal and gives an albino phenotype in heterozygous combination with viable c alleles. Female embryos carrying two copies of the deletion in chromosome 7 and also an extra fragment of chromosome 7 (with c^+) inserted into the X-chromosome (Cattanach's translocation—*see Fig.* 17.13) show only 9 per cent viability, but the ones which are viable are phenotypically more or less normal (except for coat colour variegation). A similar result is found with other seventh chromosome deletions covered by the Cattanach translocation, except that the percentage viability varies a great deal from one deletion to another, being nearly 100 per cent for some deletions. What general explanation can you suggest for the different degrees of penetrance (in supplying the deleted functions) of the translocated segment in the different cases? (Glucksohn-Waelsch, 1979, *Cell* **16**, 225.)

7 A 49-base sequence (S) is found tandemly repeated, with variations, within a few kilobases of the 'downstream' side of the J_H segments (*see Fig.* 17.8), and again at a similar distance from the 'upstream' side of each C_H segment in the mouse immunoglobulin heavy-chain gene complex. On this basis, two alternative mechanisms for heavy-chain switching have been proposed—the 'looping-out' model and the 'sister-strand exchange' model. Sketch the mechanisms which seem to be implied by these summary descriptions. (Honjo et al., 1980, *Cold Spring Harbor Symp. Quant. Biol.* **45**, 913.)

18 Population Variation and Evolution

18.1 | The fate of mutations

Genetic variation originates in mutations, the varied nature and possible origins of which were reviewed in Chapter 12. Mutation is a rare event at any particular locus, typical frequencies may be 10^{-5}–10^{-6} per gene or 10^{-8}–10^{-9} per DNA base-pair per generation, but is common reckoned over the whole genome. Accounting for the existing variation in populations is no problem so far as the availability of mutations is concerned; it is rather a question of the extent to which new mutants are able to persist in the population and the extent to which they affect the phenotype. We shall return to the question of the phenotypic effects of mutations later in this chapter.

The degree of persistence of mutant genes in populations depends upon two independently operating processes—**selection** and **drift**. Modern population genetics theory arose from the marriage of the Darwinian theory of evolution by natural selection and Mendelian genetics. It is based on the recognition that the genetic variation on which natural selection must act is due to individual segregating gene differences. The orthodox neo-Darwinian view was (and still is in some quarters) that new mutant genes survive and spread only if they confer improved Darwinian fitness (i.e. capacity to leave progeny) on the organisms carrying them, and that otherwise they are rapidly lost by adverse selection. An alternative view, while not denying the role of natural selection in establishing or eliminating some mutations, holds that many are selectively **neutral**, or nearly so, and that their increase or decrease from one generation to the next is determined mainly by the 'luck of the draw'—that is to say, on the random sample from the pool of gametes which happens to contribute to the next generation.

A third factor which must be considered is that of **recombination** in the broad sense which includes free reassortment of chromosome pairs as well as recombination of linked gene differences. The significance of recombination is that it greatly increases the possible number of genotypes available for selection.

We shall commence this chapter with a brief review of breeding systems, on which genetic recombination depends.

18.2 | Breeding systems

Outbreeding Versus Inbreeding

Single mutations with detectable phenotypic effects are generally deleterious. All experience suggests that it must be rare for a new 'raw' mutant to be unconditionally advantageous from its inception. This should not be surprising. Selection acts on phenotypes and not directly on genes, and most features of the phenotype of any complex organism are determined by interactions of alleles of many genes which must have been selected for 'good fit' with one another. Thus a new gene variant, even if it is potentially beneficial, may only yield its benefit in combination with an appropriate set of alleles of other genes. In order to survive

Box 18.1   Systems of outbreeding; reversions to inbreeding

Organism	Breeding system	Effects of system (in addition to prevention of self-fertilization)	Examples of reversion to a closed genetic system
Nearly all animals, some plants	Separate male and female individuals (sex chromosome)	—	Parthenogenesis in some insects
Some flowering plants (e.g. *Primula*)	Hermaphrodite, but with two self-incompatible, cross-compatible morphological types, controlled by 2 alleles of one locus (S)	—	Mutation at S conferring self-fertility
Many flowering plants	Many self-incompatible, mutually compatible types; multiple alleles at S locus	Restricts cross fertility of individuals with alleles in common; prohibits S homozygotes	Mutation at S locus allows self-fertility. Many examples of apomixis bypassing sexual process altogether
Simple algae (e.g. *Chlamydomonas*); Phycomycete and Ascomycete fungi	Two self-incompatible, cross-compatible mating types; two alleles one locus	—	In some Ascomycetes, regular inclusion of meiotic product nuclei of different mating type in same spore
Many Basidiomycetes (mushroom-like fungi)	Many mating types, all self-incompatible but cross-compatible in all combinations; multiple alleles at A locus	Restricts cross-compatibility of sister meiotic products to 50 per cent. Prohibits A homozygotes	As above (species with two-spored basidia—two nuclei in each spore)
Many Basidiomycetes (mushroom-like fungi)	Many mating types as above, but determined by multiple alleles at two loci A and B	Restricts sister compatibility to 25 per cent. Prohibits A or B homozygotes	?

For further information about animal systems, *see* White;[37] for plant systems *see* Stebbins;[38] for fungi *see* Fincham et al.[39]

by selection it may be necessary for a new mutation to get into a new gene association rather promptly. It is, of course, sexual reproduction and meiotic recombination which, in eukaryotes, makes this possible. This, presumably, is why the whole elaborate ritual has evolved.

To pursue this line of argument, sexual reproduction as a means of creating new recombinant genotypes depends on cross-breeding to make the necessary multiple heterozygotes. Persistent inbreeding, especially in its extreme forms of sib-mating and self-fertilization, leads eventually to homozygosity at nearly all loci. Sexual reproduction in a wholly homozygous line can give no new genotypes and is, in fact, genetically equivalent to vegetative (clonal) reproduction. Thus, if the production of recombinants really is the evolutionary 'purpose' of sexual reproduction we would expect to find devices for promoting outbreeding, since the more outbreeding the more heterozygosity.

This sounds like a good argument and it is reassuring to find that nature confirms it. Wherever one looks in the eukaryotic world one finds a strong tendency towards outbreeding, often enforced by elaborate genetic controls. In animals, a high degree of outbreeding generally follows automatically from their free dispersal and mixing, reinforced in some animals by modes of behaviour which reduce the likelihood of sib-mating. The existence in the great majority of animals of separate male and female sexes in itself prohibits self-fertilization, the most extreme form of inbreeding.

In the vegetable world, separate sexes are much less common than hermaphroditism (male and female organs on the same plant), and in the simplest fungi and algae there is often no differentiation between male and female gametes at all. Nevertheless self-fertilization is prevented in very many cases by systems of genetically determined **self-incompatibility**, usually unaccompanied by any morphological differences between the mutually compatible groups into which the population is divided. In flowering plants and the Basidiomycete class of fungi the systems of mating control go beyond what is necessary for mere prevention of self-fertilization, and have the effect of favouring crossing between relatively unrelated individuals rather than closely related ones. *Box* 18.1 summarizes the most important features of these various systems.

A survey of breeding systems in more or less related species indicates how finely balanced the respective advantages of outbreeding and inbreeding may sometimes be. The general trend towards outbreeding seems clear from the ubiquity and complexity of the systems that encourage it. But particular species in almost every group of plants and invertebrate animals have acquired means of circumventing the anti-inbreeding devices common in their near relatives. Many insects, for example, reproduce some or all of the time by parthenogenesis—that is by the production of eggs without meiosis, which then develop without fertilization. Many species of flowering plants show various forms of **apomixis** (seed production without fertilization), including formation of diploid eggs or dispensing with eggs altogether by development of embryo sacs from what would normally be nonreproductive diploid cells of the ovule; sexual cell fusion is dispensed with in either case. In certain fungal species, both Ascomycetes and Basidiomycetes, one finds mating-type systems short-circuited through the regular inclusion of two mutually compatible haploid nuclei in the same spore, which germinates to produce a mycelium that is self-fertile from the start. It seems that the selective advantage of assured reproduction, without the sometimes risky necessity of searching for a compatible sexual partner, can sometimes outweigh the longer-term advantage of outbreeding.

Equilibria in Randomly Outbred Populations

Hardy–Weinberg Equilibrium

In a randomly outbred population one expects the frequencies of the various genotypes, made possible by the presence of allelic variation, to conform to certain simple rules. The frequencies of homozygotes and heterozygotes with respect to allelic variation in a single gene should conform to the **Hardy–Weinberg equilibrium**, named after the mathematician and the geneticist who jointly formulated it.

Consider the simple case where there are just two alleles A and a, present in the population in frequencies p and q, respectively ($p+q=1$). The Hardy–Weinberg ratios of the two homozygotes and the heterozygotes will then be:

Genotype: AA Aa aa

Frequency: p^2 $2pq$ q^2

That the proportions $p:q$ among the gametes do indeed yield the Hardy–Weinberg ratios is easily demonstrated by a simple 'Punnett square' as follows:

			♀ gametes				
			A	a			
		Frequencies	p	q			
♂ gametes	A	p	AA p^2	Aa pq	Totals:	AA	p^2
	a	q	Aa pq	aa q^2		Aa	$2pq$
						aa	q^2

That a population conforming to the Hardy–Weinberg ratios transmits the same frequencies of alleles to the next generation is also easily shown. Allele A will be transmitted by all the gametes of AA individuals (frequency p^2) and by half of the gametes of Aa individuals (frequency $2pq \div 2 = pq$). Thus the total frequency of A gametes will be $p^2 + pq$ which equals $p(p+q)$ or, since $p+q=1$, simply p. By similar reasoning, the frequency of a gametes will be q.

The formula can be generalized to the case of n alleles, $a_1, a_2 ..., a_n$, present in frequencies $p_1, p_2, ..., p_n$ (where $p_1 + p_2 + ... + p_n = 1$). The equilibrium frequency of any homozygote $a_x a_x$ is then p_x^2, and that of any heterozygote $a_x a_y$ is $2p_x p_y$.

The presence of homozygous and heterozygous allelic combinations in ratios conforming to the Hardy–Weinberg equilibrium is a test of whether a population is randomly outbreeding. Self-fertilization or assortative mating, i.e. preferential mating between genetically related individuals, will lead to an excess of homozygotes over the frequencies predicted by the formula. Deviation from Hardy–Weinberg ratios can also occur as a result of different chances of survival of different genotypes—that is to say **selection**. Failure to demonstrate such deviations is, however, poor evidence against the occurrence of selection; a selection differential of a few per cent, which would have a very significant effect over a number of generations, would not show as a distorted ratio unless an extremely large sample was taken.

Linkage Equilibrium and Disequilibrium

Another prediction about genotype frequencies in a randomly outbred, unselected population is that all possible combinations of alleles of different genes should be present at the equal frequencies expected on the basis of random reassortment, irrespective of whether the genes in question are linked or not. This condition is called **linkage equilibrium**.

Suppose, to take a simple formal example, we have alleles A and a of one gene and alleles B and b of another, present in the population in frequencies p, q, r and s respectively ($p+q = 1$ and $r+s = 1$). Then, even if A and B are linked, the equilibrium frequencies of gamete genotypes will be:

Genotype: AB Ab aB ab
Frequency: pr ps qr qs Total: $p(r+s)+q(r+s) = p+q = 1$

Suppose the mutant allele b originally arose in a chromosome carrying A at the other locus, the alternative allele a already being present in the population. To begin with, the only double heterozygotes will be Ab/aB. But crossing-over will generate ab gametes, and the frequency of these will increase at the expense of Ab gametes until the equilibrium frequencies are attained; at that point the Ab combination will be regenerated by crossing-over in AB/ab heterozygotes at the same rate as it is being lost by crossing-over in Ab/aB heterozygotes.

The occurrence of significant linkage *dis*equilibrium can be due to one of two causes (or a combination of the two). Perhaps one of the alleles arose by mutation only recently and there has been insufficient time for linkage equilibrium to be attained. In *Drosophila* and man, extremely closely linked markers, detected as restriction site polymorphisms within DNA regions of the order of no more than a few tens of kilobases, have been found to show very pronounced linkage disequilibria. Obviously, where recombination is rare because of extremely close linkage, equilibration can take a very long time. But pairs of linked polymorphic loci separated by a number of map units (several per cent recombinants per generation) have nearly always been found to be in linkage equilibrium in *Drosophila* populations.

The second reason for linkage disequilibrium is that a particular combination of alleles at different linked loci is favoured by natural selection. The best examples of this situation are to be found in butterflies, where combinations of linked genes (sometimes called **supergenes**) are sometimes responsible for elaborate visible phenotypes (*see* Mimicry, p. 534). In such cases one would expect selection to operate to reduce recombination frequency in the chromosome region involved, and this may well occur.

18.3 | The nature and extent of variation in populations

The Concept of the Standard Wild Type

All outbreeding populations harbour a certain number of recessive alleles which, when made homozygous by chance or by inbreeding, determine abnormal and frequently subviable phenotypes. Such mutant alleles are not uncommon on a total genome basis; each human genome, for example, has a more than even chance of containing at least one deleterious recessive; this would quickly become obvious if incest became more popular.

Loss of viability following close inbreeding (**inbreeding depression**) is virtually universal in habitually outbred species. This is due to deleterious recessive mutations which are able to accumulate under a breeding system that ensures that they will nearly always be masked by dominant alleles. For any one gene, significantly deleterious recessive alleles are rare, but over a whole genome, or even one chromosome, they are common, and their homozygous effects are cumulative. Thus, in *Drosophila melanogaster* (reviewed by Simmons and Crow[35]) about a quarter of all second chromosomes sampled at random from wild populations were found to be lethal when made homozygous. Even the viable second chromosome homozygotes usually showed a substantial fitness deficit, up to 70 or 80 per cent, in competition with random outbreds.

For any one gene, however, variant alleles with obvious effects on the phenotype tend to be rare. It has thus been reasonable, for practical purposes, to postulate the existence of a standard **wild-type** allele of each gene; the concept of the wild-type allele is, indeed, very firmly embedded in the genetical literature and the *plus* superscript has been generally adopted to denote it. One of the major changes that have overtaken the subject during the past 15 years has been the undermining of the concept of the standard wild type. It is now recognized that a considerable proportion of all genes are demonstrably **polymorphic**—that is to say they exist in two or several fairly common allelic forms with no one allele enjoying a majority of more than, say, 98 per cent.

Some examples of polymorphism have, of course, long been known to naturalists. Many species of animals, and still more of flowering plants, exist in the wild in alternative colour forms, not infrequently interbreeding in the same population. It is possible to regard these cases as exceptions to prove the rule, in as much as the variation often involved a superficial aspect of the phenotype with no obvious effect on fitness. In a few other examples it was possible to explain the polymorphism as the consequence of some special kind of natural selection (*see below*, p. 532).

The Electrophoretic Detection of Polymorphism

That polymorphism is certainly not confined to gene products of peripheral importance can now be quickly and simply demonstrated by gel electrophoresis. In this technique, the proteins present in crude cell extracts are fractionated by their different rates of migration through a gel, made of starch or polyacrylamide, under the influence of an electrical field. The gel is usually buffered at pH 8·5 or thereabouts, and at this pH all carboxyl side chains and most basic side chains of the proteins are charged, with the balance generally favouring the former. Different proteins migrate towards the positively charged electrode at different rates dependent on their individual net charges and also on their sizes, since the lattice of the gels slows down the passage of each molecule to a degree depending on its size.

Most enzymes are not inactivated by the conditions in the gel and in many cases their positions following electrophoresis can be revealed by staining the gel with reagents that are converted to insoluble dyes by the particular enzymes it is desired to detect. For example, esterases, phosphatases and peptidases can each be stained by artificial substrates that are hydrolysed to yield coloured derivatives directly. Dehydrogenases which oxidize their specific substrates and reduce pyridine nucleotide coenzymes (NAD or NADP) can be stained by a more complex procedure. The staining mixture contains the enzyme substrate, the appropriate coenzyme NAD or NADP, phenazine methosulphate and an initially colourless tetrazolium salt. The dehydrogenase oxidizes the substrate and reduces the NAD(P), and the reduced coenzyme reduces the phenazine which in turn reduces the tetrazolium to a brilliantly coloured (red or purple) insoluble dye. Some proteins, such as haemoglobin, are intrinsically coloured and do not need to be stained.

The first application of electrophoretic screening was to the detection of unusual haemoglobins in human populations. Numerous electrophoretic variants were discovered (and later explained as single amino acid replacements, *Fig.* 12.4) but all

Fig. 18.1

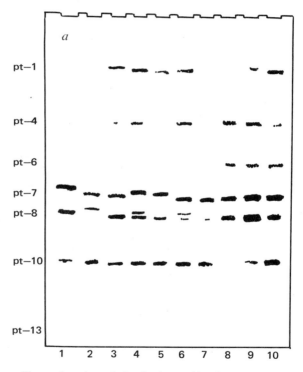

a. Electrophoretic analysis of polymorphism in *Drosophila pseudo-obscura* with respect to larval lymph proteins. Seven different proteins are resolved in the electrophoresis gel; of these two (7 and 8) show polymorphism with two alleles in each case. Larvae numbers 1, 4 and 5 were homozygous for the 'slow' allele of the gene-determining protein 7, and the others were homozygous for the 'fast' allele. With respect to the gene-determining protein 8, larva number 2 was homozygous for the 'slow' allele, numbers 1, 3, 5, 7, 8, 9 and 10 were homozygous for the 'fast' allele, and numbers 4 and 6 were heterozygotes. Note that the heterozygotes showed only two bands, indicating that protein 8 is monomeric (i.e. contains only one polypeptide chain per protein molecule). From the photograph of Singh et al.[4]

were at low frequency. Screening of human population samples for electrophoretic enzyme variants, however, showed that about one-third of the enzymes examined showed polymorphism, with the variant enzyme types in each case occurring in 2 per cent or more of the population (*Table* 18.1). In a number of instances there was, in fact, no clear majority allele. Polymorphisms characterized by similarly high frequencies of heterozygotes were also found with respect to the genes coding for certain non-enzymic proteins present in high enough concentrations to be distinct when electrophoresis gels were generally stained for protein. Some *Drosophila* larval serum proteins came into this category (*Fig.* 18.1*a*).

Assuming that each distinct band on the gel represents the product of a different allele, one can take individuals with double bands as heterozygotes. The assumption would not be a safe one if only one individual was observed, or if *all* the individuals investigated were double banded, since some proteins, even if uniform in primary structure, can undergo secondary modifications. However, the observation that some individuals (putative homozygotes) are single banded strengthens the interpretation. To be quite sure, it is necessary to undertake controlled crosses and analysis of progeny, so as to confirm that the supposed

Fig. 18.1 *contd*

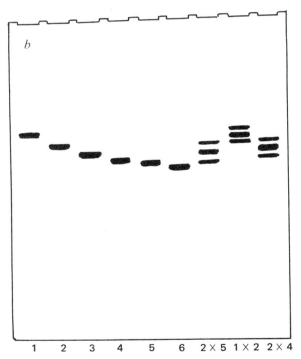

b. Electrophoretic analysis of six forms of esterase-5 in *D. pseudo-obscura* determined by six alleles of one gene. Lanes 1–6 show the stained enzyme bands from the six different kinds of homozygote. Lanes 7–9 show the band patterns obtained from three different heterozygotes: 2 × 5 in lane 7, 1 × 2 in lane 8 and 2 × 4 in lane 9. Note the three-banded pattern in the heterozygotes. The intermediate band in each case is due to a hybrid enzyme dimer with one polypeptide chain of each allelic type; the homozygote enzymes are dimers of identical chains.
Redrawn from the photographs of Singh et al.[4]

Table 18.1   **Average heterozygosity in populations of *Drosophila pseudo-obscura* and man for different genes***

D. pseudo-obscura		Man	
Enzyme/protein coded for	*Heterozygosity*	*Enzyme/protein coded for*	*Heterozygosity*
12 invariant	0·000	51 invariant	0·000
Acetaldehyde oxidase	0·012	Peptidase C	0·02
Protein-7	0·063	Peptidase D	0·02
Protein-13	0·070	Glutamate–oxaloacetate	
Malate dehydrogenase	0·102	transaminase	0·03
Octanol dehydrogenase	0·109	Leucocyte hexokinase	0·05
Leucine aminopeptidase	0·155	6-Phosphogluconate	
Protein-10	0·229	dehydrogenase	0·05
Protein-12	0·234	Alcohol	
α-Amylase-1	0·353	dehydrogenase-2	0·07
Xanthine dehydrogenase	0·492	Adenylate kinase	0·09
Protein-8	0·513	Pancreatic amylase	0·09
Esterase-5	0·741	Adenosine deaminase	0·11
		Galactose 1-phosphate	
		uridyl transferase	0·11
		Acetylcholine esterase	0·23
		Mitochondrial malic	
		enzyme	0·30
		Phosphoglucomutase-1	0·36
		Peptidase A	0·37
		Phosphoglucomutase-3	0·38
		Pepsinogen	0·47
		Alcohol	
		dehydrogenase-3	0·48
		Glutamate–pyruvate	
		transaminase	0·50
		Red blood cell acid	
		phosphatase	0·52
		Placental alkaline	
		phosphatase	0·53

*From Lewontin;[36] data from Prakash et al.[1] for *D. pseudo-obscura* and Harris and Hopkinson[2] for man.

heterozygotes show the expected allelic segregation at meiosis. This has been done in enough cases to make one fairly confident in taking a mixture of single and double-banded individuals in a population as good evidence for allelic variation. In an outbreeding population, the proportions of homo- and heterozygotes are expected to conform to those predicted by the Hardy–Weinberg formula, and they usually do so in cases of enzyme polymorphism.

One feature which can be observed in *Fig.* 18.1*b*, and which is very frequently observed in heterozygotes exhibiting two allelic variants of an enzyme, is the presence of a third band intermediate in position between the two bands corresponding to the homozygote enzyme forms. This is an indication that the enzyme is a **dimer**—that is, with each native protein molecule containing two polypeptide chains which, in homozygotes, are identical. In heterozygotes, the two chains in each dimer are drawn from a pool of two different kinds and therefore contain a proportion (50 per cent if assembly of dimers occurs randomly) of mixed dimers which will be of intermediate electric charge.

How many Alleles?

Electrophoresis will not separate all protein variants. The only differences which can be guaranteed to be detectable by this method are those that affect net charge, although certain shape (conformational) differences also affect mobility through a gel. For instance, in the size range where the sieving effect of the gel is just significant, a spherical molecule will be slowed less than an ellipsoidal molecule of the same mass. Whether or not a particular amino acid substitution affects conformation may depend on the pH, and whether or not any change in conformation is detected will depend on the pore size of the gel, which can be controlled by the acrylamide concentration. For some proteins it has been shown that many additional allelic variants can be detected when these conditions are varied.

Fig. 18.2 shows how what originally appeared to be a single allele, determining xanthine dehydrogenase of *D. pseudo-obscura*, was subclassified into nine distinct alleles by using four different conditions of electrophoresis instead of just one. Another way of revealing such 'hidden' allelic variation is by measuring the rate of enzyme inactivation at some suitable elevated temperature; many *D. pseudo-obscura* xanthine dehydrogenase alleles have been distinguished by this criterion. It is not totally implausible to suggest that so far as the *Drosophila* xanthine

Fig. 18.2

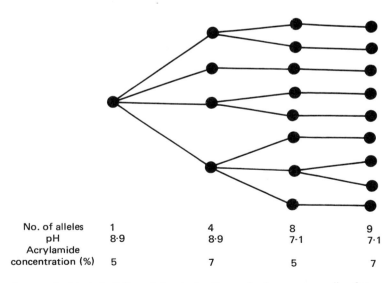

No. of alleles	1	4	8	9
pH	8·9	8·9	7·1	7·1
Acrylamide concentration (%)	5	7	5	7

Increase in detected allelic variation in the *D. pseudo-obscura* gene coding for xanthine dehydrogenase by the application of additional tests. The initial testing of 146 inbred lines in one electrophoretic system (5 per cent acrylamide, pH 8·9) distinguished six 'alleles'; three further electrophoretic systems increased the number of distinguishable alleles to 37. The diagram shows the progressive subdivision of one set of inbred lines which, on the first test, appeared to share the same allele.
From Jones,[3] quoting Singh et al.[4]

dehydrogenase (*rosy*) gene is concerned, heterozygosity is nearly universal, with any two alleles randomly sampled from a population likely to be different.

The rather extreme suggestion made in the previous sentence appears not to be applicable to all genes. For example, the *D. melanogaster adh* (alcohol dehydrogenase) gene has two common alleles ('fast' and 'slow' enzyme distinguished by electrophoresis) and attempts to show heterogeneity within these two by application of the other criteria mentioned above have failed.[6] Among the many enzymes and other proteins which have been surveyed in animal populations, some classes show relatively little polymorphism. In one study, all the soluble proteins of *Drosophila melanogaster* that were sufficiently abundant to be detected as proteins without enzyme staining were surveyed, and among these only 6 out of 54 were polymorphic.[5] Among enzymes, those involved in central areas of metabolism, for instance in the major glycolytic pathway, seem to be significantly less polymorphic than enzymes with peripheral or occasional functions. The limitation of this kind of observation is that we do not know whether the lack of observed variability in a particular protein reflects a real uniformity of amino acid sequence. It may be that the apparent uniformity is due to a strongly stabilized three-dimensional protein structure which is unaffected by most amino acid substitutions and/or to a general inviability of those substitutions that bring about charge changes. In other words, allelic variation may be much easier to observe in some proteins than in others. Only complete sequence determination on apparently identical genes or enzymes isolated from different inbred lines, a laborious and unattractive project, would provide really satisfactory evidence on this point.

Nobody doubts, however, that allelic variation must be constrained by the need to conserve function and it is evident that some proteins are more tightly constrained by conservative selection than others.

Chromosomal Variation

Structural heterozygosity is generally rare in wild populations, no doubt because of the reduction in fertility which it causes. However, for reasons explained in Chapter 6, paracentric inversions in flies do not carry this penalty, and they are found in abundance in many *Drosophila* populations. Dobzhansky made an extensive study of the distribution of inversions in *D. pseudo-obscura*; the results are summarized and discussed in detail in his book and that of Lewontin, listed at the end of this chapter.

Although the paracentric inversions found in *Drosophila* populations are fully viable, there is little to indicate that heterozygosity with respect to them actually increases fitness. In one group of flowering plants, a section of the genus *Oenothera*, one must suppose that structural heterozygosity *has* been selected for.

In this group of species the immediate products of meiosis receive one or other of two balanced haploid chromosome sets, one differing from the other through segmental interchanges affecting all the chromosome arms. The only viable pollen grains are those containing one set, while the embryo sac always develops from a meiotic product containing the other set. Consequently, all seed is structurally heterozygous. The way that the two chromosome complexes are paired and separated at the first metaphase and anaphase of meiosis (*see* p. 158, *Fig.* 6.6) ensures that, with rare exceptions, the complexes maintain their integrity from generation to generation. This is a system of **permanent hybridity** (to use Darlington's apt description) and it must be presumed that the hybridity is conducive to fitness—otherwise it is impossible to imagine how such an elaborate system could have become fixed in a whole group of species. At the same time it is a *conservative* system, ensuring the maintenance of an already well-adjusted genotype rather than encouraging the generation of new variation, which is the normal effect of sexual reproduction.

18.4 Selection

To a biologist brought up in the Darwinian tradition it is natural to attribute the presence in a population of common allelic variants to the action of natural selection. There are two ways in which one may seek to account for polymorphism by selection. The first involves postulating a steady selection of a new allele at the expense of the old one. In this case we expect the polymorphism to be transient. The second possibility is that natural selection may favour either heterozygosity or phenotypic variation as such. This kind of situation can result in stable polymorphism.

Replacement of One Allele by Another

Box 18.2 summarizes the algebra which relates the rate of change of allele frequencies to the **selection coefficient**, which is defined as the fractional deficit in fitness in the less fit phenotype. **Fitness**, in the Darwinian sense used in population genetics, means merely the ability to leave fertile progeny. Thus a selection coefficient, or fitness deficit, of $-s$ implies that individuals of the less fit phenotype leave, on average, $1 - s$ times as many fertile progeny as those of the fitter genotype. The fitness deficit will apply both to one of the homozygotes and to the heterozygote when a disadvantageous allele is completely dominant and to the homozygote only when it is completely recessive. As *Box* 18.2 shows, a new mutation with a recessive advantage is selected extremely slowly at first because it is hardly ever homozygous while at low frequency. New mutations with dominant advantageous effects are expected to become frequent very much more rapidly.

Industrial Melanism in Moths

Probably the best documented examples of population polymorphism, due to selection of a new mutant allele, concern dark (melanic) forms of originally pale-patterned moths which have become established near (or downwind of) industrial towns in Britain. In the species *Biston betularia* (peppered moth), the melanic variant has been shown to be due to a single dominant gene mutation. The original pale-speckled wild type, which was the only form recorded prior to 1848, is excellently camouflaged when resting on lichen-covered trees, but is disastrously visible to predatory birds against a soot-blackened background. It has been shown by experiment that, in sooty areas (with the lichens killed off by the pollution), the melanic form is at a strong selective advantage through its superior camouflage. In unpolluted countryside, on the other hand, the melanic form hardly occurs at all; here all the advantage lies with the pale form. In either unpolluted or very sooty environments there is no basis for polymorphism, since one type or other is decisively favoured by selection. Presumably in the years during which the

Box 18.2   **Effect of selection on allelic frequencies**

Selection coefficient s, such that AA homozygotes have a fitness of 1 and aa and Aa both have fitnesses of $1-s$. Allele frequencies: p for A, q for a with $p+q=1$.

	Genotype		
	AA	Aa	aa
Initial frequencies	p^2	$2pq$	q^2
Relative fitnesses	1	1	$1-s$
Totals after selection	p^2 +	$2pq$ +	$q^2(1-s)$ $.=1-q^2s$

Frequency of a allele after selection,

$$q' = \frac{2pq+2q^2(1-s)}{2(1-q^2s)} = \frac{pq+q^2-sq^2}{1-sq^2}$$

Since $p+q=1$, $q' = \frac{q-sq^2}{1-sq^2}$

Change in q over one generation of selection:

$$q'-q = \Delta q = \frac{(q-sq^2)-q(1-sq^2)}{1-sq^2}$$

$$\Delta q = \frac{sq^2(q-1)}{1-sq^2}$$

Time course of selection

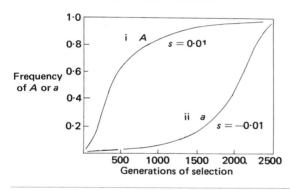

In (i) A starts at frequency 0·05 and the dominant phenotype enjoys a 1 per cent selection advantage

In (ii) a starts at frequency 0·05 and the recessive phenotype has the 1 per cent advantage

Curves as calculated by Fristrom and Spieth.[40]

melanic form was spreading, populations in industrial areas must have been transiently polymorphic. Today, polymorphism is found in populations occupying areas intermediate between the most polluted and the unpolluted; here, presumably, neither form enjoys a clear and consistent advantage, and some mixing must in any event occur by migration (*Fig.* 18.3).

Fig. 18.3

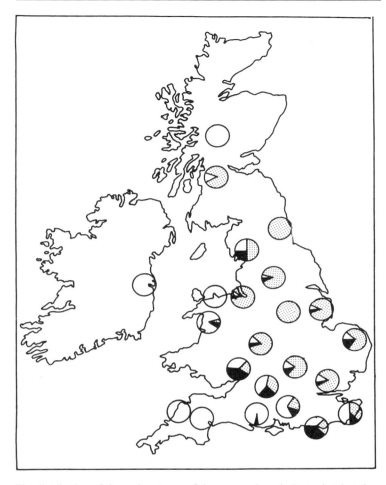

The distribution of three phenotypes of the peppered moth *Biston betularia* in Britain.
The forms *carbonaria* and *insularia* are each melanic and due to dominant alleles at different loci, the former type being the most consistently dark. The areas with the high frequencies of melanic forms are in general either industrial areas or to the east of such areas—the prevailing wind is westerly.
Redrawn from Parkin,[41] after A. Kettlewell.
○, Typical; ◉, *carbonaria*; ●, *insularia*.

Selectively Maintained Clines and the Ecological Significance of Electrophoretic Variation

A situation, such as that seen in industrial melanism, of having one allele selected in one type of environment and a second allele selected in another, commonly results in a steady change in allele frequencies along a geographical line extending between territories in which different alleles are selected. Such a genetic gradation is called a **cline**. It occurs typically in cases where there is a continuous change in environment correlating with the change in allele frequencies, but it can also occur where the environmental change is relatively abrupt but there is some diffusion of alleles by migration and interbreeding.

A good example, especially interesting because it involves a gene of known biochemical function, was reported by R. K. Koehn and his colleagues,[7] who studied populations of a marine bivalve mollusc, *Mytilus edulis* or mussel, at a series of sites along the northern shore of Long Island, New York. Relatively mature mussels showed a pronounced east–west cline with respect to two alleles, Lap^{94} and Lap^{96}, of the gene coding for leucine aminopeptidase. Both alleles, are functional, but the Lap^{94} form of the enzyme has the higher activity. The frequency of Lap^{94} correlates very closely with the salinity of the water; both are at a maximum at the north-east corner of the island facing the Atlantic Ocean and both decline steadily over the 22 miles into Long Island Sound covered by the survey.

That natural selection was operating was strongly implied by a striking relationship between allele frequency and age of the mussels. The larval stage is free swimming and can be carried long distances in the sea before settling down to a sedentary life. In the smallest settled mussels the allele frequencies were much more uniform over the 22-mile range than they were in older animals. It was inferred that larval migration largely homogenized the young population and that strong differential mortality, affecting many individuals with the 'wrong' *Lap* allele for the site in which they found themselves, restored the cline in the older survivors.

The correlation between allele frequency and salinity is not, of course, in itself good evidence that salinity is the true selective agent. Any other factor which, like salinity, was strongly correlated with position along the coastline, might fill the bill. There is, however, an independent reason for thinking that leucine aminopeptidase may be particularly important in relation to salinity. Physiological studies have related tolerance of high salt to intracellular amino acid concentration, and this may well depend on peptidase activity.

One other case in which there is at least some understanding of the selective forces which may be operating is that, mentioned above (p. 528), of the alcohol dehydrogenase polymorphism in *Drosophila*. The electrophoretically 'fast' allele adh^F codes for a more active enzyme variety than that of the 'slow' allele adh^S. Culture of larvae in alcoholic medium is strongly selective for adh^F and this allele appears also to be favoured in alcoholic 'natural' environments, such as wineries.[29] There is a significant north–south cline, with adh^S favoured in southern latitudes;[30] this has been attributed to the greater stability at high temperature of the adh^S enzyme *in vitro* but it is not clear whether temperature variations in the physiological range affect the adh^S enzyme stability in the intact fly or larva.

An attractive way of explaining apparently stable transient polymorphisms is to postulate that polymorphism *per se* is being maintained by selection, either because heterozygotes are fitter than homozygotes or because the population as a whole is fittest when it contains a mixture of different genotypes. We will deal with the former possibility first.

Heterozygote Advantage

Consider a situation in which the fitnesses of the two homozygotes and the heterozygote with respect to a pair of alleles are as follows:

$$AA \quad Aa \quad aa$$
$$1-s \quad 1 \quad 1-t$$

It can be shown that, with both homozygotes disadvantaged in comparison with the heterozygote, the proportion of heterozygotes in the population reaches an equilibrium value

$$H_{eq.} = \frac{2st}{(s+t)^2}$$

and the equilibrium allele frequencies will be $t/(s+t)$ for A and $s/(s+t)$ for a.

Attractive as this model is, it has proved to be extremely difficult to find clear examples to which it applies. The general phenomenon of hybrid vigour (**heterosis**, manifested in improved performance of F_1 hybrids as compared with inbred parent strains) is consistent with it, but can also be explained, rather more plausibly, as due to complementary action of dominant alleles of different genes. This latter explanation would account for heterosis but not easily for polymorphism unless the complementary genes were tightly linked and held in virtually permanent linkage disequilibrium (see p. 522). Otherwise, reassortment or recombination could in principle bring all the advantageous alleles together in a single haploid genome.

When the prevalence of hybrid forms of dimeric or oligomeric enzymes in heterozygotes first became apparent it appeared possible that some of these hybrid proteins might occasionally have unique advantageous properties not found in the homozygous (homodimeric or homo-oligomeric) forms. The discovery of allelic complementation (see p. 365) and its explanation had shown that hybrid protein advantage was a real possibility. However, no actual examples have so far come to light and this particular explanation of the maintenance of polymorphism has been further discredited by the finding that polymorphism is just as frequent in genes coding for monomeric proteins and in haploid fungi where heterozygote advantage would be confined to a very brief part of the life cycle.

There is, however, one very clear example of heterozygote advantage in man. In regions of southern Europe, equatorial Africa, Arabia and India, where malaria of the type caused by *Plasmodium falciparum* is, or has been, endemic, the 'sickle' allele of the gene coding for β-globin is widespread, occurring with a frequency of up to 15 per cent in parts of Kenya and Mozambique. When homozygous, this allele is extremely disadvantageous, producing a severe and crippling anaemia. Its high frequency in malarial parts of the world is undoubtedly due to the resistance of blood cells of heterozygous individuals to infection by the plasmodium. The disadvantage of the homozygous standard allele lies in malaria susceptibility, and that of the sickle homozygote in anaemia; only the heterozygote confers optimal fitness in the malarial environment; the cost is the segregation of the less fit or near-lethal homozygotes in each generation. Where malaria is eradicated, the sickle allele is unconditionally disadvantageous, and will be expected to be slowly selected out of the population (see Box 18.2).

Although it is of great importance in itself, the example of sickle-cell haemoglobin is hardly sufficient to sustain the conclusion that heterozygote advantage is the main reason for the maintenance of polymorphism. It seems, in fact, to be rather a special case.

Frequency-dependent Selection

A polymorphism will be stably maintained if the selective advantage of each of the alternative phenotypes is inversely related to its frequency. There are various ways in which this situation can come about.

Niche Selection

Suppose, for example, that the habitat is divided into territories within which different conditions prevail—what the ecologists call **niches**—and that the different morphs of a species are each well adapted to one niche and less well adapted to the others. Of course, if each morph spends its entire life in its preferred niche the species will be effectively split, and we will have two or more separate populations rather than a single polymorphic one. The situation envisaged, however, is one in which there is free intermingling during the breeding phase but a partial separation at some other essential stage of the life cycle—perhaps, in the case of flies, that of egg laying and larval development. So long as a given morph is few in number compared with the capacity of its preferred niche it will be favoured by selection, but after the niche has become saturated a negative selection coefficient will come into operation and will increase as the frequency of the corresponding morph increases. To take a simple hypothetical example, if individuals of genotypes AA (or Aa, A being completely dominant) and aa are respectively suited to niches X and Y but each totally unsuited to the other's preferred niche, and if niche X is three times as capacious as niche Y, the population will come to equilibrium, with all genotypes equally fit, when A and a are at equal frequencies and the corresponding phenotypes are in a 3 : 1 ratio. In a real situation the difference between two morphs in adaptation to two niches would be very unlikely to be of such an all-or-none character. Nor would there be likely to be such a clear-cut difference between the niches; there might very well be intergradations. Nevertheless, the general principle of frequency-dependent selection based on spatial variation in the environment seems likely to be applicable to some cases of polymorphism.

Apparently Uniform Environment

Frequency-dependent selection may occur even in the absence of obvious environmental patchiness. M. Tosic and F. J. Ayala[8] studied two alleles determining 'fast' and 'slow' electrophoretic forms of malate dehydrogenase (Mdh) in *Drosophila pseudo-obscura*. They started cultures in standard laboratory food vials with different population densities of eggs of mixed Mdh genotypes in a range of different proportions. The parental flies were drawn from forty different independently isolated wild-type stocks in such a way as to randomize as much as possible the genetic background of the Mdh alleles. From the numbers of flies of the different Mdh gentoypes which emerged in vials that started with different ratios of genotypes among the eggs, it was concluded that at high (though not at low) population densities the minority allele, whichever one it was, tended to be associated with somewhat higher fitness in terms of numbers of flies emerging.

One must remember that there may be different ways of exploiting even an apparently uniform environment. For instance, larvae of two different genotypes might preferentially deplete two different components of available food, so that each was in more direct competition with others of its own kind than with larvae of the other genotype. Such possibilities are plausible and can be tested, though not easily.

Mimicry

For many species the most powerful selective pressures come from other species—predators or fellow victims. Concealment from or discouragement of predators is obviously of prime importance, particularly in showy insects such as butterflies. Many butterfly species achieve some protection from predatory birds by being strongly distasteful. A certain number of the members of a distasteful species have to be sacrificed in the training of each fledgeling bird, but experienced birds take the characteristic colour and pattern of the species as a warning, and keep away. Once birds have been trained in this way there is an obvious advantage to another species to share the protection by acquiring the same warning pattern, and such **mimicry** is very widespread among butterflies, especially in the tropics.

Two kinds of mimicry are recognized. In the first, called **Müllerian mimicry**, two or more species, both distasteful, approximate to the same pattern, the advantage to each species being that a part of the sacrifice involved in the education of naive birds is borne by the other. This kind of mimicry is not expected to be associated with polymorphism and it generally is not. The situation is quite different in the second kind of mimicry, called **Batesian**, in which the mimic is *not* itself distasteful and acquires protection through bluff. The selection in favour of this kind of mimicry is frequency-dependent for obvious reasons; if the palatable mimic becomes too frequent in proportion to its distasteful model the birds will not be educated to recognize the pattern as a warning but will instead learn to treat it as an invitation. The advantage of each type of mimic is at a maximum when it is very rare and decreases as it becomes more common. One might predict that a palatable species in an area inhabited by several different distasteful species would tend to become polymorphic, dividing its population into a number of different types, each mimicking a different distasteful model. This does turn out to be the case. *Fig. 18.4* shows mimetic forms of two different species of East African butterfly, taking advantage of three different monomorphic distasteful species inhabiting the same territory. The genetic basis of the polymorphism in each case is mainly allelic variation at a single locus. The 'locus' here is probably a closely linked cluster of genes (sometimes called a 'supergene' by ecological geneticists), the 'alleles' being clusters of mutually adapted alleles of the different genes held in more or less permanent linkage disequilibrium through suppression of crossing-over. Modifier mutations at other (not closely linked) loci appear also to contribute to the accuracy of the imitation in several cases.

It will be appreciated that this rather well-documented set of examples of frequency-dependent selection is closely analogous to the niche selection postulated above (p. 534), the 'niches' here being the noxious model species rather than any corner of the inanimate environment. Examples of frequency-dependent niche selection in its more obvious sense have been hard to demonstrate even though it seems very likely that they must exist.

Mating Preferences

Another possible basis for frequency-dependent selection which should be mentioned is mate preference. A rather bizarre example of this has been reported in *Drosophila melanogaster*[9] where it seems that, in a mixed population of wild-type flies and those displaying the *sepia* mutant phenotype (controlled by a recessive mutant allele), males enjoy success in mating in inverse proportion to the frequency of their phenotype in the population. The implication is that the females prefer to copulate with the most unusual type of male, a degree of sophistication perhaps unexpected in an insect. It is not clear how common this type of situation is in animals.

Fig. 18.4

Mimetic forms of East African butterflies.

Three species which are noxious to predators are shown in the upper rows. The second row shows three mimetic forms of the swallowtail butterfly *Papilio dardanus*, each mimicking the species shown immediately above. The taillessness of these is due to homozygosity for a mutation *t*, and the alternative patterns are controlled by a series of 'alleles' (probably each actually a set of alleles of a number of closely linked genes in linkage disequilibrium) at another locus. The bottom row shows the non-mimetic form of *P. dardanus* flanked by two mimetic forms of another species, *Hypolimnas dubius*, mimicking two of the same model species. All insects shown are female. Dense stippling represents buff colour and light stippling yellow; other areas are black and white. After Sheppard.[42]

The Cost of Selection—an Objection to Selective Maintenance of Polymorphism

Selection implies that some proportion of the population, equivalent to the selection coefficient multiplied by the frequency of the phenotype to which it applies, is *not* being selected. This is no problem for the species if only one or a few genes are subject to selection at one time; the reproductive capacity will generally be more than enough to stand the sacrifice of the non-selected individuals. But there is much more difficulty if it is supposed that all of the many enzyme polymorphisms now known to coexist in virtually all species are maintained simultaneously by selection. Consider the model situation postulated on p. 532, in which the heterozygote with respect to a pair of alleles is fully fit and the two homozygotes subject to a fractional fitness deficit of s and t, respectively. If we take values for s and t of 0·01, and if one-half of the population belongs to one or other of the disfavoured homozygous classes (which is a minimum proportion), then about 0·5 per cent of the population in each generation must be eliminated on account of this one polymorphic gene. If there are 10^4 genes (a low estimate for a eukaryote) and 10 per cent of them are polymorphic, then every individual must take the risk of not being selected one thousand times over. The individual's chance of selection, if all polymorphisms were sustained by 1 per cent selection coefficients, would be no more than $(0·995)^{1000}$ which equals approximately 0·006—six chances in a thousand. This seems an unrealistic burden, considering all the other accidental hazards to which wild populations are subjected.

The number of polymorphic genes and the magnitude of the selection coefficients were, of course, chosen quite arbitrarily in the numerical exercise of the preceding paragraph. The estimate of load will clearly be lower if, as is possible in view of the divergent estimates obtained so far (*Table* 18.2), the proportion of polymorphic genes is lower than 10 per cent. The selection coefficients could, in principle, be much lower than 0·01, given a large and relatively stable population size and an environment which maintained the selection over long periods.

Table 18.2		Different frequencies of polymorphism and heterozygosity in genes determining different classes of protein in *Drosophila pseudo-obscura* (from Leigh-Brown and Langley[5])		
		Protein class		
		*All abundant proteins**	*Enzyme class I†*	*Enzyme class II‡*
Number of genes {	monomorphic	48	8	3
	polymorphic	6	3	7
Percentage polymorphic		11·1	27	70
Estimated average heterozygosity		0·04	0·04	0·24

* All of the proteins present in sufficient abundance to be seen by protein staining after two-dimensional separation on gels—electrophoresis in one direction and 'isoelectric focusing' at right angles.
† Class I enzymes are those involved in essential intracellular metabolism using internally generated substrates.
‡ Class II enzymes are those acting extracellularly on external substrates.

The problem of genetic load does not arise in quite the same way when a polymorphism is maintained by frequency-dependent selection. Here, provided that the genotypes are at their equilibrium frequencies, all are equally fit and survival for all is optimized. This makes frequency-dependent selection the most attractive hypothesis for the selective maintenance of polymorphism.

Considering the complexity of the relationships between genes and phenotypes, we should not be surprised to find it difficult to attribute functional significance to genetic polymorphisms. At the same time, the fact that it *is* difficult does leave the door open to the possibility that many polymorphisms are not attributable to selection at all. We consider this in the next section.

18.5 Drift

Even without selection there would still be some change in allele frequencies from one generation to the next in a finite population, since the gametes going to make the zygotes for the next generation are samples subject to ordinary sampling variation.

If in one generation two alleles are present in proportion p and $1-p$, the variation to be expected in the following generation can be expressed as the standard sampling error $[p(1-p)]/2N_e$, where N_e is the **effective population size**. The latter quantity is usually less than the actual population size to an extent that depends on the amount of inbreeding. In a perfectly mixed population in which any possible pairing is as likely as any other, the probability of a homozygote being formed as the result of the coming together of two alleles, descended from the same allele in the preceding generation, is $1/2N$, where N is the population size (number of alleles in each *diploid* generation is $2N$). In a population with some degree of inbreeding this probability is $1/2N_e$, N_e being less than N.

Given values, based on as good guesses or estimates as possible, for N_e and the frequency of mutation per gene per generation (symbolized here by μ) it is possible to estimate the degree of heterozygosity that would result from mutation and drift without any selection. In the steady state, which each population will eventually attain, heterozygosity will be created by new mutation at the same rate as it is being lost by drift. Considering a single gene, heterozygosity will decrease in each generation by a fraction $1/2N_e$ (this, as stated above, being the probability per zygote of the union of two alleles which are both descendants of the same allele in the previous generation). If H is the fraction of the population that is heterozygous, the fraction becoming homozygous through drift will be $H/2N_e$. This must be balanced by the chance of an individual that would otherwise have been homozygous becoming heterozygous by mutation, which is $2\mu(1-H)$. Solving the equation $H/2N_e = 2\mu(1-H)$ we obtain: $H = 4N_e\mu/(1+4N_e\mu)$.

This formula is widely quoted but it is not, in fact, very useful because of the uncertainties about the magnitudes of N_e and μ. One is a very large and the other a very small number[10] and that is generally all that can be said with assurance. The former, if not the latter, must vary greatly between species. One would expect the degree of heterozygosity found in different species to be very different, reflecting their undoubtedly very diverse population sizes. As R. C. Lewontin has commented, it is surprising to find so many values for various genes in various species falling in the range of 5–20 per cent. One interpretation is that what is being observed is not the true level of heterozygosity (which could be close to 100 per cent for many genes), but rather the proportion of viable mutations which bring about observable (charge-change or conformational) changes in the protein products (*see* p. 527).

At all events, there is no doubt that a part of the heterozygosity, and hence polymorphism, observed in populations *could* be due to allelic differences which are 'neutral' in the sense that the changes in their frequencies due to selection (s) are small compared with those due to random drift ($1/N_e$). The dispute concerns the *proportion* of observed polymorphism which is due to this cause; to many selectionists it has seemed implausible to suppose that many mutations could have that small an effect on fitness. On the neutralist side it may be said that rather few effects on fitness of common allelic variation have actually been demonstrated. A counter to this argument is to point out the wide range of possible values of selection coefficients that are large compared with $1/N_e$ but too small to be readily demonstrable by observation or experiment. We return to the controversy below (p. 544) in connection with fixation of mutations.

Even if many or most polymorphisms are effectively neutral they are not necessarily devoid of long-term effects on the course of evolution. In the first place, an allelic difference without significant effect under one set of environmental conditions could acquire an important selection differential if the environment changed. Secondly, even if all the alternative alleles common in a population were equivalent in all environments, they would still provide a variety of different take-off points for new mutation. For example, if a particular position in a polypeptide chain were always occupied by valine (codon GUU), six other amino acids would be accessible through a single-base change in the codon. If the population in question were polymorphic with respect to that amino acid residue, with alternatives valine (GUU), isoleucine (AUU), leucine (CUU) and phenylalanine (UUU), any of which neutral hydrophobic residues might be acceptable at a particular position, then the population as a whole would have available the possibility of single-step mutation to any of the other sixteen amino acids. Thus polymorphism, even if without immediate consequence, does confer a certain potential for change. In the very long term, mutation rates and protein structures that allowed a high level of cryptic variation to be accumulated might thus be selectively favoured.

18.6 Fixation of mutations

Evolutionary change depends on allele replacement. When we compare the structures of homologous macromolecules, nucleic acids or proteins, of different species we see similarities and differences, and the extent of the difference generally correlates well with the distance of the species relationship as assessed on other grounds. We may ask whether the molecular differences are there because of natural selection, contributing to the undoubtedly adaptive morphological and physiological features distinguishing the species, or whether they have merely accumulated through genetic drift.

Rates of Replacement under Drift and Selection

At any time an interbreeding diploid population has $2N$ copies of each gene, where N is the population size. Each of these copies can be regarded as the founder of a potential line of descendants through subsequent generations. In each generation, however, some of the lines will be extinguished by the accidents of sampling, and eventually only one will be left; one of the founding gene copies will have taken over the population. In the absence of selection, all the gene lines have the same chance, $1/2N$, of eventually monopolizing the population. It follows that any newly mutated allele will have a $1/2N$ chance of becoming 'fixed' in the population and, with new mutations arising at rate $2N\mu$, then mutations destined to be fixed must arise at the rate μ per gene per generation. It has also been shown that the average time taken from mutation to fixation, without any selection, is $4N_e$ generations.

The argument relating to mutations that confer selective advantage is much more complex, and it must suffice here to state the result. A new mutant allele with selection coefficient s as compared with the progenitor allele has a chance $2s$ of going to fixation. The reason why even an advantageous mutant allele is not assured of fixation is that, for several generations after its origin, it has a considerable chance of being wiped out by drift. Only after it has established itself in large numbers does this chance become small.

The Molecular Evolutionary 'Clock'

If the amino acid and nucleotide replacements which have occurred during macromolecular evolution are largely the result of chance fixation of selectively neutral mutations, they are expected to accumulate in proportion to time. The proportionality of replacements to time is, even granted the assumption of neutrality, subject to the reservation that since the replacement rate (*see above*) is equal to the mutation rate per gene *per generation* (leaving aside the difficult question of the relationship of real and effective population sizes) it will depend on the time occupied by each generation. It would be surprising, therefore, if the clock ran as fast for elephants as for insects.

At all events, when the extent of divergence in the amino acid sequence between various pairs of animal species is compared with the estimated times which have elapsed since the divergence of their respective lines of descent from a common ancestor, the proportionality is remarkably close. *Table* 18.3 summarizes a number of comparisons with respect to α and β-globin sequences. The estimate of millions of years since divergence is, of course, somewhat uncertain in that it depends on the interpretation of the fossil record (on which there is at best a consensus opinion) as well as on the dating of geological strata (now an increasingly exact science). The numbers given in the table are those generally agreed by palaeontologists.

Similar comparisons have been made for a number of other ubiquitous proteins, and it is apparent that some change faster than others. Cytochrome c seems to be a particularly 'conservative' protein, which presumably means that comparatively few of its amino acid residues can be changed without effect on function. If each protein is a molecular clock, the clocks do not all run at the same rate. This means that mutation frequency cannot be the only factor determining the rate of molecular divergence; the structures and functions of the gene products also come into it. Some molecules are less free than others to drift because they are more constrained by conservative selection.

Table 18.3   **The relation between amino acid difference in α and β-globins and estimated times of evolutionary divergence (from Jukes[11])**

Animals	Differences per 100 codons		Average $\alpha + \beta$	Corrected*	$10^{-6} \times$ No. of years
	α-chain	β-chain			
Placental mammals compared with each other	16·1	16·7	16·4	17·9	100
Kangaroo compared with placental mammals†	21·7	26·9	24·3	27·8	160
Chicken versus mammals‡	29·6	31·7	30·6	36·5	215
Viper versus warm-blooded animals	39·2			49·6	290
Amphibians versus terrestrial animals	46·7	48·9	47·8	65·0	380
Bony fish§ versus tetrapods‖	49·3	49·6	49·5	68·3	400
Shark versus bony vertebrates¶	57·5	63·8	60·6	93·0	545

* Correction taking account of successive changes in the same codon.
† Human, loris, mouse, rabbit, dog, cow.
‡ Including both placental mammals and kangaroo.
§ Carp and goldfish.
‖ Tetrapods include reptiles, amphibians, mammals.
¶ Shark and other primitive fishes are the only non-bony vertebrates.

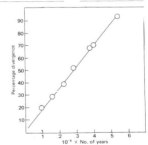

Restraints on Change in DNA Sequences

Now that many genes have had their complete DNA base sequences determined, including in a number of cases introns and flanking regions as well as the coding sequences, it is possible to make some illuminating comparisons between species, and between duplicate genes within one species, to show which parts of the DNA are most subject to change. *Table* 18.4 summarizes some of the results of the comparison between the β-globin genes of rabbit and mouse. It is evident that some parts of the genes have changed much more extensively than others and it looks as if fixation of mutations that affect the nature of the gene product has been minimized.

Table 18.4   **Comparison of the nucleotide sequences of rabbit and mouse β-globin genes (from van den Berg et al.[12])**

Part of gene	No. of nucleotides compared rat/mouse	Gaps insertions	Base-pair substitutions (b.p.s.) Transitions	Transversions	Total differences	B.p.s. as percentage of maximum number possible
Coding sequences	441/441	0	49	35	81	19
Causing replacements		0	16	19	35	10*
'Silent'		0	33	16	49	62*
5' Non-coding part of messenger	53/52	1	8	4	13	25
3' Non-coding part of messenger	95/135	41	16	7	64	47
Small intron	126/115	19	26	13	58	45
Ends (10 bp at each end)	10/10	0	4	1	5	25
Middle	106/95	19	23	12	53	48
Large intron	189/185	30	42	29	101	50
Ends	20/20	0	3	1	4	20
Middle	169/165	30	40	28	98	54

* There are three possible base-pair substitutions possible for each of the 441 base-pairs of the gene. Some will cause amino acid replacements and some not. The number of changes observed are expressed in each case as a percentage of the maximum number possible.
Points to note: If the sequence became completely randomized each base-pair would have a 75 per cent chance of ending up changed (assuming equal frequencies of all four base-pairs). 'Silent' (i.e. synonymous) changes in codons, and changes in 3' non-coding end of the messenger sequence and in the interiors of the introns have gone most of the way towards randomization in each species relative to the other. In comparison, the ends of the introns and the 5' untranslated part of the messenger are relatively highly conserved, and the amino acid sequence encoded is even more so.

Table 18.5   **Mutational differences between human globin genes (data from Efstradiatis et al.[26])**

Percentage base-pairs different*

Comparison	Amino acid replacement changes	'Silent' changes†
α—β	46	90
α—δ	47	100
α—G$_\gamma$	51	103
α—ε	51	91
G$_\gamma$—A$_\gamma$	0·35	0
β—δ	3·7	32
G$_\gamma$—ε	10	61
β—G$_\gamma$	18	74
δ—G$_\gamma$	19	75
β—ε	16	61
δ—ε	17	62

* Corrected upwards to allow for successive changes in the same base-pair (see p. 546).
† 'Silent' changes are those which result in no amino acid replacement—compare Table 18.4.

The inner sequences of the intervening sequences (introns) show many differences, including not only base-pair substitutions but also numerous deletions and insertions of material. The margins of the introns, on the other hand, are relatively conserved, in keeping with the idea, for which there is good evidence (*see* p. 414), that the 'consensus' sequences found in these regions are important for the correct cutting and splicing of the mRNA precursor.

Within the coding regions of the genes there are no deletions or insertions—most of these would, if they occurred, cause frameshifts which would effectively destroy the gene product. The base-pair substitutions within codons have a strong tendency to be those that make no difference to the amino acid coded for. Moreover, the amino acid replacements that have become fixed include a disproportionately high number of relatively conservative substitutions, such as leucine for valine or aspartic acid for glutamic acid, which would not be expected to have large effects on protein function. In fact, the code is so constructed that a high proportion of random single-base substitutions are of this kind, but this cannot easily be the whole explanation for the dearth of radical amino acid replacements—selection for preservation of function also seems to be operating.

That part of the gene transcribed into the 5' non-coding part of the mRNA outside the coding regions seems rather conservative, presumably because it has specific functions to perform (*see* p. 409), but the 3' non-coding end has diverged rather freely.

An extreme interpretation of data such as those of *Table* 18.5 is that mutations that do not affect transcription, processing of the transcript or amino acid coding are free to accumulate without selective constraints. What we really need, in order to assess this interpretation, is a yardstick of uselessness—a DNA sequence which we are sure has no function whatever and which must therefore be subject to unrestrained change. To assume that such sequences exist is to some extent begging the question, but some which seem to meet the requirements are the so-called **pseudogenes** of which there are now several examples in animals, particularly in the globin gene clusters (cf. *Fig.* 15.10, p. 439). These really do look useless, having been so damaged by mutation as to be untranslatable, and we can ask whether they show a higher rate of base substitution than, say, the third positions of codons which are members of synonymous sets of four. The answer to the question, at least so far as the mouse α-globin pseudogene is concerned, seems to be that even 'synonymous' substitutions in codons, though much more freely occurring than substitutions causing amino acid replacements, are still significantly restrained in comparison with the rate of change in a pseudogene.[13] Supposing that this is the case, why should it be?

It is reasonable to suggest that even synonymous codons are not necessarily equally conducive to optimal gene function. The relevant data here are the relative frequencies of use of different codons in various genes in various organisms. As *Table* 14.1 (p. 403) showed, codon usage can be very non-random, with differences between organisms and perhaps even between sparingly and abundantly transcribed genes in the same organism.

An obvious interpretation is that there is selection pressure on heavily used codons to fit the most plentiful tRNA species. It will be recalled (p. 44), that except in mitochondria, codons with a purine in the third position are usually efficiently recognized by one species of tRNA and synonymous codons with a pyrimidine in the third position by another. A relationship between the usage of each codon and the relative abundance of the corresponding tRNA has been shown for *E. coli*[23] and we may expect it to apply generally.

Another constraint on the establishment of certain mutations may be imposed at the level of DNA sequence itself. Mammalian DNAs in general have a remarkable shortage of 5'-CG-3' as compared with all other two-base sequences. Whatever the selective forces that led to the establishment and maintenance of this situation (and perhaps the proneness to methylation of 5'-C when adjacent to 3'-G has something to do with it—*see* p. 511), they presumably tend to prevent the establishment of any mutation to the disfavoured sequence.

The Neutralist–Selectionist Controversy

So-called neutralists, among whom M. Kimura has been the most prominent, contend (*a*) that most of the observed variation within populations is due to mutations with little or no effect on fitness, and (*b*) that most allelic substitutions are due to the random fixation of such variation. It would be possible to agree to the first of these propositions and not to the second, since, even if most polymorphism is due to inconsequential mutation and drift, it is still possible that the fixation of evolutionarily important mutations tends to be more or less assisted by selection. However, acceptance of the first part of the neutralists' position certainly makes it easier to accept the second.

The stronger selectionists hold (*a*) that practically all polymorphism is due to selection, and (*b*) that mutant alleles become fixed only when they confer some selective advantage. The second part of the selectionists' position follows from the first, since polymorphism is a necessary stage on the way to fixation. More compromising selectionists would concede that fixation may sometimes occur by drift without positive selection, especially in small populations, but would still dispute that many or most allelic replacements occur in this way.

What does the molecular evidence tell us? Very little, so far as the polymorphism question is concerned. But it does at least narrow the range of possible opinions on the neutrality or otherwise of amino acid or nucleotide replacements in macro-molecular evolution. It is clear that some mutations are very much less neutral (i.e. much less free to drift into the population) than others. Taking the rate of change of pseudogenes as a yardstick for neutrality (and a strong selectionist might object that the neutrality even of pseudogenes has not been proved), we can conclude that changes in the interior parts of introns are also neutral, or nearly so, and that 'synonymous' changes affecting third positions of codons are subject to, at most, relatively weak counterselection. The available information on non-coding DNA falling *between* genes (the vast majority of the genome in higher eukaryotes) shows that at least some of it is conserved. Repetitive sequences seem sometimes to be constrained in some way to preserve their uniformity, a form of constraint which will usually be conservative, but may occasionally hasten radical changes ('concerted evolution'—*see* p. 540).

It is easy to point to harmful amino acid substitutions in mutant proteins, but very difficult to point to beneficial ones. We would expect most natural selection of coding sequences to act to preserve structure—not to change it. The relative infrequency of fixation of mutations that change amino acid coding, compared with those that are 'silent' (*Table* 18.4), bears out this expectation. This has sometimes been regarded as a point against the neutralists. However, the neutralist position is not that selection has no role, but rather that its effect, on the whole, is to *inhibit* change, and that random drift and chance fixation are mainly responsible for the changes that become established.

In the opinion of the author, the neutralists have a very reasonable case, at least on the molecular ground on which the argument has recently been conducted. But, if so, their victory leaves us in the extremely unsatisfactory position of admitting that the molecular evidence tells us nothing at all about 'real' evolution—the kind in which genuinely advantageous changes *are* selected and in which we are compelled to believe by the obviously adaptive nature of many of the features which distinguish one species from another. Unfortunately, these obvious adaptations are mostly seen at the morphological, physiological and behavioural levels, and their connections with specific gene products are complex and difficult to trace. The molecular differences which we *can* relate directly to genes may well have adaptive significance but it is not usually obvious. For example, the β-globins of man and rabbit differ in 14 amino acid residues out of 142, but we cannot yet say that a man would not be just as well off with the rabbit molecule and vice versa. The differences could, so far as we know, be merely due to random fixation without functional effect.

Meiotic Drive

There is another way in which genetic change may occur without reference to the fitness of the organism. Whereas drift is change by chance, **meiotic drive** is something of even more problematical value—change through fratricidal competition among the products of meiosis. Here we will consider three examples. In each case the driving effect is spectacular, but its mechanism not at all understood.

In a world-wide collection of specimens of the fungus *Neurospora intermedia*, a group of isolates from Nigeria were found to carry an allele called *spore-killer* (SK).[31] Asci formed in a cross between an SK strain and one carrying the 'normal' allele sk^+ each contained two aborted spore pairs and two viable ones; the viable ascospores always contained SK. The SK allele appears to have no deleterious effect other than that on sk^+ ascospores. When such an allele arises in a population it would be expected to 'drive' itself to fixation.

In *Drosophila melanogaster*, a second-chromosome mutation called *segregation distorter* (SD) has been found repeatedly in certain wild populations. SD/sd^+ heterozygous males produce both SD and sd^+ spermatozoa, but only those carrying SD are functional.[32] SD cannot achieve fixation as it is recessive lethal, but its drive in heterozygous males is sufficient to maintain it at a high frequency in the population.

Mutations at the T locus of mouse (*Mus musculis*) show a range of complex effects, including shortening of the tail and other abnormalities of the spinal column.[33] Certain recessive lethals, recognized as t alleles through the short-tail phenotypes which they give in heterozygotes with viable t mutations, are found at high frequency in many wild populations. In these populations, the functional spermatozoa of heterozygous t^+/t males are predominantly those carrying t.

A different form of meiotic drive could be based on conversion rather than suppression of meiotic products. It was seen in Chapter 3 (p. 92) that, at least in some fungi, the frequencies of 3 : 1 and 1 : 3 tetrads may be very unequal in certain wild × mutant crosses. A mutation favoured by such an inequality could drive itself slowly but surely to fixation, provided that it was not markedly disadvantageous in other respects.[34] This must, one would think, be a very potent mechanism in lower eukaryotes with high conversion frequencies. Perhaps many standard wild-type alleles achieved their predominance by such means, but positive evidence is lacking.

18.7 Phylogenetic relationships deduced from molecular structure

We conclude this chapter with a rather brief consideration of 'real' evolution—that is to say the origin of species and higher-order groups of organisms.

Phylogenetic 'trees' showing the hypothetical divergent lines of descent of different groups of organisms have been constructed ever since the idea of evolution became acceptable. Until recently, the criteria of similarity and dissimilarity were morphological and there was an element of subjective judgement in the assessment of the degree of divergence of different species and groups. The number of mutations that would have been necessary to convert one pattern of morphological development into another could hardly be guessed at. Now that so much information is available on the amino acid sequences of proteins and the base sequences of genes, it has become possible to compute the **mutational distance** separating homologous molecules in two different species and, if reliance can be placed on the 'molecular clock', such information can be used to assess the length of time since the two lines of descent diverged from a common ancestral stem.

Where the molecules being compared are nucleic acids of the same length, it is straightforward to count the number of single base-pair mutations in the DNA needed to convert one molecule into the other. There are two complications. Firstly, when the proportion of base changes is large one must take account of the likelihood that some of the changes are the result of two successive mutations rather than just one. A correction can be made for 'double hits' making use of the Poisson distribution; the proportion of unchanged bases, if mutation is hitting the bases at random, will be e^{-m}, where m is the mean number of mutational hits per base. The estimated number of single hits will then be me^{-m} and the number of double hits $(m^2/2)e^{-m}$; the possibility of triples and higher numbers is usually neglected. The best estimate of mutational distance is the estimated *total* number of mutations, which will be very appreciably larger than the number of changed bases when the amount of change is greater than 30 or 40 per cent. The second complication arises when the two sequences being compared are different in length. Where the homology between the two is clear, it will usually be possible to see that one could have been derived from the other through one or more deletions or insertions of limited extent. One can then include the insertion/deletion events in the count when computing the mutational distance. But when homology is weak there may be doubt as to how best to align the two sequences for comparison, and the method then becomes much more problematical.

Much the same applies to comparisons of amino acid sequences of proteins. Here we cannot, of course, see the effects of mutations directly in nucleic acid base sequence, but we can easily deduce the *minimum* number of base substitutions in the mRNA (and hence the minimum number of base-pair substitutions in the coding sequence of the gene) which would make the necessary changes in the amino acid codons. It is not possible by looking at amino acid codons to deduce anything about 'silent' base changes, i.e. those (predominantly in third codon positions) which make no difference to the amino acid coded for.

Having obtained estimates of mutational distances between homologous genes in different organisms, one can construct a scheme (a 'tree') to show relationships according to the **minimal mutation** criterion. The principle used here is that the branching structure should be such as to minimize the total number of postulated mutations summed over all the branches. Thus, other things being equal, one accounts for changes shared by two species by postulating a mutation occurring once in a common stem rather than coincidentally and in parallel in two separate branches. It is usual to assume equal rates of mutation fixation in all branches; thus two related sequences are shown as having diverged from a common ancestor—not with one directly ancestral to the other.

The phylogenies deduced from sequence evidence can span vast tracts of evolutionary time when based on a very slow-running evolutionary clock, such as is provided by ribosomal RNA,[16] tRNA[14] or by the nearly ubiquitous and very conservative protein cytochrome *c*.[15] On the other hand, a fast-running clock based, for example, on the rather freely diverging satellite DNA of *Drosophila* species[18] can provide information about the relationships of closely similar species which have diverged within the last million years.

Most trees constructed from molecular sequence data merely serve to confirm relationships already deduced using non-quantitative, but nevertheless convincing, comparative morphology. Trees based on different macromolecules also tend, reassuringly, to agree with each other.[43] The most distinctive contribution of the molecular approach has been in the field of prokaryote evolution, where morphology alone is comparatively unhelpful. The prime example is a very extensive survey of 16-S ribosomal RNA sequences,[16] which has led to the recognition of a major prokaryote group (the Archebacteria, including methane bacteria and a collection of diverse forms found in extreme environments such as saline lakes and hot springs) which appear to be as remote from the remainder of the prokaryotes as either prokaryotic group is from the eukaryotes. The 16-S rRNA data have also produced information relevant to the controversial question of the possible prokaryotic origin of eukaryote organelles (cf. p. 178). There is clear sequence homology between the 16-S RNA of the blue-green 'algae' (a large group of photosynthetic prokaryotes) and plastids. The affinities of mitochondrial 16-S RNA remain rather obscure; it seems remote both from prokaryotic and cytoplasmic eukaryotic 16-S RNA. The molecular evidence bearing on the evolutionary origins of organelle DNA has recently been reviewed.[44]

Gene Duplication and Divergence

Since the method under discussion focuses on genes rather than on whole organisms, it can be applied to the diversification of a gene *within* an organism or line of descent as well as to measuring the distances *between* organisms. This is an important application from the point of view of evolutionary theory since, if one grants that complex organisms with more genes have evolved from simpler organisms with fewer, one must suppose that new gene functions have evolved by diversification of old ones. The only way this could happen is through gene **duplication**, initially leading to a redundancy of gene function. Subsequently the function of one of the two duplicates could be lost, after which either of two things could happen. Either the inactivated gene could survive as a pseudogene (cf. p. 439)—a 'rotting hulk', to use one apt description—or it might, by a fortunate combination of mutations, acquire a new function. The second outcome is much less probable than the first, but could be strongly selectable.

Fig. 18.5

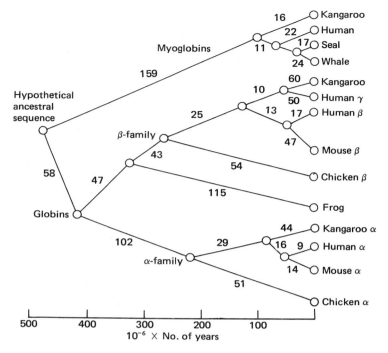

The 'tree' obtained from comparisons of the amino acid sequences of globins and related proteins.
This is a simplified version, with numerous species omitted, of the diagram of Goodman et al.[25] The numbers on the different sections of the tree are numbers of mutations, corrected for successive mutations in the same base-pair (*see* p. 546). The time-scale is based on a calibration of the 'molecular clock' by reference to the fossil record.

The best example of a molecular phylogenetic tree showing gene diversification both within and between species concerns the globins and related sequences. *Fig.* 18.5 shows the significant but distant relationships, inferred from amino acid sequences, between the myoglobins, which are oxygen-storing proteins of muscle (especially important in water-dwelling mammals), and the globins of blood. Within the globin branch, a very ancient gene duplication is thought to have generated the primeval α and β-globin genes, which somehow got separated onto different chromosomes. Within both α and β-type globins, further diversification seems to have occurred via tandem duplication resulting, in each case, in a cluster confined to a short region of one chromosome (cf. *Fig.* 15.10). The evolutionary diversification of human globin genes as inferred from base sequence data summarized in *Table* 18.5 is shown in *Fig.* 18.6. The diversification of function here is not very extreme—the products of the genes within each cluster are virtually interchangeable and differ mainly in the timing of their production. Good examples of the origin of really novel functions by duplication and divergence are harder to find—not surprisingly, since a jump in function of a gene requires an imaginative jump on the part of the investigator if he is to find it—but there are a number of examples of possibly genuine homology between enzyme proteins with currently diverse functions.

Fig. 18.6

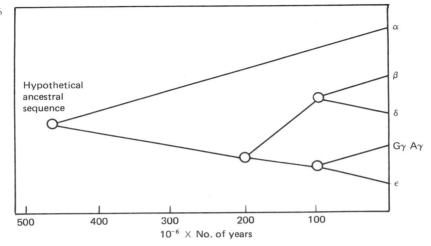

The 'tree' obtained from comparisons of the coding base sequences of human globin genes.
From Efstradiatis et al.[26]
See Table 18.5

18.8　The origin of species and higher taxonomic groups

　　How does one account for branching (sometimes called **clado-genesis**) in evolution? This is the most interesting but also the most inaccessible area of evolutionary study. To do justice to the question requires a great deal of detailed natural history for which there is no room in this book. Here I will only direct attention to a few of the main ideas and points of disagreement.

Speciation

　　In the classification of living things the most natural and least arbitrary unit is the **species**, which can be defined (*a*) as a population sharing certain characteristics which distinguish it from other species and (*b*) as a population without internal breeding barriers but with effective constraints against interbreeding with other species. The two parts of the definition, distinctive phenotype and prevention of genetic exchange with other species, are connected, since without the breeding barriers species characteristics could not be maintained. In some groups of animals, genetic isolation is considered to be a sufficient warrant for species status in itself, even without clear phenotypic difference. For example, the two *Drosophila* species, *D. melanogaster* and *D. simulans*, are virtually indistinguishable by eye and can even hybridize, but are given separate specific status because the F_1 hybrid is totally sterile and it is thus impossible for genes to be passed from one species to the other. This is an unusual case for two reasons—genetic isolation will usually lead to phenotypic divergence by drift even if not by selection, and if it does not do so the distinction between the two species is likely to go unnoticed.

The nature of the barrier to interspecies crossing varies greatly from one case to another. In some cases it is merely geographical. Among trees one may cite the fertile and vigorous F_1 hybrids formed by *Catalpa ovata* (China) and *C. bignonioides* (N. America) and by *Platanus orientalis* (Europe) and *P. occidentalis* (N. America). The plane tree widely planted in London streets is reputed to be the *Platanus* interspecific hybrid. Sometimes F_1 hybrids are sterile because the reassorted genes in their gametes are unable to function together; in other cases F_1 sterility arises because of structural chromosome changes which have come to distinguish the parent species; in yet others the chromosome sets look identical but fail to pair at pachytene. In all these cases, the incompatibility between the chromosome sets can be attributed to an accumulation of differences such as might be expected to occur by selection or drift or a combination of the two once populations have become isolated.

However it happens, once divergence has reached the point where the F_1 hybrid is sterile one might expect natural selection to operate against hybridization and its consequent waste of reproductive potential. In accordance with this prediction, we find that many pairs of apparently very similar species will not hybridize. In plants, failure of the pollen tubes of one species to grow down the style of another is the general rule, while in animals species-specific behaviour patterns are often important in preventing hybridization.

Most of those who have studied species relationships and thought about possible mechanisms of species divergence (**speciation**), consider that separation in space is an almost essential first step. It requires only a little cross-breeding between semi-isolated populations to maintain a broad similarity of their respective genetic mixes, unless exceptionally strong selective forces are operating in favour of divergence.

Probably the most realistic general scenario for the origin of new species is one in which a small population of an existing species becomes isolated, perhaps in a new habitat created by climatic change or geological upheaval. If it is a small sample of the progenitor species, it may well have an unrepresentative set of genotypes to start with (**founder effect**) and, thereafter, it will probably be subject to new selection pressures which will cause it to diverge still further from the ancestral population.

The Possible Role of Repetitive DNA Sequences

An interesting, but as yet speculative, idea[17] is that the genomic differences that cause hybrid sterility involve highly repetitive sequences. Such sequences, as we saw in Chapter 15, sometimes appear as satellites of distinctive buoyant density (reflecting distinctive base composition). Satellite DNA consists of relatively simple sequences which tend to be concentrated in tandem array in certain chromosome segments, especially adjoining centromeres. Other highly repetitive sequences, generally rather more complex, are dispersed in single copies over most or all chromosomes.

In recent years much attention has been devoted to comparisons of satellite sequences,[18] the evolution of which has already been touched upon in Chapter 15

(p. 449). Within species there are often several different 'families' of such sequences, sometimes clearly related and sometimes not. The members of each family conform, with some variations, to a consensus sequence and the same family of satellite sequences can sometimes be recognized in several related species. The variations with respect to a particular satellite sequence tend to be greater between species than within species. But the most striking feature to emerge from species comparisons is not so much the detailed differences in sequence in a particular satellite, as in the *quantity* and *distribution* of each satellite family.

An instructive study has been made[19] of the group of *Drosophila* species closely related to *D. virilis*. Preparations of three related (A + T)-rich *D. virilis* satellite families consisting of seven-base repeats (5'-ATAAACT-3', 5'-ACAAACT-3' and 5'-ACAAATT-3', respectively) were transcribed separately *in vitro* by RNA polymerase, and the transcripts, labelled with ^{32}P, were hybridized *in situ* to salivary gland chromosomes from ten different related species (for *in situ* hybridization *see Fig.* 1.18, p. 32). The chromocentric (centromeric) regions of the genomes of most of the species were labelled to some extent by transcripts from all the satellites, but the amount and exact distribution of the label varied considerably between species. One species, *D. littoralis*, appeared altogether devoid of all three satellite sequences in spite of its general similarity to the other species. Equally remarkable was the finding that, in most of the species, copies of the satellite sequences are present in small numbers at loci in the euchromatic chromosome arms, the loci being quite different from one species to another.

The general conclusions one can draw are (*a*) that the simple sequences of the *Drosophila* satellites can change in number very rapidly as compared with the slow rates of change of protein coding sequences and (*b*) that they appear to have the ability to get themselves transposed to scattered locations on all chromosomes.

Perhaps the most spectacular example of dispersed repetition is that of the *Alu* sequence in our own species[20, 24] (*see* Chapter 15, p. 449). This family of probably movable sequences has diverged to a substantial extent within human genomes (different copies may be 10–20 per cent mismatched) but there is still substantial homology between the family as a whole and corresponding families in various monkeys. It is not a satellite, since its base composition does not differ appreciably from that of total human DNA. It is not concentrated in heterochromatin but rather appears to be dispersed in single copies throughout the genome. Its total bulk amounts to about 3 per cent of the DNA.

We saw in Chapter 15 that mechanisms appear to exist for the maintenance of homogeneity among repeated sequences within a population. G. A. Dover[17] proposes concerted changes in repetitive sequences as an important cause of speciation, arguing that such changes may proceed in different directions in different isolated populations until the genomes become mutually incompatible and hybrids inviable or sterile. The argument depends on the sequences subject to concerted change being functionally important—as, indeed, some of them clearly are.

Speciation by Polyploidy

There is one well-known and clearly documented way in which a single individual can achieve what amounts to separate species status in one generation, and that is through polyploidy. As we saw in Chapter 6, a tetraploid which can arise through one failure of chromosome separation at mitosis is, from its inception, reproductively isolated from its diploid progenitor because of the almost complete sterility of the triploid progeny of tetraploid × diploid crosses. In plants, many new species or subspecies have evidently arisen in this way. Plants can often self-fertilize and, even if the diploid species had a system of self-incompatibility, this often breaks down in a derived tetraploid. Thus a newly produced tetraploid plant can usually reproduce to give tetraploid progeny. There is often little loss of fertility due to irregular disjunction of quadrivalents even in autotetraploids, while in allotetraploids there is generally none (*see* p. 160).

In animals without the capacity for self-fertilization, and with the complication of sex-chromosome systems which may fail to work in tetraploid cells, the opportunities for tetraploids are much more restricted. One important group of fish, the salmonids (including trout and salmon), appears to be entirely tetraploid. It will be recalled (p. 72) that sex differentiation in fish is minimal. It has been a matter of interest to see how far the duplicated genomes in salmonids have diverged from one another (cf. p. 547). The answer so far is that divergence appears to be minimal—most DNA sequences and gene functions probably remain duplicated.[45]

Sundry species of tetraploid insects have escaped from the difficulty of sexual reproduction by resorting to diploid parthenogenesis (p. 520). For accounts of the very significant role of polyploidy in speciation in flowering plants and ferns the interested reader is referred to the books of Stebbins and Manton.

18.9   Macroevolution: gradual or 'punctuated' by major karyotypic changes?

A long-standing division of opinion among evolutionists, which recently has surfaced once again,[21] concerns the relative importance of gradual change and relatively sudden jumps in the origin of taxonomic groups. The idea of periods of rapid change, corresponding to the major branching points on the phylogenetic trees, with long periods of near stability in between is often referred to as '**punctuated evolution**'. This model of evolution is supported by some palaeontologists on the basis of their reading of the fossil record.

So far as evolution at the species level is concerned, the gradualist view seems to serve well enough. As was argued above, it is quite plausible to attribute the divergence of closely related species to spatial isolation followed by an accumulation of differences through selection and random drift. Major innovations, such as must have accompanied the divergence of the reptiles and amphibia or reptiles and birds, or even such lesser but still spectacular changes as were involved in the great diversification of the major mammalian groups and of genera within these groups, pose a much more daunting problem. It requires something of an act of faith to believe that these could have come about by a further gradual accumulation of small changes. There have always been those who have found this too difficult and have looked for alternatives.

As to the question of what, in the 'punctuated' model, actually happened at major evolutionary branching points, little can be said. These events, assuming that they were indeed something special, occurred many millions of years apart and, though relatively sudden measured on the evolutionary scale, may each have required tens or hundreds of thousands of years to establish. The chance of obtaining any direct observational evidence is negligible. Among the ideas that have been put forward is that diversification above the level of species or genus may have required extensive restructuring of chromosome complements, bringing drastic simultaneous changes in the timing and expression of many genes, rather than the piecemeal accumulation of point mutations within individual coding sequences.

A. C. Wilson[22] has compared the degree of diversification in frogs with that in mammals—two groups of animals which, from the fossil record, seem to have originated at about the same time. During the subsequent ages the mammals appear to have been much more enterprising than the frogs. From our, perhaps biased, mammalian viewpoint, all frogs still seem to conform to rather a standard pattern, while the great variety of mammals is obvious. Yet, from comparisons of molecular structures, as revealed by electrophoretic and some sequence analyses, as much variety seems to have accumulated in the one group as in the other. The 'molecular clock' seems to have been running at about the same rate in both, but it fails to reflect the much greater rate of morphological, physiological and behavioural innovation in the mammals. What turns out to correlate much better with the rate of 'real' evolution is the rate of change in chromosome structure. Frogs are rather uniform in chromosome number and centromere positions, whereas mammals show a great range of chromosome number and individual chromosome morphology.

Interspecific comparisons of the karyotypes of related species sometimes suggest how changes in chromosome number can occur. In many species groups the number of chromosome *arms* is constant, or nearly so, while the number of *centromeres* varies. For example, in the insect order Orthoptera, the locust (*Locusta migratoria*) has 11 acrocentric autosomes, while many grasshopper species (for example of the genus *Chorthippus*) have 5 acrocentrics and 3 metacentrics; each of the 6 arms of the latter is similar in size to one of the larger locust acrocentrics. We can easily picture the origin of a metacentric chromosome in a very unequal interchange (a **Robertsonian translocation**) between the long arm of one acrocentric chromosome and the very short arm of another, the breaks being close to the respective centromeres. The very small amount of material left attached to the first chromosome centromere might be genetically inconsequential and subject to loss by accidental non-disjunction without ill-effect.

The reduction of the number of chromosome arms as well as centromeres has occurred in some species to an extreme degree *without* striking changes in phenotype. An extraordinary example is provided by the deer genus *Muntiacus*, which contains a few species rather closely similar in phenotype. *M. reevesii*, the Chinese muntjac, has $2n = 46$, much the same number as many other deer and other mammals. *M. muntjac*, the Indian muntjac, has $2n = 6$. The comparison is given in *Fig. 18.7*. In spite of this enormous difference in chromosome architecture, the single-copy DNA sequences have been estimated (by molecular hybridization) to have diverged between the two species only to the extent of 2 per cent. The main apparent difference is in the amount of middle repetitive sequence, which is somewhat reduced in the Indian muntjac. To judge from their close phenotypic similarity, virtually all the functional genes are controlled in the same way in the two species.

The reshuffling of whole chromosome arms and centromeres will undoubtedly lead to sterility in F_1 hybrids from crosses between chromosomally distinct strains and to this extent will contribute to the building of barriers to gene transfer between incipient species. But there is nothing to show that, in themselves, they

Fig. 18.7

M. reevesii

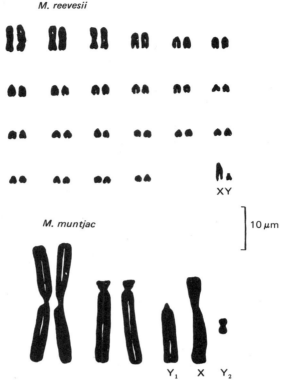

10 μm

M. muntjac

The karyotypes of two species of muntjac deer. Mitotic metaphase
chromosomes are arranged in homologous pairs.
Drawn from the figures of Schmidtke[27] for *Muntiacus muntjac* (the Indian
muntjac) and Wurster and Benirschke[28] for *M. reevesii* (the Chinese muntjac).
Both to the same scale. Note that *M. muntjac* has two 'Y'-chromosomes; these
pair at meiosis with different arms of the X and are always segregated together
into the male-determining sperm.

have any important effects on the phenotype. Neither is there much evidence at
present for large effects resulting from rearrangements *within* arms. The extensive
studies, referred to above (p. 528), of the widespread paracentric chromosomal
inversions in wild *Drosophila* populations, have not revealed any phenotypic
effects. In the absence (*see* p. 155) of any fertility deficit resulting from heterozygosity for paracentric inversions, the variant chromosomes are all functionally wild
type.

    The preceding paragraphs are not intended to imply that chromosomal
rearrangements never have important phenotypic effects. In *Drosophila*, the
placing of genes relative to heterochromatin may be important (p. 508), while in
mammals it seems that cell transformation follows from the bringing of
'oncogenes' into proximity with retrovirus promoters (p. 505). We reviewed in
Chapter 15 (p. 454) one case in which a massive magnification of one chromosome
segment brings about both a rapid and spectacular increase in chromosome size
and an appearance of a highly selectable phenotypic character—methotrexate

resistance—based on an increased level of one enzyme, dihydrofolate reductase. Whether chromosome restructuring could, in itself, ever have been responsible for major innovations in morphogenesis is at present impossible to say.

Until there is much deeper understanding of how the developmental programmes of major groups of organisms actually work, and what the differences between them are in terms of genome structure, we will not be able to assess the relevance of the evolution of the karyotype to that of the phenotype.

18.10 Summary and perspectives

In this chapter, we dealt both with variation within populations and with evolution which, in the broad sense, is taken to mean any genetic change in a population. The two topics are inextricably linked, since variation implies the possibility of change and change depends on variation.

To the extent that allelic variation in populations is essentially 'neutral' with regard to fitness, population change will inevitably occur by fixation of variants through drift at a frequency equal to the mutation rate. To the extent that the variation does affect fitness, fixation of alleles will be either promoted or inhibited by selection.

In the neutralist model, some polymorphic variation is inevitably fixed. On the selectionist view, some allelic variation is selectively maintained as polymorphism while other allelic variants are selected to fixation—either type of selection can occur without the other. In spite of this clear difference in theory, the relative effects of drift and selection, either in polymorphism or fixation, have been extremely difficult to evaluate. Even if we concede the likelihood that drift is important, there is no way in which selection can be discounted. Evidence from molecular structure testifies convincingly to the power of selection in a conservative sense. It certainly acts to inhibit deleterious change, but its effect as an innovating force is not so evident from the molecular data. Nevertheless, mutations improving fitness do occasionally arise (we have some very clear examples, including melanism in moths and drug resistance in bacteria), and natural selection will inevitably tend to fix them.

Changes involving only one or a few genes come under the heading of **microevolution**. They produce organisms altered in detail, sometimes in quite a striking way, but still clearly belonging to the original species or a very closely related one (depending on the zeal of the taxonomists working with the group in question). In this chapter we proceeded from microevolution, about which we have some direct evidence, to **macroevolution**—phylogeny on the grand scale. This is indeed an enormous step and in making it we move into a vastly greater area of enquiry while leaving behind most of the possibilities for obtaining evidence.

Using evidence from nucleic acid or protein sequences in living organisms, one can make inferences about phylogeny expressed in the form of 'trees', but this gives us no insight into the kinds of change underlying the drastic morphological and developmental restructuring which must have occurred in the diversification of the major groups of organisms. Could macroevolution be merely an extension of microevolution? It is hard to imagine, but it is also difficult to comprehend, the vast amounts of time, not to mention the area and volume of available habitat on the surface of the globe, over which natural selection has been at work. We envisage evolution as the consequence of a large number of highly selective decisions taken in succession, with each one dependent upon the previous ones. R. A. Fisher made the point well when he described natural selection as a mechanism for generating a high degree of improbability.

Chapter 18 — Selected Further Reading

Crow J. F. and Kimura M. (1970) *An Introduction to Population Genetics Theory*, 591 pp. New York, Harper and Row.

Dobzhansky T. (1941) *Genetics and the Origin of Species*, 2nd ed., 445 pp. New York, Columbia University Press.

Dover G. A. and Flavell R. B. (ed.) (1982) *Genome Evolution*, 382 pp. London, Academic Press. (Especially relevant in the context of this chapter are articles by W. F. Doolittle on selfish DNA; by C. J. Bostock and C. Tyler-Smith on methotrexate-resistant cells; by D. Gillespie et al. on the evolution of primate DNA; by D. J. Finnegan et al. on transposable DNA sequences in eukaryotes; by A. J. Jefferies on evolution of globin genes; and by H. Rees et al., R. B. Flavell, H. C. Macgregor and G. A. Dover et al., all on genome restructuring in relation to species divergence.)

Ford E. B. (1971) *Ecological Genetics*, 3rd ed., 410 pp. London, Chapman and Hall.

Lewontin R. C. (1974) *The Genetic Basis of Evolutionary Change*, pp. 346. New York, Columbia University Press.

Manton I. (1950) *Cytology and Evolution in the Pteridophyta*. Cambridge, Cambridge University Press.

Parkin D. T. (1979) *An Introduction to Ecological Genetics*, 223 pp. London, Edward Arnold.

Sheppard P. M. (1958) *Natural Selection and Heredity*, 312 pp. London, Hutchinson.

Simmons M. J. and Crow J. F. (1977) Mutations affecting fitness in *Drosophila* populations. *Annu. Rev. Genet.* **11**, 49–78.

Stebbins G. L. (1950) *Variation and Evolution in Plants*, 643 pp. Oxford, Oxford University Press.

White M. J. D. (1973) *Animal Cytology and Evolution*, 2nd ed., 961 pp. Cambridge, Cambridge University Press.

Problems

18.1   Two of the commonest systems of self-incompatibility in flowering plants, both dependent on the presence in the population of many S alleles, are exemplified by the genera *Nicotiana* and *Brassica*. In the first, pollination fails if the pollen grain and style have an S allele in common. In the second it fails if the style and the plant which *produced* the pollen grain have an S allele in common. Compare the effectiveness of the two systems in restricting the fertility of crosses between plants which are (*a*) full sibs, (*b*) half-sibs and (*c*) essentially unrelated.

18.2   In a population of the snail *Helix pomatia*, a leucine aminopeptidase was found in three alternative electrophoretic forms, which we may call 1, 2 and 3. Individual snails could possess any one of the three or any combination of two of them. Suppose that in a classification of 200 animals the following frequencies were found: type 1, 88; type 2, 11; type 3, 3; 1 + 2, 59; 1 + 3, 30; 2 + 3, 9. Are these numbers consistent with the hypothesis that the three enzyme forms are determined by three alleles in Hardy–Weinberg equilibrium?

18.3   What interpretations could be placed on the observation (*a*) of an excess and (*b*) of a deficiency of heterozygotes as compared with the Hardy–Weinberg expectation in a polymorphic population?

18.4   Assuming that individuals homozygous for the human sickle-cell allele s have zero reproductive fitness (which may not be quite true) and that the frequency of the allele is maintained at an equilibrium frequency of 10 per cent (which has been true for some parts of West Africa), calculate the selection coefficient disfavouring SS as compared with Ss individuals.

18.5   A recent study (Voelker et al., 1980, *Genetics* **94**, 961) of spontaneous mutation frequency in *Drosophila melanogaster* has yielded an estimate, for mutation to electrophoretically distinguishable functional enzyme variants, of 1.3×10^{-6} per gene per generation. Assuming (rather a large assumption) that all functional electrophoretic alleles are selectively neutral, and given (cf. *Table* 18.1) that 0.15 is a typical level of heterozygosity with respect to electrophoretic polymorphism in *Drosophila* species, what effective population size does this estimate imply?

In the same study the mutation rate to null (enzyme-negative) alleles was estimated as 3.9×10^{-6} per gene per generation. If nulls are lethal when homozygous but neutral when heterozygous with functional alleles (another large assumption), what should be the equilibrium frequency of null alleles in the population?

18.6   The variety *carbonaria* of the moth *Biston betularia* increased from being a rare variant to constituting 98 per cent of the population over a period of about 50 years in some parts of Britain in the nineteenth century. Assuming that, at the time it had achieved a frequency of 50 per cent, the *carbonaria* allele was increasing at the rate of about 4 per cent per annum, calculate the approximate selection coefficient favouring the *carbonaria* variant colour form of the moth during the establishment of the *carbonaria* variant. (Refer to *Box* 18.2.)

18.7   Mutations in 'silent' third codon positions, in the inner parts of intervening sequences and in 'pseudogenes', are clearly less subject to selective constraints than are mutations in positions which affect amino acid coding, but are they necessarily neutral? In what ways do you think that they *might* be of significance for fitness?

18.8

Man, chimpanzee	GDVEKGKKIFIMKCSQCHTVEK...
Rabbit, whale, kangaroo	GDVEKGKKIFVQKCAQCHTVEK...
Chicken, turkey	GDIEKGKKIFVQKCSQCHTVEK...
Turtle	GDVEKGKKIFVQKCAQCHTVEK...
Frog	GDVEKGKKIFCQKCAQCHTCEK...
Drosophila	GVPAGDVEKGKKLFVQRCAQCHTVEA...
Yeast	TEFKAGSAKKGATLFKTRCELCHTVEK...
Neurospora	GFSAGDSKKGANLFKTRCAECHGEGG...
Wheat	ASFSEAPPGNPDAGAKIFKTKCAQCHTVDA...
Cauliflower, mung bean	ASFNEAPPGNSKAGEKIFKTKCAQCHTVDK...

The list above shows the N-terminal sequences of cytochrome *c* from a number of different organisms, lined up to maximize homology. Discuss these sequences from the point of view of evolution (more information is, of course, conveyed by the rest of the cytochrome sequence which is too long to reproduce here).
NB One letter symbols for amino acids: A = Ala, C = Cys, D = Asp, E = Glu, F = Phe, G = Gly, H = His, I = Ile, K = Lys, L = Leu, M = Met, P = Pro, Q = Gln, R = Arg, S = Ser, T = Thr, V = Val, W = Trp, Y = Tyr, N = Asn. (Dickerson, 1972, *Sci. Am.* **226**, 58.)

18.9   Slightom et al. (*Cell*, 1980, **21**, 627) sequenced three cloned isolates of human γ-globin genes— the *A*γ and *G*γ genes from one chromosome and the *A*γ gene from a homologous chromosome. They found in all three genes a region of repetitious simple sequence near the middle of the long intervening sequence, and were surprised to observe that, whereas to the left of this sequence the two *A*γ genes were nearly identical and different at 13 positions from the *G*γ gene, to the right of it the *A*γ and *G*γ genes from the same chromosome were nearly identical and differed at 19 positions from the *A*γ gene from the other chromosome. What novel process do you think is implied by this observation? How may it be important in molecular evolution?

18.10 Gottlieb (*Biochem. Genet.* 1973, **9**, 97) studied the multiple electrophoretic forms of glutamate–oxaloacetate transaminase (GOT) found in different species of the plant genus *Stephanomeria*. Two diploid species were each found to be polymorphic with respect to alleles of a gene controlling one kind of GOT; homozygotes gave single-banded electrophoretic patterns and heterozygotes gave three-banded patterns, the middle band in each case presumably representing a hybrid dimer. An allotetraploid species, thought to have originated through chromosome doubling in a hybrid between the two diploid species, showed the same three-banded pattern in all plants examined. The allotetraploid species reproduces predominantly by self-pollination. Explain this situation. Why is it of possible interest in connection with gene evolution?

19 Human Genetics and its Applications

19.1 Introduction

The science of genetics, as well as offering understanding of the way living things work, is today credited with great powers for influencing human affairs. The clearest impact of the subject is in plant and animal breeding, where very great progress has been made, partly, it is true, through application of empirical procedures which owe little to basic genetics, but helped also in no small measure by the application of genetical theory. There is also a significant and increasing application of 'genetic engineering' to the micro-organisms used in the pharmaceutical industry. The greatest interest, however, tends to be aroused by the actual and conjectural applications of genetics directly to man himself, both in the medical and, to some extent, in the social context.

As more and more diseases due to infections are brought under control, at least in the medically advanced countries, the proportion of human disease which can be attributed, wholly or in part, to inherited defective genes becomes greater. It becomes an increasingly important part of clinical medicine to find ways of avoiding or treating 'genetic' disease.

At the same time, as more and more of the political and social obstacles to human equality are removed or at least come under challenge, the extent to which inequalities in human abilities are genetically determined becomes an increasingly sensitive issue. This chapter will be concerned predominantly with medical applications of genetics, but we will conclude with a briefer consideration of the social and philosophical implications of the subject.

19.2 Progress in the formal cytogenetics of man

The human species is becoming a very favourable one for genetic analysis. Modern medical services have brought large populations under detailed surveillance and a great deal of allelic variation, some of it clinically relevant and some of it symptomless, has come to light. The relatively small human family size and the impossibility of making experimental matings have restricted genetic analysis of the conventional kind, but this disadvantage has been to a large extent offset by the development of parasexual analysis based on fusion of cells in culture. We saw in Chapter 9 how the study of man–mouse and man–hamster cell hybrids has led to the assignment of many human genes to chromosomes or even to specific segments of chromosomes.

At the DNA level, the discovery of polymorphisms with respect to endonuclease-susceptible (restriction) sites has added a great many potential new markers which will undoubtedly be used in the future for fine-structure gene analysis. We see an example of the use of such a marker below (p. 575).

The impressive progress made up to 1980 in the mapping of the longest human chromosome is summarized in *Fig.* 19.1.

Fig. 19.1

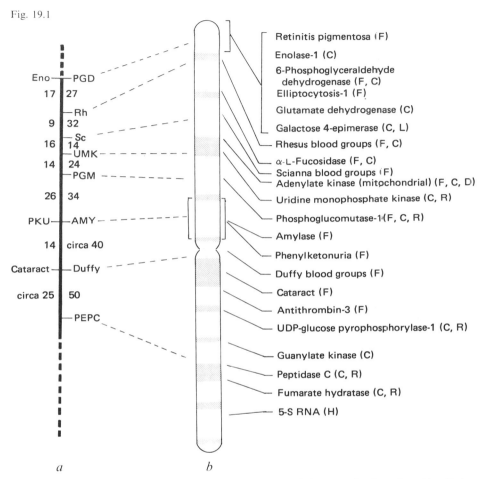

Chromosome-1 of the human genome and the genes that have been identified in it.

a. The linkage group corresponding to chromosome-1. The numbers on the left and right of the map are the recombination frequencies found in male and female meiosis; the frequencies in the female are the higher. The broken lines correspond to the terminal chromosome regions within which genetic markers have not so far been available.

b. The chromosome, showing G-bands (cf. *Fig.* 1.20*a*) and the centromere constriction. The genes listed on the right of the figure have been assigned to different regions of the chromosome by the criteria indicated in parentheses; F, family linkage studies; C, segregation in cell hybrids (cf. Chapter 9); R, correlation of loss of marker from cell hybrids with radiation-induced loss of chromosome segment; LD, linkage disequilibrium in populations with another mapped marker; H, *in situ* hybridization (cf. *Fig.* 1.18); D, deletion mapping. Data abstracted from McKusick.[5]

19.3   Congenital abnormalities

Included in the genetic variation present in all human populations are some phenotypes that are obviously defective, as well as a much larger number that are less clearly outside the 'normal' range. There may be disagreement about where the line should be drawn between clinical disorders and more or less harmless idiosyncrasies which we all possess in some degree. Nevertheless, there can be no dispute that the clearly disabling conditions at the extreme end of the range are the proper concern of the medical profession and call for prevention or treatment.

Clinical disorders which are partly or wholly genetically determined can be divided into those which can be attributed to single gene mutations and those which are multifactorial, in that they have to be explained as due to an unlucky combination of alleles of several or many genes. The conditions which are clearly due to single genes are actually a minority of all conditions with a genetic component in their causation, but they are the simplest to describe and many of them are important. We shall deal with them first.

Single-gene Defects

Single mutations that cause disease can be classified into autosomal recessives, autosomal dominants, sex-linked recessives and sex-linked dominants. The last is an uncommon class, since a dominant deleterious X-linked gene in the hemizygous male is likely to be lethal.

Fig. 19.2

Examples of pedigrees showing different kinds of inheritance of single-gene disorders. Males shown as squares, females as circles, affected individuals as filled symbols (except in *d*).

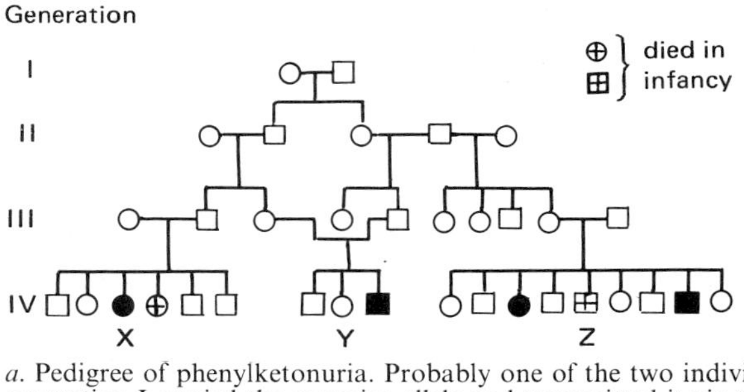

a. Pedigree of phenylketonuria. Probably one of the two individuals in generation I carried the recessive allele and transmitted it via the second and third generations to families X and Y in generation IV. The mother of family X must have been a carrier through coincidence. The appearance of the disease in family Y is not so unexpected, since the parents were first cousins who each had a $\frac{1}{4}$ chance of having inherited the phenylketonuria allele from the carrier in generation I. Family Z was another example of bad luck, since there was no reason to expect the father to be a carrier, though he evidently was; the mother presumably inherited the phenylketonuria allele transmitted from generation I. The individuals marked with a cross died young, and were most probably affected by the disease.
After Folling, Mohr and Ruud, cited by Stern.[6]

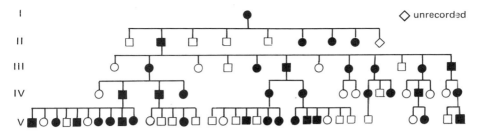

b. Pedigree of brachydactyly (short fingers) determined by a dominant allele. After Farabee, cited by Stern.[6]

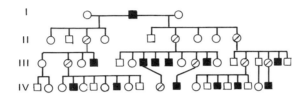

c. An example of recessive sex-linked inheritance—haemophilia. Only sons are affected (filled squares), with 50 per cent probability. The daughters each had a 50 per cent chance of carrying the mutant allele—those who definitely did so are marked by a diagonal bar. The pedigree also shows that haemophilia, though it may be severely disabling in the absence of treatment, was not necessarily lethal even in the nineteenth century. The affected man in generation I survived to father two different families.
After Bulloch and Fildes, cited by Stern.[6]

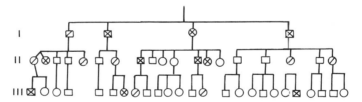

d. Pedigree of hypophosphataemic (vitamin-D resistant) rickets, due to a dominant mutant allele with incomplete penetrance to the skeletal phenotype, but always detectable through its effect in reducing the level of phosphate in the blood serum. Low phosphate with and without skeletal abnormality indicated by diagonal cross and single diagonal bar, respectively. The spouses of the affected parents, who would have been normal in each case, are not indicated. Families with one affected parent are expected to show a 1:1 segregation of affected and normal children. Overall (omitting the first generation, where other sibs, presumably normal and possibly numerous, are not recorded) the ratio is 14:31, representing a significant deficiency of affected individuals, more pronounced in generation III (for reasons not apparent—possibly mere chance). The deficiency may be due to prenatal or infantile mortality.
After McKusick.[7]

The single-gene nature of a genetic disease is usually apparent from the pattern of inheritance. *Fig.* 19.2 shows some examples of pedigrees which clearly indicate the segregation of single-gene differences. By collecting such pedigrees and pooling the data derived from them, the well-known Mendelian ratios (3:1 from mating between two heterozygotes and 1 : 1 from mating of a heterozygote to homozygous recessive) can often be demonstrated, though, where the condition under study is severely deleterious *in utero*, there may be a shortage of the handicapped phenotype as a result of unrecorded miscarriages.

In testing for goodness of fit to Mendelian ratios in man, one must remember to discount the individual (sometimes called the **propositus**) whose phenotype directed attention to the family in the first instance. Thus, if the propositus shows a condition suspected of being due to a recessive mutant gene, with the implication that both parents were heterozygous, one would expect to find a 3 : 1 ratio among the propositus's sibs (or, rather, among the sibs of a considerable number of like cases added together). To include propositi in the count would obviously bias the data in favour of the abnormal phenotype.

Table 19.1 lists some of the more frequent genetic disorders that come into the single-gene category. Very many others are known, but nearly all are very rare. Why should some be so much commoner than others? There are three possible answers: different frequencies of effective mutation, different intensities of contrary selection and positive selection in favour of the heterozygote. Of these possibilities, the first is quite conjectural, but may at least be a factor in some cases (some genes may be larger targets for mutation than others or their protein products more sensitive to amino acid replacement, so that a larger proportion of mutations have pronounced effects). The second possibility must fairly obviously be the explanation of the high frequency of some conditions, such as sex-linked red–green colour blindness, which are hardly disadvantageous at all. The third explanation

Table 19.1	Approximate frequencies of some of the most common recessive deleterious mutant alleles in man		
Condition	Frequency, per million births	Frequency of carriers, %	Frequency of allele, %
	Autosomal recessives		
Fibrocystic disease	400	4	2
Phenylketonuria	100	2	1
Albinism	100	2	1
Tay–Sachs disease	10	0·6	0·3
Galactosaemia	5	0·4	0·2
	Autosomal dominant		
Huntington's chorea	100	0·02	0·01
	X-linked recessives		
Duchenne's muscular dystrophy	200	0·01 (females)	0·01
Haemophilia	100	0·005 (females)	0·005

Abridged from Bodmer and Cavalli-Sforza.[4]

(heterozygous advantage) is true of sickle-cell anaemia (p. 533) and perhaps of other haemoglobin defects (some forms of thalassaemia) that, in heterozygotes, may also confer some resistance to malaria. Heterozygous advantage has also been considered as a possible way of explaining the extraordinarily high frequency of the recessive allele determining cystic fibrosis, which is present at frequencies of up to 4 per cent in European populations. However, no advantage of the heterozygote in this case has been clearly identified.

For obvious reasons, serious disorders due to dominant alleles are rare. If severely crippling, one would expect a dominant mutation to be removed from the population by selection in one generation. There are two circumstances in which such a mutation might persist. The first is that in which the mutant allele shows **incomplete penetrance**—that is to say it sometimes fails to show its effect, either because the effect depends also on the rest of the genotype (**modifier** genes or 'genetic background') or because it only becomes manifest in certain environmental circumstances (**genotype–environment interaction**). *Fig.* 19.2*d* shows one example in which the penetrance is incomplete at the level of obvious effect, but complete in the determination of a cryptic biochemical trait.

The second situation in which a dominant severely deleterious gene can persist is where it has its effect relatively late in life after reproduction is likely to have already occurred. A notorious example is **Huntington's chorea**, which is characterized by progressive mental deterioration making its appearance on average at about 40 years of age.

Sex-linked recessive alleles with deleterious effects also tend to occur in low frequencies because they are always subject to counterselection in males, who have no second X-chromosome to mask the effect of one carrying a defective gene. The incidence of X-chromosome disorders in males is of the same order of frequency as autosomal recessive disorders in both sexes, but it is very rare indeed in females. The best known example is **haemophilia**, a failure of blood clotting and a consequent tendency to uncontrolled bleeding. The most famous pedigree in which haemophilia appears is that founded by Queen Victoria of England, who transmitted the determining allele to one of her sons, who suffered from the disease, and to at least one of her daughters. The daughter, who became Queen of Spain, in turn passed it on to two of her three sons and to a daughter and hence to two out of four grandsons. The mutation in this case is presumed to have arisen either in Queen Victoria or in her mother.

A relatively harmless trait, already referred to, is the commonest form of red–green colour blindness, which affects about 8 per cent of the British male population but only about six in every thousand females. The responsible mutant allele is sex linked and recessive, and the different frequencies in the two sexes are those expected on the basis of a gene frequency of 0·08. In order to show the phenotype, a female needs to have inherited the allele from both parents, with probability $(0·08)^2 = 0·0064$.

Multifactorial Conditions—the Threshold Concept

There are many clinical conditions which have a strong tendency to 'run in families', but not in the regular and predictable way which one would expect if a single-gene difference were responsible. Some of these can, on the basis of examination of pedigrees, be formally ascribed to single-gene effects with

Fig. 19.3

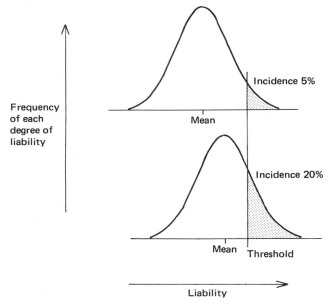

The 'threshold' model for the determination of multifactorial genetic disease (after Falconer[1]).

Two populations are shown as frequency distributions, with variation about a mean liability to a disease. An individual exceeding a threshold liability value exhibits the disease. The upper distribution is that of the population as a whole while the lower distribution is that of an especially susceptible group, for example of brothers and sisters of sufferers from the disease. The shaded areas under the curves indicate the affected fraction of the population in each case. The incidence of 5 per cent in the general population is, of course, higher than is found for any actual disease, and is chosen for the sake of clarity in the diagram.

incomplete penetrance, but very often it is impossible to identify any single gene with a major effect and any consistent degree of penetrance; one is then forced to invoke the cumulative small effects of allelic differences in several or many genes.

There is an apparent contradiction in the attempt to account for a clinical condition, which may have an all-or-nothing manifestation, in terms of a polygenic model which would normally be expected to generate a continuously graded series of phenotypes (*see* Chapter 8, p. 212). D. S. Falconer has escaped from the paradox by invoking a continuously variable physiological parameter, **liability**, underlying the all-or-nothing difference in the overt phenotype. It is postulated that when the liability exceeds a certain threshold value the clinical condition becomes fully apparent, while if the liability falls below the threshold, little or no abnormality is seen. The model is illustrated in *Fig.* 19.3. The two hypothetical bell-shaped curves represent, respectively, the distributions of liability in the population as a whole and among people with a particular degree of relatedness, for example sibship, with sufferers of the disease. The ratio of the two populations with respect to the incidence of the disease (in the figure, $\frac{20}{4} = 5$) can be used in a calculation of the heritability of liability. This, in turn, can be used to predict the incidence of the disease among relatives of other kinds. For example, the potential risk to nieces and nephews could be predicted from data on incidence in sibs.

Chromosomal Aberrations

Autosomal Aneuploidy

A number of well-defined abnormal phenotypes in man are due not to point mutations but rather to unbalanced chromosome constitutions. The most common of these is **Down's syndrome**, which is most often caused by an extra copy of (i.e. trisomy for) chromosome 21. This condition results from non-disjunction at meiosis in the oocyte. The blame is placed on the egg, rather than on the sperm cell which fertilizes it, because of the extreme dependence of the frequency of the condition on maternal age. Mothers aged 45 or over run a risk of about one in fifty of their baby exhibiting the syndrome, which is marked by rather distinct facial features and mental retardation. With mothers of under 30 the risk is less than one in a thousand. Meiotic non-disjunction is evidently much more prone to occur in older eggs. The age of the father seems to make no difference. It is likely that sperms carrying an extra chromosome 21 are non-functional or very inefficient in competition with chromosomally balanced ones, and so hardly ever fertilize eggs.

A second, though considerably less frequent, cause of Down's syndrome is the presence in the mother of a chromosomal translocation in which a major part of chromosome 21 is joined to one of the other chromosomes (generally a middle-sized one). This form of the syndrome, which is little different in phenotype from the commoner trisomy-21, is not dependent on maternal age. It does not depend on any unusual accident of non-disjunction, but is the inevitable consequence of the presence at meiosis of an association of three chromosomes which can disjoin two-and-one in two almost equally probable ways. *Fig.* 19.4 explains how eggs carrying a substantial extra fragment of chromosome 21 are produced, with an approximately 50 per cent probability for each egg, by a woman carrying the translocation. The translocation in itself produces no symptoms, provided that the translocated segment is not duplicated.

A few very rare conditions due to trisomy with respect to larger chromosomes have been reported. The disabilities associated with these are severe and lethal; no doubt many of these trisomies are lost by spontaneous abortion. Triploidy, which can be regarded as trisomy for the whole of the chromosome complement at the same time, is unconditionally lethal in man (and all mammals so far as is known), but triploid embryos are in fact rather common and account for nearly 10 per cent of all miscarriages. They must originate from diploid eggs or possibly sperm formed after failure of one of the divisions of meiosis.

Sex Chromosome Abnormalities

The numerical chromosome aberrations that are the most easily tolerated in man involve the sex chromosomes. Presumably this is because human development is adapted to cope with a variation in X-chromosome number by inactivation of any Xs in excess of one, and the Y-chromosome is (except for male determination) in any case mainly inert. Non-disjunction of X-chromosomes at the first division of meiosis in the female gives occasional XX eggs (the eggs with no X at all being inviable), while non-disjunction of X and Y in the male can result in sperm devoid of any sex chromosome and sperm carrying both X and Y (both types can function). The XX eggs on fertilization by normal sperm give XXX females and XXY males; the two kinds of non-disjunctional sperm, on fertilizing normal eggs, give XO females (O signifying the lack of a second sex chromosome)

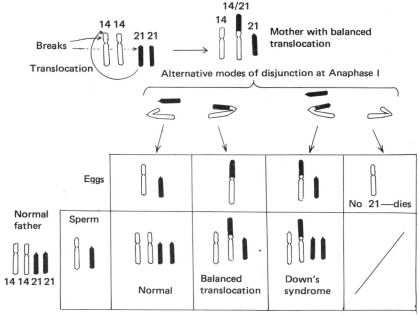

a. The origin of partial trisomy for chromosome-21, causing Down's syndrome, from a 21→14 translocation in the mother.

and XYY males. All four kinds of sex chromosome abnormality are well known, and all except the XO condition, which is rather rare, occur with frequencies of the order of one or a few per thousand of the population.

The XXX condition is virtually symptomless. It has been shown that the nuclei of XXX females have two inactivated X-chromosomes (seen as 'Barr bodies') instead of the normal one inactive X (cf. p. 509), and so it is understandable that the effect of the extra chromosome on the phenotype is minimal.

XO females exhibit a condition called **Turner's syndrome**, characterized by infertility and certain minor physical peculiarities. The marked effect of the absence of the second X shows that it must have some function in normal females. It is probably important during the several cell cleavages that precede X-chromosome inactivation.

XXY males are recognized as cases of **Klinefelter's syndrome**, with inadequate sexual development, sterility and some tendency to mental retardation. The generally male phenotype of these individuals demonstrates the positive male-determining effect of the Y-chromosome. As in normal females, one of the two X-chromosomes is inactivated in every cell and it is thus evident that, even with this inactivation, the second X-chromosome has some effect.

The XYY condition for a long time remained unrecognized. It confers very little, if any, clear abnormality of phenotype. However, it has become the subject of much attention in recent years through its discovery in the unexpectedly high proportion of 7 out of 190 men detained in a Scottish hospital for psychiatric

Fig. 19.4. contd

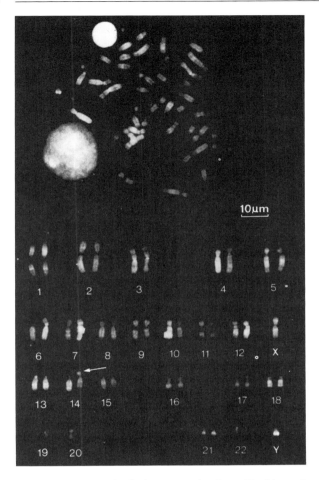

b. A photomicrograph of a karyotype showing a 21→14 translocation as well as two normal chromosomes 21. Quinacrine fluorescence staining. Photograph by courtesy of Professor H. J. Evans.

patients with criminal tendencies. These individuals were of above average height and were judged to be aggressive. It seems likely that the extra Y-chromosome in the male does, on average, have some physical and psychological effect, but it is clear that the great majority of XYY males, estimated to number about one in 2000 of all men, blend with the rest of the population without attracting any special attention to themselves.

Balanced Segmental Interchanges

A very large number of human karyotype determinations have been made in recent years, and many symptomless chromosome variations have been detected. These are mostly balanced segmental interchanges, with no overall gain or loss of chromosomal material. Meiosis in an interchange heterozygote is liable to lead to a proportion of inviable products—67 per cent if the association of four chromosomes formed at meiotic prophase disjoins two-and-two at random at first anaphase (cf. *Fig.* 6.2). The reduction of fertility due to duplication/deficiency karyotypes in gametes may well pass unnoticed; the altogether excessive fertility of the average healthy human being can be reduced a good deal and still remain adequate for the achievement of the family sizes that most people desire. Some segmental interchanges, however, are more sinister in their associations. Certain forms of cancer are rather consistently associated with specific chromosome rearrangements arising during development (*see* p. 502, Chapter 17). It seems possible that certain rearrangements, or the breaks leading to them, at least predispose cells to certain kinds of malignant transformation, even if they are not the immediate cause. The opposite interpretation—that the malignant transformation occurs first and in some way confers a selective advantage during tumour growth on the segmental interchange—is not ruled out.

19.4   Intervention for the avoidance of genetic disease

Genetic Counselling

There are a number of ways in which our knowledge of genetic diseases can, or could, be used in attempts to reduce their incidence. Genetic counselling, for couples who believe they may be at risk of producing a defective child, is becoming a routine aspect of medical practice. The advice given may lead to voluntary abstention from parenthood or to selective abortion of fetuses either strongly suspected of being defective or, by the use of **antenatal diagnosis** (*see* p. 574), definitely shown to be so.

Calculating the Odds

The usual situation which calls for counselling is that in which there is some family history of disease or disability that is suspected of having a genetic basis. Potential parents may well want to know what risk they run of having a sick or handicapped child. The counsellor can usually at least calculate the odds.

The simplest situation is that in which the couple seeking advice have already had one defective child and wish to know how likely it is that a further child would be affected in the same way. If the particular type of disability can be identified as one resulting from an autosomal recessive allele in homozygous condition, then the simple answer is that each subsequent child will have a one in four chance of being affected and a three in four chance of being phenotypically normal (though with a two-thirds chance of carrying the defective allele).

Sometimes the counsellor can say that the condition in question has no known genetic basis and is probably due to some accident at birth or other environmental misfortune which is unlikely to be repeated. In this case the parents can be reassured that their next child will have as good a chance of normal health as anyone else. Perhaps the commonest situation is that in which there is thought to be some genetic basis for the disability, but not a simple one—for example the determination may be polygenic, or due to one 'major' gene with incomplete penetrance. Penetrance variation may depend on genetic modifiers or on environmental variation or on both. In complex situations of these kinds the counsellor has no simple logical formula for calculating the risk but must resort to the published medical statistics, which may tell him how frequently, in the population as a whole, a sib of a child showing the specific phenotype will itself be affected by the same condition. Everything depends on the correct identification of the disease, which is why most genetic counsellors are clinicians in the first instance and geneticists only second.

A somewhat more complex calculation has to be made if a relative of one of the potential parents is affected by a genetic disease. Suppose that the causative factor of the condition shown by the relative is known to be a single recessive allele, and that the relative is a brother or sister (i.e. uncle or aunt of the potential child). Then the person seeking advice can be told that he or she has a two-thirds chance of being a heterozygous carrier of the recessive mutation and a one-third chance of being homozygous normal. His or her spouse has a very much smaller chance, though still a calculable one, of also being a carrier. To take the worst case, if the disease in question is fibrocystic disease of the pancreas (cystic fibrosis), the frequency of heterozygotes among the general population is about $\frac{1}{22}$, and so the probability that *both* potential parents will be heterozygotes is $\frac{2}{3} \times \frac{1}{22} = \frac{1}{33}$. The chance that any child born to such a couple will suffer from the disease will be $\frac{1}{4} \times \frac{1}{33} = \frac{1}{132}$. A risk of this magnitude may or may not be regarded as worth taking depending on temperament and individual and social circumstances.

Genetic diseases determined by dominant mutations present problems to the genetic counsellor when they are characterized either by incomplete penetrance or by late onset of effect. In both cases calculation of the odds has to be based on medical statistics. Thus, if, from the family history (an affected parent or sib, for example) the person seeking advice is known to have had a 50 per cent chance of inheriting the deleterious dominant, but does not show its effects, he or she may either have had the good luck not to inherit it (chance $\frac{1}{2}$) or may be carrying the mutant allele without its effects penetrating to the overt phenotype—a chance of $\frac{1}{2}(1-x)$ where x is the degree of penetrance of the allele. Hence the probability that the person is a carrier will be $\frac{1}{2}(1-x)/[\frac{1}{2} + \frac{1}{2}(1-x)] = (1-x)/(2-x)$. This estimate of risk may be in error if the degree of penetrance is itself inheritable; the value of x obtained from the general population may not then apply to the family in question.

The most notorious example of the difficulty arising from late onset of a dominant genetic disease is that of **Huntington's chorea**, referred to above. A relatively young person whose father or mother suffered from the disease has, at birth, a 50 per cent chance of having inherited the responsible allele. The older he or she gets without showing symptoms the more grounds there are for optimism. But because the age of onset averages about 40, and often is delayed until beyond the age of 50, there can be no reasonable assurance that the mutation is not present until it is probably too late to start a family in any case. One is put in mind of the ancient Greek saying, quoted by Herodotus: 'Call no man fortunate until he is dead.' A biochemical test which would allow an early identification one way or the other would be a great blessing—at least for those who received a favourable verdict.

Heterozygote Detection

A part of the uncertainty in genetic counselling can be eliminated if a method is available that will allow the identification of heterozygotes with respect to recessive deleterious alleles. Such a method is in principle available whenever the homozygous abnormal allele is associated with a more or less complete deficiency of an enzyme activity which can be quantitatively assayed. It has to be remembered that the terms 'dominant' and 'recessive' are relative to the aspect of the phenotype which is being looked at. So far as enzyme activity is concerned, dominance is generally absent—that is to say, the heterozygote has an enzyme activity more or less exactly intermediate between the activities of the two homozygotes or 50 per cent of normal. But at the level of the physiology and development of the whole organism, a 50 per cent drop in the level of a single enzyme can usually be accommodated without obvious effect, and dominance of the normal allele will appear to be complete. Given the availability of a good clinical biochemical laboratory (a big proviso), a person suspected of being a carrier of the recessive gene for galactosaemia (galactose 1-phosphate uridyl transferase deficiency) or for the Lesch–Nyhan syndrome (HGPRT deficiency), to name but two clear-cut enzyme-based genetic diseases, can in principle be given a definite answer one way or the other at the expense of the donation of a sample of blood or a few skin cells.

In some cases a biochemical abnormality consists not in the absence of a protein gene product but in the production of a mutationally altered form of it. The heterozygote then produces two different forms of the protein and these can sometimes be separated by electrophoresis. For example, heterozygous red blood cells carrying the alleles both for normal and for sickle-cell haemoglobin give two clearly distinct red bands when their lysates are subjected to electrophoresis.

In cases where the suspected heterozygosity is with respect to chromosome structure, diagnosis is now a rather routine procedure—again, of course, providing that the laboratory facilities and expertise are available. A small sample of blood is taken and the leucocytes are stimulated to divide in culture by the addition of phytohaemagglutinin, a cell-agglutinating protein from beans. The dividing cells are arrested in mitotic metaphase by the mitotic inhibitor colchicine (cf. p. 160), flattened on a microscope slide, stained to show the chromosomes and photographed. For analysis, the images of the chromosomes are usually cut out separately from the photograph of a single clear metaphase and matched up in pairs. In this way segmental translocations and interchanges become evident (*Fig. 19.4b*).

One of the most important applications to genetic counselling is in cases where a mother has already given birth to a child with Down's syndrome and wants to know whether she carries a chromosome-21 translocation or whether (which is considerably more probable) she was the victim of a non-disjunctional accident at meiosis which is relatively unlikely to be repeated. It is the mother rather than the father who is relevant here since an extra translocated chromosome-21 fragment is hardly ever transmitted through the sperm, which appears to be much more vulnerable than the egg to chromosome imbalance.

Antenatal Diagnosis

When a pregnant woman is known to have a considerable chance of bearing a child with a genetic defect it is possible in many cases to diagnose the condition in the fetus and to carry out an abortion if this is desired. Antenatal diagnosis depends on the recovery of living cells from the fetus. This can be achieved by a procedure called **amniocentesis**, which consists of the insertion of a hypodermic syringe needle into the amniotic cavity and withdrawing a few millilitres of the fluid surrounding the fetus. The amniotic fluid contains free cells of fetal origin, and these can be cultured and tested, in various ways, for the

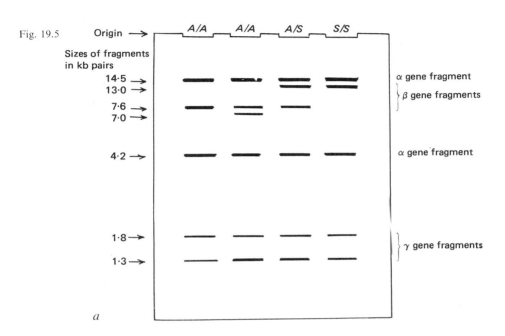

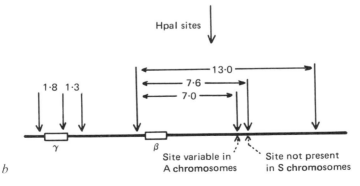

Restriction site polymorphism linked to the β-globin gene of human chromo-some 11 (after Kan and Dozy[2]).
a. HpaI restriction fragments hybridized to β-globin cDNA probe. Note that the probe hybridizes to other globin gene sequences as well as to the β-globin gene sequence.
b. Restriction site map of the β-globin/δ-globin chromosome region. One of the two A/A individuals analysed in (a) turned out to be heterozygous with respect to the HpaI 7·0/7·6 restriction fragment difference explained in (b).

enzymes which they can (or cannot) produce, for their karyotype and even for restriction-site patterns in their DNA. W. Bodmer and L. L. Cavalli-Sforza[4] listed 35 recessive genetic diseases which can, or could in principle, be recognized before birth in this way. That list is now considerably longer.

A new dimension has been added to antenatal screening by the discovery that human populations are replete with DNA polymorphisms, not, in most cases, with any effects on essential genes, but adding or removing sites for restriction endonuclease action and hence altering the sizes of particular DNA restriction fragments. A first and classic example of the use of DNA polymorphism for following the transmission of a defective gene was due to Y. Kan and his colleagues in San Francisco.[2] They found that a restriction fragment, detected on Southern transfers by a radioactive probe for α- and β-globin genes, varied within the US population. Moreover, at least in a first sample, all sufferers from sickle-cell anaemia were homozygous with respect to one size, all homozygous normals were homozygous with respect to the other size, and heterozygous carriers of the sickle-cell allele had restriction fragments of both sizes. These findings had two implications: first, that the polymorphic restriction site was very closely linked to the β-globin gene and second, that all of the sickle-cell alleles present in the first sample must have descended from the same original mutation which presumably occurred many hundreds of years ago in Africa in a chromosome which was marked by a restriction site not present in most of the other chromosomes of the population. Over a long period one would expect the globin mutation and the restriction site difference to come into linkage equilibrium (cf. p. 522), but the distance between them is so short and crossing-over so rare that linkage disequilibrium would be expected to persist for a very long time. Subsequent samples from the American 'black' population have shown that a substantial number of sickle-cell alleles are *not* any longer linked (if they ever were) to the distinguishing restriction site, and so it is not after all possible to use the DNA analysis on fetal cells as an infallible indication of their globin gene status. However, given some tests of both globin and DNA pattern, on the parents and perhaps other members of the family as well, it would often be possible to use the close linkage of the DNA marker and the β-globin gene for antenatal diagnosis of sickle-cell disease (*see Fig. 19.5*).

Considerable effort is now being put into the mapping of DNA polymorphisms in the human genome in the hope that it will eventually be possible to find restriction-site differences that could be used as closely linked markers for many genes of interest. This kind of knowledge would not permit antenatal diagnosis in every case, since a DNA marker would only be of use if it happened to be segregating in the family under investigation. Nevertheless, this approach does hold promise for the identification of many kinds of harmful alleles in the fetus. It would not be necessary for these alleles to be detectable through their protein products; indeed, they need not be expressed in the fetus at all.

There are a few examples in which the presence of a defective gene can be detected antenatally not by assays for the protein gene product, or even through direct analysis of the DNA, but through indirect biochemical effects. Such a case is **spina bifida**, a severe failure in the development of the central nervous system such that the main neural tube never becomes properly closed. This condition is, in its milder forms, susceptible to at least partial correction by postnatal surgery, but the success of this is very limited in the more severe cases. A somewhat similar but often even more severe malformation is **anencephaly**, in which brain development is affected. These two conditions together occur with a frequency of approximately three per thousand births. Where one child in a family is affected by either, there is a three or four per cent probability that a subsequent child of the same parents will be affected also, a finding which makes it very probable that there is some genetic component in the causation of the defect. It is therefore desirable to have some means of antenatal diagnosis. This is provided by the discovery that a protein, α-fetoprotein, is characteristically present in the amniotic fluid when the fetus has a severe lesion of the type associated with spina bifida or anencephaly, but not when the fetus is normal. Amniocentesis can reveal the presence of this protein and thus permit the detection of the most severe cases. The α-fetoprotein test may, however, miss cases of medium severity which are actually a greater problem in as much as they are less likely to be spontaneously aborted.

The whole purpose of antenatal screening is to allow termination of pregnancies which threaten to lead to the birth of severely deformed babies. If abortion is not regarded as acceptable there is no point in carrying out amniocentesis, especially since the procedure, though relatively safe for the pregnant woman, is not absolutely so.

Therapy for Genetic Diseases

To show that a disease has a genetic basis is not the same as pronouncing it incurable. We have seen earlier in this book how auxotrophic mutants of micro-organisms can be made practically equivalent to wild type by supplying them with the nutrients that they are unable to synthesize for themselves. Simple auxotrophy is not known in man, presumably because no human being lives on a diet equivalent to minimal medium. There is, however, an analogous situation in diabetes mellitus, a condition which, as we saw above, has some genetic basis although not a simple one. Here what would, if left untreated, be a disabling and often lethal condition can be virtually totally ameliorated by insulin injection.

Simple mutational blocks in metabolic pathways in man have their disabling effects more frequently through accumulation of intermediates than by deficiency of end-product. For example, the two kinds of arginine-specific *Neurospora* auxotroph which accumulate citrulline and argininosuccinate, respectively (*see* p. 350), have their analogues in man, the accumulated intermediates appearing in high concentrations in the urine. The conditions citrullinuria and argininosuccinicuria are each determined by rare recessive alleles, and the obvious phenotypic manifestation in each case is mental retardation rather than a dietary requirement for arginine, which is always present in the diet in any case. The effects on mental development seem likely to be due to the accumulated intermediates.

A more frequently occurring condition resulting from the accumulation of a metabolite behind a metabolic block is phenylketonuria (PKU). This is one of several effects of mutation in man on different steps in the breakdown of phenylalanine (*Fig.* 19.6), and specifically involves the conversion of phenylalanine to tyrosine, a reaction normally catalysed by phenylalanine hydroxylase. When, as a result of homozygosity for a recessive mutation, this enzyme is missing, phenylalanine, which is constantly produced by hydrolysis of dietary protein, is converted by a side reaction (transamination) to phenylpyruvic acid which has a disastrous effect on the development of the nervous system. Fortunately, no damage has occurred at the time of birth and, provided that the condition can be promptly diagnosed, as is easily possible by a test for phenylketonuria in the urine, at least the worst consequences can be avoided by provision of a special diet which is as free as possible of phenylalanine. There are indications, though as yet no

Fig. 19.6

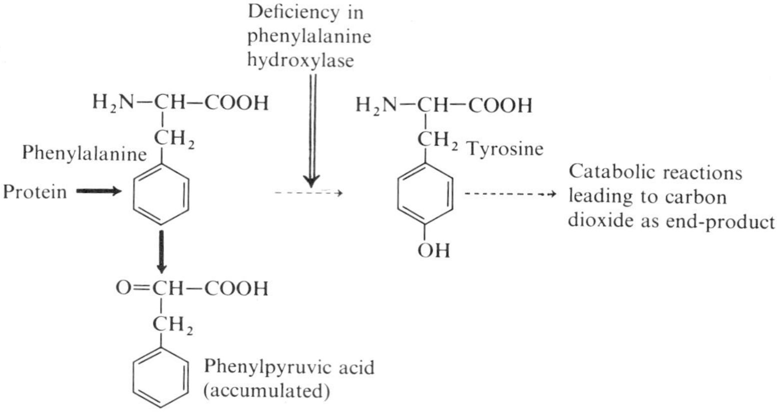

The metabolic block in phenylketonuria.

certainty, that at the age of about 3 years the sensitive period of neurological development is over, and that it may be safe at this time to transfer the child to a normal diet. This is not necessarily the end of the problem, however, as there is at least a suspicion that an affected woman who has escaped the consequences of her own phenylpyruvate accumulation, may nevertheless damage her unborn child should she become pregnant on a normal protein-containing diet.

As more understanding of the metabolic basis of genetic diseases is achieved it is to be expected that the possibilities of treatment by dietary restriction or by hormonal injections will increase. As we have already observed, some abnormalities with a genetic component can even be treated, sometimes with quite good success, by surgery.

Restriction of Mating between Relatives

One aspect of deliberate human behaviour which prevails in practically all human societies, and certainly affects the incidence of genetic disease, is the avoidance of mating between brothers and sisters, fathers and daughters and mothers and sons. The concept of incest as something forbidden quite often also includes less close relationships, such as that between first cousins or between uncles and nieces, though, with respect to these, customs are more variable.

The effects of mating between close relatives on the frequencies of genetic diseases can be easily calculated where the diseases in question are determined by single-gene mutations and where the frequencies in the population of the mutant alleles are known. Phenylketonuria (PKU) can be taken as an example. The incidence of this disease at birth is of the order of one in 10^4, implying that about 4×10^{-4} of all marriages are between individuals heterozygous for the PKU allele (the probability for each child from such a marriage of suffering from the disease is $\frac{1}{4}$). The frequency of individuals in the population carrying the gene, assuming that mating is at random, must then be about 2×10^{-2} (the square root of 4×10^{-4}), that is 2 per cent. Now if a person is heterozygous with respect to a recessive autosomal allele, the probability that a sister or brother is also heterozygous is $\frac{1}{2}$, since each sib will have the same 50 per cent chance of having inherited that allele from the parent carrying it. Thus, in the case of a brother–sister mating, the chance of *both* being heterozygous with respect to PKU would not be 2×10^{-2} squared but rather $2 \times 10^{-2} \times 0.5 = 10^{-2}$ or 1 per cent. The children resulting from such a union would each run a $\frac{1}{4}$ per cent risk of PKU, which is 25 times greater than that in the population at large.

The risk is increased by a lesser factor in the case of a more frequent deleterious recessive, such as that determining cystic fibrosis; the frequency of heterozygotes in this case is about 4 per cent, accounting for the observed frequency of the disease at birth of $\frac{1}{4} \times \frac{4}{100} \times \frac{4}{100} = \frac{4}{10\,000}$, or one in about 2500 births. The risk to children of a brother–sister union would be $\frac{1}{4} \times \frac{4}{100} \times \frac{1}{2} = 0.5$ per cent—rather more than ten times the general population risk. The general formula is:

$$\text{Chance of disease in each child} = \tfrac{1}{4} \times h \times p$$

where h is the frequency of heterozygotes with respect to the deleterious recessive allele and p is the probability of the two related parents *both* carrying a particular infrequent allele given that *one* of them does so. The formula assumes that h is fairly small so that the frequency of homozygotes for the deleterious allele can be ignored, and also that h is small compared with p. The value of p is $\frac{1}{2}$ for sibs, $\frac{1}{2}$ for parents and offspring, $\frac{1}{4}$ for the uncle–niece relationship, $\frac{1}{8}$ for first cousins and smaller fractions for more distant relationships. The factor by which the risk is increased, as compared with the general outbred population, by the relatedness of the parents is $\frac{1}{4}hp/\frac{1}{4}h^2 = p/h$.

In our society the only frequent degree of close relationship between parents is that of cousinhood. From the above formula we see that if one person in fifty is heterozygous for the PKU allele (which is approximately the case), the chance that the children of first cousin marriages will have the disease is increased by a factor of $\frac{50}{8} = 6.25$ as compared with the general population. This is in approximate agreement with observation. The enhancement of risk becomes greater than this for diseases determined by rarer deleterious recessives (PKU is one of the commoner ones) and, of course, greater for relationships closer than that of first cousin.

This would appear to constitute a rather strong argument not only against incest, as it is at present legally defined in most countries, but against cousin marriages also. This, at any rate is what one would conclude if the short-term prevention of genetic disease were the main concern. If, on the other hand, more importance were to be placed on the reduction in the frequencies of deleterious recessives one would not discourage cousins from marrying—rather the reverse, since by making a deleterious recessive allele more often homozygous one is exposing it to selection and reducing the effective fertility of the carrier individuals.

It is said that the royal families of ancient Egypt practised brother–sister mating over many hundreds of years. If this was so, strongly deleterious recessive alleles must have been virtually eliminated from that line of descent—but perhaps at some cost in terms of useful diversity of talents.

Possibilities of Replacement of Defective Genes

The most radical type of intervention in human genetics which we can conceive is the repair of genetic defect by the replacement or supplementation of the defective allele by a functional one. This would not so long ago have been regarded as totally impossible, and even now it is some way from being a practical proposition. Nevertheless, most of the basic techniques which would be required are now established in model systems, and there is no doubt that a great deal of the work going on in mammalian cell genetics is motivated by the hope that direct **gene therapy**, as it may be called, will one day become possible.

The most plausible scenario for gene therapy runs as follows. It would first be necessary to establish the precise nature of the metabolic defect from which the patient was suffering and trace it to a specific gene which could be cloned in a suitable host–vector system (at present a plasmid or lambda phage derivative with *E. coli*, but alternative systems based on mammalian virus with mammalian cells are being developed). The number of diseases which at present satisfy these requirements is very limited; there are many for which the metabolic deficiencies have been determined but few among them that correspond to genes for which the nucleic acid probes necessary for cloning can be made available (*see* pp. 386–392). Haemoglobin deficiencies provide the most obvious possibilities.

The next step would be to obtain from the patient a sample of cells in whose descendants the therapeutic gene could be expressed. In the case of globin genes, these might be the bone marrow cell precursors of red blood cells. These cells could be cultured and treated with DNA containing the cloned sequence of the normal functional gene under conditions (presence of calcium etc.—*see* p. 243) favouring uptake of the DNA. There would be some problem in selecting a suitably transformed cell line. It might in some cases be possible to select directly for expression of the desired gene in cell culture—for example for $HGPRT^+$ to cure the Lesch–Nyhan syndrome (cf. p. 574). If not, an alternative means of selecting the desired transformants might be to construct a plasmid in which the therapeutic gene was physically linked to a foreign, but harmless, sequence which *could* be selected for, and hope that the two sequences would be integrated together (cotransformation—cf. p. 243). The chances of this happening could be assessed beforehand by suitable model experiments. The period of *in vitro* cell culture would be kept to the minimum time necessary for transformation and subsequent selection, since long-term mammalian cell cultures, in so far as they survive at all, generally accumulate chromosomal abnormalities which might easily prevent them from resuming their normal functions on return to the body of the patient. The final step in the procedure would be the reintroduction of the transformed cells to the appropriate tissue where, if all went well, they would multiply, produce the normal gene product and cure or at least alleviate the patient's disability.

At least two serious problems remain to be mentioned. First, it is one thing to introduce a gene into a cell, and another to get in into the right context (in terms of neighbouring DNA sequence and perhaps higher-order chromatin structure) for something approaching normal expression. A particular DNA sequence might sometimes be integrated at its normal locus, either forming a tandem duplication or replacing the resident homologous gene (the latter is obviously the desirable outcome). But in some experiments, transforming DNA has been integrated in apparently random positions and when this happens it may come under an inappropriate system of transcriptional control or even fail to be transcribed at all. One possible solution to this problem would be to introduce, along with the coding sequence, all of the adjacent DNA necessary for transcription and normal control. It is not at present clear how large a tract of flanking DNA this would have to be.

The second problem, perhaps even more intractable, is that of getting the reintroduced cells established as a significant part of the total cell population. Unless special measures were taken, one would expect them to be greatly outnumbered by the non-cured resident cells. A proposed solution to this problem is the inclusion, in the transforming 'package' of DNA sequences, of one which confers a strong selective advantage on the transformed cells. One possibility is the Herpes Simplex virus gene coding for thymidine kinase, an enzyme activity which can be readily selected for (*see* p. 236) and which apparently boosts the growth of cells carrying it. Another possible tactic, even more drastic but more clearly feasible, would be to destroy a large part of the resident cell population by radiation or conceivably even by surgery. Obviously at some point the question whether the attempted cure was worse than the disease would have to be faced.

In spite of the formidable difficulties, gene therapy, through transformation of a proportion of certain specialized cells of an individual, may become possible for a few kinds of disorder in the foreseeable future. Similar therapy for germ cells, with the object of providing a repaired gene for the next generation, is not a realistic prospect at present. Although, as we saw on p. 245, DNA can now be introduced and integrated into mammalian eggs, such integration is uncontrolled and probably at random. There is no way of ensuring a favourable outcome for any particular egg.

For an expert and balanced review of the possibilities, the reader is referred to a recent article by R. Williamson.[8]

19.5   Differences between human races

Human racial differences and the causes of them are of importance to sociologists, anthropologists, historians and psychologists. They are, of course, also much stressed in some of the more pathological forms of politics. The science of genetics also has something to say in the matter, but not so much as is sometimes supposed.

Physical Differences

First of all, what is meant by a race? There is no doubt that, for example, almost any Bushman can be distinguished with certainty from almost any Finn, and many other ethnic groups are equally readily distinguishable from one another. Should we then regard these readily distinguishable racial types as distinct species? There are two reasons why to do so would be a great exaggeration of the separateness of human races. First, and most important, all the races are completely interfertile, and inter-racial hybrids show no disabilities resulting from their mixture of genomes (rather the contrary, in fact, if one may judge from the number of outstanding athletes to have been generated by racial mixture). Nor, so far as evidence goes, do further generations following hybridization show any segregation of unbalanced genomes. The second reason, which is both a consequence and a confirmation of the first, is that the clear differences between races tend to become blurred where their respective territories come into contact or overlap. Any attempt to assign *all* humans to one or other of a number of clearly distinct racial groups would certainly fail, and would have done so even before the recent rapid increase in mobility of the world's population.

Not a great deal can be said about the genetic basis of the obvious phenotypic differences between the 'typical' members of different racial groups. Clearly, they must have a genetic basis and, indeed, one must suppose that, to revert to our previous example, virtually all Bushmen have alleles of genes to do with skin and hair colour, hair form and perhaps physical build which are not present at all in Finns, and vice versa. Little, however, is known about the numbers of genes involved in the determination of any of these character differences. Some study has been made of the inheritance of skin colour, with the general conclusion that the number of segregating gene differences affecting this trait is fairly small—perhaps no more than four or thereabouts.[3] This conclusion is based on the degree of segregation of darker and lighter phenotypes among the second generation (F_2)

Fig. 19.7

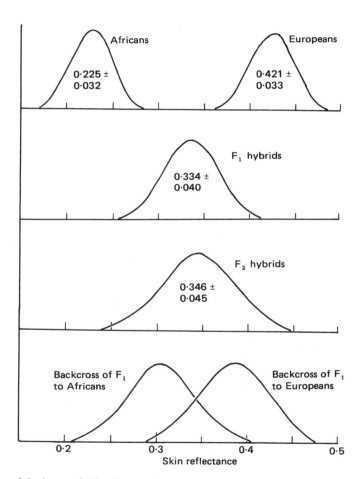

Inheritance of skin pigmentation.
The curves are based on the means and standard deviations of samples of Africans and Europeans, F_1 hybrids and F_2 and backcross progenies. The metric used is light reflected from the skin under standard conditions, transformed to antilogs (since this gave better approximations to normal curves). The spread of values in the F_2 and backcross samples, compared with the pure-race and F_1 samples, is close to what would be expected if the racial difference was determined by between three and four gene differences with additive effects. There is no appearance of dominance.
Based on Harrison and Owen.[3]

progeny following inter-race crossing (*Fig.* 19.7). There is no clear dominance of 'dark' over 'light' alleles or vice versa.

So far as two of the most conspicuous components of racial phenotype are concerned, skin colour and physical build, there are clear correlations with climate. The indigenous inhabitants of equatorial countries are all dark skinned, and there is a general tendency for pigmentation to decrease with distance from the equator. Increased pigmentation may well have been selected as a protection against excessive sunlight. There is also a significant tendency for equatorial peoples to be less heavily built than those living nearer to the poles, with a higher ratio of body surface area to body weight. A part of these differences may be attributable to nutrition, but there is little doubt that genetic factors are involved as well. It is reasonable to suggest that changes in body conformation that encourage heat loss have been naturally selected in hot countries, and that those which encourage the retention of heat have been selected in cold ones. Considerations such as these, somewhat speculative though they are, encourage the conclusion that the races of man diverged from a common stock in response to selective pressures exercised by the environment, the most important component of which was probably climate.

Mental Differences?

Much controversy has understandably been provoked by the possibility, for which there is some evidence, of differences between racial means with respect to mental qualities. The most provocative data are those which show that Americans of African descent score on average about 15 points lower on IQ tests than Americans of European or Asian descent. We reviewed in Chapter 8 some of the methodological difficulties which beset attempts to assess the relative effects of genotype and environment on continuous variation in human populations. The difficulties are all the greater when intergroup, as opposed to within-group, differences are at issue. It is impossible to equalize environments when two groups are visibly distinguishable and are regarded, both by themselves and by others, as different on account of this. It is virtually impossible to know how much allowance to make for inter-racial environmental difference, particularly its cultural and psychological components.

It may be argued—indeed, it often is argued—that there is no good evidence for *any* genetic component in the causation of any inter-racial mental differences. To some it seems that such evidence should not even be sought since it could, in any case, lead to no useful conclusions regarding social or educational policy and would only cause aggravation. Meanwhile, and in order to put the 'race–IQ' controversy into better perspective, it may be useful to stress the following points: (i) We are talking about population *means*, not absolute differences; there is no denying the broad overlap between the IQ ranges of different groups; the range *within* each group far exceeds the greatest difference between group means, (ii) High IQ is in any case not the only socially beneficial human quality, nor even the only desirable mental quality. (iii) In principle, the controversy will disappear as soon as people can learn to judge their fellow humans as individuals without reference to racial stereotypes which, even if not wholly unfounded, can never do justice to individual qualities. Meanwhile, discussion of the issue should be conducted with restraint.

Differences with Respect to Polymorphic Genes

Human races are recognized on the basis of more-or-less clear physical differences, the genetic determination of which is for the most part poorly defined. When we look, in different racial groups, at those genes which we *can* define (those coding for electrophoretically variable enzymes or determining blood group or other antigens) we see, on the whole, no all-or-none differences but only differences in allele frequencies. The classic example is that of the A/B/O blood group polymorphism. All three alleles are present in nearly all human populations, but with widely differing relative frequencies. The commonest allele almost everywhere is O (giving blood group O when homozygous), and its frequency approaches 100 per cent in many American Indian populations (especially in the southern half of the continent). In other populations, the A and B alleles occur with substantial frequencies, usually accounting together for 40–50 per cent of the total. The frequency of the B allele is the most variable, ranging from only about 5 per cent in the UK to as high as 50 per cent in parts of India. The variation occurs within as well as between nations; for example there are very significant differences, and a clear north–south trend (lower frequency of A in the north and higher in the south) within England and Scotland. There are fewer data on most other polymorphisms, but so far as they go they support the same generalization—the same alleles are distributed through all races but with different frequencies.

So far as relatively rare alleles are concerned, there *are* some racial differences. For example, phenylketonuria and cystic fibrosis, both caused by relatively rare recessives, are considerably less common in Africans and Americans of African descent than in Caucasians (a grouping including Europeans, Arabs and most Indians). The very rare Tay–Sachs disease, again caused by a single recessive allele, is most common in Ashkenazi Jews. The allele frequency is about 0·015 as compared with about 0·002 in Sephardic Jews and non-Jews. These differences presumably mean that the present incidence of the disease in question is due to rather few mutations which happen to have occurred by chance in some populations and not in others. As we have already seen, the 'sickle' allele of the β-globin gene seems to have arisen independently (but rarely) in a number of ethnic groups (Southern Europeans, Africans, Indians) and to have achieved detectable frequency only in areas in which it is, or has been, favoured by selection.

19.6   Summary and perspectives

There are several reasons for being interested in human genetics. Firstly, it has applications in medicine, mainly in genetic counselling but also in connection with therapy. Knowledge of the nature of particular hereditary diseases already enables some of them to be treated nutritionally; in the future, therapy at the DNA level may become feasible even though it may be doubted whether it will ever be applicable in more than a few special situations.

Secondly, man is becoming rather good material for the investigation of questions of gene structure and organization relevant to eukaryote genetics in general. Human genetics is, of course, greatly restricted by the lack of controlled crosses; one has to draw what conclusions one can from ready-made pedigrees and population statistics. On the other hand, the large scale of modern human population surveys means that the human geneticist has available a great wealth of information on allelic variation. As a consequence, the human species is one of those most frequently cited in discussions of genetic polymorphism and its significance. Furthermore, the development of hybrid cell techniques, together with the availability of so much naturally occurring variation (as well as new mutations selected in cell culture), has made mapping of genes to chromosomes relatively rapid and easy. With all chromosomes distinguishable by banding techniques and, to an increasing extent, by molecular landmarks, the human genome is becoming more thoroughly mapped than that of any other mammal. The molecular cloning of human genes and the complete determination of their DNA sequences, introns and all, is also well advanced.

Human genetics also raises in acute form questions of a more general and philosophical nature.

The main programme of genetics is the explanation of the phenotype in terms of developmental processes stemming ultimately from the genotype—that is to say from the DNA. As we have seen, for some aspects of the phenotype the description of the chain of causation is complete or very nearly so. Some people are very resistant to the idea that all the unique characteristics of the human species are to be explained in this way and do not like to think that they are determined by their DNA. It is not clear what else they would prefer to be determined by. It hardly seems more in keeping with the dignity of man to be moulded like putty by the environment, still less to be battered to and fro by truly indeterminate processes at the atomic level. If there is a problem in accepting genetic determinism, it is the problem of determinism in general, which is in the realm of metaphysics not of science; that is to say, there is no way in which it could be resolved by objective observation.

So, what can we say about the human genome and its connection with the unique qualities of man? So far, it has to be admitted, the answer is discouraging. What we see at the DNA level is a number of identifiable genes, very similar in organization to those of other mammals and, so far as detailed comparisons have been made, almost identical in nucleotide sequence to homologous genes in our relatives the apes. Beyond and between these we see a typical higher eukaryote organization or lack of it—long tracts of apparently meaningless sequence, much of it repetitive, with, here and there in the 'pseudogenes', signs of function long since extinct. Some of the sequences may even be parasitic—something that we have to adjust to and tolerate, as if they were a part of the environment rather than of 'ourselves'. The prospect is not at present very inspiring. As a scientifically informed clergyman was heard to remark in a recent radio talk: 'We do not find the soul entwined in the rungs of the double helix.'

The detailed examination of the genomes of man and higher animals is still, of course, in its early stages, and it may be that some insight may in time be obtained into the genetic determination of the special characteristics of the human brain. Whether it will ever be within the power of the brain to achieve a complete understanding of itself may well be questioned, but we can expect that at least some progress will be made in this area. But it seems, in any case, quite unlikely that scrutiny of DNA sequence, or of the structures and functions of macro-molecules that can be related to specific individual genes, will help very much. It seems much more reasonable to suppose that all the specifically human modes of development at the mental level occur in the area of epigenetics—that obscure, complex and self-regulating network of interactions which comes between the primary gene products and the ultimate phenotype. Genetic determination certainly exists in the sense that potentially definable mutations in the genes bring about definable changes in the phenotype, no aspect of which is immune to mutational effects. But to trace all the sequences of cause and effect through the epigenetic maze is another matter, and it is not clear that it will ever be possible.

Chapter 19   Selected Further Reading

Bodmer W. F. and Cavalli-Sforza L. L. (1976) *Genetics, Evolution and Man*, 782 pp. San Francisco, W. H. Freeman.

Emery A. E. H. (1971) *Elements of Medical Genetics*, 2nd ed., 222 pp. Edinburgh, E. & S. Livingstone.

McKusick V. A. (1969) *Human Genetics*, 2nd ed., 220 pp., Englewood Cliffs, New Jersey, Prentice-Hall.

McKusick V. A. (1980) The anatomy of the human genome. *J. Hered.* **71**, 370–391.

Stern C. (1973) *Principles of Human Genetics*, 3rd ed., 890 pp. San Francisco, W. H. Freeman.

Problems

19.1 What degree of reassurance might be given by a genetic counsellor in the following situations?

a. A woman's uncle (mother's brother) suffered from haemophilia. The woman is now pregnant and wants to know the risk to her child. How would the situation be altered if the affected uncle were the father's brother?

b. A woman's first cousin died of the effects of fibrocystic disease. If the woman were to marry and have children what would be the risk that the children would have the disease?

c. A man's elder sister has been showing neurological symptoms now diagnosed as those of Huntington's chorea. He is 40 years of age and wants to know the risk both to himself and to any further children that he might have.

19.2 Harelip occurs with a frequency of about 0·1 per cent at birth. The concordance between identical twins with respect to the condition is about 50 per cent, between ordinary sibs 3·5 per cent, between first cousins 0·7 per cent and between second cousins 0·3 per cent. These concordance values are not affected by sex. What mode of inheritance, if any, is suggested by these data?

19.3 In a classic study on twins, Newman, Freeman and Holzinger (*Twins: A Study of Heredity and Environment*, 1937, University of Chicago Press) determined correlations with respect to height, head length, intellectual achievement (using a standard test) and IQ (measured by two different tests) between (*a*) monozygotic twins reared together, (*b*) dizygotic twins (same sex) reared together and (*c*) monozygotic twins separated in infancy and reared in different homes. They reported the following correlations:

	Height	Head length	Achievement	IQ(1)	IQ(2)
a.	0·981	0·910	0·955	0·910	0·922
b.	0·934	0·691	0·883	0·640	0·621
c.	0·969	0·917	0·507	0·679	0·727

What conclusions may be drawn from these data concerning the degree of heritability of each of these measures?

19.4 Spencer et al. (*Nature*, 1964, **201**, 299–300) identified three electrophoretically distinct forms of acid phosphatase, A, B and C, in human red cells. In a sample of 178 English adults they found the following frequencies:

A only	17	B and C	9
B only	61	A and C	5
C only	none	A, B and C	none
A and B	86		

Propose a hypothesis to explain these data. Why was the class with C only not found?

Specific References

Chapter 1

1. Margulis L. (1970) *Origin of Eukaryotic Cells.* New Haven, Yale University Press.
2. Watson J. D. and Crick F. H. C. (1953) *Nature* **171**, 964–967.
3. Meselson M. and Stahl F. W. (1968) *Proc. Natl Acad. Sci. USA* **44**, 671–682.
4. Prescott D. M. and Kuempel P. L. (1973) *Methods Cell Biol.* **7**, 147.
5. Matsumoto K. et al. (1974) *J. Cell Biol.* **63**, 146–159.
6. Kriegstein H. G. and Hogness D. S. (1974) *Proc. Natl Acad. Sci. USA* **71**, 135.
7. Britten R. J. and Kohne D. E. (1968) *Science* **161**, 529–531.
8. Petes T. D., Byers B. and Fangman W. L. (1973) *Proc. Natl Acad. Sci. USA* **70**, 3072–3076.
9. Kavanoff R. and Zim B. H. (1973) *Chromosoma* **41**, 1–27.
10. Thoma F. and Koller Th. (1977) *Cell* **12**, 101–107.
11. Marsden M. P. F. and Laemmli U. K. (1979) *Cell* **17**, 849–858.
12. Paulson J. R. and Laemmli U. K. (1977) *Cell* **12**, 817–828.
13. Ryter A. (1968) *Bacteriol. Rev.* **32**, 39–54.
14. Klug A. et al. (1980) *Nature* **287**, 509–516.
15. Seidman M. M., Levine A. J. and Weintraub H. (1979) *Cell* **18**, 439–449.
16. Bak A. L., Zeuthen J. and Crick F. H. C. (1977) *Proc. Natl Acad. Sci. USA* **74**, 1595–1599.
17. Peacock W. J. et al. (1977) *Cold Spring Harbor Symp. Quant. Biol.* **42**, 1121–1135.
18. Wolff S. and Perry P. (1975) *Exp. Cell Res.* **93**, 23–30.
19. Jones K. W. (1970) *Nature* **225**, 912–915.
20. Comings D. E. (1978) *Annu. Rev. Genet.* **12**, 25–46.
21. McKnight S. L. and Miller O. L. (1979) *Cell* **17**, 551–563.
22. Laird C. D. and Chooi W. Y. (1976) *Chromosoma* **58**, 193–218.
23. Rich A. and Raj Bhandary U. L. (1976) *Annu. Rev. Biochem.* **45**, 805–860.
24. Crick F. H. C. (1966) *J. Mol. Biol.* **19**, 548–555.
25. Zickler D. (1970) *Chromosoma* **30**, 287–304.
26. Lesins K. et al. (1970) *Chromosoma* **30**, 111.
27. Bennett M. D. and Rees H. (1969) *Chromosoma* **27**, 235.
28. Gall J. D. et al. (1971) *Chromosoma* **33**, 321.
29. Young H. S. and Dhaliwal S. S. (1972) *Chromosoma* **36**, 258.
30. Singleton J. R. (1953) *Am. J. Botany* **40**, 141.
31. Hewitt G. M. and John B. (1971) *Chromosoma* **34**, 308.
32. Stent G. S. and Calendar R. (1978) *Molecular Genetics—An Introductory Narrative,* 2nd ed., pp. 206–207. New York, Freeman.
33. Watson J. D. (1975) *The Molecular Biology of the Gene,* 3rd ed., p. 287. New York, W. A. Benjamin.
34. Sodd M. A. (1976) In: Fasman G. D., ed., *Handbook of Biochemistry and Molecular Biology, Nucleic Acids,* 3rd ed., Vol. 2, pp. 423–456. Cleveland, Ohio, CRC Press.
35. Sulston J. E. and Brenner S. (1974) *Genetics* **77**, 95–104.
36. Chooi W. Y. (1971) *Genetics* **68**, 195–211.
37. Fincham J. R. S., Day P. R. and Radford A. (1979) *Fungal Genetics,* 4th ed., p. 38. Oxford, Blackwell Scientific Publications.

38. King R. C. (ed.) (1974) *A Handbook of Genetics*, Vol. 2, pp. 364–368. New York, Plenum Press.
39. White M. J. D. (1973) *Animal Cytology and Evolution*, 3rd ed., pp. 39–41. Cambridge, Cambridge University Press.
40. Swanson C. P., Merz Y. and Young W. J. (1981) *Cytogenetics—the Chromosome in Division, Inheritance and Evolution*, 2nd ed. Englewood Cliffs, New Jersey, Prentice-Hall.
41. Weisbrod S. (1982) *Nature* **297**, 289–295.
42. Robson G. E. and Williams K. L. (1979) *Curr. Genet.* **2**, 229–232.
43. Callan H. G. (1982) *Proc. R. Soc. Lond. (Biol.)* **214**, 417–448.

Chapter 2

1. Westergaard M. and von Wettstein D. (1972) *Annu. Rev. Genet.* **6**, 71–110.
2. Carpenter A. T. C. (1980) *Genetics* **92**, 511–541.
3. Gillies C. B. (1979) *Genetics* **91**, 1–17.
4. Tease C. and Jones G. H. (1978) *Chromosoma* **69**, 163–178.
5. Brown S. W. and Zohary D. (1955) *Genetics* **40**, 850–873.
6. Winge Ø. (1934) *C. R. Trav. Lab. Carlsberg* **21**, 1–49.
7. Brenner S. (1974) *Genetics* **77**, 71–94.
8. Srb A. M., Owen R. D. and Edgar R. S. (1965) *General Genetics*, 2nd ed. San Francisco, W. H. Freeman.
9. White M. J. D. (1973) *Animal Cytology and Evolution*, 3rd ed. Cambridge, Cambridge University Press.

Chapter 3

1. Whitehouse H. L. K. (1957) *J. Genet.* **55**, 348.
2. Fincham J. R. S. (1971) *Using Fungi to Study Recombination*. Oxford, Clarendon Press.
3. Freeling M. (1976) *Genetics* **83**, 701–717.

Chapter 4

1. Swanson C. P., Merz T. and Young W. J. (1981) *Cytogenetics*, 2nd ed. Englewood Cliffs, New Jersey, Prentice-Hall.
2. Harrison B. J. and Fincham J. R. S. (1964) *John Innes Institute Annual Report*. Bayfordbury, Herts.
3. King R. C. (ed.) (1975) *A Handbook of Genetics*. New York, Plenum Press.
4. Coe E. H. Jr and Neuffer M. G. (1977) The genetics of corn. In: Sprague G. F., ed., *Corn and Corn Improvement*, 2nd ed. Madison, Wisconsin, Am. Soc. Agron.
5. Fincham J. R. S., Day P. R. and Radford A. (1979) *Fungal Genetics*. Oxford, Blackwell Scientific Publications.

Chapter 5

1. Lewis E. B. (1952) *Proc. Natl Acad. Sci. USA* **38**, 953–961.
2. Hilliker A. J. and Chovnick A. (1981) *Genet. Res.* **38**, 281–296.
3. Case M. E. and Giles N. H. (1964) *Genetics* **49**, 529–540.
4. Kitani Y. and Whitehouse H. L. K. (1974) *Genet. Res.* **24**, 229–250.
5. Wagner R. Jr and Meselson M. (1976) *Proc. Natl Acad. Sci. USA* **73**, 4135–4139.
6. Holliday R. (1964) *Genet. Res.* **5**, 282–304.
7. Meselson M. and Radding C. M. (1975) *Proc. Natl Acad. Sci. USA* **72**, 358–361.
8. Fincham J. R. S. (1967) *Genet. Res.* **9**, 49–62; Brett M. et al. (1976) *J. Mol. Biol.* **106**, 1–22.

9. Sherman F., Jackson M., Liebman, S. W. et al. (1975) *Genetics* **81**, 51–73.
10. Bell L. and Byers B. (1979) *Proc. Natl Acad. Sci. USA* **76**, 3445–3449.

Chapter 6

1. Perkins D. D. (1974) *Genetics* **77**, 459–489.
2. Roberts P. A. (1976) In: Ashburner M. and Nowitski E., ed., *The Genetics and Biology of Drosophila,* Vol. 1a. New York, Academic Press.
3. Clausen R. E. (1932) *Sven. Bot. Tidskr.* **26**, 123–136.
4. Melchers G., Sacristan M. D. and Holder A. A. (1978) *Carlsberg Res. Commun.* **43**, 203–218.
5. Hassold T. et al. (1980) *Ann. Hum. Genet.* **44**, 151–178.
6. Cleland R. E. (1929) *Z. Vererbungsl.* **51**, 126–145.
7. Rick C. M. and Barton D. W. (1954) *Genetics* **39**, 640–666.
8. Sears E. R. (1953) *Am. Naturalist* **87**, 245–252.
9. Fincham J. R. S., Day P. R. and Radford A. (1979) *Fungal Genetics,* 4th ed. Oxford, Blackwell Scientific Publications.
10. Manton I. (1950) *Cytology and Evolution in the Pteridophyta.* Cambridge, Cambridge University Press.

Chapter 7

1. Jinks J. L. (1959) *J. Gen. Microbiol.* **21**, 397–409
2. Mahler H. R. et al. (1978) In: Bacila M., Horecker B. L. and Stoppani H. O. M., ed., *Biochemistry and Genetics of Yeasts,* pp. 513–547. New York, Academic Press.
3. Morimoto R. et al. (1978) *Mol. Gen. Genet.* **163**, 241–255.
4. Lewin F., Morimoto R. and Rabinowitz M. (1978) *Mol. Gen. Genet.* **163**, 257–275.
5. Wolf K. et al. (1973) *Mol. Gen. Genet.* **125**, 53–90.
6. Southern E. M. (1975) *J. Mol. Biol.* **98**, 503–517.
7. Dujon B. (1980) *Cell* **20**, 173–183.
8. Van Ommen G.-J. et al. (1980) *Cell* **20**, 173–183.
9. Anderson S. et al. (1981) *Nature* **290**, 457–470.
10. Kemble R. J. and Bedbrook J. R. (1980) *Nature* **284**, 565–566.
11. Sakano K., Kung S. D. and Wildman S. G. (1974) *Mol. Gen. Genet.* **130**, 91–97.
12. Bedbrook J. R. et al. (1977) *Cell* **11**, 739–749.
13. Bedbrook J. R. et al. (1979) *J. Biol. Chem.* **254**, 405–410.
14. Harris E. H. et al. (1977) *Mol. Gen. Genet.* **163**, 257–275.
15. Beggs J. (1978) *Nature* **275**, 104–109.
16. Herring A. J. and Bevan E. A. (1977) *Nature* **268**, 464–465.
17. Livingston D. M. (1977) *Genetics* **86**, 73–84.
18. Sonneborn T. M. (1970) *Proc. R. Soc. Lond. (Biol.)* **176**, 347–366.
19. Leaver C. J. and Gray M. W. (1982) *Annu. Rev. Plant Physiol.* **33**, 373–402.
20. Struhl K. et al. (1979) *Proc. Natl Acad. Sci. USA* **76**, 1035–1039.
21. Borst P. and Grivell L. A. (1981) *Nature* **290**, 443–444.
22. Tilney-Bassett R. A. E. (1978) The plastids. In: Tilney-Bassett R. A. E., ed., *Inheritance and Genetic Behaviour of Plastids,* 2nd ed., Part 2, pp. 251–521. Amsterdam, Elsevier-North Holland.
23. Gillham W. W. (1978) *Organelle Heredity.* New York, Raven Press.
24. Bacila M., Horecker B. L. and Stoppani A. O. M. (eds) (1978) *Biochemistry and Genetics of Yeasts.* New York, Academic Press.

25. Beale G. H. (1954) *The Genetics of* Paramecium aurelia. Cambridge, Cambridge University Press.
26. Belcour L. et al. (1981) *Curr. Genet.* **3**, 13–21.
27. Pring D. R., Levings C. S. III and Conde M. F. (1980) In: Davies D. R. and Hopwood D. A., ed., *The Plant Genome*, pp. 111–120. John Innes Symposium 4.
28. Riou G. and Delain E. (1969) *Proc. Natl. Acad. Sci. USA* **62**, 210–217.

Chapter 8

1. East E. M. (1915) *Genetics* **1**, 335–429.
2. Clayton G. A., Morris J. A. and Robertson A. (1957) *J. Genet.* **55**, 131–151.
3. Falconer D. S. (1953) *J. Genet.* **51**, 470–501.
4. Mather K. and Harrison B. J. (1949) *Heredity* **3**, 1–52.
5. Thoday J. M. (1961) *Nature* **191**, 368–370.
6. Jinks J. L. and Towey P. (1976) *Heredity* **37**, 69–81.
7. Long E. O. and Dawid I. B. (1980) *Annu. Rev. Biochem.* **49**, 727–764.
8. Tartof K. D. (1975) *Annu. Rev. Genet.* **9**, 355–385.
9. Darlington C. D. and Mather K. (1949) *The Elements of Genetics*, p. 57. London, Allen & Unwin.
10. Mather K. and Jinks J. L. (1977) *Introduction to Biometrical Genetics*. London, Chapman & Hall.
11. Falconer D. S. (1981) *An Introduction to Quantitative Genetics*, 2nd ed., p. 165. London, Longmans.

Chapter 9

1. Pontecorvo G. and Käfer E. (1958) *Advan. Genet.* **9**, 71–104.
2. Käfer E. (1960) *Nature* **186**, 619–620; Käfer E. (1961) *Genetics* **46**, 1581–1609.
3. Stern C. (1936) *Genetics* **21**, 624–630.
4. Miller O. J. et al. (1971) *Science* **173**, 244.
5. Ruddle F. H. et al. (1971) *Nat. New Biol.* **232**, 69–73.
6. Douglas G. R., McAlpine P. J. and Hamerton J. L. (1973) *Proc. Natl Acad. Sci. USA* **70**, 2737–2740.
7. McKusick V. A. (1980) *J. Hered.* **71**, 370–391.
8. Rosenkraus M. and Chasin L. A. (1975) *Proc. Natl Acad. Sci. USA* **72**, 493–497.
9. Klobutcher L. A. and Ruddle F. H. (1979) *Nature* **280**, 657–660.
10. Wigler M., Axel R. et al. (1978) *Cell* **14**, 725–731.
11. Wigler M. et al. (1979) *Cell* **16**, 777–785.
12. Constanti F. and Lacy E. (1981) *Nature* **294**, 92–94.
13. Wagner T. E. et al. (1981) *Proc. Natl Acad. Sci. USA* **78**, 6376–6380.
14. Beggs J. (1978) *Nature* **275**, 104–109.

Chapter 10

1. Kontomichalou P., Mitani M. and Clowes R. C. (1970) *J. Bacteriol.* **104**, 34–44.
2. Sharp P. A. et al. (1973) *J. Mol. Biol.* **75**, 235–255.
3. Hu S., Ohtsubo E. and Davidson N. (1975) *J. Bacteriol.* **122**, 748–763.
4. Cunningham R. P., Shibata T., DasGupta C. et al. (1979) *Nature* **281**, 191–195.
5. Gill R. E. et al. (1979) *Nature* **282**, 797–801.
6. Chou J. et al. (1979) *Nature* **282**, 801–806.
7. Alton N. K. and Vapnek D. (1979) *Nature* **282**, 864–869.

8. Ohtsubo H. and Ohtsubo E. (1978) *Proc. Natl Acad. Sci. USA* **75**, 615–619.
9. Heffron F. et al. (1981) *Cold Spring Harbor Symp. Quant. Biol.* **45**, 259–268.
10. Sherratt D. et al. (1981) *Cold Spring Harbor Symp. Quant. Biol.* **45**, 275–281.
11. Heynecker H. L. et al. (1976) *Nature* **263**, 748–752.
12. Jacob F. and Wollman E. (1961) *Sexuality and the Genetics of Bacteria.* New York, Academic Press.
13. Broda P. W. (1979) *Plasmids.* Oxford, Freeman.
14. Bukhari A. I., Shapiro J. A. and Adhya S. L. (ed.) (1977) *DNA Insertion Elements, Plasmids and Episomes.* New York, Cold Spring Harbor Laboratory.

Chapter 11

1. Doermann A. H. (1973) *Annu. Rev. Genet.* **7**, 325–341.
2. Wu R. and Taylor E. (1971) *J. Mol. Biol.* **57**, 491–511.
3. Doermann A. H. and Hill M. B. (1953) *Genetics* **38**, 79–90.
4. Broker T. R. and Lehman I. R. (1971) *J. Mol. Biol.* **60**, 131–149.
5. Streisinger G., Edgar R. S. and Denhardt G. H. (1964) *Proc. Natl Acad. Sci. USA* **51**, 775–779.
6. Benzer S. (1961) *Proc. Natl Acad. Sci. USA* **47**, 403–415.
7. Benzer S. (1957) In: McElroy W. D. and Glass B., ed., *The Chemical Basis of Heredity.* Baltimore, Johns Hopkins Press.
8. Rothman J. (1965) *J. Mol. Biol.* **12**, 892–912.
9. Bidwell K. and Landy A. (1979) *Cell* **16**, 397–406.
10. Thomas M., Cameron J. R. and Davis R. W. (1974) *Proc. Natl Acad. Sci. USA* **71**, 4579–4583.
11. Leder P. et al. (1977) *Science* **196**, 175–177.
12. Gottesman M. E. and Weisberg R. A. (1971) In: Hershey A. D., ed., *The Bacteriophage Lambda.* New York, Cold Spring Harbor Laboratory.
13. Hayes W. (1968) *The Genetics of Bacteria and their Viruses*, 2nd ed. Oxford. Blackwell Scientific Publications.

Chapter 12

1. Rucknaegel D. L. and Winter W. P. (1974) *Ann. N.Y. Acad. Sci.* **241**, 82–84.
2. Schaar R. L., Muzyczka N. and Bessman M. J. (1973) *Genetics* **73**, 137–140.
3. Glickman B., van den Elsen P. and Radman M. (1978) *Mol. Gen. Genet.* **163**, 307–312.
4. Coulondre C., Miller J. H., Farabaugh P. J. et al. (1978) *Nature* **274**, 775–780.
5. Duncan B. K. and Miller J. H. (1980) *Nature* **287**, 560–563.
6. Streisinger G. et al. (1966) *Cold Spring Harbor Symp. Quant. Biol.* **31**, 77–84.
7. Tartof K. D. (1974) *Proc. Natl Acad. Sci. USA* **71**, 1272–1276.
8. Ghosal D. and Saedler H. (1978) *Nature* **275**, 611–617.
9. Heseltine W. A. et al. (1980) *Nature* **285**, 634–641.
10. Gottesman S. (1981) *Cell* **23**, 1–2.
11. Cooper P. K. and Hanawalt P. (1972) *Proc. Natl Acad. Sci. USA* **69**, 1156–1160.
12. Müller W. et al. (1978) *J. Mol. Biol.* **124**, 343–358.
13. Drake J. W. (1970) *The Molecular Basis of Mutation.* San Francisco, Holden-Day.

14. Auerbach C. (1976) *Mutation Research*. London, Chapman & Hall.
15. Russell W. L., Russell L. B. and Kelly E. M. (1958) *Science* **128**, 1546–1550.
16. Cleaver J. E. and Bootsma D. (1975) *Annu. Rev. Genet.* **9**, 19–38.

Chapter 13

1. O'Donnell J. et al. (1975) *Genetics* **79**, 78–83.
2. Smith J. D. et al. (1970) *J. Mol. Biol.* **54**, 1–14.
3. Sherman F. et al. (1974) *Genetics* **77**, 255–284.
4. Reissig J. L. (1963) *J. Gen. Microbiol.* **30**, 317–325.
5. Yanofsky C. et al. (1966) *Proc. Natl Acad. Sci. USA* **51**, 266–272.
6. Thatcher D. R. (1980) *Biochem. J.* **187**, 875–886.
7. Weatherall D. J. (1978) *Ciba Found. Symp.* **66**, 135–174.
8. Stewart J. W. and Sherman F. (1972) *J. Mol. Biol.* **68**, 83–96; Stewart J. W., Sherman F., Jackson M. et al. (1972) *J. Mol. Biol.* **78**, 169–184.
9. Stewart J. W., Sherman F., Shipman N. A. et al. (1971) *J. Biol. Chem.* **246**, 7429–7445.
10. Sherman J. W. and Stewart J. W. (1974) In: *Biochemistry of Gene Expression in Higher Organisms*, pp. 55–86. Sydney, Austral. & N.Z. Book Co.
11. Smith M. et al. (1979) *Cell* **16**, 753–761.
12. Seidman J. G. et al. (1974) *J. Mol. Biol.* **90**, 677–689.
13. Altman S. et al. (1971) *J. Mol. Biol.* **56**, 195–197.
14. Piper P. W. (1978) *J. Mol. Biol.* **122**, 217–235.
15. Piper P. W. et al. (1976) *Nature* **262**, 757–761.
16. Crick F. H. C. (1966) *J. Mol. Biol.* **19**, 548–555.
17. Jones S. A., Arst H. N. and MacDonald D. W. (1981) *Curr. Genet.* **3**, 49–56.
18. Arst H. and Cove D. J. (1973) *Mol. Gen. Genet.* **126**, 111–141.
19. Hynes M. J. (1975) *Nature* **253**, 210–212.
20. Grove G. and Marzluf G. (1981) *J. Biol. Chem.* **256**, 463–470.
21. Paigen K. (1979) *Annu. Rev. Genet.* **13**, 417–461.
22. Fincham J. R. S. and Harrison B. J. (1967) *Heredity* **22**, 211–224.
23. Gielow L. et al. (1971) *Genetics* **69**, 289–302.
24. Sheppard D. E. et al. (1971) *J. Bacteriol.* **139**, 1085–1088.
25. Gunsalus R. P. and Yanofsky C. (1980) *Proc. Natl Acad. Sci. USA* **77**, 7117–7121.
26. Bennett G. N. and Yanofsky C. (1978) *J. Mol. Biol.* **121**, 179–192.
27. Zurawski G. et al. (1978) *Proc. Natl Acad. Sci. USA* **75**, 5938–5942.
28. Fincham J. R. S., Day P. R. and Radford A. (1979) *Fungal Genetics*, 4th ed., p. 346. Oxford, Blackwell Scientific Publications.
29. Fincham J. R. S. (1976) *Microbial and Molecular Genetics*, 2nd ed., p. 79. London, Hodder & Stoughton.
30. Altman S. (1979) In: Celis J. E. and Smith J. D., ed., *Nonsense Mutations and tRNA Suppressors*. New York, Academic Press.
31. Beckwith J. R. and Zipser D. (ed.) (1970) *The Lactose Operon*. New York, Cold Spring Harbor Laboratory.
32. Callan H. G. (1982) *Proc. R. Soc. Lond. (Biol.)* **214**, 417–448.
33. Schwartz D. (1963) *Structure and Function of the Genetic Material.* Erwin Bauer Gedachtnisvorlesungen III, pp. 201–203. Berlin, Akademie Verlag.

Chapter 14

1. Evans M. J. and Lingrel J. B. (1969) *Biochemistry* **8**, 3000–3005.

2. Maniatis T. et al. (1976) *Cell* **8**, 163–182.
3. Itakura K. and Riggs A. D. (1980) *Science* **209**, 1392–1396.
4. Maniatis T. et al. (1978) *Cell* **15**, 687–704.
5. Montgomery D. L., Hall B. D., Gillam S. et al. (1978) *Cell* **14**, 673–680.
6. Jeffries A. J. and Flavell R. A. (1977) *Cell* **12**, 1097–1108.
7. Grunstein M. and Hogness D. (1975) *Proc. Natl Acad. Sci. USA* **72**, 3961–3965.
8. Benton W. D. and Davis R. W. (1977) *Science* **196**, 180–182.
9. Ratzkin B. and Carbon J. (1977) *Proc. Natl Acad. Sci. USA* **74**, 487–491.
10. Paterson B. M. et al. (1977) *Proc. Natl Acad. Sci. USA* **74**, 4370–4374.
11. Tilghman S. M. et al. (1978) *Proc. Natl Acad. Sci. USA* **75**, 725–729.
12. Dugaiczyk A. et al. (1979) *Proc. Natl Acad. Sci. USA* **76**, 2253–2257.
13. Pribnow D. (1975) *Proc. Natl Acad. Sci. USA* **72**, 784–788.
14. Holland J. P. and Holland M. J. (1979) *J. Biol. Chem.* **254**, 9839–9845.
15. Gallwitz D. and Sures I. (1980) *Proc. Natl Acad. Sci. USA* **77**, 2546–2550.
16. Smith M. et al. (1979) *Cell* **16**, 753–761.
17. Montgomery D. et al. (1980) *Proc. Natl Acad. Sci. USA* **77**, 541–545.
18. Schaffner W. et al. (1978) *Cell* **14**, 655–671.
19. Yamawaki-Kataoka Y. (1980) *Nature* **283**, 783–789.
20. Rosenberg M. and Court D. (1979) *Annu. Rev. Genet.* **13**, 319–353.
21. Goeddel D. V. et al. (1978) *Proc. Natl Acad. Sci. USA* **75**, 3578–3582.
22. Breathach R. and Chambon P. A. (1981) *Annu. Rev. Biochem.* **50**, 349–383.
23. Corden J. et al. (1980) *Science* **209**, 1406–1414.
24. Shine J. and Delgarno L. (1974) *Proc. Natl Acad. Sci. USA* **71**, 1342–1346.
25. Lee F., Bertrand K., Bennett G. et al. (1978) *J. Mol. Biol.* **121**, 193–217.
26. Keller E. B. and Calvo J. M. (1979) *Proc. Natl Acad. Sci. USA* **76**, 6186–6190.
27. Lerner M. R. et al. (1980) *Nature* **283**, 220–224.
28. Hagenbüchle O. et al. (1981) *Nature* **289**, 643–646.
29. Edge M. D. et al. (1981) *Nature* **292**, 756–762.
30. Proudfoot N. J. (1976) *J. Mol. Biol.* **107**, 491–525.
31. Busslinger M. et al. (1981) *Cell* **27**, 289–298.
32. Ohtsubo H. and Ohtsubo E. (1978) *Proc. Natl Acad. Sci. USA* **75**, 615–619.
33. Maxam A. M. and Gilbert W. (1980) *Methods Enzymol.* **65**, 499–560.
34. Sanger F. et al. (1977) *Proc. Natl Acad. Sci. USA* **74**, 5463–5467.
35. Galas D. J. and Schmitz A. (1978) *Nucleic Acids Res.* **5**, 3157–3170.
36. Nordheim A. et al. (1981) *Nature* **294**, 417–422.
37. Weisbrod S. (1982) *Nature* **297**, 289–295.
38. Cantor G. R. (1981) *Cell* **25**, 293–295.

Chapter 15

1. Post L. E. et al. (1979) *Proc. Natl Acad. Sci. USA* **76**, 1697–1701.
2. Post L. E. and Nomura M. (1980) *J. Biol. Chem.* **255**, 4660–4666.
3. Wood W. A. and Revel H. (1976) *Bacteriol. Rev.* **40**, 847–868.
4. Botstein D. and Herskowitz I. (1974) *Nature* **251**, 584–589.

5. Abraham J. et al. (1980) *Proc. Natl Acad. Sci. USA* **77**, 2477–2481; Sanger F. et al. (1977) *Nature* **265**, 687–698; Fiddes J. C. and Godson G. N. (1979) *J. Mol. Biol.* **133**, 19–43.
6. Fiers W. et al. (1976) *Nature* **260**, 500–507.
7. Bernardi A. and Spahr P.-F. (1972) *Proc. Natl Acad. Sci. USA* **69**, 3033–3037.
8. Gralla J., Steitz J. and Crothers D. M. (1974) *Nature* **248**, 204–208.
9. Fiers W. et al. (1978) *Nature* **273**, 113–120.
10. Judd B. H. et al. (1972) *Genetics* **71**, 139–156.
11. Ashburner M. and Bonner J. J. (1979) *Cell* **17**, 241–254.
12. Bonner J. J. and Pardue M.-L. (1976) *Cell* **8**, 43–50.
13. Spradling A. et al. (1975) *Cell* **14**, 901–919.
14. Ish-Horowicz D. and Pinchin S. M. (1980) *J. Mol. Biol.* **142**, 231–245.
15. Lis J. T., Prestidge L. and Hogness D. S. (1978) *Cell* **14**, 901–919.
16. Nunberg J. H. et al. (1980) *Cell* **19**, 355–364.
17. Smith M. et al. (1979) *Cell* **16**, 753–761.
18. Holland J. P. and Holland M. J. (1979) *J. Biol. Chem.* **254**, 9839–9845.
19. Gallwitz D. and Sures I. (1980) *Proc. Natl Acad. Sci. USA* **77**, 2546–2550.
20. Nishioka Y. and Leder P. (1979) *Cell* **18**, 875–882.
21. Lauer J. et al. (1980) *Cell* **20**, 119–130.
22. MacDonald R. L. et al. (1980) *Nature* **287**, 117–122.
23. Dugaiczyk A. et al. (1979) *Proc. Natl Acad. Sci. USA* **76**, 2255–2258.
24. Cochet M. et al. (1980) *Nature* **282**, 567–574.
25. Catterall J. F. et al. (1979) *Nature* **278**, 323–327.
26. Goeddel D. V. et al. (1980) *Nature* **290**, 20–26.
27. Fritsch E. F. et al. (1980) *Cell* **19**, 959–972.
28. Kaufman R. E. et al. (1980) *Proc. Natl Acad. Sci. USA* **77**, 4229–4233.
29. Pressley L. et al. (1980) *Proc. Natl Acad. Sci. USA* **77**, 3586–3589.
30. Proudfoot N. and Maniatis T. (1980) *Cell* **21**, 537–544.
31. Schaffner W. et al. (1978) *Cell* **14**, 655–671.
32. Weatherall D. (1979) *Ciba Found. Symp.* **66**, 47–74.
33. Leder A. et al. (1981) *Nature* **293**, 196–200.
34. Church G. M. and Gilbert W. (1980) In: Scott W. A., Joseph D. R., Schultz J. et al., ed., *Mobilization and Reassembly of Genetic Information.* New York, Academic Press.
35. Church G. M., Slonimski P. P. and Gilbert W. (1979) *Cell* **18**, 1209–1215.
36. Eibel H. et al. (1980) *Cold Spring Harbor Symp. Quant. Biol.* **45**, 609–617.
37. Levis R. et al. (1980) *Cell* **21**, 581–588.
38. Potter S. S. et al. (1979) *Cell* **17**, 429–439.
39. Potter S. S. et al. (1980) *Cell* **20**, 639–647.
39a. Potter S. S. (1982) *Nature* **297**, 201–204.
40. Gehring W. and Paro R. (1980) *Cell* **19**, 897–904.
41. Wensinck P. C. et al. (1980) *Cell* **18**, 1231–1246.
42. Strobel E. et al. (1979) *Cell* **17**, 429–439.
43. Fincham J. R. S. and Sastry G. R. K. (1974) *Annu. Rev. Genet.* **8**, 15–50.
44. Houck C. M., Rinehart F. P. and Schmid C. W. (1979) *J. Mol. Biol.* **132**, 289–306.
45. Orgel L. and Crick F. H. C. (1980) *Nature* **284**, 604–607.
46. Doolittle W. F. and Sapienza C. (1980) *Nature* **284**, 601–604.

47. Dover G. A. (1980) *Nature* **285**, 618–620.
48. Bostock C. J. and Clark E. M. (1980) *Cell* **19**, 709–715.
49. Leigh-Brown A. J. and Ish-Horowicz D. (1981) *Nature* **290**, 677–682.
50. Slightom J. L., Blechl A. E. and Smithies O. (1981) *Cell* **21**, 627–638.
51. Jackson J. A. and Fink G. R. (1981) *Nature* **292**, 306–311.
52. Thompson J. N. Jr and Woodruff R. C. (1978) *Nature* **274**, 317–321.
53. Jagadeeswaran P., Forget B. G. and Weissman S. M. (1981) *Cell* **26**, 141–142.
54. Jones R. N. (1977) *Int. Rev. Cytol.* **40**, 1–100.
55. Spradling A. C. and Rubin G. M. (1982) *Science* **218**, 341–347.

Chapter 16

1. Deppe V. et al. (1978) *Proc. Natl Acad. Sci. USA* **75**, 376–380.
2. Laufer J. S., Bazzicalupo P. and Wood W. B. (1980 *Cell* **19**, 569–577.
3. Kelly S. J. (1977) *J. Exp. Zool.* **167**, 365–376.
4. Mintz B. (1967) *Proc. Natl Acad. Sci. USA* **58**, 344–351.
5. Hotta Y. and Benzer S. (1972) *Nature* **240**, 527–535.
6. Poodry C. A. et al. (1980) In: Wright T. R. F. and Ashburner M., ed., *Biology and Genetics of* Drosophila, Vol. 2d, p. 408. New York, Academic Press.
7. Postlethwaite J. H. and Schneiderman H. A. (1973) *Annu. Rev. Genet.* **7**, 381–434.
8. Lawrence P. A. (1981) *J. Embryol. Exp. Morph.* **64**, 321–323.
9. Garcia-Bellido A. (1975) *Ciba Found. Symp.* **29**, 161–178.
10. Bryant P. J. (1975) *J. Exp. Zool.* **193**, 49–77.
11. Kauffman S. A. et al. (1978) *Science* **199**, 259–270.
12. Morata G. and Lawrence P. A. (1977) *Nature* **265**, 211–216.
13. Lewis E. B. (1978) *Nature* **276**, 565–570.
14. Johnson M. H. and Ziomek C. A. (1981) *Cell* **24**, 71–80.
15. Marx J. L. (1981) *Science* **213**, 1485–1487.

Chapter 17

1. Steinbrück G. et al. (1981 *Chromosoma* **83**, 199–208.
2. van de Putte P. et al. (1980) *Nature* **286**, 218–222; van de Putte P. et al. (1980) *Cold Spring Harbor Symp. Quant. Biol.* **45**, 347–353.
3. Zieg J. et al. (1980) *Science* **196**, 170–172.
4. Silverman M. et al. (1980) *Cold Spring Harbor Symp. Quant. Biol.* **45**, 17–26.
5. Kushner P. J. et al. (1979) *Proc. Natl Acad. Sci. USA* **76**, 5264–5268.
6. Hopper A. K. and Hall B. D. (1975) *Genetics* **80**, 61–76.
7. Hicks J. et al. (1979) *Nature* **282**, 478–483.
8. Nasmyth K. A. et al. (1981) *Cold Spring Harbor Symp. Quant. Biol.* **45**, 961–982.
9. Klar A. et al. (1981) *Cold Spring Harbor Symp. Quant. Biol.* **45**, 983–990.
10. Hoeijmakers J. H. J. et al. (1980) *Nature* **284**, 78–80.
11. Lewin B. (1974) *Gene Expression*, Vol. 1, p. 358.
12. Ptashne M. et al. (1980) *Cell* **19**, 1–11.
13. Roth G. E. and Moritz K. B. (1981) *Chromosoma* **83**, 169–190.
14. King T. J. and Briggs R. (1956) *Cold Spring Harbor Symp. Quant. Biol.* **21**, 271–290.
15. Gurdon J. B. and Uehlinger V. (1966) *Nature* **210**. 1240–1241.
16. Milstein C. (1981) *Proc. R. Soc. Lond. (Biol.)* **211**, 393–412; **21**, 271–289.
17. Amzel L. M. and Poljak R. J. (1979) *Annu. Rev. Biochem.* **48**, 961–997.

18. Davis M. M. et al. (1980) *Nature* **283**, 733–739.
19. Early P. et al. (1980) *Cell* **19**, 981–992.
20. Maki R. et al. (1980) *Proc. Natl Acad. Sci. USA* **77**, 2138–2142.
21. Kataoka T. et al. (1980) *Proc. Natl Acad. Sci. USA* **77**, 919–923.
22. Hood L. et al. (1980) *Cold Spring Harbor Symp. Quant. Biol.* **45**, 887–898.
23. Sakano H. et al. (1979) *Nature* **280**, 288–294.
24. Rowley J. D. (1980) *Annu. Rev. Genet.* **14**, 17–40.
25. Cruciger J. Q. V. et al. (1976) *Cytogenet. Cell Genet.* **17**, 231–235.
26. Jonasson J. et al. (1977) *J. Cell Sci.* **24**, 255–263.
27. Lania L. et al. (1979) *Cold Spring Harbor Symp. Quant. Biol.* **44**, 597–603.
28. Swanstrom R. et al. (1981) *Proc. Natl Acad. Sci. USA* **78**, 124–128.
29. Hayward W. S. et al. (1981) *Nature* **290**, 475–480.
30. Demerec M. (1940) *Genetics* **25**, 618–627.
31. Cattanach B. (1974) *Genet. Res.* **23**, 291–306.
32. Holliday R. and Pugh J. E. (1975) *Science* **187**, 226.
33. Razin A. and Riggs A. D. (1980) *Science* **210**, 604–610.
34. Wigler M. et al. (1981) *Cell* **24**, 33–40.
35. Shen C.-K. J. and Maniatis T. (1980) *Proc. Natl Acad. Sci. USA* **77**, 6634–6638.
36. Elgin S. L. R. (1981) *Cell* **27**, 413–415.
37. McGhee J. D. et al. (1981) *Cell* **27**, 45–55.
38. Sakano H. et al. (1981) *Nature* **290**, 562–565.
39. Sonneborn T. M. (1977) *Annu. Rev. Genet.* **11**, 349–367.
40. Echols H. (1980) In: Leighton T. and Loomis W. F., ed., *The Molecular Genetics of Development*. New York, Academic Press.
41. White M. J. D. (1973) *Animal Cytology and Evolution,* 2nd ed. Cambridge, Cambridge University Press.
42. Shih C. and Weinberg R. A. (1982) *Cell* **29**, 161–169.
43. Parade L. F., Tabin C. J., Shih C. et al. (1982) *Nature* **297**, 474–478.
44. Honjo T. et al. (1980) *Cold Spring Harbor Symp. Quant. Biol.* **45**, 913–923.
45. Tabin C. J., Bradley S. M., Bargmann C. I. et al. (1982) *Nature* **300**, 143–149.
46. Weinberg R. A. (1982) *Cell* **30**, 3–4.

Chapter 18

1. Prakash S. et al. (1969) *Genetics* **61**, 841–858.
2. Harris H. and Hopkinson D. A. (1972) *J. Hum. Genet.* **36**, 9–20.
3. Jones J. S. (1980) *Nature* **288**, 10–11.
4. Singh R. S. et al. (1976) *Genetics* **84**, 609–629.
5. Leigh-Brown A. and Langley C. H. (1980) *Proc. Natl Acad. Sci. USA* **76**, 2381–2384.
6. McKay J. (1981) *Genet. Res.* **37**, 227–238.
7. Koehn R. K. et al. (1980) *Proc. Natl Acad. Sci. USA* **77**, 5385–5389.
8. Tosic M. and Ayala F. J. (1981) *Genetics* **97**, 679–701.
9. Auxdabehere D. (1980) *Genetics* **95**, 743–755.
10. Voelker R. A. et al. (1980) *Genetics* **94**, 961–968.
11. Jukes T. H. (1980) *Science* **210**, 973–978.
12. van den Berg J. et al. (1978) *Nature* **276**, 37–44.
13. Miyata T. and Yasunaga T. (1981) *Proc. Natl Acad. Sci. USA* **78**, 450–453.
14. Cedergren R. J. et al. (1980) *Proc. Natl Acad. Sci. USA* **77**, 2791–2795.
15. Dickerson R. E. (1972) *Sci. Am.* **226**, 58–72.

16. Fox G. E. et al. (1980) *Science* **209**, 457–463.
17. Dover G. A. (1981) In: Barigozzi C., Montalenti G. and White M. J. D., ed., *Mechanisms of Speciation*. New York, Alan Lis.
18. Miklos G. L. G. and Gill A. C. (1982) *Genet. Res.* **39**, 1–30.
19. Cohen E. H. and Bowman S. C. (1979) *Chromosoma* **73**, 327–355.
20. Rubin C. M. et al. (1980) *Nature* **284**, 372–374
21. Lewin R. (1980) *Science* **210**, 883–887.
22. Wilson A. C. (1975) *Stadler Genetics Symp.* **7**, 117–134.
23. Ikemura T. (1981) *J. Mol. Biol.* **151**, 389–409.
24. Houck C. M., Rinehart F. P. and Schmidt C. W. (1978) *J. Mol. Biol.* **132**, 289–306; Houck C. M. and Schmidt C. W. (1981) *J. Mol. Evol.* **17**, 148–155.
25. Goodman M. G. et al. (1975) *Nature* **253**, 603–608.
26. Efstradiatis A. et al. (1980) *Cell* **21**, 653–668.
27. Schmidtke J. et al. (1981) *Chromosoma* **84**, 187–194.
28. Wurster D. H. and Bernirschke K. (1967) *Cytologia* **32**, 273–285.
29. Hickey D. A. and McLean M. D. (1980) *Genet. Res.* **36**, 11–15.
30. Vigue C. L. and Johnson F. M. (1973) *Biochem Genet.* **9**, 213–227.
31. Turner B. C. and Perkins D. D. (1979) *Genetics* **93**, 587–606.
32. Peacock W. J. and Erikson J. (1965) *Genetics* **51**, 313–328.
33. Silver L. M. (1981) *Cell* **27**, 239–240.
34. Lamb B. C. and Helmi S. (1982) *Genet. Res.* **39**, 199–217.
35. Simmons M. J. and Crow J. F. (1977) *Annu. Rev. Genet.* **11**, 49–78.
36. Lewontin R. C. (1974) *The Genetic Basis of Evolutionary Change*. New York, Columbia University Press.
37. White M. J. D. (1973) *Animal Cytology and Evolution*. Cambridge, Cambridge University Press.
38. Stebbins G. L. (1950) *Variation and Evolution in Plants*. Oxford, Oxford University Press.
39. Fincham J. R. S., Day P. R. and Radford A. (1979) *Fungal Genetics*. Oxford, Blackwell Scientific Publications.
40. Fristrom J. W. and Spieth P. T. (1980) *Principles of Genetics*, p. 577. New York, Chiron Press.
41. Parkin D. T. (1979) *An Introduction to Ecological Genetics*. London, Edward Arnold.
42. Sheppard P. M. (1958) *Natural Selection and Heredity*. London, Hutchinson.
43. Penny D., Foulds L. R. and Hendy M. D. (1982) *Nature* **297**, 197–201.
44. Gray M. W. and Doolittle W. F. (1982) *Microbiol. Rev.* **46**, 1–42.
45. Schmidtke J. and Kandt I. (1981) *Chromosoma* **83**, 191–197.

Chapter 19

1. Falconer D. S. (1965; 1967) *Ann. Hum. Genet.* **29**, 51–83; **31**, 1–20.
2. Kan Y. W. and Dozy A. M. (1978) *Proc. Natl Acad. Sci. USA* **75**, 5631–5635.
3. Harrison G. A. and Owen J. T. (1964) *Ann. Hum. Genet.* **28**, 23–37.
4. Bodmer W. F. and Cavalli-Sforza L. L. (1976) *Genetics, Evolution and Man*. San Francisco, W. H. Freeman.
5. McKusick V. A. (1980) *J. Hered.* **71**, 370–391.
6. Stern C. (1973) *Principles of Human Genetics*, pp. 91, 99, 220. San Francisco, W. H. Freeman.
7. McKusick V. A. (1969) *Human Genetics*, 2nd ed. Englewood Cliffs, New Jersey, Prentice-Hall.
8. Williamson R. (1982) *Nature* **298**, 416–417.

Answers to Problems

Chapter 1

1.1

Species:	Rat	Xenopus	Drosophila	Neurospora	Saccharomyces
G+C, %:	42·9	64·3	38·8	54·1	39·8
T_m:	86·9	95·7	85·2	91·5	85·6

1.2 The pattern indicates semiconservative replication with diverging replication forks. The double-stranded DNA loop ('chromosome') completed one round of replication with one strand labelled with tritium and one strand unlabelled. In the second round the replication forks are generating, as they move apart, one duplex labelled in both strands (highest density of silver grains in the radiosensitive film) and one duplex labelled in one strand only.

1.3 About 20 per cent of the sequence is highly repetitive (C_0t 10^{-3}–10^{-4}), about 35 per cent is intermediate repetitive (C_0t about 1) and about 40 per cent single copy (C_0t about 10^3–10^4). Relative to the single copy the highly repetitive fraction is 10^6–10^7-fold reiterated and the middle-repetitive fraction 10^3–10^4-fold. The different behaviour of the 2000 and 400-base fragments shows that single-copy sequences are mostly within a few hundred bases of repetitive sequences, i.e. interspersed with them. The 'instantly' reannealing fraction probably consists of inversely repeated sequences which can fold back on themselves hairpin-wise.

1.4 *a.* In *Xenopus*, with circa 10^6 µm of *extended* DNA, circa 450 origins; in *Triturus*, with circa 7×10^6 µm, 850 origins.
b. In *Xenopus* $7·8 \times 10^4$ µm/h compared with maximum rate of $3·2 \times 10^5$ with all origins active bidirectionally; in *Triturus* $1·3 \times 10^5$ µm/h compared with maximum of $3·4 \times 10^5$.
c. In *Xenopus* it appears that about 24 per cent of the origins are active at any one time during the S-phase, and in *Triturus* about 38 per cent. In checking the calculation remember the $40 \times$ contraction factor between extended DNA and fibre.

1.5 *Neurospora* about 7 µm, about 2000-fold contracted at metaphase. *Tradescantia*, about 84 µm, 90 000-fold contracted (presumably largely achieved by looping as shown in *Fig.* 1.13c).

1.6 All of the third metaphase chromatids have one DNA strand with no BrdU (replicated in the third S-phase). Darkly stained chromatids have their second strand (conserved from the start of the experiment) also unlabelled with BrdU. Medium-stained chromatids have their second strand moderately labelled (replicated in the second S-phase). Lightly stained chromatids have their second strand heavily labelled (replicated in the first S-phase). Sister-strand exchange from the first mitosis is seen as switches of staining intensity from dark to light in single chromatids, the sister chromatid showing uniform medium staining. Exchanges in the second mitosis each show as a switch from dark to medium in one chromatid with a corresponding switch from medium to light in the sister chromatid. Exchanges in the third mitosis show as reciprocal switches dark/medium or light/medium between sister chromatids.

Chapters 2 and 3

1   The result shows that the phenotype of the seed is determined by the genotype of the next generation, not that of the maternal plant. In fact, the greater part of the pea seed consists of the seedling leaves (cotyledons), which in leguminous plants serve as a food store for the developing seedling.

2   In *Phycomyces*, meiosis must occur after the formation of zygospores and before the formation of spores in the sporangium—either in the zygospore or in the sporangium. The mycelium is haploid and the zygospore (and perhaps the young sporangium) diploid. (Colour and mating type segregate independently in the spores.)

3   In *Phytophthora*, the mycelium is diploid, segregation occurring in gametangium formation when meiosis must occur. Only the gametangia are haploid.

4   The endosperms show $1:1$ segregation with respect to every locus at which the parent tree was heterozygous. They are the products of meiosis and are haploid. Had the tree been entirely self-pollinated the embryos would have shown an approximately $1:2:1$ ratio of F, F/S and S types.

5   The two strains are homozygous for alleles r and R, R being incompletely dominant over r. The cross $RR \times rr$ (seed parent written first) gives RRr triploid endosperm, which is not quite so strongly pigmented as RRR endosperm in the pure-breeding purple strain. The reciprocal cross $rr \times RR$ gives rrR endosperms, which are paler still. Plants from either cross will produce approximately equal numbers of R and r embryo sacs and approximately equal numbers of R and r pollen grains. Since the embryo sac contributes two chromosome sets to the endosperm and the pollen grain one, the four equally probable kinds of random pollination ($R \times R$, $R \times r$, $r \times R$ and $r \times r$) will give endosperms of constitutions RRR, RRr, Rrr and rrr, respectively. Thus we expect a ratio of 1 dark purple : 1 less dark purple : 1 light purple : 1 white.

6   Assume that white is due to an allele a, recessive to A, black, and that each of the original black rabbits was A/a. Each of their black progeny then has a two-thirds chance of being A/a. The probability of black brother and sister both being A/a is $\frac{2}{3} \times \frac{2}{3} = \frac{4}{9}$. With a litter size of 8, a $A/a \times A/a$ mating has a chance of $(\frac{3}{4})^8 = \dfrac{6561}{65\,536} \sim \frac{1}{10}$ of producing no white offspring (each has a $\frac{3}{4}$ chance of being black). So each black $\times$ black pair has a $\frac{4}{9} \times \frac{9}{10} = \frac{4}{10}$ chance of having at least one white offspring, and a $\frac{6}{10}$ chance of having none. With three such pairs, the chance of there being no white progeny from any of them is $(0 \cdot 6)^3 = 0 \cdot 216$.

7   The free reassortment of the two segregating differences during male meiosis shows that the alleles responsible are on two different chromosome pairs (there is no crossing-over in the *Drosophila* male—cf. p.71). We expect a $9:3:3:1$ ratio of wild, dumpy, hairy and dumpy-hairy phenotypes. Crosses of F_2 dumpy $\times$ hairy could be of four different types:
(1) $dy/dy\ +/+\ \times\ +/+\ h/h$, probability 1/9, giving all wild phenotype among progeny.

(2) $dy/dy\ -/h\ \times\ +/+\ h/h$, probability 2/9, giving 1:1 wild and hairy among progeny.

(3) $dy/dy\ +/+\ \times\ dy/+\ h/h$, probability 2/9, giving 1:1 wild and dumpy among progeny.

(4) $dy/dy\ +/h\ \times\ dy/+\ h/h$, probability 4/9, giving 1:1:1:1 wild, dumpy, hairy and dumpy-hairy among progeny.

8 C/c determines coloured versus white; P/p determines purple versus red, if coloured at all; p and c are both recessive. F_1 purple was $C/c\ P/p$. The first cross between F_2 plants was $C/c\ P/p \times c/c\ p/p$; the second was $C/c\ P/p \times c/c\ P/p$. (You can check the observed numbers against the expected ratios by the χ^2 test.)

9 The v mutation is sex linked, so that all the F_1 females are v^+/v and all the F_1 males are v/Y. Thus in the F_2 there is a 1:1 segregation of non-vermilion and vermilion eye and a 3:1 segregation of normal and vestigial wing in both sexes, the latter being the normal F_2 ratio when there is no sex linkage. Hence the 3:3:1:1 ratio.

Had the original cross been reversed, with vestigial females and vermilion males, all of the F_1 flies would have been phenotypically wild type. The F_2 generation would have shown a 9:3:3:1 ratio with all of the vermilion-eyed flies male.

10 v is sex-linked and st autosomally inherited, both v^+ and st^+ are needed for the formation of brown eye pigment. We can predict the phenotypic ratio in the F_2 by means of a Punnett square:

			F_1 females $v^+/v\ st^+/st$ (wild phenotype)			
		Eggs:	v^+st^+	v^+st	vst^+	vst
F_1 males	Sperm:	$v\ st^+$	$+$	$+$	$-$	$-$
$v/Y\ st^+/st$		$v\ st$	$+$	$-$	$-$	$-$
(vermilion		$Y\ st^+$	$+$	$+$	$-$	$-$
phenotype)		$Y\ st$	$+$	$-$	$-$	$-$

$+$ indicates wild-type eye colour
$-$ indicates mutant bright-red eye colour
3:5 ratio expected in both sexes

11 st eliminates brown pigment in the eye and bw the red pigment; the double mutant phenotype has neither pigment and is white. The F_2 exhibits the 9:3:3:1 ratio expected for independent assortment of two autosomal loci with dominance. The white F_2 flies, crossed to wild type, will give the wild phenotype in the F_1 and 9:3:3:1 wild, scarlet, brown and white in the F_2. The white F_2 flies will give scarlet when crossed to scarlet and brown when crossed to brown.

12 sfo segregates only at the first division (in this sample) and so is close to the centromere—perhaps inseparable from it. Mating type shows limited crossing-over with the centromere (20 per cent—map distance 10); fl recombines freely with the centromere (approx. 67 per cent second-division segregation) and must be distant from it (at least 33 map units). Mating type and sfo must be on different chromosomes since NPD approximately equals PD and exceeds T tetrads in number; fl is either on a third chromosome or so far from the centromere on one of the first two that linkage is not detectable.

Chapter 4

4.1 From the first cross we expect coloured non-shrunken, coloured shrunken, colourless non-shrunken, colourless shrunken in a $1:1:1:1$ ratio. From the second cross we expect coloured shrunken and colourless non-shrunken, each about 45 per cent, and coloured non-shrunken and colourless shrunken, each about 5 per cent.

$$\text{First cross:}\quad \frac{C\,sh}{c\,sh}\times\frac{c\,Sh}{c\,sh}\qquad \text{Second cross:}\quad \frac{C\,sh}{c\,Sh}\times\frac{c\,sh}{c\,sh}$$

C and Sh are linked with about 10 per cent recombination (actually 9·3 per cent on the data given).

4.2 The general formula for the F_2 from the 'coupling' double heterozygote is $\frac{1}{4}\{2+(1-p)^2\}:\frac{1}{4}\{1-(1-p)^2\}:\frac{1}{4}\{1-(1-p)^2\}:\frac{1}{4}(1-p)^2$. For the 'repulsion' case it is $\frac{1}{4}(2+p^2):\frac{1}{4}(1-p^2):\frac{1}{4}(1-p^2):\frac{1}{4}p^2$. Note that both of these formulae reduce to $9/16:3/16:3/16:1/16$ when $p=\frac{1}{2}$. The use of F_2 progenies for estimating linkage is inefficient because F_2 phenotypes are not directly representative of F_1 gamete genotypes; much information about the latter is obscured by dominance and by uncertainty about the contributions of male and female gametes.

4.3 Two pairs of alleles, segregating without linkage, are expected to give a $9:3:3:1$ ratio in the F_2 generation, if one member of each pair is completely dominant. Taking the three pairs of alleles two at a time:

$a/+$ and $bt/+$ gave 110 agouti non-belted, 39 agouti belted, 38 black non-belted, 13 black belted, obviously a close fit to the 'expected' numbers for $9:3:3:1$ of 112·5, 37·5, 37·5 and 12·5.

$bt/+$ and $un/+$ gave 112 non-belted straight-whiskered, 36 non-belted wavy, 35 belted straight and 17 belted wavy, again an obviously good fit to $9:3:3:1$. Thus bt is not appreciably linked to a or un.

$a/+$ and $un/+$ gave 96 agouti straight, 53 agouti wavy, 51 black straight and no black wavy. These numbers are significantly different from $9:3:3:1$ (you can work out the χ^2 value for three degrees of freedom) but are a good fit to $2:1:1:0$, which is what is expected if the two markers are completely linked. But, in fact, they are also compatible with as much as 20 per cent recombination between a and un, for which the expectation would be only 1 per cent of the double recessive phenotype (*see* answer to problem 4.2). This illustrates the insensitivity of F_2 data from 'repulsion' crosses as a means of measuring linkage.

4.4 The map is lg—19·3—gl—21·2—B. If there were no interference, the expected frequency of double crossover products would be $0·193\times0·212=0·041$ or 4·1 per cent—that is 22·2 of a total of 543. The observed number was 11, which is significantly less than 22 ($\chi^2=5·5$ with one degree of freedom, $P<0·05$) so there is evidence of interference. The coefficient of coincidence is estimated as $11/22=0·5$.

4.5 In the F_1 generation we expect phenotypically wild-type females and males exhibiting the $w\,m\,f$ phenotype. In the F_2, on assumption (i) we expect $0·35\times0·2=7$ per cent double crossover products among the eggs, equally divided between $w+f$ and $+m+$. We expect 28 per cent (i.e. $35-7$) single crossovers between w and m, equally divided between $w++$ and $+mf$, and 13 per cent (i.e. $20-7$) between m and f equally divided between $w\,m+$ and $++f$. The remaining 59 per cent will be non-crossovers, equally $w\,m\,f$ and $+++$. On assumption (ii), the yield of double crossover products should have frequency $0·5\times7=3·5$ per cent. The frequency of single crossovers between w and m would then be $35 - 3 \cdot 5 = 31 \cdot 5$ per cent and that between m and f $20 - 3 \cdot 5 = 16 \cdot 5$ per cent.

4.6 A distance of 45 centimorgans does not imply as much as 45 per cent recombination; double and multiple crossovers will be very significant within such a long interval and the real recombination frequency will be less than 45 per cent to an extent depending on the intensity of interference. For the sake of argument assume 40 per cent which will not be far out. Then from cross (i) we will get: 30 per cent wild type, 30 per cent white, 20 per cent cinnabar, 20 per cent brown; from cross (ii) we expect 20 per cent white, 20 per cent wild type, 30 per cent cinnabar, 30 per cent brown. Cross (iii) will give a $2:1:1$ ratio of wild : cinnabar : brown. Cross (iv) will give 65 per cent wild, 10 per cent cinnabar, 10 per cent brown, 15 per cent white and cross (v) the same result as cross (iii).

4.7 For *his* and *met* there are 37 parental ditype tetrads and no non-parental ditypes; linkage is unequivocal, and the map distance is 13 (half the percentage tetratypes, since there are no NPDs). There is no evidence for linkage of either *met* or *his* with *arg* (slightly more NPDs than PDs in fact for both pairs), but it is possible that *arg* is as far away on the same chromosome as the other two since the PD:T:NPD ratio is not significantly different from $1:4:1$ for either *his/arg* or *met/arg*. The odds are in favour of *arg* being distant from the centromere on another chromosome.

4.8 Second-division segregation frequencies of *pro* and *ad* are 23 per cent and 41 per cent, respectively, giving estimates of $11 \cdot 5$ and $20 \cdot 5$ centimorgans from the centromere. The map distance between *pro* and *ad* is estimated as 11 (22 per cent tetratypes, no NPDs). Clearly the order is centromere—$11 \cdot 5$—*pro*—$11 \cdot 0$—*ad*. It is easy to see that *ad* and *pro* are in the same arm since second-division segregation for *ad* almost always accompanies second division segregation for *pro*.

4.9

$a\ b\ c$	$a\ b\ c$	$a\ b\ c$	$a\ b\ c$	$a\ b\ +$	$a\ b\ c$	$a\ b\ +$
$a\ b\ c$	$+\ b\ c$	$a\ b\ +$	$a\ +\ c$	$a\ +\ +$	$a\ +\ +$	$a\ +\ c$
$+\ +\ +$	$a\ +\ +$	$+\ +\ c$	$+\ b\ +$	$+\ b\ c$	$+\ b\ +$	$+\ b\ c$
$+\ +\ +$	$+\ +\ +$	$+\ +\ +$	$+\ +\ +$	$+\ +\ c$	$+\ +\ c$	$+\ +\ +$
64	16	16	1	1	1	1 per cent

(unordered tetrads)

Chapter 5

5.1 The three *lozenge* mutations, though showing no complementation, are separable by recombination and map in the sequence

$$ct—sn—lz^{BS}—lz^{46}—lz^{g}—v$$

with the *lz* mutations apparently spanning a map distance of about $0 \cdot 15$ centimorgans. The '*lz*' alleles generated in the experiment were double *lz* mutants reciprocally related to the wild-type recombinants and (allowing for their detectability only in males) produced with about the same frequency. The conclusion drawn from these data was that the different *lz* mutations affected functions that could only complement one another in *cis*. At first these were still thought to be *different* functions, and it was only later knowledge of gene–protein relationships obtained from molecular analysis that led to the realization that separable mutations could affect the same function in the sense of a single polypeptide chain.

5.2
The first three classes (6, 3 and 2) could have arisen either from reciprocal crossing-over between ry^5 and ry^{41} (ry^5 being on the *kar* side and ry^{41} on the *l26* side) or from conversion at one or other *ry* site with accompanying crossing-over in an adjacent interval. The fourth class (10 flies) is most likely due to conversion of ry^{41} without crossing-over and the fifth class (2 flies) to conversion of ry^5 without crossing-over. The *ry* allele marked * will be the double mutant $ry^5 ry^{41}$ if the ry^+ on the attached chromosome is due to reciprocal crossing-over; if it is due to conversion, * will be ry^5 if ry^{41} has been converted and ry^{41} if ry^5 has been converted.

5.3
The great majority of the ade^+ spores resulted either from reciprocal recombination between mutant *ade* sites or from conversion of one of these sites to wild type with crossing-over occurring in an adjacent interval. Site 8 is certainly to the right (on the *bio* side) of the other three, from the predominance of $y\,ade^+\,bio$ products in all three crosses. Taking frequency of recombination as a measure of distance, site 11 is the closest to site 8, with 16 and 20 not distinguishable on these data.

5.4
In all 14 asci cys^+ has risen by site conversion (of site *17* in 10 cases and of *64* in 4 cases). In most asci (9/14) conversion has occurred without associated crossing-over. In the 5 cases where associated crossing-over has occurred the cys^+ always has the flanking marker combination $+ +$, indicating the order *lys—17—64—ylo* if one assumes that the crossing-over occurred immediately adjacent to the conversion. One ascus shows half-chromatid conversion and postmeiotic segregation, without associated crossing-over.

5.5
Sites *4* and *17*, 2·30 per cent; sites *2* and *17*, 0·41 per cent recombination (total of $+ +$ and ab as percentage of total). In 4×17, 46 single-site conversion and 3 coconversions—6·1 per cent coconversion. In 2×17, 9 single-site conversions and 27 coconversions—75 per cent coconversion. Coconversion and recombination frequencies are inversely related since only single-site conversion produces recombination between sites. The data also suggest an inverse relationship between frequency of coconversion and frequency of reciprocal crossing-over between sites (9 in 4×17, 0 in 2×17). Sites need to be separated by a certain distance for a reciprocal crossover to occur between them, and must be within a certain distance to be coconverted.

Chapter 6

6.1
Heterozygosity with respect to an inversion covering *B—gl2* but excluding *lg* would account for the observation. Confirmation could be obtained by microscopical observation of pachytene of meiosis; an inversion loop in the appropriate region of chromosome 2 should be visible if the hypothesis is correct. The 40 per cent aborted pollen would follow from an 80 per cent incidence of crossing-over within the inversion loop, so the loop should be a large one, corresponding to at least 40 centimorgans.

6.2
The 70 per cent of spermatocytes in which no crossing-over occurs between the centromere and the breakpoint will give viable sperm assuming alternate disjunction of centromeres. On the same assumption, the 30 per cent in which there is such a crossover will give two out of four inviable cells in each tetrad with segmental duplication and deficiency. It will be the crossover products that will be inviable. Thus the viable sperm count will be reduced on account of the interchange heterozygosity by 15 per cent. Crossing-over within the centromere-breakpoint interval (though not distal to the breakpoint) will be suppressed so far as the

production of viable products is concerned. Note that viable crossover products are possible if some adjacent disjunction of centromeres occurs.

6.3 The radiation induced the transposition to another chromosome of a chromosomal segment including pyr^+. In the half-aborted asci, adjacent (as opposed to alternate) segregation of the association of four chromosomes has produced two spore pairs with lethal deficiency and two with viable duplication with respect to the pyr^+ segment; the duplication has some effect in reducing growth rate. Crosses of duplication pyr^+/pyr^+ strains to a pyr^- mutant result in trisomic segregation ratios in the meiotic tetrads. The transposed pyr^+ segment always signals itself by slow growth.

6.4 The A strain had a segmental interchange involving linkage groups 1 and 7. The breakpoint in linkage group 7 is very close to *met* (no recombination in 85 products) and that in linkage group 1 is fairly close to A/a (13 recombinants in 85, suggesting a map distance of 15, but this may be an underestimate since synapsis tends to be reduced in the vicinity of segmental exchanges). The *thi* locus is more distant from the exchange point than is *met*, as shown by the single crossover; its normal frequency of recombination with *met* is greatly reduced by the exchange.

6.5 The original sporulating culture had become tetraploid, segregating a/a, a/α and α/α ascospores.

6.6 A diploid a/a culture crossed to a standard haploid gives an $a/a/\alpha$ triploid. The further a isolate had only one copy of the mating-type chromosome but two of the chromosome carrying $ade3^+$—very probably a disomic. Crossed to $ade3$ it gives trisomic ratios for $ade3$ and normal diploid segregation for mating type.

6.7 Calling the 'purple' and 'pink' alleles P and p respectively, the three kinds of purple-flowered plant had the genotypes $P/p/p/p$, $P/P/p/p$ and $P/P/P/p$, respectively. The ratios obtained by crossing each of these to $p/p/p/p$ can be predicted, provided that it can be assumed that sister chromatids remain joined to the same centromere at the first division of meiosis, by considering the number of ways of segregating four chromosomes two-and-two. However, crossing-over with the centromere can join sister p alleles to different centromeres. If those two centromeres segregate to the same pole at anaphase I, then a p/p gamete can be produced by a $P/P/P/p$ plant. This happens only rarely, probably because the p locus is fairly close to its centromere.

Chapter 7

7.1 The information about restriction sites shows that the green/white variation correlates with the origin of the plastid DNA as would be expected if plastid phenotype were determined by the plastid genome. In *Pelargonium*, plastid DNA is inherited through both eggs and pollen; in *Mirabilis*, it appears to be excluded from the pollen.

7.2 The neutral property (resistance to killing) is dependent on the presence of a self-replicating dsRNA molecule which shows extranuclear inheritance. The maintenance of the dsRNA is, however, dependent on a maintenance gene M in the nucleus. S4, like N1, carries M but, unlike N1, it lacks the dsRNA. S1 carries the recessive allele m which is incompatible with maintenance. The m segregants from S1 × N1 possess the dsRNA initially but they lose it after a few budding cycles. If they are mated in time to S4 their dsRNA can be rescued and thereafter maintained in the diploid, subject to loss in the m haploid segregants following sporulation.

7.3   Make a heterokaryon between senescent green and vigorous *buff* hyphae and, after a short period of growth, isolate single conidia or single-nucleate hyphal fragments. An extranuclear basis for senescence is indicated if some or all of the *buff* reisolates are senescent. If senescence were based in the nucleus it should segregate with the nuclear colour marker. (In fact, the result favoured the extranuclear alternative.)

7.4   The CO_2 sensitivity is due to a virus, transmissible by injection. The virus can be visualized with the electron microscope.

7.5

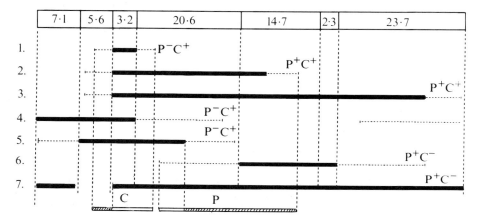

The upper part of the diagram shows the restriction site map—the two ends are to be imagined as joined to form a loop. The solid bars in the middle section show the *minimum* extents of the mtDNA sequences retained in each petite mutant; in some instances they extend a part of the way across restriction fragments that are not retained *in toto*, this being the only way of accounting for the size of the extra fragment within the bounds set by the flanking restriction sites. The broken-line extensions of the solid bars show the outside limits of the retained sequences, given the sizes of the extra fragments. The bars at the foot of the diagram show the segments within which the whole of the *C* determinant and the whole of the *P* determinant must be contained (no mtDNA outside these segments can be needed for the resistance functions). The cross-hatched subsegments must each contain at least part of the essential sequence of the resistance determinant, inasmuch as deletions that certainly extend no further than the ends of these subsegments eliminate the resistance function.

7.6   Since karyogamy is blocked by the *kar* mutation, the parental nuclei segregate vegetatively from the abortive zygotes as haploids, showing no recombination of chromosomal markers. In cross (*a*), the nucleus that came from the cir$^+$ strain remains associated with the plasmid, but half of the nuclei originating from the plasmid-negative strain become associated with it following the cell fusion. In cross (*b*) at least one of the two kinds of nuclei put into the cross remains associated with the type of plasmid with which it was originally associated. The results suggest that plasmid DNA is transmitted within the nuclei, but that nuclei initially lacking plasmid DNA can acquire it by infection from another nucleus in the same cell.

Chapter 8

8.1
$$\frac{V_G}{V_P} = \frac{V_P - V_E}{V_P} = \frac{36\cdot0 - 4\cdot8}{36\cdot0} = \frac{31\cdot2}{36\cdot0} = 0\cdot87$$

Note that the F_1 variance provides the best estimate of the environmental component of the F_2 variance; the parental variances are not comparable because of the very different means. To estimate the narrow-sense heritability one needs an estimate of the dominance component of the F_2 variance, and this would require data from backcross progenies.

8.2 The difference is just significant at the 5 per cent level. Variance of mean 1 is 3·07, variance of mean 2 is 2·96, variance of the difference between means is 6·03 and the standard error of the difference 2·45. Thus the difference, 5·3, is just greater than twice its standard error. Doubling the numbers would have halved the variances of the means and reduced the standard error of the difference by the factor $\sqrt{2}$, that is to 1·74; in relation to this, a difference of 6·03 is strongly, not just barely, significant.

8.3 The heritability is about 0·43.

8.4 The regression (slope of the best-fit straight line) is 0·33, giving an estimate of heritability of 0·66. (The data were taken from a real experiment but, to make the exercise easier, the father–son pairs were selected so as to give an obvious fit by eye to a straight line. The complete data showed much more scatter, with a standard error for the regression of $\pm 0\cdot13$.)

8.5 There will be $2^{20} = 1048\,576$ possible RI lines. With each generation of sib mating the probability of each initially heterozygous locus remaining heterozygous is halved. After n generations the probability of heterozygosity at any one locus is $(\frac{1}{2})^n$. With 20 loci the chance of heterozygosity at at least one of them is $20 \times (\frac{1}{2})^n$. To make this probability less than $\frac{1}{20}$, $(\frac{1}{2})^n$ must be less than $\frac{1}{400}$. Thus nine generations are required (counting the F_2 as the first generation), since $2^8 = 256$ and $2^9 = 512$.

Chapter 9

9.1 y is on one chromosome and w and Acr are on another, both in the same arm with w closer to the centromere.

9.2 Selection was being made for homozygosity with respect either to *ade3* or to *ade6*—*mal* is not subject to selection. Homozygosity for *ade6* is always associated with homozygosity for *ade3* and *mal*, indicating that the latter two markers are distal to *ade6* in the same chromosome arm. Homozygosity for *ade3* is associated in only 11 out of 16 cases with homozygosity for *mal*, indicating that *mal* is closer to the centromere than *ade3*. The linkage order is centromere—*ade6*—*mal*—*ade3*. All of the isolates can be explained by single mitotic crossovers in one or other interval.

9.3 The *PGK* gene may very well be sex linked to judge from these data (in fact it is). All the other markers are autosomal, and all associated with different autosomes.

9.4 One control would be to use a non-reverting deletion mutant as recipient. Another would be to use a donor of distinctive enzyme type. To determine whether the transforming DNA is integrated at the normal *am* locus, cross a transformant to wild type. If integration has occurred at another locus, recombination will yield recombinants in which am^+ is either missing or duplicated. If integration is into a different chromosome, normal am^+, transformed am^+, duplicated am^+ and am^- should be segregated in a $1:1:1:1$ ratio.

9.5   Mutants with increased penicillin production could be sought in different strains already carrying mutually complementary colour (or nutritional) markers. Heterokaryons could then be constructed by pairwise combinations of mutant strains, and diploid hybrids could be selected as single conidia showing complementation of the parental markers. Haploids could then be selected from the diploid hybrids as sectors or single conidial colonies showing segregation of the parental markers. If different mutations to increased penicillin production, present in the parental strains, are on different chromosomes, one-quarter of the haploid segregants should carry both mutations. Further parasexual hybridization of the double mutant to a third mutant could lead to the isolation of a triple mutant haploid segregant. The effects of the different mutations on penicillin production would not necessarily be additive, but double or multiple mutants would quite probably show increased penicillin yield as compared with single mutants. The procedure might well be more efficient than accumulating mutations by successive rounds of mutagenesis within a single strain.

Chapters 10 and 11

1   The determinant of sulphonamide resistance (sul^r) was transferred from the leu^- his^+ strain to the leu^+ his^- with relatively high efficiency, but the determinant of colicin production (col^+) is not transferred at all in the opposite direction; neither is there any transfer of the chromosomal markers leu and his. However, after acquiring sul^r, the leu^+ his^- strain could not only transfer this determinant but, at a lower frequency, col^+ as well. An explanation consistent with the evidence is that sul^r and col^+ are functions of different plasmids, the first determining its own transmission and the second dependent for its transmission on the presence of another self-transmissible plasmid. ColE1 is an example of a plasmid of this second type.

2   The colony harboured an F′ factor carrying the chromosomal markers $thyA^+$ and $argA^+$, i.e. a chromosomal fragment amounting to something over 1 per cent of the chromosome by the time map. *E. coli* cells can be 'cured' of F′ factors, as they can of F, by acridine orange. Most of the cells treated with acridine orange were cured; in some cases recombination between plasmid and chromosome occurred before curing. In the control the plasmid was lost in only one colony; again a certain amount of chromosome–plasmid recombination is seen.

3   The sequence of markers (*ile–arg–met–pro*) is the same in all three species. Hfr1 and Hfr3 transfer in the same direction (anticlockwise) from an origin, or origins, on the clockwise side of *pro*. Hfr2 transfers in the opposite direction from an origin on the anticlockwise side of *ile*. The two *Salmonella* Hfrs transfer at about the same rate, which is little more than half the rate in *E. coli* on which the standard time map is based.

4   The purpose is to ensure that all the selected colonies had actually received the Hfr pho^a sequence (*pho* enters from this Hfr before the selected markers thr^+ leu^+). The pho^+ frequencies therefore reflect recombination frequencies given that both homologues are present to recombine. The best order using recombination frequencies may be

U12–U3–E1–U7–U18, but there is some ambiguity about the order of E1 and U7. *Lac* is a sufficiently closely linked marker to serve as a third point in a three-point cross, and could provide an unequivocal ordering of pairs of *pho* sites on the basis of the very unequal frequencies of single and triple exchanges (cf. problem 6). The 0·18 per cent *pho*$^+$ recombinants from U12 × U18 (corresponding to 0 36 per cent recombination when the undetected double-mutant recombinants are taken into account) corresponds to 0·036 min on the time map, or about 4×10^{-4} of the whole genome, or about 1·6 kb.

5 The genes *leu* and *thr* are cotransduced with a frequency of only 3 per cent, indicating that they are separated by almost the maximum amount of DNA that can be encapsidated in a P1 particle. If length of DNA is proportional to distance on the time map, this maximum amount will be a little more than 2 per cent of the *E. coli* genome—that is a little more than 80 kilobase pairs. This is, in fact, close to the size of the P1 genome, and so it seems that transducing P1 particles have phage DNA replaced by an approximately equivalent amount of *E. coli* DNA.

6 The order is *trpA–trpB–trpE*.

7 Prophage is induced to lytic growth when it enters a non-immune F-recipient cell from a lysogenic Hfr. The substantial drop in recombinants at (or somewhat before) the time of entry of *gal*$^+$ is consistent with the lambda integration site being close to *gal*. In the case of 424 the prophage site enters much later—between *trp* and *his*. The unexplained feature in both cases is the small percentage of cells which must have received the prophage (if it was stably associated with the Hfr chromosome) and yet survived. Some kind of abortive induction may account for these.

8 At high multiplicities of infection nearly every transduced cell will have received a normal lambda genome as well as λdgal, and the former will supply the functions which the latter needs for replication and packaging after induction. At low multiplicities, the λdgal prophage will usually be without this help and will be unable to form phage because of its deleted functions. Superinfection with a range of lambda strains with different mutations will show which functions the λdgal prophage needs help with.

9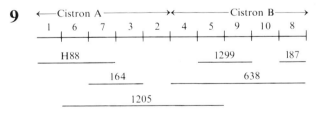

10 The presence of a foreign sequence (of any size up to 15 kb) inserted in the EcoRI site within the *lac* gene is signalled by the formation of a white rather than a blue plaque.

11 Cut both plasmids with EcoRI, mix the fragments, allow them to anneal end to end through their short EcoRI-generated single-stranded tails, seal with ligase and use the ligated mixture to transform *E. coli*, selecting for kanamycin resistance. From among the resistant transformants, select those which produce colicin I. Isolate DNA from these and check the plasmid component for size. Eliminate one EcoRI site by *in vitro* mutagenesis.

12   The most likely possibility is that Tn5 has transposed from lambda into many sites in the chromosome, causing auxotrophy whenever the site of insertion is within the coding sequence for an enzyme necessary for the synthesis of a molecule which can be supplied in the growth medium. It should be possible to map genetically both the neomycin resistance and the auxotrophic mutation; if the hypothesis is correct they should be in the same place.

13   The Charon 4A 'arms', common to both clones, are about 17·2 and 10·0 kb, respectively. The *Neurospora* inserts are about 10·7 kb in the left-hand track, and 9·2 + 12·1 kb in the right-hand track; in the latter case, two EcoRI fragments have been ligated into the vector end-to-end, as sometimes happens.

Chapters 12 and 13

1   The experiment showed that the mutations must have arisen sponta-neously at different times during the growth of the cultures *before* exposure to bacteriophage, and were not induced by the exposure. In the latter case there would have been no difference in variation between plates due to fractionation of the culture used to seed them.

2   The amino acid sequence could be coded by the highly repetitious sequence AGGCUGGCUGGCUGGCA. The DNA corresponding to this could well be liable to 'slippage' during replication leading to change in repeat number and consequent frameshift.

One expects two classes of frameshifts—additions and deletions. It might be expected that the addition class, with one or more extra repeats of the GGCT, would in turn be highly mutable, since it could readily lose the extra repeats by replication slippage within the now still more repetitious sequence. The deletion class might be considerably more stable since it has fewer repeats. The paper of Farabaugh et al. verified this interpretation.

3   1 and 4: Probably G-C→A-T transition.
2: Deletion.
3 and 6: Frameshift.
5: Probably A-T→G-C transition.
7 and 8: Possibly transversions, 7 most likely to give a G-C base-pair and 8 to give an A-T base-pair.

4   The observations on revertants are compatible with either UAA or UAG as the chain-terminating codon in the original mutant. The effect of nitroquinoline oxide is more consistent with UAG.

5   Type (ii) will be the least frequent since it requires change in one specific DNA base-pair. Types (iv) and (v) will be more frequent; (iv) requires a specific change in any one of what are usually reiterated copies of a tRNA gene, while (v) can often occur as the result of compensating change in any one of a number of sites in the originally mutated gene. Type (i) is probably still more common, since change in any one of a large number of positions can usually inactivate an enzymic protein. Type (iii) could be most common of all, since inactivation of any one of two or more enzymes functioning in the competing pathway might result in phenotypic suppression.

6 Hydroxylamine specifically attacks cytosine and induces G-C→A-T transitions. Hence reversion of the aspartic acid (5′ GA $\frac{U}{C}$ 3′) codon to the asparagine (5′ AA $\frac{U}{C}$ 3′) codon could not be induced in the 5′ GA $\frac{T}{C}$ 3′ DNA sequence but only in the complementary sequence 3′ CT $\frac{A}{G}$ 5′ on the other strand. We would thus expect reversions from treatment of only one of the strands. The strand that responded to the mutagen would be the transcribed strand, complementary to the mRNA. The experiment would give mosaic clear/turbid plaques if incorporation of wild-type A instead of mutant G occurred at the first replication after infection opposite the modified C, while the complementary strand, with unmodified G, replicated to propagate the mutant pair G–C. The finding of entirely turbid plaques would suggest that some correction mechanism was replacing the G opposite the modified C with A, and perhaps equally probably the modified C with T, *before* replication (the second mode of correction would restore the original mutation and not be observed). So far as the author is aware this particular experiment has not been done.

7 The wild-type codon sequence is:
AUG UCU AAC CUU CCC UCC GA . . . ., the methionine coded for by the initiating AUG being cleaved off and replaced by acetyl.

In the primary mutant there was a single base-pair insertion changing UCC in the sixth codon to UCCX, thus throwing the reading frame out of phase from the seventh codon onwards. The revertant had a compensating deletion of either the first or the second base in the second codon, giving: AUG $\frac{U}{C}$ UA ACC UUC CCU CCX GA . . . ., with the correct reading frame restored. The N-terminal methionine is evidently not removed when flanked to the right by a leucine residue.

8 The *hisD* and *hisB* genes are subject to a common control by the *O* segment. The control is *cis*-acting; the O^c mutation affects only those *his* genes which are coupled to it on the same chromosome, and for those genes the effect is dominant. An obvious hypothesis (confirmed in other ways) is that the *his* cluster is an operon, or unit of transcription, and that *O* controls the access of RNA polymerase.

9 The recombination is needed for setting up the *cis–trans* comparison:

$$\frac{Gus\text{-}r^r\ Gus\text{-}s^f}{Gus\text{-}r^n\ Gus\text{-}s^s} \quad \text{versus} \quad \frac{Gus\text{-}r^r\ Gus\text{-}s^s}{Gus\text{-}r^n\ Gus\text{-}s^f}$$

(*r* and *n* indicate responsive and non-responsive to hormone; *s* and *f* slow and fast in electrophoresis)

If *Gus-r* is a *cis*-acting control element, like an operator or promoter, only the fast variant will respond to the hormone in the first diploid and only the slow variant will do so in the second. If *Gus-r* can act in *trans* by, for example, coding for a diffusible activator or repressor protein, the enzyme variants will be equally expressed in both diploids.

10 The *arg-1* complementation map is suggestive of a single gene because of the predominance of non-complementing mutants and the continuous series of overlaps in the map. The *ad-3* is strongly suggestive of two independently acting genes because of the clean division of the great majority of the mutants into two mutually complementing groups.

Chapters 14 and 15

1

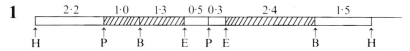

The shaded fragments hybridize with the probe. The coding sequence corresponding to the cDNA probe is interrupted by an intervening sequence.

2 The data suggest three stages of processing. The protection of the full-length gene (1290 b approximating to 1250) indicates the presence of some full-length unprocessed transcript. The removal of the first intron seems to proceed in two 'bites'; the first removes the greater part of it from the 5′ end, separating the 146-base 5′ coding sequence (approx. 141 b) from an approximate 1085-base unprocessed 3′ sequence, and the second removing the remainder of the small intron leaving a 1018-base (approx. 1000 b) 3′ unprocessed sequence. The third step removes the large intron without detectable intermediates, effectively separating the middle and 3′ coding sequences (indistinguishable by size—approximating to 222 b).

3 Yeast $1 \cdot 1 \times 10^4$, *Drosophila* $1 \cdot 0 \times 10^5$, mouse $1 \cdot 8 + 10^6$. Note that there will be twice as many clones containing only a *part* of the 5-kb sequence.

4 The ATG, 10 from the 3′ end is the only possible translation initiation codon in the sequence shown; 8–12 bases upstream of this putative initiation codon is a typical Shine–Delgarno sequence (AGGAG), likely to be involved in ribosome binding; 74–80 bases upstream from the ATG is the sequence TAAAATT which looks like a Pribnow 'box' and a likely transcription start signal, and 14–24 bases further upstream again is the sequence CTTGACACCTT, which is similar to the '−30' sequence implicated in transcription initiation in the *lac* operon (cf. *Fig.* 14.8). The putative gene product would begin Met-Ala-Thr-Val . . . .

5 In each case the poly(A) is preceded by G-C, and then by a U-rich sequence which would be capable of forming a short length of double-stranded structure by pairing in reverse orientation with an A-rich sequence less than ten nucleotides away on the 5′ side. The A-rich sequence in each case includes AAUAAA, which appears to be a polyadenylation 'signal' in many eukaryote messenger RNAs.

6 The two kinds of cDNA could each be used as a probe to identify those parts of the genomic DNA clone from which the mRNA had been transcribed. This would involve restriction site mapping of the DNA clone and molecular hybridization of the probes to transferred restriction fragments. The genomic DNA clone and the cDNA clones could also be sequenced to show that the two cDNAs corresponded to different combinations of sequences present in the genomic DNA. The pattern of homology between the cloned genomic DNA and each of the cDNAs could also be shown by heteroduplex analysis, using the electron microscope or single-strand-specific DNAase (to define the DNA segments protected from digestion by hybridization to each kind of mRNA).

7 It does not look like an operon because there is no known basis for the complete elimination of two or more enzyme activities by a single-point mutation when these enzyme activities are due to separate protein (polypeptide) molecules. The polar effects of chain-terminating mutations in bacterial operons are not as extreme as this, and in any case the close coupling of transcription and translation thought to account for such polar effects cannot be used as an explanation in eukaryotes. The situation is best explained as one in which a *single* polypeptide chain is responsible for a number of *different* enzyme activities which are rather independent of one another and can be lost by mutation one at a time. Chain-terminating mutations obviously eliminate all activities dependent on that part of the chain which fails to get synthesized. The fact that classes (*c*), (*d*) and (*e*) lack all activities individually and yet can complement certain class (*a*) mutants suggests that they cannot form stable enzyme by themselves (many are incomplete chains) but can do so by aggregation with the complete mis-sense chains formed by (*a*) mutants. These speculations have been confirmed by biochemistry.

8 The cDNA clone could be used as a probe for cloning the normal *sh* gene from the maize genome and also the unstable mutant gene from a plant in which the transposable element appears to be at the *sh* locus. Comparison of the two will reveal whether the latter has an inserted sequence within it. If so, the inserted element could be cut out and cloned separately, and then used as a probe for the element elsewhere in the genome.

9 Several possible definitions: (i) DNA sequence transcribed to give a single primary RNA transcript. (ii) DNA sequence(s) whose transcript or translation product ultimately appears in a single type of functional macromolecule—polypeptide chain, rRNA or tRNA. On this definition the gene is often discontinuous in the genomic DNA. (iii) The unit of transcription (alternative (i) above) plus any adjacent sequences necessary for correct transcription initiation or termination, or for *cis*-acting control of the level of transcription. There is no agreement on the use of the term at the DNA level.

Chapters 16 and 17

1 The observation suggests that X-inactivation is clonal, with the state of activity or inactivity of the X transmitted as stably during growth and development as are the two different genomes in a chimaera. It also suggests that stable X-inactivation has already occurred in the 16-celled blastocyst.

2 The experiment shows that changing the genotype from h^+/h to h/h later than 8 hours prior to pupation cannot alter a decision already made that a wing cell will develop a chaeta rather than a trichome; h^+ was necessary for that decision but it does not have to stay around to enforce it. That yellow sectors can be observed both on thorax and (in h/h flies) on wing following later irradiation shows that potential wing progenitor cells remain capable of mitotic crossing-over later than 8 hours before pupation.

3   The first observation shows that compartments are not clones each stemming from a single cell, but must arise through common decisions taken by groups of cells (polyclones). The further observations show that the compartment boundaries separating different thoracic segments, or the posterior/anterior compartments within each segment, are established earlier than the dorsal/ventral and left/right boundaries within each segment.

4   Control of leg movement can be mapped to six foci, one for each leg, separate from, but near to, the foci of the legs themselves, mapped by colour. The behaviour foci probably correspond to the cells destined to form the ganglia of the nerve network.

5   The mating type is fixed anew, with an apparently random choice between two alternatives, whenever the macronucleus is regenerated. Immediately after macronuclear regeneration from the micronucleus both choices are still open, but the daughter macronuclei formed in the first mitotic division become differentiated independently of one another. Irreversible DNA rearrangement at or immediately following that division seems likely, though it has not been directly demonstrated.

6   At the time of X-chromosome inactivation a gene on an autosomal fragment inserted into the X is also inactivated to an extent which depends (*a*) on the random proportion of cells, in the tissues in which the gene is active, which retain the activity of the normal X rather than the one with the insertion, and (*b*) on the proportion of cell clones in which the insertion X-chromosome has undergone inactivation but the inactivation has not reached the locus of the inserted gene. A gene in the middle of the insert, relatively far from the heterochromatized X chromatin, is likely to escape inactivation in some clones. Thus the ability of the insertion to compensate for a deletion will depend on the positions of the necessary genes in relation to the ends of the inserted segment, as well as on the level of gene activity that is needed for viability of the mouse.

7

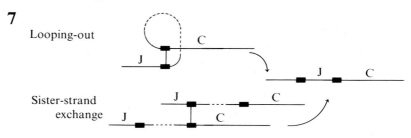

Looping-out

Sister-strand exchange

Chapter 18

18.1   *Nicotiana* system: fertility is 50 per cent on average in sib mating ($\frac{1}{4}$ crosses sterile, $\frac{1}{2}$ 50 per cent fertile, $\frac{1}{4}$ fully fertile) and 75 per cent on average in half-sib matings ($\frac{1}{2}$ crosses fully fertile and $\frac{1}{2}$ 50 per cent fertile). *Brassica* system: fertility is 25 per cent on average in sib matings ($\frac{3}{4}$ crosses sterile and $\frac{1}{4}$ fully fertile) and 50 per cent in half-sib matings ($\frac{1}{2}$ crosses sterile and $\frac{1}{2}$ fully fertile). Crosses between unrelated plants usually fully fertile in both systems.

18.2   Yes indeed. The numbers fit almost exactly—they were obviously invented.

18.3 An excess could be a consequence of increased fitness of heterozygotes as compared with homozygotes. A deficiency *could* be due to heterozygote *un*fitness, but is much more likely to be a consequence of inbreeding (in plants, self-fertilization).

18.4 $s = -0.11$ (from formula on p. 533, with $t = 1$).

18.5 Using the formula $H = 1 - 1/(4N\mu + 1)$ and the values 0.15 for H and 1.3×10^{-6} for μ, N may be calculated as 3.4×10^4. A frequency of a null allele of 1.97×10^{-3} would give a frequency of null homozygotes of 3.9×10^{-6}, sufficient to balance through loss of the null allele the new appearances of it by mutation.

18.6 $s = 0.30$, calculated from the formula in *Box* 18.2, with $\Delta q = -0.04$ and $q = 0.5$.

18.7 Some ideas: If in third codon positions they might make translation of a codon dependent on a different tRNA, which might be present in greater or lesser abundance. They could create sequences especially prone to methylation which could affect gene activity. They could create mutational 'hot-spots' which might affect fitness in the long run.

18.8 Extensions of the chain at the N terminus distinguish major groups (insects, vertebrates, fungi and green plants) from one another. In the common part of the chain, the vertebrates, as a group, clearly share features absent in the other organisms; within vertebrates the complete match between man and chimpanzee and between the two birds is indicative of evolutionary relationship. The two flowering plants (from very distinct families) share many substitutions relative to the others; the two fungi are rather different from plants and remarkably different from each other, throwing doubt on whether the Ascomycetes (in which yeast and *Neurospora* are both placed) are a natural group. The degree of homology between all species is striking; cytochrome *c* is a very conservative protein.

18.9 It seems that the $A\gamma$ and $G\gamma$ genes on the same chromosome have acquired greater homology with one another by the substitution into $A\gamma$ of a sequence derived from $G\gamma$, starting at the repetitious simple sequence in the long intron. This transfer of sequence information was not accompanied by crossing-over between chromatids, since this would have resulted in change in repeat number to either triplication or single copy. A process of intrachromosome sequence conversion is suggested; the repetitious sequence could assist the initiation of such a process by providing a region of ready hybridization between complementary DNA strands from the two genes. The occurrence of sequence transfer between homologous but non-allelic repeated gene copies could be the explanation for the maintenance of a high degree of homology between members of repetitious gene families within a species.

18.10 The implication is that one allele has been fixed at one of the duplicated loci and another allele at the other. Thus both enzyme forms, and the hybrid derivative of them, are present in all plants. This is a situation in which the functions of the two fixed alleles might diverge, and the hybrid protein might evolve distinctive and advantageous properties of its own. Duplication and subsequent divergence must certainly be important in gene and protein evolution, though not always involving polyploidy (globin genes and globins are obvious examples).

Chapter 19

19.1 *a.* The woman has a 1 in 4 chance of carrying the haemophilia allele; each of her daughters will have a 1 in 8 chance of receiving the allele but will show no effects herself; each of her sons will have a 1 in 8 chance of suffering from the disease. If the affected uncle had been the father's brother the woman's children would be at no risk, assuming that her father did not show the disease.

b. The woman has a 1 in 4 chance of carrying the recessive allele which was homozygous in the first cousin. Her spouse would have the same chance of carrying the allele as the general population—about 1 in 25. Therefore the chance that both parents will be carriers is 1 in 100, and the risk to each of their children 1 in 400.

c. The man had, at birth, a 1 in 2 chance of having inherited the allele causing the disease. Had he inherited it he would by now have a certain chance (say 50 per cent from the medical statistics) of beginning to show symptoms. The fact that he does not do so decreases the chance that he has inherited the allele to 1 in 3. The present risk to any of his children is a half of this, 1 in 6.

19.2 The clear correlation between concordance with respect to the condition and degree of relationship indicates some degree of genetic causation. The 50 per cent concordance between identical twins reinforces this conclusion, but also shows that genetic determination is far from complete. The large difference in degree of concordance between twins and ordinary sibs strongly suggests multifactorial inheritance; if the genetic causation lay in a single recessive allele, the concordance of ordinary sibs should be one-quarter that of identical twins, while if a dominant allele were responsible, ordinary sibs should have half as much concordance as identical twins, provided that the non-genetic component of causation did not include environmental variables which were more important for ordinary sibs than for the twins.

19.3 A high degree of heritability is indicated by similarity between (*a*) and (*c*) combined with pronounced difference between (*a*) and (*b*). An environmental basis for the observed variation is supported by similarity between (*a*) and (*b*) combined with a difference between both and (*c*). Head length is evidently highly heritable and little influenced by those environmental differences that distinguish the homes of separated twins. 'Achievement' is clearly greatly influenced by environment and has relatively low heritability (although the genetic component is still apparent in the difference between (*a*) and (*b*)). Height and IQ fall somewhere in between these two—higher heritability than 'achievement' but lower than head length. The first kind of IQ test appears to be more influenced by environment than the second.

19.4 The data are consistent with the determination of the three enzyme varieties by three alleles at one locus with estimated frequencies ph^A 0.35, ph^B 0.61 and ph^C 0.04. The observed frequencies of the enzyme phenotypes are in reasonable agreement with Hardy–Weinberg equilibrium (AA 0.095, compared with the expected $0.35^2 = 0.122$; BB 0.343, compared with $0.61^2 = 0.372$; AB 0.483, compared with $2 \times 0.35 \times 0.61 = 0.427$; AC 0.028, compared with $2 \times 0.35 \times 0.04 = 0.028$; BC 0.05, compared with $2 \times 0.61 \times 0.04 = 0.049$). The expected frequency of the C only (CC) phenotype is $0.04^2 = 0.0016$, which corresponds to only 0.28 in a sample of 178; finding none is the most likely result.

Index